MODERN MICROWAVE TRANSISTORS

MODERN MICROWAVE TRANSISTORS

Theory, Design, and Performance

FRANK SCHWIERZ
JUIN J. LIOU

A JOHN WILEY & SONS, INC., PUBLICATION

The cover photograph, a 100 nm metamorphic GaAs HEMT with mushroom gate, has been supplied courtesy of the Fraunhofer Institut für Angewandte Festkörperphysik, Freiburg, Germany.

This text is printed on acid-free paper. ∞

Copyright © 2003 by John Wiley & Sons, Inc. All rights reserved.

Published by John Wiley & Sons, Inc., Hoboken, New Jersey.
Published simultaneously in Canada.

No part of this publication may be reproduced, stored in a retrieval system or transmitted in any form or by any means, electronic, mechanical, photocopying, recording, scanning or otherwise, except as permitted under Section 107 or 108 of the 1976 United States Copyright Act, without either the prior written permission of the Publisher, or authorization through payment of the appropriate per-copy fee to the Copyright Clearance Center, Inc., 222 Rosewood Drive, Danvers, MA 01923, (978) 750-8400, fax (978) 750-4744, or on the web at www.copyright.com. Requests to the Publisher for permission should be addressed to the Permissions Department, John Wiley & Sons, Inc., 111 River Street, Hoboken, NJ 07030, (201) 748-6011, fax (201) 748-6008, e-mail: permreq@wiley.com.

Limit of Liability/Disclaimer of Warranty: While the publisher and author have used their best efforts in preparing this book, they make no representation or warranties with respect to the accuracy or completeness of the contents of this book and specifically disclaim any implied warranties of merchantability or fitness for a particular purpose. No warranty may be created or extended by sales representatives or written sales materials. The advice and strategies contained herein may not be suitable for your situation. You should consult with a professional where appropriate. Neither the publisher nor author shall be liable for any loss of profit or any other commercial damages, including but not limited to special, incidental, consequential, or other damages.

For general information on our other products and services please contact our Customer Care Department within the U.S. at 877-762-2974, outside the U.S. at 317-572-3993 or fax 317-572-4002.

Wiley also publishes its books in a variety of electronic formats. Some content that appears in print, however, may not be available in electronic format.

Library of Congress Cataloging-in-Publication Data

Modern microwave transistors : theory, design, and performance / Frank Schwierz, Juin J. Liou.
    p. cm.
    "A Wiley-Interscience publication"
    Includes index.
    ISBN 0-471-41778-5 (cloth : alk. paper)
    1. Microwave transistors.   I. Liou, Juin J.

TK7871.96.M53 M64 2003
621.381'3—dc21
                                        2002027230

Printed in the United States of America.

10 9 8 7 6 5 4 3 2 1

CONTENTS

Preface		xi
1	**Background on Microwave Transistors**	**1**
	1.1 Introduction	1
	1.2 Microwave Transistor Figures of Merit	4
	1.2.1 The Concept of Two-Port Networks	4
	1.2.2 The Problem of Stability	6
	1.2.3 Power Gain Definitions	7
	1.2.4 The Characteristic Frequencies f_T and f_{max}	9
	1.2.5 Minimum Noise Figure and Associated Gain	11
	1.2.6 Output Power and Power-Added Efficiency	12
	1.3 Historical View of Microwave Transistors	12
	1.3.1 The Early Years	12
	1.3.2 Development of Microwave Transistors with Heterostructures	15
	1.3.3 Recent Developments	20
	1.4 State of the Art of Microwave Transistors in the Year 2001	21
	1.4.1 III–V FETs	21
	1.4.2 BJTs and HBTs	28
	1.4.3 Wide Bandgap Transistors	34
	1.4.4 Si MOSFETs	37
	1.5 Application Aspects	38
	1.5.1 Civil Applications of Microwave Systems	38
	1.5.2 Other Applications of Transistors with GHz Capabilities	40
	1.5.3 Microwave Transistors for Wireless Communications up to 2.5 GHz	41
	1.6 Summary and Outlook	43
	References	52
2	**Basic Semiconductor Physics**	**61**
	2.1 Introduction	61
	2.2 Free-Carrier Densities	63

	2.2.1	Band Diagrams and Band Structure	63
	2.2.2	Carrier Statistics	67
	2.2.3	Approximations for the Carrier Densities	69
2.3	Carrier Transport	77
	2.3.1	Introduction	77
	2.3.2	Classical Description of Carrier Transport	78
	2.3.3	Nonclassical Description of Carrier Transport	91
2.4	PN Junctions	96
2.5	Schottky Junctions	106
2.6	Impact Ionization	108
2.7	Self-Heating	112
References	117

3 Heterostructure Physics — 121

3.1	Introduction	121
3.2	Band Diagrams	125
	3.2.1 The Anderson Model	125
	3.2.2 Built-In Voltage and Thickness of the Space–Charge Region	127
	3.2.3 Bandgaps and Band Offsets	130
3.3	Carrier Transport Across Heterojunctions	134
	3.3.1 Currents Across Spike Heterojunctions	134
	3.3.2 Currents Across Smooth Heterojunctions	139
3.4	Carrier Transport Parallel to Heterojunctions and Two-Dimensional Electron Gas	140
	3.4.1 Band Diagram	140
	3.4.2 2DEG Sheet Density	142
	3.4.3 2DEG Mobility	154
	3.4.4 Spontaneous and Piezoelectric Polarization Effects	157
References	160

4 MESFETs — 163

4.1	Introduction	163
4.2	DC Analysis	166
	4.2.1 The PHS Model	167
	4.2.2 The Cappy Model	176
4.3	Small-Signal Analysis	180
	4.3.1 Small-Signal Equivalent Circuit	180
	4.3.2 Modeling the Equivalent Circuit Elements Based on the PHS Model	184
	4.3.3 Modeling the Equivalent Circuit Elements Based on the Cappy Model	187
	4.3.4 Small-Signal Parameters and Gains	189
4.4 Noise Analysis	194

	4.4.1	Noise Mechanisms	194
	4.4.2	Noise Modeling Using the PHS Noise Model	197
	4.4.3	Noise Modeling Using the Cappy Noise Model	199
	4.4.4	Minimum Noise Figure	199
4.5	Power Analysis		204
4.6	Issues of GaAs MESFETs		208
	4.6.1	Transistor Structures	208
	4.6.2	Low-Noise GaAs MESFETs	210
	4.6.3	Power GaAs MESFETs	216
4.7	Issues of Wide Bandgap MESFETs		222
	4.7.1	Transistor Structures	222
	4.7.2	SiC MESFETs	223
	4.7.3	GaN MESFETs	225
References			226

5 High Electron Mobility Transistors — 231

5.1	Introduction		231
5.2	DC Analysis		237
	5.2.1	PHS-Like HEMT Model	237
	5.2.2	Cappy-Like HEMT Model	246
5.3	Small-Signal Analysis		253
	5.3.1	Introduction	253
	5.3.2	Modeling the Circuit Elements Based on the PHS-Like Model	254
	5.3.3	Modeling the Circuit Elements Based on the Cappy-Like Model	259
	5.3.4	The Concept of Modulation Efficiency	259
	5.3.5	The Concept of Delay Times	262
5.4	Noise and Power Analysis		265
5.5	Issues of AlGaAs/GaAs HEMTs		266
	5.5.1	Transistor Structures	266
	5.5.2	Low-Noise AlGaAs/GaAs HEMTs	267
	5.5.3	Power AlGaAs/GaAs HEMTs	267
5.6	Issues of GaAs pHEMTs		270
	5.6.1	Transistor Structures	270
	5.6.2	Low-Noise GaAs pHEMTs	271
	5.6.3	Power GaAs pHEMTs	273
5.7	Issues of GaAs mHEMTs		273
	5.7.1	Transistor Structures	273
	5.7.2	Performance of GaAs mHEMTs	275
5.8	Issues of InP HEMTs		276
	5.8.1	Transistor Structures	276
	5.8.2	Low-Noise InP HEMTs	278
	5.8.3	Power InP HEMTs	280

	5.9 Issues of AlGaN/GaN HEMTs	280
	5.9.1 Transistor Structures	280
	5.9.2 AlGaN/GaN HEMT Performance	282
	References	284

6 MOSFETs   292

  6.1 Introduction   292
  6.2 Two-Terminal MOS Structure   295
      6.2.1 Qualitative Description   295
      6.2.2 Derivation of the Threshold Voltage   299
  6.3 DC Analysis   302
      6.3.1 Introduction   302
      6.3.2 PHS-Like MOSFET Model   303
      6.3.3 Effective Mobility   307
      6.3.4 Modifications of the Threshold Voltage   310
  6.4 Small-Signal Analysis   313
      6.4.1 MESFET/HEMT-Like Equivalent Circuit   313
      6.4.2 Transmission Line Model   316
      6.4.3 Compact Models   319
  6.5 Noise and Power Analysis   320
  6.6 Issues of Small-Signal Low-Noise MOSFETs   321
      6.6.1 Transistor Structures   321
      6.6.2 Si MOSFET Performance   322
  6.7 Issues of Power MOSFETs   324
  References   327

7 Silicon Bipolar Junction Transistors   334

  7.1 Introduction   334
  7.2 DC Analysis   344
      7.2.1 First-Order Model Development   344
      7.2.2 Extensions of the First-Order Model   348
  7.3 Small-Signal Analysis   371
      7.3.1 Small-Signal Equivalent Circuit   371
      7.3.2 Delay Time Analysis   376
      7.3.3 Cutoff Frequency and Maximum Frequency of Oscillation   378
  7.4 Noise Analysis   379
  7.5 Power Analysis   384
  7.6 Issues of Si BJTs   387
      7.6.1 Transistor Structures   387
      7.6.2 BJT Performance   390
      7.6.3 Low-Noise BJTs   394
      7.6.4 Power BJTs   395
  References   397

8 Heterojunction Bipolar Transistors — 400

    8.1 Introduction — 400
    8.2 DC Analysis — 405
        8.2.1 Minority Carrier and Bandgap Narrowing Parameters — 405
        8.2.2 HBTs with Smooth Emitter–Base Heterojunction — 408
        8.2.3 HBTs with Spike Emitter–Base Heterojunction — 413
        8.2.4 HBT Structures with a Reduced Spike — 415
        8.2.5 Other Issues Related to HBT DC Behavior — 417
    8.3 Small-Signal, Noise, and Power Analysis of HBTs — 422
    8.4 Self-Heating of HBTs — 424
        8.4.1 Temperature-Dependent Collector Current in Multifinger HBTs — 424
        8.4.2 Temperature Dependence of Current Gain — 427
        8.4.3 Current Gain Collapse — 430
    8.5 Issues of GaAs-Based HBTs — 432
        8.5.1 Transistor Structures — 432
        8.5.2 GaAs-Based HBT Performance — 440
    8.6 Issues of InP-Based HBTs — 443
        8.6.1 Transistor Structures — 443
        8.6.2 InP HBT Performance — 448
    8.7 Issues of SiGe HBTs — 450
        8.7.1 Transistor Structures — 450
        8.7.2 SiGe HBT Performance — 457
    References — 461

Appendixes

A.1 Frequently Used Symbols — 469
A.2 Physical Constants and Unit Conversions — 473
A.3 Microwave Frequency Bands — 476
A.4 Two-Port Calculations — 477
A.5 Important Material Properties of Selected Materials — 481

Index — 483

PREFACE

Microwave electronics are advancing at an incredibly fast pace. In the past, microwave electronics have mainly been related to defense applications. This changed dramatically during the 1990s when wireless communications created the first consumer mass markets for microwave systems. Cellular phones have become common place and are now sold in millions of units annually. Furthermore, new applications are under development or already on the market. Examples are Bluetooth, wireless local area networks (WLANs), or civil satellite communication.

Microwave transistors are the backbone of all these microwave systems. In recent years, they have also undergone an impressive evolution. Continuous efforts in research and development have made transistors faster and more powerful. Performance of traditional GaAs FETs and HBTs has been improved and new classes of microwave transistors, such as wide bandgap FETs and metamorphic GaAs HEMTs, have been introduced. Furthermore, Si saw a renaissance in microwave electronics. SiGe HBTs are now available commercially and Si MOSFETs, which formerly were considered slow devices not suited for microwave operation, are gaining popularity for applications in the lower GHz range.

The purpose of this book is to thoroughly provide the material necessary for understanding the physics, operation, design, and performance of microwave transistors. This includes different transistor structures, analysis of transistor behavior, and design guidelines. The book also covers the relevant basics of semiconductor physics so that a course in this field is not a prerequisite.

The beginning of the book is rather unorthodox. Commonly, device books begin with theoretical chapters on semiconductor lattices or quantum mechanics. We thought that a practice-related start would be more inspiring for the reader. Thus, Chapter 1 serves as kind of an appetizer, giving an overview of the existing field of microwave transistors. It deals with the evolution of microwave transistors and highlights important past and future trends of their development. Basic transistor structures and commonly used figures of merit are introduced, and the status of microwave transistors in the early 2000s is reviewed. Chapters 2 and 3 provide the theoretical background that is indispensable for understanding the operation and analysis of transistors. This includes both basic semiconductor device physics and different aspects of heterojunctions, which are widely used in advanced microwave transistors. Relevant material properties and parameters are also given to place the reader in a position to really calculate device characteristics and design microwave transistors. Chapters 4 to 8 address different types of microwave transistors currently in use or under development. Subsequently MESFETs, HEMTs, silicon MOSFETs, conventional bipolar transistors (BJTs), and heterostructure bipolar transistors (HBTs) are discussed in a unified manner.

The five transistor chapters have identical structure and organization, thus enabling the reader to compare and assess the differences and relative merits of different transistor types. In each chapter, first the basic structure and operation of the corresponding transistor type are explained. Then, dc, small-signal, noise, and power behaviors are analyzed. The focus of the book is on the physical understanding of operation and design of microwave transistors. Thus the main emphasis is physics-based modeling while the field of compact modeling is intentionally excluded. Each chapter closes with advanced transistor structures, design issues, and a description of the state-of-the-art performance. We are aware that the latter is only a snapshot showing the situation in early 2002. The performance of several transistor types is expected to improve considerably in the near future, while other types have already reached or even exceeded their zenith. Nevertheless, the performance data compiled in this book represent a valuable synopsis. Plots of transistor figures of merit as a function of design parameters (e.g., cut off frequency versus gate length or base width) permit the estimation of performance limits and a serious comparison of different transistor types.

Many people contributed to this book in one way or another. We would like to thank our undergraduate and graduate students whose many questions incessantly led us to think more deeply about how device physics can be explained correctly and understandably, at the same time. We are grateful to our editor at Wiley, George Telecki, and to our production editor, Lisa Van Horn, for their continuous encouragement and support. Furthermore, we wish to thank our families for their support and understanding during the course of manuscript preparation.

One of us (F.S.) would like to thank U. König (DaimlerChrysler) for many helpful discussions and providing lots of experimental material, M. Schlechtweg (IAF Freiburg) for numerous fruitful discussions and providing the cover photo, Mark Rodwell (UCSB) and M. Paßlack (Motorola) for helpful comments, V. Polyakov (TU Ilmenau) for delivering the results of Schrödinger-Poisson simulations presented in this book, his Ph.D. students M. Kittler, R. Granzner, and J. Geßner for the helpful discussions about the manuscript, Ms. G. Müller and Ms. A. Schwierz for preparing a large part of the figures, and his wife Sabine Schwierz and her team at impuls GmbH Ilmenau for continuous technical assistance and preparation of scans.

FRANK SCHWIERZ
JUIN J. LIOU

Ilmenau, Germany
Orlando, Florida
September 2002

MODERN MICROWAVE TRANSISTORS

CHAPTER 1

BACKGROUND ON MICROWAVE TRANSISTORS*

1.1 INTRODUCTION

Since the invention of the bipolar transistor in 1947, semiconductor electronics has been advancing and evolving at an enormous pace. This can be attributed mainly to the dramatic reduction of the device dimensions and therefore the integration of more and more transistors onto a single Si chip. Thanks to these advances, microprocessors now contain hundreds of millions of transistors and Gbit DRAMs are commercially available. This field of semiconductor electronics is called Si VLSI (Very Large Scale Integration) and has become a multibillion dollar industry. The trend of ever increasing integration levels and decreasing device dimensions is expected to continue at least for the next several years [7, 8]. Figure 1.1 shows the evolution of the memory capacity and the minimum device feature size up to the year 2001, as well as the targets of the International Technology Roadmap for Semiconductors (ITRS) for the next several years. In past three decades, the minimum feature size of production stage Si ICs decreased by a factor of about 0.7 and the capacity of DRAMs increased by a factor of four every three years. This trend results in an almost linear slope of the curves in the semilogarithm plot shown in Fig. 1.1, known as Moore's Law.

Besides Si VLSI, however, there are other emerging fields in semiconductor electronics. Despite the fact that their current market share is much smaller than that of Si VLSI, some of these fields are in a state of dynamic growth. Among them, microwave (including radio frequency) electronics with microwave transistors as its basic building blocks is likely the most prominent one. In the context of this book, the term "microwaves" means electromagnetic waves with frequencies around and above 1 GHz.

The development of microwave transistors went almost unnoticed until 1980 because, unlike Si VLSI, there were no mass consumer markets for microwave systems. Most applications of microwave transistors had been military or exotic scientific projects. Examples of military microwave applications are equipment for secure communications, electronic warfare systems, missile guidance, control elec-

---

*The data given in this chapter have been compiled from around 1000 references in the technical literature [1]. Part of the material has been presented by the authors in invited papers in scientific journals and at international conferences, see, e.g., [2–6].

1

2 BACKGROUND ON MICROWAVE TRANSISTORS

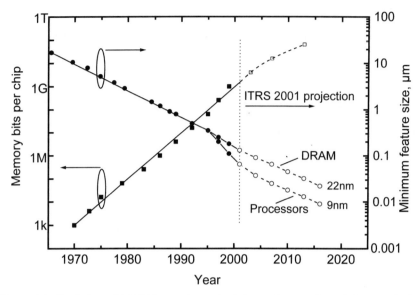

Figure 1.1. Evolution of DRAM capacity and minimum feature size of Si MOS devices from 1970 to 2001. The targets of the International Technology Roadmap for Semiconductors (ITRS) beyond 2001 are also shown.

tronics for smart ammunition, and radar systems. In the 1980s, satellite television using low-noise transistors operating around 12 GHz in the receiver front-ends was the first civil application of microwave transistors with a market volume worth mentioning.

During the 1970s and 1980s, when military applications were dominant in microwave electronics, performance was most important, while economic aspects played only a minor role. The situation changed dramatically in the 1990s. The new global political situation has led to considerable cuts in military budgets. Furthermore, a shift to consumer applications took place, and consumer applications clearly became dominant. Thus, the design philosophy for many microwave systems changed from "performance at any price" to "sufficient performance at lowest cost."

Currently, we are witnessing far-reaching upheavals in civil communication technology, which have created mass consumer markets for microwave systems. Modern communication systems including mobile communication (e.g., cellular phones and other advanced communication services) and the Internet will have an impact on human society at least as significant as personal computers had in the past. These new communication systems transmit, process, and receive great amounts of data in very short time intervals and operate in the GHz range. Microwave transistors are the backbone of these modern communication systems, and the widespread use of mobile phones during the 1990s created the first real mass market for microwave transistors. In 1998, for the first time more mobile phones than PCs had been produced. The 1990s were a decade of unprecedented growth in the microwave electronics industry.

The year 2001, however, has been disastrous for the entire semiconductor industry worldwide and has frequently been called the worst year in semiconductor history. Chip sales decreased dramatically and most semiconductor manufacturers had to announce layoffs. The crisis concerned not only traditional semiconductor products such as memories and microprocessors, but microwave electronics as well. This, however, does not alter the fact that the future of microwave electronics looks bright. We are still in the beginning of the information age, and new communication services requiring faster circuits and devices will be introduced. In the near future, the markets for microwave electronics will continue to grow considerably.

The evolution of microwave electronics followed a path that differs somewhat from that of the Si VLSI. As indicated by the name, the only semiconductor material used in Si VLSI is silicon. Furthermore, only two basic types of transistors are widely used in the Si VLSI: MOSFETs (Metal–Oxide Semiconductor Field-Effect Transistors) and BJTs (Bipolar Junction Transistors). The basic operation principles of MOSFETs and BJTs have remained essentially the same over the past three decades; only the device structures changed gradually over time.

For microwave electronics, on the other hand, a large variety of different semiconductor materials (Si, SiGe, GaAs, InP, further III–V compounds, and wide bandgap materials) have been employed. Furthermore, various types of microwave transistors exist, including

- MESFETs (Metal–Semiconductor Field-Effect Transistors)
- HEMTs (High Electron Mobility Transistors)
- MOSFETs
- BJTs
- HBTs (Heterojunction Bipolar Transistors)

Microwave transistors are used in a large number of different circuits such as low-noise amplifiers, power amplifiers, mixers, frequency converters and multipliers, attenuators, and phase shifters. Although the requirements on transistor performance differ from application to application, microwave transistors in principle can be distinguished into two groups as small-signal low-noise transistors and power transistors.

Any discussion of the principles, properties, and performance of microwave transistors inevitably leads to the so-called figures of merit, which are commonly used to describe and assess transistor performance. In the following section the most important figures of merit of microwave transistors are explained.

1.2 MICROWAVE TRANSISTOR FIGURES OF MERIT

To assess the capabilities and the performance of electronic devices, figures of merit are often used. Figures of merit are numbers or quantities that enable device and circuit engineers to estimate device performance and to compare the merits of different types of devices. The commonly used figures of merit for microwave transistors are introduced and discussed below.

1.2.1 The Concept of Two-Port Networks

Before getting to the figures of merit themselves, it is necessary to discuss the two-port concept of electronic devices. Any active device such as a microwave transistor can be treated as a two-port network. As shown in Fig. 1.2, the input of such a two-port network is connected to a signal source (often called the generator) consisting of a voltage source v_G and a source impedance Z_G, and the output is connected to a load with a load impedance Z_L. The input and output currents of the two-port network are i_1 and i_2, respectively, whereas the input and output voltages are denoted v_1 and v_2. The voltages and currents are in lower case to indicate that they are ac small-signal quantities. In Fig. 1.2, the two-port network containing the microwave transistor is a black box and does not give any information on transistor performance. This information can be represented by a number of small-signal parameter sets such as Y, H, or Z parameters. The following discussion is restricted to Y parameters but is in principle applicable to other parameter sets.

The relationship between Y parameters and the currents and voltages of the circuit in Fig. 1.2 is given by

$$Y \text{ parameters } \begin{bmatrix} i_1 \\ i_2 \end{bmatrix} = \begin{bmatrix} y_{11} & y_{12} \\ y_{21} & y_{22} \end{bmatrix} \begin{bmatrix} v_1 \\ v_2 \end{bmatrix} \quad \begin{array}{l} i_1 = y_{11}v_1 + y_{12}v_2 \\ i_2 = y_{21}v_1 + y_{22}v_2 \end{array} \quad (1\text{-}1)$$

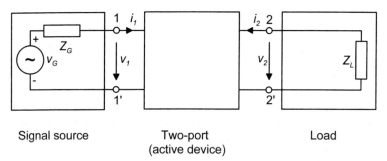

Figure 1.2. A microwave transistor (active device) as a two-port network with signal source and load.

The four Y parameters are defined under ac short circuit conditions by

$$y_{11} = \left.\frac{i_1}{v_1}\right|_{v_2=0} \quad \text{input admittance}$$

$$y_{12} = \left.\frac{i_1}{v_2}\right|_{v_1=0} \quad \text{reverse transfer admittance}$$

$$y_{21} = \left.\frac{i_2}{v_1}\right|_{v_2=0} \quad \text{forward transfer admittance} \quad (1\text{-}2)$$

$$y_{22} = \left.\frac{i_2}{v_2}\right|_{v_1=0} \quad \text{output admittance}$$

Figure 1.3 shows the equivalent circuit of the active device using Y parameters. At frequencies up to about 100 MHz, the external voltages and currents of the two-port network (v_1, i_1, v_2, i_2) can be easily measured, and the Y parameters are determined based on Eq. (1-2).

For operating frequencies in the microwave range, however, the measurement of the external voltages and currents and the realization of the required short-circuit conditions become more complicated. Therefore, another set of parameters, the so-called S or scattering parameters, is commonly used. S parameters are not defined as quotients of currents and voltages but as ratios of the powers of travelling waves:

$$S \text{ parameters} \quad \begin{bmatrix} b_1 \\ b_2 \end{bmatrix} = \begin{bmatrix} s_{11} & s_{12} \\ s_{21} & s_{22} \end{bmatrix} \begin{bmatrix} a_1 \\ a_2 \end{bmatrix} \quad \begin{array}{l} b_1 = s_{11}a_1 + s_{12}a_2 \\ b_2 = s_{21}a_1 + s_{22}a_2 \end{array} \quad (1\text{-}3)$$

As in the case of Y parameters, the subscripts 1 and 2 designate the input and the output of the two-port network, respectively, whereas a and b are the powers of incoming (or incident) and outgoing (or reflected) waves. Fig. 1.4 shows the two-port network with the incident (a_1, a_2) and reflected (b_1, b_2) waves.

Despite the fact that Y parameters cannot be measured in the microwave range,

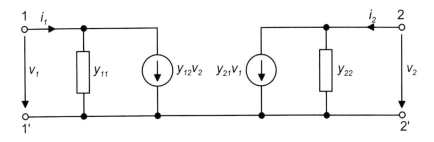

Figure 1.3. Equivalent circuit of an active device based on Y parameters.

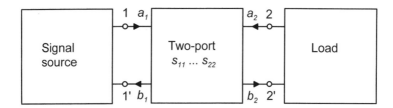

Figure 1.4. Two-port network characterized by S parameters.

they are widely used for discussing the properties of microwave transistors. Y parameters are more closely related to device physics and are more interpretable for device engineers than S parameters. Y parameters can be converted into S parameters, and vice versa, using simple formulas. In Appendix 4 the conversion rules between the different sets of small-signal parameters are given. All parameters introduced above depend on the specific bias conditions and on the operating frequency of the two-port network. In general, large values for the 21-parameters (parameters with subscript 21, e.g., y_{21} and s_{21}) are desirable. The 21-parameters describe the transistor's ability to control and amplify input signals.

1.2.2 The Problem of Stability

In general, a microwave transistor is capable of power amplification or sustained oscillation. Whether the transistor in a circuit will oscillate or not depends on the transistor itself (i.e., on the values of its small-signal parameters at the operating frequency under the particular bias conditions), and on the source and load impedances.

If the transistor is to be used for power amplification, oscillation is not desired. Therefore, it is important to understand the conditions under which oscillations may occur. In other words, we have to deal with the stability of the transistor. A transistor is said to be unconditionally stable if it does not oscillate regardless of the values of the signal source and load impedances or any additional passive components connected to the transistor's input or output. If, however, certain source and load impedances can cause oscillations, then the transistor is conditionally stable. The stability behavior of a transistor can be described by the stability factor k as introduced by Rollett [9]:

$$k = \frac{2Re(y_{11})Re(y_{22}) - Re(y_{12}y_{21})}{|y_{12}y_{21}|} \qquad (1\text{-}4)$$

where Re denotes the real part of the quantity in parantheses. For $k > 1$, the transistor is unconditionally stable and for $k < 1$ it is conditionally stable and unintended oscillations may occur.

1.2.3 Power Gain Definitions

A key feature of a transistor is its ability to amplify currents and voltages, and thus to deliver larger amounts of power to the load than received from the signal source. This property is called the power gain. In general, the power gain is the ratio of the power P_2 delivered from the transistor output to the load to the power P_1 delivered from the signal source to the transistor input. In practice, however, the problem is more complex. The matching conditions between the signal source and transistor and between the transistor and load influence the power transfer. Furthermore, only a stable non-self-oscillating transistor can be used as an amplifier. Thus, there are several different power gain definitions commonly used to characterize microwave transistors (see, e.g., [10]) which will be discussed briefly below.

If a transistor is to achieve the maximum power gain, then power matching is required. For conditions at which the transistor is unconditionally stable, i.e., for $k > 1$, power matching is obtained when both the input and output of the transistor are conjugately impedance-matched to the signal source and the load, respectively. The power gain obtained under these matching conditions is the maximum available gain, MAG, and can be calculated by

$$MAG = \left|\frac{y_{21}}{y_{12}}\right|(k - \sqrt{k^2 - 1}) \tag{1-5}$$

If, however, the transistor is conditionally stable ($k < 1$), auxiliary external admittances y_1 (at the input) and y_2 (at the output) have to be connected to the transistor as shown in Fig. 1.5 to suppress its tendency to oscillate. Thus, an overall stability factor K of the network consisting of the transistor itself and of the admittances y_1 and y_2 of equal or greater unity can be obtained. The overall stability factor is given by

$$K = \frac{2[Re(y_{11}) + Re(y_1)][Re(y_{22}) + Re(y_2)] - Re(y_{12}y_{21})}{|y_{12}y_{21}|} \tag{1-6}$$

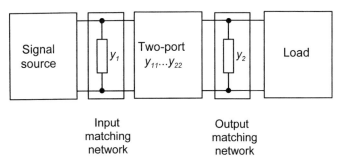

Figure 1.5. Two-port network including signal source, load, and auxiliary external admittances y_1 and y_2.

If the admittances y_1 and y_2 are chosen so that at the operating frequency the overall stability factor is exactly equal to unity, and if the input and output of the whole network (including y_1, transistor, and y_2) is conjugately impedance-matched to signal source and load, then the so-called maximum stable gain MSG defined as

$$MSG = \left| \frac{y_{21}}{y_{12}} \right| \qquad (1\text{-}7)$$

is achieved. In other words, MAG is the maximum gain an unconditionally stable transistor can achieve without any external network. MSG, on the other hand, is the maximum gain obtainable from a microwave transistor in combination with external matching impedances under the condition of $K = 1$ at the operating frequency. For MSG, there is no restriction concerning the operating frequency. This means that at the operating frequency, the transistor may be either unconditionally stable or conditionally stable. A comparison between Eqs. (1-5) and (1-7) shows that MAG is always lower than MSG except for the case in which the stability factor k of the transistor is equal to unity.

The unilateral power gain U as defined by Mason [11] is another frequently used measure for the maximum gain attainable from a microwave transistor. In general, U is the gain of a two-port network having no output-to-input feedback, but with input and output conjugately impedance-matched to signal source and load, respectively. A zero output-to-input feedback would mean that the output is completely isolated from the input. Because all microwave transistors possess a nonzero feedback from output to input, a lossless network must be added to cancel such a feedback. The resulting network (i.e., the transistor in combination with the lossless network canceling the feedback) will not oscillate unintentionally since only an output-to-input feedback not equal to zero can cause oscillations. Thus, the unilateral power gain is defined over the whole frequency range irrespective of the value of the stability factor k of the transistor. U can be calculated using

$$U = \frac{|y_{21} - y_{12}|^2}{4[Re(y_{11})Re(y_{22}) - Re(y_{12})Re(y_{21})]} \qquad (1\text{-}8)$$

Power gains such as MAG, MSG, and U are commonly given in decibels (dB):

$$\text{Power Gain}[dB] = \frac{P_2}{P_1}[dB] = 10 \log\left(\frac{P_2}{P_1}\right) \qquad (1\text{-}9)$$

It should be noted that the stability factors and all gains discussed above are calculated from small-signal parameters. Therefore, these quantities strictly speaking are applicable only for the small-signal case. Nevertheless, MSG, MAG, and U are also used as figures of merit for power transistors under large-signal operations.

The abovementioned power gains require certain matching conditions. Because these matching conditions are typically not fulfilled in practical circuits, MSG,

MAG, and U define the upper limits of the power gain a microwave transistor can achieve.

1.2.4 The Characteristic Frequencies f_T and f_{max}

The cutoff frequency f_T and the maximum frequency of oscillation f_{max} are the most important figures of merit for the frequency characteristics of microwave transistors. The cutoff frequency, often also designated as the gain–bandwidth product, is related to the short-circuit current gain h_{21}. This current gain is defined as the ratio of the small-signal output current to input current of the transistor with the output short-circuited. Such a current gain is frequency dependent, and its magnitude rolls off at higher frequencies at a slope of –20 dB/decade for any transistor. Frequently, the slope is given by –6 dB/octave. It can easily be checked that both slopes are identical. The octave is a factor of two in frequency and the decade is a factor of 10. The cutoff frequency is the frequency at which the magnitude of h_{21} equals unity (or 0 dB).

The maximum frequency of oscillation f_{max} is the frequency at which the unilateral power gain U equals unity. Therefore, f_{max} is the maximum frequency at which the transistor still provides a power gain. The somewhat misleading notation of maximum frequency of oscillation stems from the fact that it is also the highest frequency at which an ideal oscillator would still be expected to operate [12]. Like the short-circuit current gain h_{21}, U rolls off at –20 dB/dec. Figure 1.6 shows the measured current gain and unilateral power gain of a microwave GaAs MESFET with a gate length of 0.2 µm [13]. Also shown in the figure are the values of f_T and f_{max}, which are not measured directly but determined by extrapolating the measured h_{21} and U with the –20 dB/dec slope. This practice is not only convenient but in some cases inevitable because of the frequency limits of the measurement equipment. In his review of the history of GaAs FETs, P. Greiling stated: "For those of us associated with this technology, this measurement problem always seems to exist. We are in a catch 22 situation in which we are developing circuits for instruments that are needed to measure the circuits we are developing" [14]. One could argue that the value of f_{max} is only of academic interest because at this frequency the transistor looses its ability to amplify. This, however, is not true and the knowledge of f_{max} is also of practical importance. From the known –20 dB/dec slope of the unilateral power gain we find

$$U(f) = -20 \log f + 20 \log f_{max} \qquad (1\text{-}10)$$

Thus, if the extrapolated maximum frequency of oscillation of a certain transistor is given, one can easily estimate the maximum power gain this transistor will show at a certain frequency f. It should be noted that MSG rolls off at –10 dB/dec, whereas MAG shows no definite roll-off slope.

Sometimes in the literature, f_{max} is referred to as the frequency at which the maximum available gain MAG, rather than the unilateral gain U, decreases to unity.

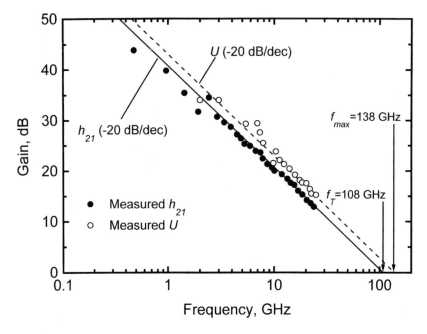

Figure 1.6. Measured current gain h_{21} and unilateral power gain U, and extrapolated f_T and f_{max} of a GaAs MESFET with a gate length of 0.2 μm (after [13]).

This is not entirely correct. It is true that the frequencies at which U and MAG become unity in most cases are not significantly different from each other, but f_{max} extrapolated from U can be different from that extrapolated from MAG. This problem will be discussed in more detail in Section 4.3.3.

The value of f_{max} for a specific microwave transistor may be either larger or smaller than the value of f_T. Transistors with $f_{max} > f_T$ can have useful power gains at frequencies above f_T and up to f_{max}. A simple explanation for this is that the current gain smaller than unity is compensated by a voltage gain greater than unity in the frequency range between f_T and f_{max}. Thus a power gain > 1 is possible. Transistors with $f_{max} < f_T$, however, can achieve power gain only at frequencies up to f_{max} and cannot be used as power amplifiers at frequencies between f_{max} and f_T.

A frequently asked question is which of the two characteristic frequencies, f_T or f_{max}, is more important for microwave transistors. There is no unequivocal answer. The commonly cited statement that f_T is a more important figure of merit for digital circuits, whereas for analog applications f_{max} is more significant is far too simple. The importance of f_T and f_{max} depends on the specific application of the transistor. Certainly, both characteristic frequencies are desired to be as high as possible. Manufacturers of microwave transistors often strive for $f_T \approx f_{max}$ so that the devices are useful for a large number of different applications.

Another important issue is what is the maximum operating frequency f_{op} of a microwave transistor having certain f_T and f_{max}. Again, there is no definite answer, and we need to discuss the issue separately for small-signal and power applications. Concerning small-signal transistors, a rather conservative rule of thumb is that the f_T of the transistor should be at least ten times the operating frequency of the system in which the transistor is to be used, and that f_{max} should be at least equal to f_T [15]. Low-noise transistors in particular benefit from high f_T's and f_{max}'s. A less stringent requirement is that the operating frequency of a microwave system should be not higher than half the cutoff frequency of the transistors used [16]. For power transistors, f_T should be at least 1 to 1.5 times f_{op}, and f_{max} should be 2 to 3 times f_{op} [17, 18]. If no transistor fulfilling this requirement is available, then, naturally, a transistor with inferior frequency performance has to be used. In general, the requirements on f_T and f_{max} differ from application to application.

1.2.5 Minimum Noise Figure and Associated Gain

A microwave transistor used as an amplifying device receives the signals and noises at its input terminals. Because the transistor cannot distinguish between the signal and noise, both will be amplified. Besides the amplified signals and amplified external noises, an additional component appears at the output of the transistor: the intrinsic noise generated in the transistor. For front-end amplifiers in which the signal-to-noise ratio is small, the intrinsic noise must be kept as small as possible. A figure of merit describing the amount of intrinsic noise produced in microwave transistors is the noise figure NF. The noise figure (commonly given in decibels) is defined by

$$NF[dB] = 10 \log \frac{P_{Si}/P_{Ni}}{P_{So}/P_{No}} \quad (1\text{-}11)$$

Here P_{Si} and P_{So} are the signal powers at the input and the output, and P_{Ni} and P_{No} are the noise powers at input and output, respectively. An ideal noiseless transistor would show a noise figure of 1 (or 0 dB). The magnitude of NF is dependent on the matching conditions at the input of the transistor, bias condition, and frequency. Unfortunately, the matching and the bias conditions required for minimum noise figure NF_{min} are different from those for maximum power gain. Therefore, if the transistor is to be operated for minimum noise, it will possess a power gain considerably lower than MAG, MSG, and U. The power gain obtained from the transistor biased and matched for minimum noise is called the associated gain G_a.

Another occasionally used figure of merit to describe the noise characteristics of a microwave transistor is the noise temperature T_N. It is related to the noise figure by

$$NF[dB] = 10 \log\left(1 + \frac{T_N}{T_0}\right) \quad (1\text{-}12)$$

where T_0 is the room temperature. If the noise figure is specified as a pure ratio and is not in dB, then the noise temperature can be calculated by

$$T_N = T_0(NF - 1) \qquad (1\text{-}13)$$

1.2.6 Output Power and Power-Added Efficiency

The output power P_{out} and the power-added efficiency PAE are two figures of merit relevant to power transistors. For power transistors, the amount of microwave power that can be delivered to the load is of primary importance, whereas the noise figure is of no concern. P_{out} is dependent on the frequency and type of the amplifier circuit, e.g., class-A amplifier or class-B amplifier. The output power can be given either directly in watts (W) or in dBm according to

$$P_{out}[dBm] = 10 \log P_{out}[mW] \qquad (1\text{-}14)$$

The output power is frequently reported in terms of power density, and commonly used figures of merit are the output power per mm gate width for FETs and the output power per μm^2 emitter area for bipolar transistors. Despite the fact that these figures of merit are not measures for the total output power of a certain transistor, they give a general idea of the power handling capability of the transistor and allow for the comparison of power performance for different transistors or even different types of transistors.

In power amplifiers where heat dissipation or battery power is of concern, the power-added efficiency is an important figure of merit. It is defined by

$$PAE = \frac{P_{out}(hf) - P_{in}(hf)}{P_{in}(dc)} \qquad (1\text{-}15)$$

where $P_{out}(hf)$ and $P_{in}(hf)$ are the transistor's microwave output and input powers, respectively, and $P_{in}(dc)$ is the dc power delivered by the power supply and dissipated into heat in the transistor. For microwave power transistors, high output power and power-added efficiency are desirable.

1.3 HISTORICAL VIEW OF MICROWAVE TRANSISTORS

1.3.1 The Early Years

In the first half of the 20th century, vacuum tubes were used exclusively as active devices in microwave applications. These tubes were bulky, unreliable, and consumed large amounts of power. Since the invention of the transistor, engineers have spent a lot of efforts to increase their operating frequencies for high-speed applications and to replace vacuum tubes. Ge BJTs developed in 1958–1959 were the first transistors operating above 1 GHz. As in the rest of semiconductor electronics, however, the dominance of Ge transistors in microwave applications declined very quickly. By 1963, Si BJTs became competitive and in 1970 almost all microwave transistors were Si BJTs [19]. Figure 1.7 demonstrates the output power and minimum noise figure obtained from the state-of-art microwave transistors reported in 1970.

It became clear in the early 1960s that Si is not the optimal semiconductor for microwave transistors. GaAs, having a six-fold electron mobility and a higher maximum electron drift velocity compared to Si, is a far better material for high-speed transistors. Attempts to develop GaAs microwave BJTs were fruitless, however. The work on GaAs FETs, on the other hand, was successful and led to a new type of microwave transistor, revolutionizing the microwave electronics industry. In 1966, Mead presented the first GaAs MESFET [20]. This transistor operated like a junction FET and consisted of an n-type active layer with a Schottky contact as the control electrode. Despite the fact that this device was not designed for microwave applications originally, it marked a major milestone in the development of the current microwave field-effect transistors. Since then, the GaAs MESFET evolved rapidly. The first GaAs MESFET with practical microwave performance was reported in 1967 and showed a f_{max} of 3 GHz [21]. In 1970, the record f_{max} of GaAs MESFETs increased to 30 GHz, which clearly exceeded the microwave performance of all Si BJTs at that time [22].

Si BJTs and GaAs MESFETs were the two only microwave transistor types in use in the 1970s and early 1980s. During that period, rapid improvements of GaAs MESFET performance had been obtained while progress in the more mature Si

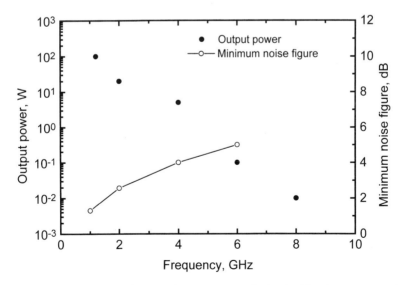

Figure 1.7. Minimum noise figure and output power obtained with microwave transistors (state of the art as of 1970, after [19]).

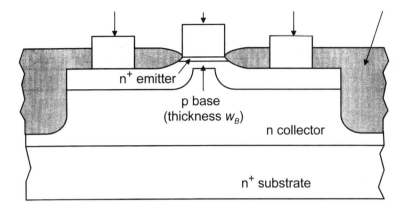

Figure 1.8. Schematic of a microwave Si BJT (after [23]). The vertical extension of the p base beneath the emitter is the base thickness w_B.

technology proceeded only gradually. Figures 1.8 and 1.9 show the basic structures of an ion-implanted Si microwave BJT and an epitaxial GaAs MESFET of that period. The critical dimensions to obtain good microwave performance are the base thickness w_B in BJTs and the gate length L for GaAs MESFETs, both of which should be as small as possible.

At frequencies below 4 GHz, Si BJTs were commonly used, whereas in the frequency range between 4 and 18 GHz the GaAs MESFET was the device of choice. Figure 1.10 shows the noise performance of the microwave BJTs and MESFETs in 1980, together with the noise performance of earlier MESFETs. GaAs MESFETs (gate length ranging from 0.25 to 0.5 μm) with noise figures below 2 dB at frequencies up to 12 GHz and Si BJTs with noise figures around 2 dB between 2 and 4 GHz could be achieved. Also, it is clear from the figure that good progress on noise performance in GaAs MESFET was made from 1967 to 1980.

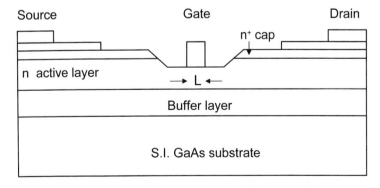

Figure 1.9. Cross section of a typical GaAs MESFET (after [24]).

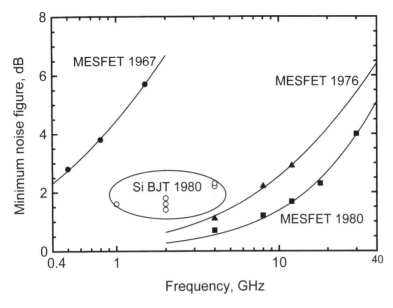

Figure 1.10. Minimum noise figure versus frequency obtained from GaAs MESFETs and Si BJTs (state of the art as of 1980).

The output power and the maximum frequency of oscillation of Si BJTs and GaAs MESFETs reported in 1980 are listed in Table 1.1.

1.3.2 Development of Microwave Transistors with Heterostructures

Despite the fact that aggressive device scaling (shrinking of FET gate length and BJT base thickness) always plays an important role in transistor engineering, only the use of heterostructures after 1980 offered the opportunity of tremendous progress toward improved high-frequency performance of microwave transistors. Since then, heterostructures have been a key component in modern high-performance microwave transistors such as HEMTs and HBTs.

Table 1.1. Output power and maximum frequency of oscillation of GaAs MESFETs and Si BJTs reported in 1980

Transistor type	P_{out}	f_{max}
Si BJT	60W @ 2 GHz	35 GHz
	6 W @ 5 GHz	
GaAs MESFET	10 W @ 10 GHz	up to 100 GHz

A heterostructure is a combination of at least two layers of different semiconductors with distinct bandgaps. The use of heterostructures in high-speed devices resulted from the progress made in the epitaxial growth based on molecular beam epitaxy (MBE) [25]. Using MBE, it is possible to grow layers with a thickness of only a few nanometers and with sharp interfaces between the adjacent layers. Another epitaxial growth technique used to produce high-quality heterostructures for microwave transistors is metal–organic chemical vapor deposition (MOCVD) [26].

During the late 1970s, intensive work was done at Bell Labs to grow and characterize sequences of thin GaAs and AlGaAs layers, called superlattices. Measurements of the Hall mobility in such layer sequences consisting of n-type AlGaAs and undoped GaAs showed mobilities clearly exceeding those of doped bulk GaAs or AlGaAs at room and lower temperatures [27]. It was logical then to utilize these enhanced mobilities for fast field-effect transistors, and research groups at different labs around the world started to create transistors taking advantage of the AlGaAs/GaAs heterostructure. The first device of this kind came from Fujitsu [28] and was called HEMT. Shortly afterward, researchers from other labs such as Bell Labs, University of Illinois, Cornell University, and Thomson-CSF also reported transistors based on the same concept but with different names such as, selectively doped heterostructure field-effect transistor (SDHT), modulation doped field-effect transistor (MODFET), and two-dimensional electron gas field-effect transistor (TEGFET).

The heterostructure physics exploited in AlGaAs/GaAs HEMTs is shown in Fig. 1.11. The different bandgaps of AlGaAs and GaAs cause a bandgap difference ΔE_G that results in band offsets ΔE_C and ΔE_V in the conduction and valence bands, respectively, at the heterointerface. In HEMTs, a large ΔE_C is desired, which stimulates the transfer of electrons from the n-type AlGaAs (conduction band at a higher energy level) to the lightly doped or undoped GaAs (conduction band at a lower energy level). The transferred electrons are confined to a region only a few nanometers thick in the GaAs layer near the heterointerface, called the two-dimensional electron gas (2DEG). Because the 2DEG electrons are spatially separated from the donors, ionized impurity scattering is suppressed and the electron mobility in the channel is in-

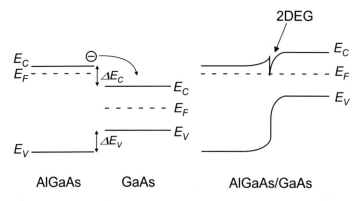

Figure 1.11. Energy band diagram of the heterostructure in an AlGaAs/GaAs HEMT.

creased. The term 2DEG stems from the fact that the electrons can move freely only in the two spatial directions parallel to the interface but not across it.

The second type of microwave transistors using heterostructures is the HBT. The idea of the HBT is almost as old as the bipolar transistor itself. In 1948, W. Shockley outlined the advantage of incorporating a heterostructure into a bipolar transistor [29]. In 1957, H. Kroemer formulated the basic HBT theory [30]. Because during that time high-quality heterostructures could not be grown, it was not until the early 1980s when practical HBTs, first using the AlGaAs/GaAs system, could be successfully fabricated [31].

The heterostructure effect exploited in HBTs is shown in Fig. 1.12. The key part of a HBT is the emitter-base heterojunction with the bandgap of the emitter being larger than that of the base. Because of the different badgaps and the resulting bandgap difference ΔE_G, electrons moving from the emitter to the base encounter a smaller energy barrier to be surmounted than holes travelling from base to emitter. Thus hole injection from the base into the emitter is strongly suppressed, and higher current gains compared to those in homojunction BJTs can be obtained. The emitter injection efficiency G_e of an HBT describes the ratio of the desired electron injection to the undesired hole injection at the emitter-base junction. It is connected to the current gain and obeys the relation

$$G_e \propto \frac{N_{DE}}{N_{AB}} \exp\left(\frac{\Delta E_G}{k_B T}\right) \qquad (1\text{-}16)$$

for the HBT from Fig. 1.12. N_{DE} and N_{AB} are the emitter and base doping concentrations, k_B is the Boltzmann constant, and T is the absolute temperature in the device. Due to the exponential term in Eq. (1-16), it is possible to dope the base much higher than the emitter without significantly decreasing the current gain. Strictly speaking, a large valence band offset ΔE_V is desirable for HBTs. This differs from HEMTs in which a large conduction band offset is beneficial. Using a high base

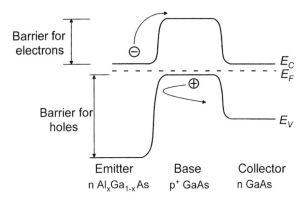

Figure 1.12. Energy band diagram of the heterostructure in an AlGaAs/GaAs HBT (smooth emitter-base heterojunction.

doping density produces the following two positive effects on the transistor performance: (1) a low base resistance resulting in higher f_{max} and lower NF_{min}, and (2) a very thin base (without running the risk of base punch-through), leading to a short base transit time and thus a high cutoff frequency f_T. Furthermore, by varying the composition of the base a graded bandgap (decreasing bandgap toward the collector) in the base can be engineered. This gives rise to a field aiding the transport of free carriers in the base and further increases f_T.

Besides the conventional AlGaAs/GaAs heterojunction, a number of other material combinations are being used in microwave transistors. When considering the lattice constants of the semiconductors, three different heterostructure types can be found: lattice matched, pseudomorphic, and metamorphic heterostructures. The first HEMTs and HBTs made use of the material system AlGaAs/GaAs (on GaAs substrates). The lattice constants of AlAs and GaAs are nearly the same. Therefore $Al_xGa_{1-x}As$ layers of any Al content x can be grown *lattice matched* on GaAs substrates. Because of the existence of so-called DX centers (deep-level states in doped AlGaAs), however, the maximum Al content in AlGaAs/GaAs heterostructures in typical microwave transistors is restricted to about 0.35. Figure 1.13 shows the bandgap of commonly used semiconductors as a function of lattice constant. As can be seen, another lattice matched system is $Al_{0.48}In_{0.52}As/In_{0.53}Ga_{0.47}As$/InP (on InP substrates). This system offers larger ΔE_G and ΔE_C values than the AlGaAs/GaAs counterpart, which makes heterostructures on InP substrates more promising for high-performance microwave transistors. Recently, the lattice matched system $AlAs_{0.56}Sb_{0.44}/In_{0.53}Ga_{0.47}As$/InP (not shown in Fig. 1.13) received attention in research because it offers an extraordinary large conduction band offset of around 1.75 eV. The drawbacks of fragility, availability of only small diameter wafers and high price, however, currently hamper the use of InP substrates in high-volume commercial applications.

At the beginning of heterostructure research it was believed that only materials with nearly the same lattice constant could result in heterostructures useful for electron devices. It turned out, however, that it is also possible to grow good quality heterostructures from materials with different lattice constants, provided the thickness of the grown layer does not exceed a certain critical value t_c. If the grown layer is thinner than t_c, its crystalline structure accommodates to that of the substrate material. This causes a lattice deformation in the grown layer, and a *pseudomorphic* layer, frequently also called a strained layer, is created. As an example, Fig. 1.14 shows the schematic of a pseudomorphic AlGaAs/InGaAs/GaAs heterostructure in which the pseudomorphic InGaAs layer is under compressive strain. The amount of strain depends on the lattice mismatch between substrate and layer, and on the layer thickness. When the thickness exceeds t_c, the grown layer will relax, causing a large number of dislocations at the interface. The critical thickness is about 40 nm, 20 nm, and 13 nm for x = 0.1, 0.2, and 0.3 in the $In_xGa_{1-x}As$/GaAs system [32].

Since 1986, pseudomorphic AlGaAs/InGaAs/GaAs heterostructures with In contents in the range of 15 to 25% were successfully grown on GaAs substrates and used in pseudomorphic HEMTs (GaAs pHEMT) [33]. At about the same time,

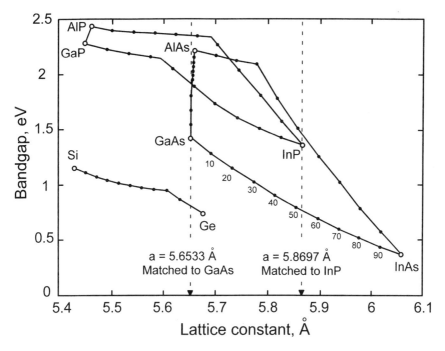

Figure 1.13. Bandgap versus lattice constant plots for commonly used semiconductors. The numbers at the curve between GaAs and InAs designate the In percentage in the InGaAs alloy (for example, 20 corresponds to $In_{0.2}Ga_{0.8}As$).

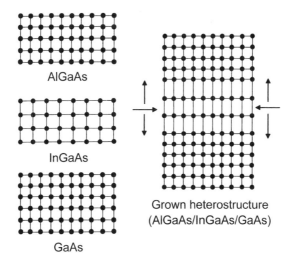

Figure 1.14. Pseudomorphic heterostructure (on the right) consisting of the layer sequence AlGaAs-InGaAs-GaAs (on the left). The arrows indicate the compressive strain occuring in the pseudomorphic InGaAs layer.

pseudomorphic heterostructures using $In_xGa_{1-x}As$ layers with $x > 0.53$ grown on InP substrates also became popular for pHEMTs [34]. Another pseudomorphic heterostructure frequently used in microwave transistors is strained SiGe grown on Si. It is used in SiGe HBTs [35].

The third and newest kind of heterostructures currently used in microwave transistors is the so-called *metamorphic* type. The basic concept is to use a substrate material (e.g., GaAs) and to overgrow a graded buffer layer (e.g., InAlAs) with a thickness much greater than t_c [36]. The buffer layer serves as a relaxed pseudosubstrate for the actual device layer. Because the buffer is extremely thick, dislocations arising at the interface substrate/buffer barely influence the electrical properties of the device layer on top of the buffer. Metamorphic $In_xGa_{1-x}As$-channel/InAlAs-buffer/GaAs structures with In contents x in the channel above 0.5 have been demonstrated in metamorphic HEMTs (mHEMTs) [37]. The main advantage of the metamorphic approach is that inexpensive GaAs substrates can be used to obtain high ΔE_C values, and thus InP-HEMT-like performance can be attained with GaAs.

1.3.3 Recent Developments

Two new directions in microwave transistor research that occurred in the 1990s are worth mentioning. The first is the application of the standard device of Si VLSI, i.e., the Si MOSFET, as a microwave device. Despite the fact that in the past the Si MOSFET had not been considered seriously for microwave applications due to its relatively low speed, the continuous FET scaling and increasing maturity of Si MOS technology in recent years has led to MOSFETs becoming a strong candidate. In fact, the topic of microwave CMOS was frequently discussed at all major device conferences around the world in late 1990s. Meanwhile, CMOS microwave circuits with operating frequencies up to 5 GHz have been reported [38, 39]. Si power MOSFETs transformed into well-accepted microwave transistors at frequencies up to about 2.5 GHz are widely used in base stations of wireless communication systems [40, 41].

The second direction is the investigation of wide bandgap semiconductors, such as SiC and III-nitrides, for power transistors with large output powers in the GHz range [42]. The wide bandgap of these materials also allows operating temperatures far exceeding those for Si and III–V transistors. This could open up new applications for microwave transistors, e.g., in automobiles and aircraft.

Figure 1.15 summarizes the major milestones in the development of microwave transistors in the past 50 years. The frequency limits of microwave transistors have been extended considerably during this period. It is worth mentioning that both bipolar and field-effect transistors made from III–V compound semiconductors with cutoff frequencies exceeding 300 GHz and maximum frequencies of oscillation above 600 GHz were reported in late 1990s. In Fig. 1.16 the evolution of the record values for f_T and f_{max} obtained from the most advanced microwave transistors in the period from 1960 to 2000 is shown. The record values of f_T are evidently lower than those of f_{max}. In 2000, InP-based transistors showed the highest f_T and

1.4 STATE OF THE ART OF MICROWAVE TRANSISTORS IN THE YEAR 2001

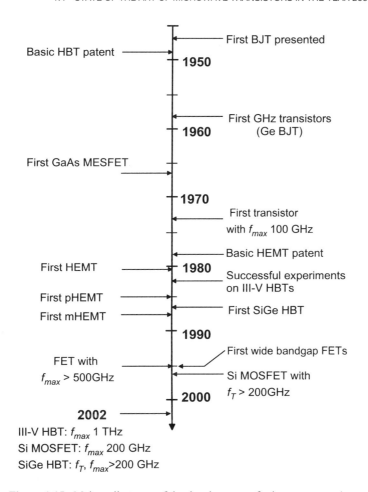

Figure 1.15. Major milestones of the development of microwave transistors.

f_{max} among all semiconductor devices. The highest f_{max} in excess of 1 THz has been obtained from an InP transferred substrate HBT [58], whereas the highest f_T of 396 GHz came from an InP HEMT [56].

1.4 STATE OF THE ART OF MICROWAVE TRANSISTORS IN THE YEAR 2001

1.4.1 III–V FETs

Both the HEMT and GaAs MESFET are widely used microwave devices due to their simple structure and superior high-frequency performance. Up to the year

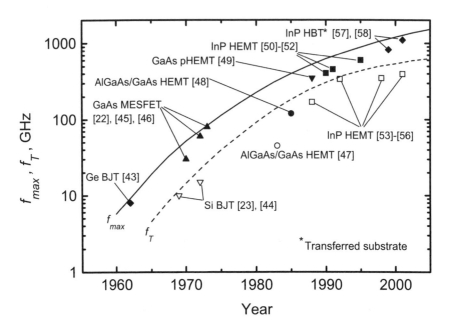

Figure 1.16. Evolution of the frequency limits (record f_T and f_{max}) of microwave transistors from 1960 to 2001.

2001, 0.12 μm gate-length GaAs MESFETs with f_T and f_{max} of 121 GHz and 160 GHz, respectively, as well as minimum noise figures of 0.73 dB at 12 GHz and 0.9 dB at 18 GHz have been realized [59]. A record f_T of 168 GHz has been obtained from a 0.06 μm gate-length GaAs MESFET [60] and a record f_{max} of 177 GHz from a 0.12 μm gate transistor [61]. At 26 GHz, a 0.11 μm gate-length MESFET showed a minimum noise figure of 0.78 dB [62]. A power module containing four GaAs MESFETs delivered an output power of 125 W at 2.2 GHz [63]. In terms of the GaAs MESFET output power density, no significant progress has been made since 1980. The highest output power density reported for GaAs MESFETs is 1.4 W/mm at 8 GHz [64].

The most important developments in the field of III–V FETs resulted from HEMTs. Figure 1.17 shows the cross section of a typical HEMT. All early HEMTs consisted of GaAs substrate, buffer, channel, and n-type AlGaAs barrier layers. A characteristic feature of many microwave FETs (both MESFETs and HEMTs) is the cross section of the gate. When the gate length is reduced down to the deep submicron range, the cross section of the gate decreases. Thus, the resistance of the small gate strip becomes large, which has a negative influence on the gain and noise behavior at high frequencies. Therefore, so-called mushroom gates, as shown in Fig. 1.17, are frequently used to achieve a short gate length and a small gate resistance.

The high free-carrier mobilities in the 2DEG channel of AlGaAs/GaAs heterostructures raised much hope that AlGaAs/GaAs HEMTs could surpass both the

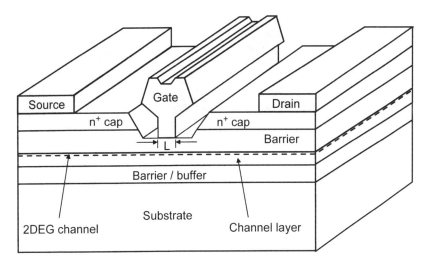

Figure 1.17. Cross section of an advanced HEMT (after [65]).

noise behavior and the frequency limits of GaAs MESFETs. Experimental work during the 1980s revealed, however, that these expectations had been too optimistic. A high 2DEG mobility in HEMT channels does not automatically lead to superior device performance. Moreover, the 2DEG must have a high electron sheet density n_s and good electron confinement. These two properties require a conduction band offset larger than that found in the AlGaAs/GaAs system.

Therefore, since the late 1980s, the focus of HEMT research has been shifted to systems that offer large conduction band offsets, such as pseudomorphic heterostructures on GaAs as well as lattice-matched and pseudomorphic structures on InP, all using InGaAs channel layers. Another added advantage of such systems is that the mobility in the $In_xGa_{1-x}As$ channel layer is in many cases higher than that in GaAs. Tables 1.2 and 1.3 compare the properties of different III–V heterostructures frequently used in HEMTs. The much higher electron sheet densities in heterostructures on InP substrates and in pseudomorphic heterostructures on GaAs substrates result from the larger conduction band offset and from the fact that two heterojunctions contribute to the 2DEG in these structures. In the case of an $In_{0.52}Al_{0.48}As/In_{0.53}Ga_{0.47}As/InP$ heterostructure, for example, the upper $In_{0.52}Al_{0.48}As/In_{0.53}Ga_{0.47}As$ heterojunction ($\Delta E_C \approx 0.52$ eV) and the lower $In_{0.53}Ga_{0.47}As/InP$ heterojunction ($\Delta E_C \approx 0.27$ eV) form the boundaries of the channel. In the case of the conventional AlGaAs/GaAs HEMT, only one heterojunction with an ΔE_C of around 0.22 eV existed. As can be seen from Table 1.3, pseudomorphic heterostructures on InP possess sheet concentrations n_s and low-field mobilities μ_0 about two to three times higher than those in conventional AlGaAs/GaAs structures, thus resulting in a far superior microwave performance in InP HEMTs than in GaAs HEMTs.

During the 1990s, considerable experimental work was done in the field of meta-

Table 1.2. Bandgap discontinuities and conduction band offsets for different III–V heterojunctions

Heterojunction type	ΔE_G, eV	ΔE_C, eV	Substrate
$Al_{0.3}Ga_{0.7}As/GaAs$ (lm)*	0.38	0.22	GaAs
$Al_{0.3}Ga_{0.7}As/In_{0.2}Ga_{0.8}As$ (pm)	0.582	0.407	GaAs
$In_{0.52}Al_{0.48}As/In_{0.53}Ga_{0.47}As$ (lm)	0.714	0.52	InP
$In_{0.53}Ga_{0.47}As/InP$ (lm)	0.616	0.271	InP
$In_{0.52}Al_{0.48}As/In_{0.65}Ga_{0.35}As$ (pm)	0.768	0.660	InP

*Lattice matched and pseudomorphic heterojunctions are denoted lm and pm, respectively.

morphic HEMTs (mHEMTs) grown on GaAs substrates. The aim was to achieve InP-HEMT-like performance but to avoid expensive InP substrates. In 1999, a 3 in diameter InP substrate cost about $700 compared to $170 for a 4 in diameter GaAs substrate [37]. Table 1.4 summarizes the record f_T and f_{max} values reported for the different III–V HEMTs.

In Figs. 1.18 and 1.19, the upper limits of the cutoff frequencies and maximum frequencies of oscillation for GaAs MESFETs, GaAs HEMTs, and InP HEMTs are shown as a function of gate length. The upper limit curves demonstrate the state of the art of microwave MESFETs and HEMTs (December 2001) and have been obtained from the best reported f_T and f_{max} values of the corresponding transistor type at different gate length levels. Clearly, the characteristic frequencies f_T and f_{max} depend on the gate length. For gate lengths ranging from about 0.3 to 1 μm, the f_T increase is almost linear in the log–log plot. This behavior corresponds to the L^{-a} dependence (a is a constant for a certain transistor type) of f_T mentioned in various recent FET review papers. For FETs with extremely short gates (0.1 μm and below), f_T increases only slightly with decreasing gate length. The reason is that parasitic effects, such as the gate fringing capacitances and gate bonding pad capacitances, become comparable to the intrinsic gate capacitance of the FET. The maximum frequency of oscillation shows a saturation behavior at short gate lengths as well.

In general, the upper limit plots of the different FET types look similar. As expected from the superior properties of InP heterostructures (see Tables 1.2 and 1.3), InP HEMTs are the best III–V FETs for microwave applications in terms of both f_T

Table 1.3. 2DEG mobilities and electron sheet concentrations in HEMTs

Heterojunction type	μ_0, cm^2/Vs	n_s, cm^{-2}	Substrate
$Al_{0.3}Ga_{0.7}As/GaAs$ (lm)	5400	1.4×10^{12}	GaAs
$Al_{0.25}Ga_{0.75}As/In_{0.15}Ga_{0.85}As$ (pm)	6400	2.2×10^{12}	GaAs
$Al_{0.25}Ga_{0.75}As/In_{0.22}Ga_{0.78}As$ (pm)	5300	3.5×10^{12}	GaAs
$In_{0.52}Al_{0.58}As/In_{0.53}Ga_{0.47}As$ (lm)	10000	3.0×10^{12}	InP
$In_{0.52}Al_{0.48}As/In_{0.8}Ga_{0.2}As$ (pm)	12700	3.6×10^{12}	InP
$In_{0.35}Al_{0.65}As/In_{0.65}Ga_{0.35}As$ (pm)	15000	4.6×10^{12}	InP

Table 1.4. Record cutoff frequencies and maximum frequencies of oscillation of experimental HEMTs

FET Type	f_T, GHz	L, μm	Ref.	f_{max}, GHz	L, μm	Ref.
AlGaAs/GaAs HEMT (lm)*	113	0.1	[66]	151	0.24	[67]
GaAs HEMT (pm)	152	0.1	[68]	350	0.15	[49]
GaAs HEMT (mm)	204	0.18	[69]	400	0.1	[70]
InP HEMT (lm)	396	0.025	[56]	455	0.15	[51]
InP HEMT (pm)	340	0.05	[71]	600	0.1	[52]

*The notations lm, pm, and mm denote lattice matched, pseudomorphic, and metamorphic.

and f_{max}. Surprising trends are found when comparing the cutoff frequencies obtained from GaAs MESFETs, AlGaAs/GaAs HEMTs, and GaAs pHEMTs. As expected, for long-gate devices, GaAs MESFETs show lower f_T than the GaAs-based HEMTs. For shorter gates, however, the cutoff frequency of MESFETs is actually higher than that of AlGaAs/GaAs HEMTs and GaAs pHEMTs. This behavior might be due to the decreased influence of mobility on device speed in short-gate FETs due to the effect of drift velocity saturation.

The III–V FET noise performance is shown in Fig. 1.20. The minimum noise

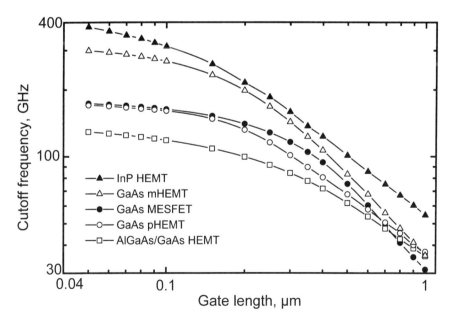

Figure 1.18. Upper f_T limits for different types of microwave III–V FETs. The upper limits have been empirically constructed from experimental data published up to the year 2001.

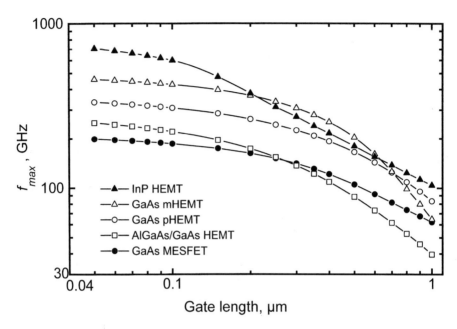

Figure 1.19. Upper f_{max} limits for different types of microwave III–V FETs. The upper limits have been empirically constructed from experimental data published up to the year 2001.

figures of Si MOSFETs are also included for comparison. Experimental results published in the literature have been used to construct the lower limit NF_{min} versus frequency curves shown in Fig. 1.20. Clearly, InP HEMTs possess by far the best noise behavior of all III–V FETs. This can be attributed to the excellent transport properties (i.e., extremely high mobility and peak velocity) of the 2DEG channel with high In content. Minimum noise figures are clearly below 1 dB up to more than 60 GHz, and are around 1 dB at 94 GHz. Also, GaAs mHEMTs showed very low noise figures up to 24.5 GHz. Although the noise figures of GaAs mHEMTs are slightly higher than those of InP HEMTs, the inexpensive GaAs substrate makes this device attractive. Despite the fact that GaAs pHEMTs have higher noise figures than InP HEMTs and GaAs mHEMTs, these transistors are widely used in low-noise amplifiers at frequencies up to about 50 GHz, and noise figures as low as 1.2 dB have been obtained at this frequency. The reasons for the popularity of GaAs pHEMTs are that their fabrication is easier and less expensive compared to InP HEMTs and GaAs mHEMTs and that GaAs pHEMT technology already reached a high level of maturity. The best reported noise figures of AlGaAs/GaAs HEMTs are slightly lower than those of GaAs MESFETs and only a little higher than those of GaAs pHEMTs.

Useful power amplification up to 200 GHz using monolithically integrated InP HEMT amplifiers has been reported [72]. The output powers reported for GaAs

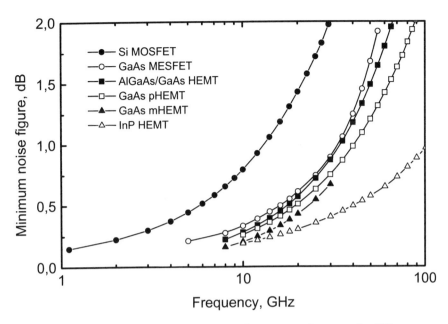

Figure 1.20. Lower limit of the minimum noise figure versus frequency for different types of microwave FETs.

MESFETs, AlGaAs/GaAs HEMTs, InP HEMTs, GaAs pHEMTs, and wide bandgap FETs are shown in Fig. 1.21. The following relation holds between the upper limit output power and operating frequency:

$$P_{out} \times f_{op}^2 = c_p \qquad (1\text{-}17)$$

where c_p is a constant describing the state of power FET technology and design cleverness at a particular time. Over the years, c_p increased continuously and reached a value around 4×10^{21} WHz^2 for III–V FETs in the year 2000.

In principle, the output power at low frequencies can be increased simply by increasing FET periphery, i.e., by using a larger gate width W. At high frequencies, however, there is an upper limit for W beyond which the power performance deteriorates. Around 2 GHz, the highest output powers reported are 240 W for AlGaAs/GaAs HEMTs* [73] and 125 W for GaAs MESFETs* [63]. At 94 GHz, GaAs pHEMTs and InP HEMTs are the only two FET types delivering useful power gain. The output powers, however, are orders of magnitude lower than those at 2 GHz.

---

*The data are for a module consisting of four single transistors.

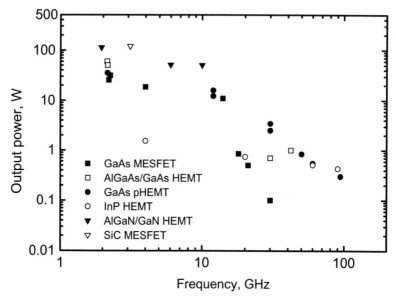

Figure 1.21. Output power versus frequency for experimental III–V and wide bandgap power FETs. Included are only single-chip devices.

1.4.2 BJTs and HBTs

Si BJTs are still used in microwave systems, and some efforts have been made to improve the high-frequency performance of these devices. Currently, the highest values for f_T and f_{max} obtained with experimental Si BJTs are 100 GHz and 101 GHz, respectively [74], but there is a clear trend to shift research and development to HBTs for microwave applications.

The first successfully realized HBTs for microwave applications were based on GaAs. Much work on InP and SiGe HBTs has also been done recently. Figure 1.22 shows the cross section of a GaAs HBT consisting of an AlGaAs emitter and GaAs base and collector regions. In the figure, the dashed lines separate the intrinsic part (within the dashed lines) and the extrinsic part (outside the dashed lines) of the HBT. The extrinsic HBT does not have any useful effect on transistor operation, but rather it deteriorates the power gain, the maximum frequency of oscillation, and the minimum noise figure due to the external components, such as the collector-base capacitance C_{cb} and the base resistance R_B. The influence of R_B and C_{cb} on f_{max} can be clearly seen from the expression

$$f_{max} \approx \sqrt{\frac{f_T}{8\pi R_B C_{cb}}} \qquad (1\text{-}18)$$

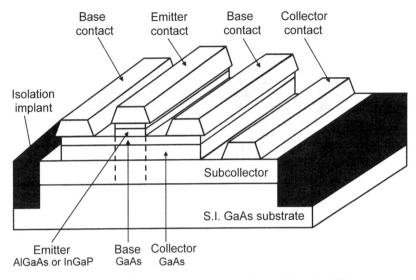

Figure 1.22. Cross section of a typical GaAs HBT (after [75]).

GaAs HBTs with AlGaAs and InGaP emitters became commercially available in the 1990s and are commonly used in modern microwave power amplifiers, whereas the technology for InP HBTs is less mature. Table 1.5 summarizes the state of the art of GaAs and InP HBTs in terms of f_T and f_{max}. These transistors have a structure similar to that shown in Fig. 1.22.

Both GaAs and InP HBTs show record f_T and f_{max} values of more than 150 GHz, but with InP HBTs having an edge over GaAs HBTs when both high f_T and f_{max} are required. Moreover, double heterojunction InP HBTs with InP collectors have a higher breakdown voltage compared to GaAs HBTs. The record f_T and f_{max} of GaAs and InP HBTs, as well as of SiGe HBTs, which will be discussed later, are summarized in Figs. 1.23 and 1.24, respectively. Note that the frequency performance is plotted versus the base thickness w_B, which is of similar importance to the bipolar transistors as the gate length to the FETs. The solid lines in Figs. 1.23 and 1.24 are

Table 1.5. Reported f_T and f_{max} of GaAs and InP HBTs

Transistor type	f_T, GHz	f_{max}, GHz	Year	Ref.
GaAs	156	255	1998	[76]
GaAs	138	275	1997	[77]
GaAs	60	350	1992	[78]
InP	270	300	2000	[79]
InP	341	238	2001	[80]

the empirical upper limits for f_T and f_{max}. Both f_T and f_{max} increase with decreasing base thickness down to about 20–30 nm. For thinner base layers, f_{max} reaches a maximum followed by a pronounced falloff. This effect can be attributed to the increasing base resistance R_B and its influence on f_{max} [see Eq. (1–18)]. The trend of the f_T versus w_B curve is similar to that of f_{max}. The falloff at small base widths, however, is weaker.

Recently, an interesting and novel InP HBT made from a substrate transfer process has been reported [81]. In these transistors, the size of the extrinsic device part is dramatically reduced, thus leading to a very small external collector–base capacitance and resulting in an extraordinarily high f_{max}. Transferred substrate HBTs with record f_{max} of 820 and 1080 GHz have been successfully fabricated [57, 58]. These are the highest f_{max} ever reported for a three-terminal device to date. Another InP transferred substrate HBT showed high values of $f_T = f_{max}$ = 295 GHz [82]. It is expected that by further optimization of the design of transferred substrate HBTs, f_T and f_{max} can be increased to 500 GHz and 1500 GHz, respectively. Despite the fact that transferred substrate HBTs are not suitable for cost-effective mass production, they offer the possibility of realizing useful power amplification at extremely high frequencies.

The major disadvantage of InP HBTs in general is, as discussed before, the brittle and expensive InP substrate. Also, the technology of InP HBTs is relatively immature compared to that of GaAs HBTs.

A main advantage of III–V HBTs compared to III–V FETs is that high f_T and

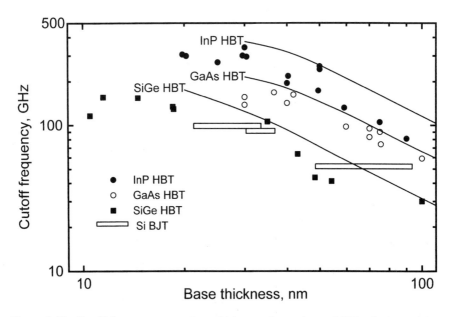

Figure 1.23. Cutoff frequency versus base thickness of experimental SiGe, GaAs, and InP HBTs. Representative f_T values for conventional Si BJTs are included for comparison.

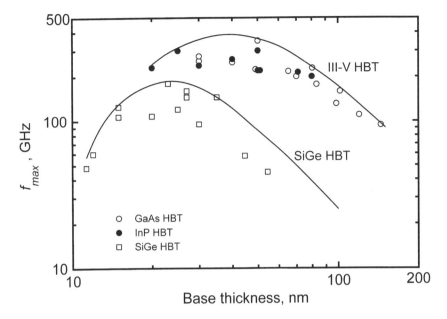

Figure 1.24. Maximum frequency of oscillation versus base thickness of experimental SiGe, GaAs, and InP HBTs.

$f_{\max}$ can be obtained without the limitation of photolithography. The lateral minimum device dimension in HBTs is usually around or above 1 μm, whereas in FETs with comparable frequency limits the gate length is 0.25 μm or less (see Table 1.4). In bipolar transistors (conventional homojunction bipolar transistors and HBTs), electrons travel perpendicular to the wafer surface from emitter to collector. Therefore, the device dimensions critical for transistor speed are not lateral but vertical and are nearly independent of the photolithography process.

III–V HBTs have inferior high-frequency noise behavior compared to III–V FETs. Nonetheless, the use of HBTs in low-noise amplifiers in wireless communication systems operating at frequencies up to 2.5 GHz has become an interesting option. State of the art noise figures of both GaAs and InP HBTs (state of the art as of 2001) are listed in Table 1.6. Worth mentioning is the fact that between 1989 and 2001, no significant improvement of the noise behavior of InP HBTs could be obtained.

The main application of III–V HBTs, however, is in microwave power amplifiers. HBTs offer much higher power densities than III–V FETs, which makes impedance matching easier and leads to smaller chip sizes. Table 1.7 shows a compilation of output power, power density, gain, and power-added efficiency of III–V HBTs operating at microwave frequencies. In the table, P_D in W/mm is the output power density per mm emitter length and P_D in W/μm² is the output power density per μm² emitter area.

Table 1.6. Minimum noise figures of III–V HBTs (in dB)

HBT Type	NF_{min} @ 2 GHz	NF_{min} @ 6 GHz	NF_{min} @ 12 GHz	NF_{min} @ 18 GHz	Year	Ref.
GaAs	0.9	1.1	1.2	1.6	1998	[83]
InP	0.46	1.09	2.7	3.3	1989	[84]
InP	—	1.55	1.9	—	1999	[85]

Another increasingly popular and important heterostructure device is the SiGe HBT. The basic concept of this transistor is similar to that of the III–V HBTs, but the SiGe HBT possesses a huge advantage over the III–V HBTs in the sense that the SiGe HBT can be fabricated with the existing Si CMOS technology with only a few more steps added. The layer sequence of SiGe HBTs from the bottom up is Si substrate, n^+-Si subcollector, n-Si collector, p-strained-SiGe base, and n-Si emitter. There are two heterojunctions in SiGe HBTs, one at the emitter-base junction and the other at the base–collector junction. The first experimental SiGe HBT was reported in 1987 [89]. Since then, SiGe HBTs have evolved from a laboratory curiosity to well-accepted microwave devices and achieved commercial status in the late 1990s. Experimental devices showing f_T and f_{max} of 210 GHz and 285 GHz, respectively, have been reported [90, 91]. Theoretical investigations led to the conclusion that optimally designed SiGe HBTs showing both f_T and f_{max} around 300 GHz should be possible [92]. Unlike III–V HBTs, SiGe HBT noise figures are quite low. Minimum noise figures of 0.18 dB, 0.55 dB, and 1.18 dB at 4, 8, and 12 GHz, respectively, were reported in [93], and 1.0 dB and 1.7 dB at 12 GHz and 18 GHz were reported in [94]. Figure 1.25 shows the best minimum noise figures of different HBTs reported in the literature. For comparison, the noise performances of the state-of-art Si BJTs [95] and GaAs MESFETs are also included. In the lower frequency range (2 to 5 GHz), the noise figures of SiGe HBTs are comparable with those of GaAs MESFETs. At higher frequencies, especially above 10 GHz, SiGe HBTs become noisier than GaAs MESFETs. GaAs HBTs show nearly constant noise figures of around 1 dB between 1 and 10 GHz.

A power SiGe HBT delivering an impressive 230 W output power at 2.8 GHz has been reported in [96]. The main problem of Si and SiGe bipolar power transistors is the relatively low breakdown voltage. Typical GaAs HBTs show

Table 1.7. Power performance of III–V HBTs

HBT Type	f, GHz	P_{out}, W	P_D, W/mm	P_D, mW/μm^2	Gain, dB	PAE, %	Ref.
GaAs	7.5	2.52	12.6	4.2	3.4	31.8	[86]
GaAs	10	0.60	30.0	10.0	7.1	60	[86]
GaAs	26	0.74	6.4	4.0	—	—	[87]
InP	18	1.17	4.88	2.44	7.3	54	[88]

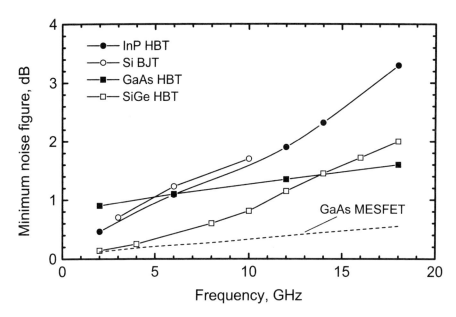

Figure 1.25. Reported minimum noise figure versus frequency for GaAs HBTs, InP HBTs, SiGe HBTs, and Si BJTs.

collector–emitter and collector–base breakdown voltages BV_{CE0} and BV_{CB0} around 15 V and 20 to 25 V, respectively, thus allowing a safe operation in a 6 V power supply system. In the case of Si BJTs and SiGe HBTs, BV_{CE0} and BV_{CB0} are only 3 to 5 V and 10 to 15 V, respectively, which is too low for several practical applications [97].

From the economical perspective, the main advantages of SiGe HBTs compared to III–V HBTs are (1) large diameter and inexpensive Si substrates, and (2) existing Si technology available for the production of SiGe HBTs. A cost study for HBTs yields $0.12/mm² for SiGe, $0.5/mm² for GaAs, and $1.2/mm² for InP based on the use of 6″ SiGe, 4″ GaAs, and 3″ InP wafers and technology available in 1998 [98].

Figure 1.26 summarizes the state of the art of bipolar power microwave transistors. Of these, the GaAs HBT is the dominating microwave bipolar power device with numerous commercial applications. A comparison of Figures 1.26 and 1.21 reveals one interesting detail. Although the output power density of bipolar transistors is much higher compared to FETs (see Fig. 1.29 in Section 1.4.3 and Table 1.7), their total output per chip is frequently lower than that of FETs. The lower output power density of FETs is compensated by the successful realization of large area devices resulting in high total chip output powers of FETs.

Finally, Fig. 1.27 shows the power added efficiency *PAE* of HBTs and III–V FETs as a function of frequency. At low GHz-frequencies HBTs show the highest *PAE* while at very high frequencies (above 50 GHz) InP HEMTs are superior.

34 BACKGROUND ON MICROWAVE TRANSISTORS

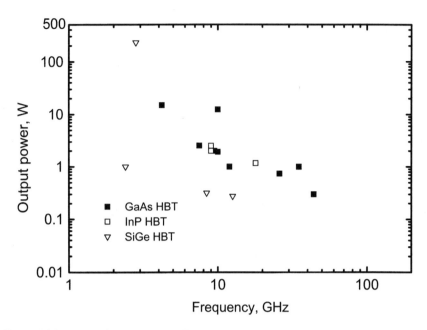

Figure 1.26. Reported output power of single-chip power HBTs as a function of frequency.

1.4.3 Wide Bandgap Transistors

Since 1990, wide bandgap semiconductors such as SiC and group III nitrides (AlN, GaN, and AlGaN) have received increasing attention for high-power microwave applications due to their high breakdown field and their high electron peak and saturation velocities. In the case of SiC, there is no unique crystal type, as in the case of Si or GaAs, but rather more than 100 different so-called polytypes with different crystal structures. The 4H and 6H SiC polytypes are commonly used for microwave FETs, with 4H SiC being the favorite material. The group III nitrides exist in two crystal types, namely the wurtzite (hexagonal) and zincblende (cubic) crystals. Un-

Table 1.8. Properties of Si, GaAs, and the wide bandgap materials 6H SiC, 4H SiC, and GaN

Material	Si	GaAs	6H SiC	4H SiC	GaN
E_G, eV	1.12	1.42	2.86	3.2	3.4
E_{br}, V/cm	5.7×10^5	6.4×10^5	3.0×10^6	3.3×10^6	3.8×10^6
μ_0, cm²/Vs	710	4700	350	610	1200
v_{peak}, 10^7cm/s	1	1.8	2	2	2.46
κ, W/Kcm	1.3	0.5	2.9	2.9	1.2
					0.4*

*Data for sapphire substrate.

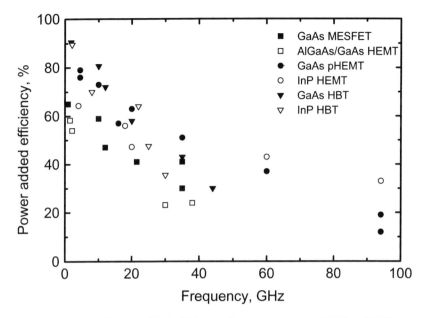

Figure 1.27. Reported power added efficiency of microwave power HBTs and FETs.

til now, only the wurtzite type has been used in microwave transistors. Table 1.8 summarizes important properties (bandgap E_G, breakdown field E_{br}, electron low-field mobility μ_0, electron peak velocity v_{peak}, and thermal conductivity κ) of 4H and 6H SiC, wurtzite GaN, and, for comparison, Si and GaAs (all materials n-type with a doping concentration of 10^{17} cm^{-3}).

The stationary electron drift velocity versus field characteristics of the wide bandgap materials 4H SiC and wurtzite GaN, and, for comparison, GaAs are shown in Fig. 1.28. At high fields above 30 kV/cm, the wide bandgap materials show much higher drift velocities than GaAs. On the other hand, GaAs has a higher low-field drift velocity, which leads to a higher low-field electron mobility. To obtain high output powers at low supply voltages (e.g., 3 V), a high mobility resulting in a low on-resistance of the transistor is important. For high voltage operation, on the other hand, a high breakdown voltage of the transistor is imperative together with a

Table 1.9. f_T and f_{max} of experimental wide bandgap FETs

FET type	L, μm	f_T, GHz	f_{max}, GHz	Year	Ref.
SiC	0.45	22	50	1998	[105]
SiC	0.5	13.2	42	1996	[106]
AlGaN/GaN	0.05	110	140	2000	[107]
AlGaN/GaN	0.12	101	155	2001	[108]
AlGaN/GaN	0.25	50	100	2000	[109]

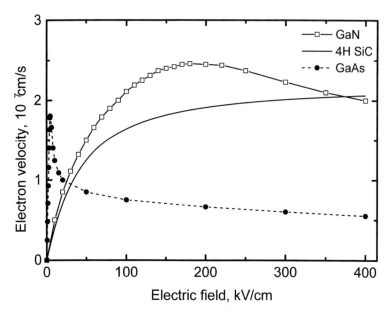

Figure 1.28. Stationary velocity-field characteristics of GaAs, 4H SiC, and GaN. Data taken from [99–102].

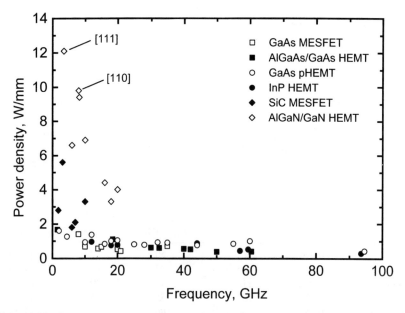

Figure 1.29. Output power density versus frequency for different types of power field-effect transistors.

large carrier velocity at high electric fields. The large breakdown fields in SiC and GaN result in extremely high breakdown voltages for SiC- and GaN-based FETs.

The first commercial SiC MESFET was announced in 1999 [103]. Another class of wide bandgap FETs is AlGaN/GaN HEMTs. Because there are still no GaN substrates, GaN transistors are realized on sapphire or SiC substrates. Sapphire substrates offer the advantage of lower cost, but SiC substrates possess a much higher thermal conductivity, which is important for high-power transistors. Another promising option is to grow GaN on Si substrates [104]. AlGaN/GaN transistors are still being developed in research labs and were not yet available commercially as of 2001. Table 1.9 presents recently published f_T and f_{max} values for SiC MESFET and AlGaN/GaN HEMTs.

Although power amplification is the targeted application of AlGaN/GaN HEMTs, the high-frequency noise performance of such transistors has also been investigated. Minimum noise figures less than 1 dB up to 18 GHz have been reported for AlGaN/GaN HEMTs with a gate length of 0.12 μm [108].

The most impressive feature of SiC MESFETs and AlGaN/GaN HEMTs is the high output power per mm gate width. This combined with the high thermal conductivity of SiC substrates will lead to new microwave power FET designs. Typically, III–V FETs show much lower output power densities compared to wide bandgap FETs but can operate at higher frequencies. A compilation of reported output power densities of these types of microwave power FETs is shown in Fig. 1.29. The 9.8 W/mm and 12.1 W/mm output powers at 8 GHz and 3.5 GHz, respectively, obtained from an AlGaN/GaN HEMT is the highest output power density ever reported for a field-effect transistor [110, 111].

1.4.4 Si MOSFETs

The advances in CMOS processing, continuous scaling of gate length, progress in SOI (silicon on insulator) technology, and development of Si LDMOSFETs (laterally diffused MOSFET) stirred serious discussions on the suitability of MOSFETs and CMOS for microwave applications. Despite some continuing controversy in the debate, the prospects for CMOS with operating frequencies in the lower GHz range are promising. Microwave CMOS circuits operating at frequencies up to 5 GHz have been realized and CMOS chips for different GHz applications are commercially available. For example, a single-chip solution for 2.4 GHz Bluetooth containing transceiver and analog and digital circuitry including memory came to market in 2001 [112]. Furthermore, LDMOSFETs used as power transistors for frequencies up to 2.5 GHz have been commercially established as well.

Experimental work on low-noise microwave CMOS has been carried out with both conventional Si bulk technology and SOI technology. The SOI concept seems to be more promising because of the ease of integration with other high-performance microwave components also fabricated on insulators. A cutoff frequency of 245 GHz for a 60 nm gate bulk Si MOSFET [113] and 150 GHz f_T for a 125 nm gate SOI MOSFET [114] have been reported in 2001 and 1997, respectively. For a long time short-gate, high-f_T MOSFETs suffered from relatively low power gains at high frequencies and low maximum frequencies of oscillation. Up to the year 2000, the record f_{max} value remained as low as 66 GHz, obtained with a 0.5 μm gate length SOI MOSFET [115]. The main reasons for this were the high specific resistance of the gate material (polysilicon) and the fact that mushroom gates usually used in short-gate III–V FETs were not yet available in standard Si CMOS technology. A lot of work has been done to reduce the gate resistance by depositing silicides or metal on top of the polysilicon and by employing metal gates. These efforts finally led to improved power gain and f_{max} behavior, and in 2001, a 0.18 μm gate length bulk MOSFET with an f_{max} of 150 GHz [116], a 80 nm gate SOI MOSFET with an f_{max} of 185 GHz [117], and a 50 nm gate SOI MOSFET with a record f_{max} of 198 GHz [118] have been reported.

The lowest minimum noise figures reported for Si MOSFETs (state of the art as of 2001) are below 0.25 dB at 2 GHz and below 1 dB up to 8 GHz [116] (see Fig. 1.20). These noise figures are comparable to those of GaAs MESFETs in 1980. The noise performance of Si MOSFET will always be inferior to that of advanced III–V FETs because of their poorer electron transport properties. Another problem of microwave Si MOSFETs is the fact that good f_T, f_{max}, and NF_{min} can only be obtained with extremely scaled MOSFETs. The breakdown voltages of such MOSFETs may be too low for many practical applications. However, in the mass consumer markets, where cost and the ability for integration are of major concern, microwave Si MOSFETs have clear advantages over other microwave devices.

For high-power amplifiers operating at frequencies around 2 GHz, on the other hand, no deep submicron gates are required. Here LDMOSFETs optimized for high breakdown voltage are used. Commercial transistors of this kind delivering a maximum output power of 220 W at 2 GHz are commercially available [40].

1.5 APPLICATION ASPECTS

Discussions of microwave transistor applications in this section will be confined to electronics systems used in the civil domain. This field is evolving fast, and the market literally exploded during the 1990s. In the following, civil microwave applications that currently have large market volumes or that are expected to create mass markets in the near future are described. Furthermore, applications of different microwave transistor types in wireless communications are addressed. The focus will be placed on low-noise and power transistors for use in handsets and base stations for mobile or wireless communication systems.

1.5.1 Civil Applications of Microwave Systems

The operating frequencies of civil microwave applications range from a few hundreds MHz to 100 GHz. Currently, most systems having real mass markets operate at frequencies below 5 GHz. The number of units sold in these markets is in the order of millions per year.

At the lower bound of this frequency spectrum, a mass market is pagers operating at frequencies from 200 to 900 MHz. Microwave chips for identification purposes (so-called RF ID) operate at frequencies from the MHz range up to lower GHz frequencies. Cellular phones and the global system for mobile communications (GSM) running around 900 MHz and 1.8 GHz represent another large market for microwave transistors. More advanced mobile communications services, such as DECT (Digital European Cordless Telecommunications), PCS (Personal Communications System), PHS (Japanese Personal Handy-Phone System), DCS (Digital Communications Systems), and UMTS (Universal Mobile Telecommunications Services) operating in the frequency range from 1.8 to 2.4 GHz are also on the rise. A new system called the wireless local area network (WLAN) occupies the frequency band around 2.4 GHz. WLAN had been expected to be produced in large volumes in the 1990s but the acceptance of traditional inexpensive coaxial-based systems by the majority of users prevented the breakthrough of WLANs [119]. During the early 2000s, a number of large companies made another attempt using Bluetooth. Bluetooth has a somewhat different system concept and architecture compared to the traditional WLAN and is frequently called WPAN (Wireless Personal Area Network). It is designed to wirelessly link devices like desktop and notebook computers, printers, cellular phones, digital cameras, etc., over short distances around 10 m. New standards for high-speed data transfer like HIPERLAN (operating frequency 5.2 GHz) and HIPERLAN 2 (operating frequency 17 GHz) and the associated hardware were being developed in the early 2000s.

Examples for civil communications systems in the higher frequency range are direct-to-satellite communication (20 GHz downlink, 30 GHz uplink [119]), LDMS (Local Multipoint Communication Services, 27.5 to 29.5 GHz [120]), millimeter-wave digital radio systems (23 and 38 GHz, and possibly 12, 15, 18, and 26 GHz [121]).

Another field of civil microwave applications is automobiles. The envisaged applications include the GPS operating at 1.8 GHz, collision avoidance radar (77 GHz), vehicle identification, traffic management, and others [122]. GPS and collision avoidance radar are market segments with expected large volumes for microwave transistors.

Certain radar and sensor applications will operate around 94 GHz. When InP HEMTs became operable above 100 GHz in late the 1990s, discussions started on using microwave transistors also in frequency bands in the range from 140 GHz to 220 GHz. There are low-attenuation atmospheric windows at 140 to 165 GHz and 200 to 220 GHz for these applications [72].

In general, the civil microwave markets in the different frequency ranges mentioned above are expected to grow quickly during the next few years. Several market segments are not saturated yet, and it is certain that new applications will create even larger consumer markets. There is, however, no guarantee that a new and useful microwave product will automatically find commercial acceptance. Examples of this include the delay in the large volume production of WLAN systems and the economic disaster of the direct-to-satellite communication system IRIDIUM during the years 1999 and 2000.

1.5.2 Other Applications for Transistors with GHz Capabilities

Besides the "pure" microwave applications discussed above, there are two other important areas for transistors with the capability of GHz operation. The first of these is microprocessors, which comprise a large share of silicon VLSI electronics. Since the invention of the microprocessor in 1971, processor chip complexity has increased dramatically. Microprocessors are traditionally realized in CMOS technology, and transistors used are n- and p-channel silicon MOSFETs. The critical dimension in these MOSFETs is the gate length, which has decreased at a rate of approximately 0.87/year in the past. A reduction of critical device dimensions reduces the device switching time and consequently enhances the clock frequency of the microprocessor. The ITRS roadmap expectations for the feature size and clock frequency of microprocessors is listed in Table 1.10 [8].

The on-chip clock can be considered as the lowest frequency limit for transistor operation. In the year 2000, the on-chip clock was on the order of 1 GHz, and a speed of more than 10 GHz is expected for future high-performance processors. Thus the frequency limits of silicon MOSFETs to be employed in microprocessors must be in the GHz range. Although these transistors operate as digital devices, their frequency response has to be that of typical microwave transistors. In other words, there is an urgent need to design and produce silicon MOSFETs with high-frequency properties that in the past could be achieved only by III–V transistors.

The second "nonpure" microwave area requiring transistors with microwave performance is mixed-signal circuits and systems-on-chip (SoC). A SoC is the integration of almost all aspects of a system design on a single chip. This means that different components such as memory, high-performance/low-power logic, analog, and microwave circuit parts have to be realized side by side on one chip. The microwave part and the interface between it and other parts of the system must be able to handle signals in the microwave range. Table 1.11 shows the expected key performance requirements for transistors to be applied in the microwave part of future SoCs according to the ITRS 2001 edition [8]. Some remarks are required for the interpretation of the data in Table 1.11.

First, for SoCs it is imperative that all technologies be integrated into the standard CMOS process with minimum cost. Thus, the use of III–V components is out of question, and all microwave transistors must be silicon-based transistors, i.e., Si BJTs, SiGe HBTs, and Si MOSFETs. The trend is being driven by the fact that the integration levels for future SoCs are extremely large (millions of transistors per chip).

For the definition of the f_{max} roadmap target of bipolar transistors it was assumed

Table 1.10. Expected future minimum feature size and clock frequency for production-stage high-performance microprocessors

Year of production	2001	2004	2007	2010	2013	2016
Gate Length, nm	65	37	25	18	13	9
On-chip local clock, GHz	1.7	4	6.7	11.5	19.3	28.7

Table 1.11. Targeted microwave performance of transistors to be used in future mixed signal circuits and SoCs

Year of production	2001	2004	2007	2010	2013	2016
Physical gate length (ASIC), nm	90	53	32	22	16	11
System operating frequency, GHz	0.5–10	0.5–20	0.5–30	0.5–50	0.5–75	0.5–100
Bipolar f_T, GHz	45	60	80	95	110	125
Bipolar f_{max}, GHz	90	120	160	190	220	250
nMOSFET f_T, GHz	132	225	372	541	744	1082
nMOSFET f_{max}, GHz	160	175	190	200–230	230–260	260–290

that f_{max} should be approximately 5–10 times the transmit/receive frequency of the system. From Table 1.11, we see that the required cutoff frequency of bipolar transistors is only 50% of f_{max}. This is different from the $f_T \approx f_{max}$ guideline discussed in Section 1.2. The cutoff frequency roadmap target of nMOSFETs is related to the carrier transit time from source to drain. For the channel length (i.e., the effective source-to-drain distance) the physical gate length of low-power ASICs has been assumed together with an average electron velocity along the channel of about 7.5 × 10^6 cm/s. This f_T definition is also different from that typically used by device engineers. Nevertheless, the roadmap targets clearly indicate that considerable research efforts will be needed to meet the requirements of future mixed-signal solutions and SoCs.

The need for fast silicon transistors in microprocessors, SoCs, and low-cost communication applications, all of which will continue to enjoy a mass consumer market for many years to come, has put more emphasis on the development of silicon-based microwave components. Anyhow, a large portion of microwave electronics, especially those operating at higher GHz frequencies, will still be dominated by III–V transistors.

1.5.3 Microwave Transistors for Wireless Communications up to 2.5 GHz

Wireless communication systems such as cellular phones can be divided into the user part (handset) and the infrastructure (base stations). The user part consists of a receiver and a transmitter. The design considerations to build these parts are discussed below.

In the receiver part of a handset, low-noise microwave transistors are used to amplify the incoming signals. As in any low-noise amplifier, a low minimum noise figure of the transistors is desired. The noise requirements of the microwave devices for this application are, however, not as stringent as in satellite communications. In wireless communications, the receiver "feels" the noise of the whole environment, which is interference-dominated, whereas in satellite communications the signal comes from the sky with less background noise [97]. Consequently, for wireless communications, the noise produced intrinsically in the microwave devices is somewhat less important because of the noisy environment. Besides GaAs MESFETs and Si BJTs as the traditional low-noise transistors used in wireless communications, GaAs pHEMTs are widely used today. In the early 2000s, commercial SiGe HBTs have also been introduced as low-noise devices for receiver front ends. As discussed in Section 1.4, the minimum noise figures of SiGe HBTs are comparable to GaAs MESFETs up to about 5 GHz. Despite the fact that GaAs HBTs are relatively noisy (best experimental devices with minimum noise figures around 1 dB and commercial devices with NF_{min} around 1.7 dB between 1.8 and 2.45 GHz), they fulfill the requirements for several wireless systems [41]. The use of Si CMOS, possibly merged with SiGe HBTs in a low-cost BiCMOS process, in the receiver has been a heavily discussed issue in industry during the early 2000s [94, 123, 124].

Another requirement for the handset is the reduction of power consumption. To this end, the supply voltage of handsets has been reduced continuously during the past years. In the year 2000, a supply voltage of 3 V was established as a standard. To deliver a high output power combined with a high efficiency at a limited supply voltage of 3 V, microwave power transistors possessing a large on-current and a low on-resistance are required in the transmit section of the handset. In this application, the GaAs HBT, and GaAs pHEMTs to a lesser extent [41], have replaced almost entirely the formerly dominating GaAs MESFETs and Si BJTs [125].

GaAs HBTs have two main advantages compared to MESFETs or HEMTs for power applications. First, the power density of GaAs HBTs is much larger, thus allowing more compact transistor designs with smaller device periphery. This makes matching between transistor and surrounding circuit much easier. Second, the GaAs HBT is a normally off device and requires only a single-polarity voltage supply. Typically, GaAs MESFETs and HEMTs are designed as depletion-mode (normally on) devices requiring a positive drain voltage and a negative gate voltage to control the drain current. Because handsets operate with a single-polarity power supply only, additional circuitry for providing the negative gate voltage is needed in FET power amplifiers [119]. A solution would be to use enhancement mode (normally off) FETs, but these devices offer lower on-currents and power densities than their depletion mode counterparts [125].

In 2001, the dominant power microwave transistor used in base stations of wireless communications systems with operating frequencies up to 2.5 GHz was the Si LDMOSFET, which in the last several years has replaced all other competing Si and GaAs transistors. Si LDMOSFETs combine the advantages of moderate cost, high reliability, and extremely high output power. To obtain the required output powers, the supply voltage in base stations (more than 10 V) is much higher than that in handsets (3 V). The very high breakdown voltage of the Si LDMOSFET allows this device to operate under a drain voltage of more than 60 V.

Progress in the development of wide bandgap transistors offers the possibility of high-power SiC MESFETs and AlGaN/GaN HEMTs. These devices have already demonstrated power densities clearly surpassing those of all other types of FETs. In principle, these devices have the potential for power amplifiers used in base stations.

1.6 SUMMARY AND OUTLOOK

The aim of this section is to summarize some important trends of the evolution of microwave transistors discussed in the preceding sections and to give a condensed assesment of the different types of transistors for microwave applications. It should be noted that the following discussions reflect the personal opinion of the authors and are based on the level of knowledge in early 2002.

During the past decades, worldwide research and development activities led to a continous improvement of the performance of microwave transistors, and further progress will be achieved in the future. The evolution was governed by the following general trends.

Trend 1

The first trend, and probably the most important one, is that during the last years the application areas of microwave systems have shifted more and more from defense and space purposes to commercial mass markets. Until the 1980s, defense and space applications clearly dominated while nowadays microwave electronics for the most part is becoming mainstream and commercial products using microwave components are part of the daily life. The market volumes of commercial microwave products grew considerably since 1990. The main impetus came from the introduction of wireless communications systems. Cordless and cellular phones were the most prominent products for mass markets, and new applications for microwave devices have steadily been developed and introduced during the 1990s. Regardless of short-term fluctuations at the global markets, which temporarily may affect the microwave electronics industry, this trend will continue for many years to come. Beside the commercial applications, however, also in the future part of the microwave products will be defense-related. While the operating frequencies of most commercial microwave products are in the lower GHz range (up to about 6 GHz), the different defense- and space-related microwave systems cover the entire frequency range up to more than 100 GHz.

Trend 2

Over the years, the performance of microwave transistors has been improved continuously. The improvents were related to the operating frequencies and the frequency limits, the noise figures, and the figures of merit associated with power amplification. The progress has mainly been achieved by:

- Scaling of the intrinsic device dimensions related to transistor speed, i.e., the gate length in field-effect transistors and the base thickness in bipolar transistors.
- Minimizing the undesired parasitic components of the transistors. Most important for field-effect transistors is the reduction of the parasitic source and

gate resistances, while in bipolar transistors the reduction of the base and emitter resistances and of the external collector-base capacitance is critical.
- Developing heterostructures and introducing of transistors utilizing heterostructures (i.e., HEMTs and HBTs).
- Using new materials and material systems offering better carrier transport properties (higher mobilities and velocities), better carrier confinement, and/or higher breakdown fields.

Trend 3

This trend is connected to the growing role of commercial mass markets for microwave systems. While for the traditional defense applications the transistor performance was (and is) the most critical issue, for mass applications low cost is extremely important. This led to the development of low cost microwave technologies since the early 1990s. The focus on economics resulted in considerable efforts to develop fast Si-based transistors such as SiGe HBTs and microwave Si MOSFETs.

The current status of the different semiconductor technologies for microwave transistors together with their preferred applications (low-noise and power) are compiled in Table 1.12. The maximum operating frequency $f_{\text{op,max}}$ is related to the frequency requirements of the *currently* targeted applications and markets.

In general, there is no optimum single transistor type, or semiconductor technology, for the variety of different microwave applications. For applications at the lower end of the frequency range (up to about 3 GHz), all technologies listed in Table 1.12 show sufficient performance. Here, the Si-based technologies offer a clear cost advantage. Currently, especially Si BJTs and Si LDMOSFETs are very successful, and SiGe HBTs will gain ground in the near future. Between 3 GHz and 20 GHz GaAs-based transistors are most important because the performance of most Si-based transistors (BJT, MOSFET) is no longer satisfying the requirements of many applications. Only SiGe HBTs compete successfully with GaAs transistors in this frequency range. Possibly Si MOSFETs will become competitive as well due to further scaling and the increasing experience and cleverness in designing microwave MOSFETs. By then, BiCMOS integrated circuits combining SiGe HBTs (in the high-speed section) and Si MOSFETs (in the lower-speed sections) is a promising option. The frequency range above 20 GHz is currently the domain of GaAs- and InP-based transistors. For very-high frequency applications, InP transistors are the devices of choice.

Table 1.13 compares the different types of microwave transistors in terms of operating frequency, breakdown voltage, threshold voltage uniformity, output power density, low-frequency noise ($1/f$ noise), high frequency noise, possibility of a single power supply, lithography requirements, maturity, and cost.

In general, bipolar transistors show a better threshold voltage uniformity than MESFETs and HEMTs. The threshold voltage of bipolar transistors is mainly determined by the bandgap (i.e., by a material property) and is almost independent of

Table 1.12. Semiconductor technologies used in microwave electronics together with the estimated maximum operating frequency $f_{op,max}$, maximum operating voltage $V_{op,max}$, current status, and preferred application

Technology (Transistor Type)	$f_{op,max}$, GHz	$V_{op,max}$	Status	Preferred Application
Si CMOS	5	low	R&D, P	LN, possibly PA
Si LDMOS	3	high	P	PA
Si BJT	5	low–medium	P	LN, PA
SiGe HBT	50	low	P, R&D	LN, possibly PA
GaAs MESFET	20	medium	P	LN, PA
GaAs HEMT	60	medium	P	LN, PA
GaAs HBT	30	medium	P	PA, possibly LN
InP HEMT	200	low	R&D, P[a]	LN, PA
InP HBT	150	low–medium	R&D, P[a]	PA
SiC MESFET	10	very high	R&D, P	PA
AlGaN/GaN HEMT	20	very high	R&D	PA, possibly LN

R&D: research and development, P: production, LN: low-noise, PA: power.
[a]Low-volume production in early 2000s.

possible variations in the fabrication process. In the case of MESFETs and HEMTs, on the other hand, the threshold voltage critically depends on the thickness and doping of the semiconductor layers underneath the gate.

The low-frequency $1/f$ noise of bipolar transistors is considerably lower (lower corner frequency) than that of field-effect transistors. In bipolar transistors the collector current flows mainly perpendicular to the semiconductor surface and is hardly influenced by the properties of the semiconductor surface and by surface effects, such as traps. Such traps originate the $1/f$ noise. Thus, bipolar transistors are bulk devices. In FETs, on the other hand, the drain current flows in close vicinity to the surface and is strongly affected by surface effects.

Finally, Table 1.14 shows the preferred microwave transistor technology for important defense/space and commercial applications. The remark "future" in parantheses is not a prediction but rather means a possible future option.

In the following, the status of the different microwave transistors is briefly reviewed. During the 1990s, the research and development activities on Si BJTs and GaAs MESFETs essentially stopped. This fact, however, should not seduce to the misleading assertion that Si BJTs and GaAs MESFET do no longer play an important role in modern microwave electronics. Instead, these transistors are still widely used in different wireless communication systems [127, 128]. Their market share, however, is declining. The traditional applications of Si BJTs become more and more the target of SiGe HBTs, and GaAs MESFETs have to compete with SiGe HBTs, GaAs HBTs, and GaAs pHEMTs. In Fig. 1.30 the evolution of the share of GaAs MESFETs, GaAs HEMTs and GaAs HBTs on the total GaAs microwave market (integrated circuits) between 1992 and 2003 is shown. The trend is based on data taken from [129-132].

Table 1.13. Comparison of the different microwave transistor types

Transistor Type	f_{op}	BV	V_{th} (Uniformity)	P_D	LF Noise	HF Noise	Single Power Supply	Lithography Requirements	Maturity	Cost
Si CMOS	−	−	0	0	0	−	yes	−	0	0/+[a]
Si LDMOS	−	+	0	0	−	−	yes	+	+	+
Si BJT	−	−	+	+	+	−	yes	0	+	+
SiGe HBT	0	−	+	0	+	0	yes	0	0	+
GaAs MESFET	0	0	0	0	−	0	no	−	+	0
GaAs HEMT	0/+	0	−	0	−	+	no	+	+	0
GaAs HBT	0	+	+	+	+	−	yes	−	+	0
InP HEMT	+	−	−	−	−	+	no	+	−	−
InP HBT	+	0	+	+	+	−	yes	+	−	−
AlGaN HEMT	0	+	−	+	−	0	no	−	−	−
SiC MESFET	−	+	−	+	−	−	no	−	−	−

f_{op}: operating frequency, BV: breakdown voltage, V_{th}: threshold voltage uniformity, P_D: output power density, LF noise: low-frequency noise ($1/f$ noise), HF noise: high frequency noise, single power supply: Is a single-polarity power supply in the circuitry sufficient?, lithography requirements: Is deep submicron lithography necessary to fabricate transistors for the typical targeted operating frequency?, maturity and cost: describe the status of the typical fabrication technology.
+ good, 0 avearge, − poor
[a]Depending on the gate length level

Table 1.14. Preferred HEMT technologies for different defense/space and commercial applications.

Frequency	Defense/Space	Commercial	Low-noise	Power
0.85-2.5 GHz		Wireless	Si BJT	Si BJT(hs)
			SiGe HBT	GaAs HBT (hs)
			GaAs MESFET	GaAs HEMT (hs)
			GaAs pHEMT	Si LDMOSFET (bs)
				SiC MESFET (future bs),
				AlGaAs HEMT (future bs)
2.5-6 GHz		Wireless	GaAs MESFET	GaAs pHEMT (hs)
			GaAs pHEMT	GaAs HBT (hs)
			SiGe HBT	AlGaAs/GaAs HEMT (bs)
			InP HEMT (future)	SiC MESFET (future bs)
			InP HBT (future)	AlGaAs HEMT (future bs)
12 GHz	Phased array radar	DBS	GaAs pHEMT	GaAs pHEMT
				GaAs HBT
20 GHz	Satellite downlinks		GaAs pHEMT	AlGaN/GaN (future)
27-35 GHz	Missile seekers	LMDS	GaAs pHEMT	GaAs pHEMT
				GaAs HBT
44 GHz	Satellite ground terminals	MVDS	InP HEMT	GaAs pHEMT
60 GHz	Satellite cross links	Wireless LAN	InP HEMT	GaAs pHEMT
			InP HEMT	GaAs pHEMT
77 GHz		CAR	InP HEMT	GaAs pHEMT
			InP HEMT	
94 GHz	FMCW radar		InP HEMT	InP HEMT
> 100 GHz	Radio astronomy		InP HEMT	
Digital 10 Gbit/s		Fiber optic communication		GaAs pHEMT, SiGe HBT
Digital 40 Gbit/s		Fiber optic communication		nP HEMT, InP HBT, SiGe HBT
Digital > 40 Gbit/s		Fiber optic communication		InP HEMT, InP HBT

bs: base station, hs: handset, DBS: Direct-Broadcast Satellite, LMDS: Local Multipoint Distribution System, MVDS: Multipoint Video Distribution System, CAR: Collision Avoidance Radar, LAN: Local Area Network, FMCW: Frequency-Modulated Continuous Wave (radar). Part of the content has been taken from the overview given by Chavarkar and Mishra [126].

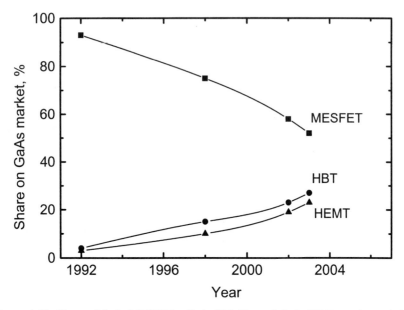

Figure 1.30. Share of GaAs MESFETs, GaAs HEMTs, and GaAs HBTs on the total GaAs microwave IC market.

During the 1990s, the SiGe HBT became a well-accepted microwave device and in the early 2000s it reached breakthrough to widespread commercial applications. The growing interest of the industry in SiGe technology is demonstrated in Table 1.15 showing the chipmakers involved in research, development, and fabrication of SiGe HBTs in the early 1990s and, 10 years later, in the early 2000s. The frequency limits of SiGe HBTs, i.e., the cutoff frequency f_T and the maximum frequency of oscillation f_{max} have been enhanced continuously over the years as shown in Fig. 1.31.

Si LDMOSFETs are widely used as power amplifiers in base stations of wireless communication systems. The technology of these devices is well established and

Table 1.15. Semiconductor companies involved in the SiGe HBT business

Early 1990s		Early 2000s	
IBM	IBM	Texas Instruments	Conexant
Daimler Benz	Agere	Hitachi	SiGe Microsystems
	Atmel	Motorola	Communicant
	Infineon	NEC	AMS
	STM	Analog Devices	Mitel
	Philips	TSMC	XFAB
	Maxim	UMC	...

After [133].

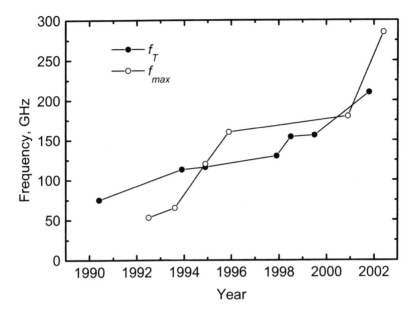

Figure 1.31. Evolution of the frequency performance of SiGe HBTs in terms of f_T and f_{max}.

not expensive. The maximum operating frequency of LDMOSFETs, however, is quite low and is not expected to considerably exceed 3 GHz.

Currently, the assessment of the prospects of microwave CMOS is difficult. Small-signal microwave Si MOSFETs show promising performance. During the last two years, considerable progress has been made in improving the maximum frequency of oscillation (which is a measure of the power gain obtainable at a certain frequency). Si MOSFETs with both f_T and f_{max} between 100 and 200 GHz and with reasonable noise figures in the lower GHz range have been reported. Furthermore, CMOS offers a notable cost advantage compared to the other competing technologies. For these reasons, the proponents of microwave CMOS (including the authors) expect the application of Si MOSFETs in low-end products for the cost-driven mass markets in the near future. On the other hand, many experts doubt that microwave CMOS can successfully compete with SiGe and GaAs transistors even at low GHz frequencies. An impression on the different opinions on the prospects of microwave CMOS can be found in [134].

GaAs HEMTs are widely used commercially. Most popular is the GaAs pHEMT while conventional AlGaAs/GaAs HEMTs gain less attention. The technology of AlGaAs/GaAs HEMTs and GaAs pHEMTs can be considered as mature and no further significant improvement of the performance is expected. GaAs mHEMTs, on the other hand, are still under development. Their high-frequency performance clearly exceeds that of AlGaAs/GaAs HEMTs and GaAs pHEMTs. More work needs to be done, however, in the field of reliability.

GaAs HBTs are matured and widely used for power amplification. Most interest is now focused on InGaP/GaAs HBTs, although the conventional AlGaAs/GaAs HBT is still successful. In contrast to GaAs MESFETs, the share of GaAs-based HEMTs and HBTs on the total GaAs market is growing (see Fig. 1.30).

For many years researchers tried to grow GaAs on Si wafers. The motivation of this work was to use inexpensive and large diameter Si wafers for the cost-effective production of GaAs devices. Although several experimental GaAs transistors on Si substrates have been reported, the results were not satisfying. In the early 2000s, however, a breakthrough in this area has been announced by Motorola [135].

InP transistors offer the best high-frequency performance among all types of microwave transistors. InP HEMTs show the lowest noise figures at all GHz frequencies. Their frequency limits are higher than those of all other microwave field-effect transistors. InP HBTs outperform the other microwave bipolar transistors in terms of f_T and f_{max}. Transferred substrate InP HBTs with an estimated f_{max} above 1 THz have been fabricated. This is the highest f_{max} of any three-terminal semiconductor device reported so far. Another advantage of InP HBTs is the low turn-on voltage which makes these devices a promising candidate for ultra-low-power applications. The major drawback of InP transistors is the still less matured and expensive technology. Up to now this prevented the use of InP transistors in large volume consumer applications.

SiC MESFETs and AlGaN/GaN HEMTs offer extremely high output power densities which are far beyond the limits of any other type of field-effect transistors. First SiC MESFETs are in the market while AlGaN/GaN HEMTs are still under development. Because currently no commercial GaN substrates exist, AlGaN/GaN HEMTs are fabricated on sapphire, SiC, or Si substrates. AlGaN/GaN HEMTs on SiC substrates offer the best power performance. The major drawback of these devices is the high substrate cost. In early 2002, the price of a 50-mm SiC wafer with epitaxially grown GaN was about $10 000 [136]. Other problems are related to the reliability of AlGaN/GaN HEMTs. Since considerable funding is spent on these devices and a growing number of research groups from academia and the industry is working intensively on AlGaN/GaN HEMTs, further progress is expected.

We conclude with a list summarizing the major merits (M) and drawbacks (D) of the different microwave transistor types, together with some comments (C).

Si CMOS
M   In general matured and inexpensive technology (spin-off of Si VLSI)
    Excellent potential for integration
D   Deep submicron lithography necessary
    Low breakdown voltage
C   Still controversial discussion on the suitability and prospects of microwave CMOS

Si LDMOS
M   High output power and breakdown voltage
    Low cost, relaxed requirements on lithography

D Operating frequency limited to about 3 GHz
C Dominating device for power amplification in base stations up to 2.5 GHz

Si BJT
M Matured and inexpensive technology
   Low 1/f noise, lower emitter turn-on voltage than GaAs HBT (0.7 V vs. 1.4 V)
D Relatively low operating frequency
   Low breakdown voltage
C Still widely used in the lower GHz range but importance is declining

SiGe HBT
M Meanwhile matured and relatively inexpensive technology
   Combines low 1/f noise with reasonable high-frequency noise, low emitter turn-on voltage
D Low breakdown voltage
C Commercially available, importance is increasing

GaAs MESFET
M Matured technology, simple device, low cost (especially ion implanted GaAs MESFET).
D Not well suited for operation above 20 GHz.
C Still important in both low-noise and power applications

GaAs HEMT
M Meanwhile matured technology
   High operating frequency, low noise figure, reasonable output power
D Commonly a normally-on device (needs a negative gate voltage to switch off)
C Most popular are the GaAs pHEMTs, conventional AlGaAs HEMTs are less important

GaAs HBT
M Relaxed requirements on lithography (no lateral submicron structures necessary)
   High linearity, high output power density
D Application in very low-power circuits problematic because of high emitter turn-on voltage
C Popular device for power amplification in cellular phones

InP HEMT
M Very high operating frequencies and frequency limits
   Lowest noise figure of all microwave transistor types
D Expensive technology
C Application still limited to low-volume high-performance markets (early 2000s)

InP HBT
M Highest operating frequencies and frequency limits of all bipolar transistors

Low emitter turn-on voltage (less than Si BJT, important for very-low-power applications)
D Expensive technology
C Application still limited to low-volume high-performance markets (early 2000s)

Wide bandgap FETs (AlGaN/GaN HEMT, SiC MESFET)
M Highest output power density of all microwave field-effect transistors
   High operating temperatures, high breakdown voltage
D Expensive and immature technology
   Reliability still an open question
C First SiC MESFETs commercially available, AlGaN HEMTs not yet commercially available
   Targeted application: power amplification at GHz frequencies.

REFERENCES

1. F. Schwierz, Microwave Transistors: State of the Art in the 1980s, 1990s, and 2000s. A Compilation of 1000 Top References, TU Ilmenau 2002, Unpublished.
2. F. Schwierz, Microwave Transistors—The Last 20 Years, *Proc. 3rd ICCDCS,* Circuits, and Systems, pp. D28.1–8, 2000.
3. F. Schwierz and J. J. Liou, Semiconductor Devices for Wireless Communications and High-Speed Internet, *Proc. SSGRR-2000,* pp. 331.1–10, 2000.
4. F. Schwierz and J. J. Liou, Semiconductor Devices for RF Applications: Evolution and Current Status, *Microelectron. Reliab. 41,* pp. 145–168, 2001.
5. J. J. Liou and F. Schwierz, Evolution and Current Status of RF Semiconductor Devices, *Extended Abstracts, Solid State Devices and Material Conference,* pp. 74–75, 2001.
6. F. Schwierz and J. J. Liou, Development of RF Transistors: A Historical Prospect, *Proc. 6th International Conference on Solid-State and Integrated-Circuit Technology,* pp. 1314–1319, 2001.
7. *The National Technology Roadmap for Semiconductors,* Semiconductor Industry Association 1992, 1994, 1997.
8. *ITRS—The International Technology Roadmap for Semiconductors,* Semiconductor Industry Association 1999, 2001.
9. J. M. Rollett, Stability and Power-Gain Invariants of Linear Twoports, *IRE Trans. Circuit Theory, CT-9,* pp. 29–32, 1962.
10. H. Beneking, *High Speed Semiconductor Devices—Circuit Aspects and Fundamental Behaviour,* Chapman & Hall, London 1994.
11. S. J. Mason, Power Gain in Feedback Amplifier, *IRE Trans. Circuit Theory, CT-1,* pp. 20–25, 1954.
12. S. Tiwari, *Compound Semiconductor Device Physics,* Academic Press, San Diego, 1992.
13. K. Onodera, M. Tokumitsu, M. Tomizawa, and K. Asai, Effects of Neutral Buried p-Layer on High-Frequency Performance of GaAs MESFET's, *IEEE Trans. Electron Devices, 38,* pp. 429–436, 1991.

14. P. Greiling, The Historical Development of GaAs FET Digital IC Technology, *IEEE Trans. Microwave Theory Tech., 32,* pp. 1144–1156, 1984.
15. J. Finol, J. C. Durec and D. K. Lovelace, Building Blocks for Digital Wireless Communications in Sub-Micron Technologies: An Overview, *Proc. 2nd ICCDCS,* pp. 113–126, 1998.
16. A. B. Vollmer, Gate Arrays Break Performance Records with 75-GHz SiGe Process, *Electronic Design, 47,* pp. 36–42, May 3, 1999.
17. H. Q. Tserng, B. Kim, P. Saunier, H. D. Shih, and M. A. Khatibzadeh, Millimeter-Wave Power Transistors and Circuits, *Microwave Journal, 32,* pp. 125–135, April 1989.
18. F. Ali, A. Gupta, and A. Higgins, Advances in GaAs HBT Power Amplifiers for Cellular Phones and Military Applications, *IEEE Microwave and Millimeter-Wave Monolithic Circuits Symp. Dig.,* pp. 61–66, 1996.
19. H. F. Cooke, Microwave Transistors: Theory and Design, *Proc. IEEE, 59,* pp. 1163–1181, 1971.
20. C. A. Mead, Schottky Barrier Gate Field Effect Transistor, *Proc. IEEE, 54,* pp. 307–308, 1966.
21. W. W. Hooper and W. I. Lehrer, An Epitaxial GaAs Field-Effect Transistor, *Proc. IEEE, 55,* pp. 1237–1238, 1967.
22. K. Drangeid, R. Sommerhalder, and W. Lehrer, High-Speed Gallium-Arsenide Schottky-Barrier Field-Effect Transistors, *Electron. Lett., 6,* pp. 228–229, 1970.
23. J. A. Archer, Low-Noise Implanted-Base Microwave Transistors, *Solid-State Electron., 17,* pp. 387–393, 1974.
24. K. Kamei, S. Hori, H. Kawasaki, and T. Chigira, Quarter Micron Gate Low Noise GaAs FET's Operable up to 30 GHz, *IEDM Tech. Dig.,* pp. 102–105, 1980.
25. For reviews on MBE see (a) K. Ploog, Molecular Beam Epitaxy of III–V Compounds, in *Crystals: Growth, Properties, and Applications,* H. C. Freyhardt (ed.), vol. 3, pp. 73–162, Springer Verlag, New York 1980; (b) A. Y. Cho (ed.), *Molecular Beam Epitaxy (Key Papers in Applied Physics),* Springer Verlag, New York 1994.
26. For reviews on MOCVD see (a) R. D. Dupuis, L. A. Moudy, and D. P. Dapkus, Preparation and Properties of $Ga_{1-x}Al_xAs$-GaAs Heterojunctions Grown by Metal–Organic Vapor Deposition, Gallium Arsenide and Related Compounds 1978, *Inst. Phys. Conf. Ser. 45,* pp. 1–9, 1979; (b) G. B. Stringfellow, *Organometallic Vapor Phase Epitaxy: Theory and Practice,* Academic Press, Boston 1989.
27. R. Dingle, H. L. Störmer, A. C. Gossard, and W. Wiegmann, Electron Mobilities in Modulation-Doped Semiconductor Heterojunction Superlattices, *Appl. Phys. Lett., 33,* pp. 665–668, 1978.
28. T. Mimura, S. Hiyamizu, T. Fujii, and K. Nanbu, A New Field-Effect Transistor with Selectively Doped GaAs/n-$Al_xGa_{1-x}As$ Heterojunctions, *Jpn. J. Appl. Phys., 19,* pp. L225–L227, 1980.
29. W. Shockley, U.S. Patent 2 569 347, Filed 26 June 1948, Issued 25 Sept. 1951.
30. H. Kroemer, Theory of a Wide-Gap Emitter for Transistors, *Proc. IRE, 45,* pp. 1535–1537, 1957.
31. H. T. Yuan, W. V. McLevige, and H. D. Shih, GaAs Bipolar Digital Integrated Circuits, in N. G. Einspruch and W. R. Wisseman (eds.), *VLSI Electronics Microstructure Science,* vol. 11, pp. 173–213, Academic Press, Orlando, 1985.

32. L. D. Nguyen, L. E. Larson, and U. Mishra, Ultra-High-Speed Modulation-Doped Field-Effect Transistors, *Proc. IEEE, 80* pp. 494–518, 1992.
33. A. Ketterson, W. T. Masselink, J. S. Gedymin, J. Klem, W. Kopp, H. Morkoc, and K. R. Gleason, Characterization of InGaAs/AlGaAs Pseudomorphic Modulation-Doped Field Effect Transistors, *IEEE Trans. Electron Devices, 33,* pp. 564–571, 1986.
34. J. M. Kuo, B. Calvic, and T. Y. Chang, New Pseudomorphic MODFET Utilizing $Ga_{0.47-u}In_{0.53+u}/Al_{0.48+u}In_{0.52-u}$ Heterostructures, *IEDM Tech. Dig.,* pp. 460–463, 1986.
35. D. L. Harame, J. H. Comfort, J. D. Cressler, E. F. Crabbe, J. Y.-C. Sun, B. S. Meyerson, and T. Tice, Si/SiGe Epitaxial-Base Transistors - Part I: Materials, Physics, Circuits, *IEEE Trans. Electron Devices, 42,* pp. 455–468, 1995.
36. K. Inoue, J. C. Harmand, and T. Matsumo, High-Quality $In_xGa_{1-x}As/InAlAs$ Modulation-Doped Heterostructures Grown Lattice-Mismatched on GaAs Substrates, *J. Crystal Growth, 111,* pp. 313–317, 1992.
37. W. E. Hoke, P. J. Lemonias, J. J. Mosca, P. S. Lyman, A. Torabi, P. F. Marsh, R. A. McTaggart, S. M. Lardizabad, and K. Helzar, Molecular Beam Epitaxial Growth and Device Performance of Metamorphic High Electron Mobility Transistor Structures Fabricated on GaAs Substrates, *J. Vac. Sci. Technology, B17,* pp. 1131–1135, 1999.
38. H. Samavati, H. R. Rategh, and T. H. Lee, A 5-GHz CMOS Wireless LAN Receiver Front End, *IEEE J. Solid-State Circuits, 35,* pp. 765–772, 2000.
39. L. E. Frenzel, RF, Wireless, and Optical Technologies Become the Hot Topics, *Electronic Design, 49,* pp. 95–100, Febr. 21, 2001.
40. J. Browne, More Power Per Transistor Translates into Smaller Amplifiers, *Microwaves & RF, 40,* pp. 132–136, Jan. 2001.
41. G. Heftman, Wireless Semi Technology Heads into New Territory, *Microwaves & RF, 39,* pp. 31–40, Feb. 2000.
42. J. C. Zolper, Wide Bandgap Semiconductor Microwave Technologies: From Promise to Practice, *IEDM Tech. Dig.,* pp. 389–392, 1999.
43. D. N. McQuiddy, J. W. Wassel, J. B. Lagrange, and W. R. Wisseman, Monolithic Microwave Integrated Circuits: An Historical Perspective, *IEEE Trans. Microwave Theory Tech., 32,* pp. 997–1008, 1984.
44. N. J. Gri, Microwave Transistors—From Small Signal to High Power, *Microwave J., 14,* pp. 45–50, Febr. 1971.
45. W. Baechtold, W. Walter, and P. Wolf, X and Ku Band GaAs MESFETs, *Electron. Lett., 8,* pp. 35–37, 1972.
46. W. Baechtold, K. Daetwyler, T. Forster, T. O. Mohr, W. Walter, and P. Wolf, Si and GaAs 0.5 μm Gate Schottky-Barrier Field-Effect Transistors, *Electron. Lett., 9,* pp. 232–234, 1973.
47. P. C. Chao, T. Yu, P. M. Smith, S. Wanaga, J. C. M. Hwang, and W. H. Perkins, Quarter-Micron Gate Length Microwave High Electron Mobility Transistor, *Electron. Lett., 19,* pp. 894–896, 1983.
48. P. C. Chao, S. C. Palmateer, P. M. Smith, U. K. Mishra, K. H. G. Duh, and J. C. M. Hwang, Millimeter-Wave Low-Noise High Electron Mobility Transistors, *IEEE Electron Device Lett., 10,* pp. 531–533, 1985.
49. L. F. Lester, P. M. Smith, P. Ho, P. C. Chao, R. C. Tiberio, K. H. G. Duh, and E. D. Wolf, 0.15 μm Gate-Length Double Recess Pseudomorphic HEMT with F_{max} of 350 GHz, *IEDM Tech. Dig.,* pp. 172–175, 1988.

50. P. C. Chao, A. J. Tessmer, K. H. G. Duh, P. Ho, M.-Y. Kao, P. M. Smith, J. M. Ballingall, S.-M. Liu, and A. A. Jarba, W-Band Low-Noise InAlAs/InGaAs Lattice-Matched HEMT's, *IEEE Electron Device Lett., 11,* pp. 59–62, 1990.

51. P. Ho, M. Y. Kao, P. C. Chao, K. H. G. Duh, J. M. Ballingall, A. T. Allen, A. J. Tessmer, and P. M. Smith, Extremely High Gain 0.15 μm Gate-Length InAlAs/InGaAs/InP HEMTs, *Electron. Lett., 27,* pp. 325–327, 1991.

52. P. M. Smith, S.-M. J. Liu, M.-Y. Kao, P. Ho, S. C. Wang, K. H. G. Duh, S. T. Fu, and P. C. Chao, W-Band High Efficiency InP-Based Power HEMT with 600 GHz f_{max}, *IEEE Microwave and Guided Wave Lett. 5,* pp. 230–232, 1995.

53. U. K. Mishra, A. S. Brown, S. E. Rosenbaum, C. E. Hooper, M. W. Pierce, M. J. Delaney, S. Vaughn, and K. White, Microwave Performance of AlInAs-GaInAs HEMT's with 0.2- and 0.1-μm Gate Length, *IEEE Electron Device Lett., 9,* pp. 647–649, 1988.

54. L. D. Nguyen, A. S. Brown, M. A. Thompson, and L. M. Jelloian, 50-nm Self-Aligned-Gate Pseudomorphic AlInAs/GaInAs High Electron Mobility Transistors, *IEEE Trans. Electron Devices, 39,* pp. 2007–2013, 1992.

55. T. Suemitsu, T. Ishii, H. Yokoyama, Y. Umeda, T. Enoki, Y. Ishii, and T. Tamamura, 30-nm-Gate InAlAs/InGaAs HEMTs Lattice-Matched to InP Substrates, *IEDM Tech. Dig.,* pp. 223–226, 1998.

56. Y. Yamashita, A. Endoh, K. Shinohara, M. Higashiwaki, K. Hikosaka, T. Mimura, S. Hiyamizu, and T. Matsui, Ultra-Short 25-nm-Gate Lattice-Matched InAlAs/InGaAs HEMTs within the Range of 400 GHz Cutoff Frequency, *IEEE Electron Device Lett., 22,* pp. 367–369, 2001.

57. Q. Lee, C. Martin, D. Mensa, R. P. Smith, J. Guthrie, and M. J. W. Rodwell, Submicron Transferred-Substrate Heterojunction Bipolar Transistors, *IEEE Electron Device Lett., 20,* pp. 396–398, 1999.

58. M. J. W. Rodwell, M. Urteaga, T. Mathew, D. Scott, D. Mensa, Q. Lee, J. Guthrie, Y. Betser, S. C. Martin, R. P. Smith, S. Jaganathan, S. Krishnan, S. I. Long, R. Pullela, B. Agarwal, U. Bhattacharya, L. Samoska, and M. Dahlstrom, Submicron Scaling of HBTs, *IEEE Trans. Electron Devices, 48,* pp. 2606–2624, 2001.

59. H. Hsia, Z. Tang, D. Caruth and M. Feng, Direct Ion-Implanted 0.12-μm GaAs MESFET with f_T of 121 GHz and f_{max} of 160 GHz, *IEEE Electron Device Lett., 20,* pp. 245–247, 1999.

60. M. Tokumitsu, M. Hirano, T. Otsuji, S. Yamaguchi, and K. Yamasaki, A 0.1-μm Self-Aligned-Gate GaAs MESFET with Multilayer Interconnection Structure for Ultra-High-Speed IC's, *IEDM Tech. Dig.,* pp. 211–214, 1996.

61. D. C. Caruth, R. L. Shimon, M. S. Heins, H. Hsia, Z. Tang, S. C. Shen, D. Becher, J. J. Huang, and M. Feng, Low-Cost 38 and 77 GHz CPW MMICS Using Ion-Implanted GaAs MESFETs, *IEEE MTT-S Dig.,* pp. 995–998, 2000.

62. K. Onodera, K. Nishimura, S. Aoyama, S. Sugitani, Y. Yamane, and M. Hirano, Extremely Low-Noise Performance of GaAs MESFET's with Wide-Head T-Shaped Gate, *IEEE Trans. Electron Devices, 46,* pp. 310–319, 1999.

63. K. Ebihara, T. Takahashi, Y. Tateno, I. Igarashi and J. Fukaya, L-Band 100-Watts Push-Pull GaAs Power FET, *IEEE MTT-S Dig.,* pp. 703–706, 1998.

64. H. M. Macksey and F. H. Doerbeck, GaAs FETs Having High Output Power Per Unit Gate Width, *IEEE Electron Device Lett., 6,* pp. 147–148, 1981.

65. P. C. Chao, Gate Formation Technologies, in R. L. Ross, S. S. Svensson and P.

Lugli (eds.), *Pseudomorphic HEMT Technology and Applications*, Kluwer, Dordrecht 1996.

66. A. N. Lepore, H. M. Levy, R. C. Tiberio, P. J. Tasker, H. Lee, E. D. Wolf, L. F. Eastman, and E. Kohn, 0.1 μm Gate Length MODFETs with Unity Current Gain Cutoff Frequency Above 110 GHz, *Electron. Lett., 24,* pp. 364–366, 1988.

67. I. Hanyu, S. Asai, M. Nukokawa, K. Joshin, Y. Hirachi, S. Ohmura, Y. Aoki, and T. Aigo, Super Low-Noise HEMTs with a T-Shaped WSi_x Gate, *Electron. Lett., 24,* pp. 1327–1328, 1988.

68. L. D. Nguyen, P. J. Tasker, D. C. Radulescu, and L. F. Eastman, Characterization of Ultra-High-Speed Pseudomorphic AlGaAs/InGaAs (on GaAs) MODFETs, *IEEE Trans. Electron Devices, 36,* pp. 2243–2248, 1989.

69. D. C. Dumka, W. E. Hoke, P. J. Lemonias, G. Cueva, and I. Adesida, Metamorphic $In_{0.52}Al_{0.48}As/In_{0.53}Ga_{0.47}As$ HEMTs on GaAs Substrate With f_T Over 200 GHz, *Electron. Lett., 35,* pp. 1854–1856, 1999.

70. M. Zaknoune, Y. Cordier, S. Bollaert, D. Ferre, D. Theron, and Y. Crosnier, 0.1 μm High Performance Metamorphic $In_{0.32}Al_{0.68}As/In_{0.33}Ga_{0.67}As$ HEMT on GaAs Using Inverse Step InAlAs Buffer, *Electron. Lett., 35,* pp. 1670–1671, 1999.

71. S. E. Rosenbaum, B. K. Kormanyos, L. M. Jelloian, M. Matloubian, A. S. Brown, L. E. Larson, L. D. Nguyen, M. A. Thompson, L. P. B. Katehi, and G. M. Rebeiz, 155- and 213-GHz AlInAs/GaInAs/InP HEMT MMIC Oscillators, *IEEE Trans. Microwave Theory Tech., 43,* pp. 927–932, 1995.

72. S. Weinreb, T. Gaier, R. Lai, M. Barsky, Y. C. Leong, and L. Samoska, High-Gain 150–215-GHz MMIC Amplifier with Integral Waveguide Transistions, *IEEE Microwave and Guided Wave Lett., 9,* pp. 282–284, 1999.

73. K. Inoue, K. Ebihara, H. Haematsu, T. Igarashi, H. Takahashi, and J. Fukaya, A 240 W Push-Pull GaAs Power FET for W-CDMA Base Stations, *MTT-S Dig.,* pp. 1719–1722, 2000.

74. Y. Kiyota, E. Ohue, T. Onai, K. Washio, M. Tanabe, and T. Inada, Lamp-Heated Rapid Vapor-Phase Doping Technology for 100-GHz Si Bipolar Transistors, *Proc. BCTM,* pp. 173–176, 1996.

75. F. A. Myers, Advanced Gallium Arsenide Circuits, *GEC Review, 13,* pp. 86–97, 1998.

76. T. Oka, K. Hirata, K. Ouchi, H. Uchiyama, T. Taniguchi, K. Mochizuki, and T. Nakamura, Advanced Performance of Small-Scaled InGaP/GaAs HBT's with f_T over 150 GHz and f_{max} over 250 GHz, *IEDM Tech. Dig.,* pp. 653–656, 1998.

77. T. Oka, K. Hirata, K. Ouchi, H. Uchiyama, K. Mochizuku, and T. Nakamura, InGaP/GaP HBT's with High-Speed and Low-Current Operation Fabricated Using WSi/Ti as the Base Electrode and Burying SiO_2 in the Extrinsic Collector, *IEDM Tech. Dig.,* pp. 739–742, 1997.

78. W. J. Ho, N. L. Wang, M. F. Chang, A. Sailer, and J. A. Higgins, Self-Aligned, Emitter-Edge-Passivated AlGaAs/GaAs Heterojunction Bipolar Transistors with Extrapolated Maximum Oscillation Frequency of 350 GHz, Dev. Res. Conf., paper IVA-1, 1992. See also *IEEE Trans. Electron Devices, 39,* p. 2655, 1992.

79. M. W. Dvorak, O. J. Pitts, S. P. Watkins, and C. R. Bolognesi, Abrupt Junction InP/GaAsSb/InP Double Heterojunction Bipolar Transistors with f_T as High as 250 GHz and $BV_{CEO} > 6$ V, *IEDM Tech. Dig.,* pp. 178–181, 2000.

80. M. Ida, K. Kurishima, N. Watanabe, and T. Enoki, InP/InGaAs DHBTs with 341-

GHz f_T at High Current Density of Over 800 kA/cm², *Tech. Dig. IEDM,* pp. 776–779, 2001.
81. Q. Lee, B. Agarwal, D. Mensa, R. Pullela, J. Guthrie, L. Samoska, and M. J. W. Rodwell, A > 400 GHz f_{max} Transferred-Substrate Heterojunction Bipolar Transistor IC Technology, *IEEE Electron Device Lett., 19,* pp. 77–79, 1998.
82. Y. Betser, D. Scott, D. Mensa, S. Jaganathan, T. Mathews, and M. J. Rodwell, InAlAs/InGaAs HBTs with Simultaneously High Values of F_T and F_{max} for Mixed Signal Analog/Digital Applications, *IEEE Electron Device Lett., 22,* pp. 56–58, 2001.
83. H. Dodo, Y. Amamiya, T. Niwa, M. Mamada, N. Goto, and H. Shimawaki, Microwave Low-Noise AlGaAs/InGaAs HBTs with p+-Regrown Base Contacts, *IEEE Electron Device Lett. 19,* pp. 121–123, 1998.
84. Y.-K. Chen, R. N. Nottenburg, M. B. Panish, R. A. Hamm, and D. A. Humphrey, Microwave Noise Performance of InP/InGaAs Heterostructure Bipolar Transistors, *IEEE Electron Device Lett., 10,* pp. 470–472, 1989.
85. S. S. H. Hsu and D. Pavlidis, Low Noise, High-Speed InP/InGaAs HBTs, *Dig. GaAs IC Symp.,* pp. 188–191, 2001.
86. B. Bayraktaroglu, J. Barette, L. Kehias, C. I. Huang, R. Fitch, R. Neidhard, and R. Scherer, Very High-Power-Density CW Operation of GaAs/AlGaAs Microwave Heterojunction Bipolar Transistors, *IEEE Electron Device Lett., 14,* pp. 493–495, 1993.
87. S. Tanaka, High-Power, High-Efficiency Cell Design for 26 GHz HBT Power Amplifier, *IEEE MTT-S Dig.,* pp. 843–846, 1996.
88. R. S. Virk, M. Y. Chen, C. Nguyen, T. Liu, M. Matloubian, and D. B. Rensch, A High-Performance AlInAs/InGaAs/InP DHBT K-Band Power Cell, *IEEE Microwave and Guided Wave Lett., 7,* pp. 323–325, 1997.
89. S. S. Iyer, G. L. Patton, S. L. Delage, S. Tiwari, and J. M. C. Storck, Silicon-Germanium Base Heterojunction Bipolar Transistors by Molecular Beam Epitaxy, *IEDM Tech. Dig.,* pp. 874–877, 1987.
90. S. J. Jeng, B. Jagannathan, J.-S. Rieh, J. Johnson, K. T. Schonenberg, D. Greenberg, A. Stricker, H. Chen, M. Khater, D. Ahlgreen, G. Freeman, K. Stein, and S. Subbanna, A 210-GHz f_T SiGe HBT with a Non-Self-Aligned Structure, *IEEE Electron Device Lett., 22,* pp. 542–544, 2001.
91. B. Jagannathan, M. Khater, F. Pagette, J.-S. Rieh, D. Angell, H. Chen, J. Florkey, F. Golan, D. R. Greenberg, R. Groves, S. J. Jeng, J. Johnson, E. Mengistu, K. T. Schonenberg, C. M. Schnabel, P. Smith, A. Stricker, D. Ahlgreen, G. Freeman, K. Stein, and S. Subbanna, Self-Aligned SiGe NPN Transistors with 285 GHz f_{MAX} and 207 GHz f_T in a Manufacturable Technology, *IEEE Electron Device Lett. 23,* pp. 258–260, 2002.
92. J. Geßner and F. Schwierz, Vertical and Lateral Design of SiGe HBTs for High-Frequency Operation, Unpublished, 2002.
93. H. Schumacher, U. Erben, and W. Durr, SiGe Heterojunction Bipolar Transistors—The Noise Perspective, *Solid-State Electronics, 41,* pp. 1485–1492, 1997.
94. C. A. King, M. R. Frei, M. Mastrapasqua, K. K. Ng, Y. O. Kim, R. W. Johnson, S. Moinian, S. Martin, H.-I. Cong, F. P. Klemens, R. Tang, D. Nguyen, T.-I. Hsu, T. Campbell, S. J. Molloy, L. B. Fritzinger, T. G. Ivanov, K. K. Bourdelle, C. Lee, Y.-F. Chyan, M. S. Carroll, and C. W. Leung, Very Low Cost Graded SiGe Base Bipolar Transistors for a High Performance Modular BiCMOS Process, *IEDM Tech. Dig.,* pp. 565–568, 1999.

95. J. Böck, H. Knapp, K. Aufinger, M. Wurzer, S. Boguth, R. Schreiter, T. F. Meister, M. Rest, M. Ohnemus, and L. Treitinger, 12 ps Implanted Base Silicon Bipolar Technology, *IEDM Tech. Dig.*, pp. 553–556, 1999.
96. P. A. Potyraj, K. J. Petrowsky, K. D. Hobart, F. J. Kub, and P. E. Thompson, A 230 Watt S-Band SiGe HBT, *1996 IEEE MTT-S Digest*, pp. 673–676, 1996.
97. N.-L. L. Wang, Transistor Technologies for RFICs in Wireless Applications, *Microwave Journal, 41*, pp. 98–110, Feb. 1998.
98. U. Konig, SiGe and GaAs as Competitive Technologies for RF-Applications, *Proc. BCTM*, pp. 87–92, 1998.
99. B. Carnez, A. Cappy, A. Kaszynski, E. Constant, and G. Salmer, Modeling of a Submicrometer Gate Field-Effect Transistor Including Effects of Nonstationary Electron Dynamics, *J. Appl. Phys., 51*, pp. 784–790, 1980.
100. M. V. Fischetti, Monte Carlo Simulation of Transport in Technologically Significant Semiconductors of Bulk Diamond and Zinc-Blende Structures—Part I: Homogeneous Transport, *IEEE Trans. Electron Devices, 38*, pp. 634–649, 1991.
101. J. Kolnik, I. H. Oguzman, K. F. Brennan, R. Wang, P. P. Ruden, and Y. Wang, Electronic Transport Studies of Bulk Zincblende and Wurtzite Phases of GaN Based on an Ensemble Monte Carlo Calculation Including a Full Zone Band Structure, *J. Appl. Phys., 78*, pp. 1033–1038, 1995.
102. M. Roschke and F. Schwierz, Electron Mobility Models for 4H, 6H, and 3C SiC, *IEEE Trans. Electron Devices, 48*, pp. 1442–1447, 2001.
103. J. Browne, SiC MESFET Delivers 10-W Power at 2 GHz, *Microwaves & RF, 38*, pp. 138–139, Oct. 1999.
104. E. M. Chumbes, A. T. Schremer, J. A. Smart, Y. Wang, N. C. Mac Donald, D. Hogue, J. J. Komiak, S. J. Lichwalla, R. E. Leoni III, and J. R. Shealy, AlGaN/GaN High Electron Mobility Transistors on Si(111) Substrates, *IEEE Trans. Electron Devices, 48*, pp. 420–426, 2001.
105. S. T. Allen, R. A. Sadler, T. S. Alcorn, J. Sumakeris, R. C. Glass, C. H. Carter, Jr., and J. W. Palmour, Silicon Carbide MESFET's for High-Power S-Band Applications, *Mat. Sci. Forum, 264-268*, pp. 953–956, 1998.
106. S. T. Allen, J. W. Palmour, and C. H. Carter, Jr., Silicon Carbide MESFET's with 2W/mm and 50% P. A. E. at 1.8 GHz, *IEEE MTT-S Dig.*, pp. 681–684, 1996.
107. M. Micovic, N. X. Nguyen, P. Janke, W.-S. Wong, P. Hashimoto, L.-M. McCray, and C. Nguyen, GaN/AlGaN High Electron Mobility Transistors with f_T of 110 GHz, *Electron. Lett., 36*, pp. 358–359, 2000.
108. W. Lu, J. Yang, M. A. Khan, and I. Adesida, AlGaN/GaN HEMTs on SiC with over 100 GHz f_T and Low Microwave Noise, *IEEE Trans. Electron Devices, 48*, pp. 581–585, 2001.
109. N. X. Nguyen, M. Micovic, W.-S. Wong, P. Hashimoto, L.-M. McCray, P. Janke, and C. Nguyen, High Performance Microwave Power GaN/AlGaN MODFETs Grown by RF-Assisted MBE, *Electron. Lett., 36*, pp. 468–469, 2000.
110. Y.-F. Wu, D. Kapolnek, J. P. Ibbetson, P. Parikh, B. P. Keller, and U. K. Mishra, Very-High Power Density AlGaN/GaN HEMTs, *IEEE Trans. Electron Devices, 48*, pp. 586–590, 2001.
111. J. W. Palmour, S. T. Sheppard, R. P. Smith, S. T. Allen, W. L. Pribble, T. J. Smith, Z. Ring, J. J. Sumakeris, A. W. Saxler, and J. W. Milligan, Wide Bandgap Semiconductor

Devices and MMICs for RF Power Applications, *IEDM Tech. Dig.*, pp. 385–388, 2001.
112. J. Browne, Chip Contains Full Bluetooth Solution, *Microwaves & RF, 40*, p. 139, April 2001.
113. H. S. Momose, E. Morifuji, T. Yoshimioto, T. Ohguro, M. Saito, and H. Iwai, Cutoff Frequency and Propagation Delay Time of 1.5-nm Gate Oxide CMOS, *IEEE Trans. Electron Devices, 48*, pp. 1165–1174, 2001.
114. C. Wann, F. Assaderaghi, L. Shi, K. Chan, S. Cohen, H. Hovel, K. Jenkins, Y. Lee, D. Sadana, R. Viswanathan, S. Wind, and Y. Taur, High-Performance 0.07μm-CMOS with 9.5-ps Gate Delay and 150 GHz f_T, *IEEE Electron Device Lett., 18*, pp. 625–627, 1997.
115. R. A. Johnson, P. R. de la Houssaye, C. E. Chang, P.-F. Chen, M. E. Wood, G. A. Garcia, I. Lagnado, and P. M. Asbeck, Advanced Thin-Film Silicon-on-Sapphire Technology: Microwave Circuit Applications, *IEEE Trans. Electron Devices, 45*, pp. 1047–1054, 1998.
116. L. F. Tiemeijer, H. M. J. Boots, R. J. Havens, A. J. Scholten, P. H. W. de Vreede, P. H. Woerlee, A. Heringa, and D. B. M. Klassen, A Record High 150 GHz f_{max} Realized at 0.18 μm Gate Length in an Industrial RF-CMOS Technology, *IEDM Tech. Dig.*, pp. 223–226, 2001.
117. T. Hirose, Y. Momiyama, M. Kosugi, H. Kano, Y. Watanabe, and T. Sugii, A 185 GHz f_{max} SOI DTMOS with a New Metallic Overlay-Gate for Low-Power RF Applications, *IEDM Tech. Dig.*, pp. 943–945, 2001.
118. S. Narashima, A. Ajmera, H. Park, D. Schepis, N. Zamdmer, K. A. Jenkins, J.-O. Plouchart, W-H. Lee, J. Mezzapelle, J. Bruley, B. Doris, J. W. Sleight, S. K. Fung, S. H. Ku, A. C. Mocuta, I. Yang, P. V. Gilbert, K. P. Muller, P. Agnello, and J. Welser, High Performance Sub-40nm CMOS Devices on SOI for the 70nm Technology Node, *IEDM Tech. Dig.*, pp. 625–628, 2001.
119. D. Halchin and M. Golio, Trends for Portable Wireless Applications, *Microwave Journal, 40*, pp. 62–78, Jan. 1997.
120. M. K. Siddiqui, A. K. Sharma, L. G. Callejo and R. Lai, A High Power and High Efficiency Monolithic Power Amplifier for Local Multipoint Distribution Service, *IEEE MTT-S Dig.*, pp. 569–572, 1998.
121. L. Raffaelli, MMW Digital Radio Front Ends: Market, Application and Technology, *Microwave Journal, 40*, pp. 92–96, Oct. 1997.
122. H. Bierman, Personal Communications, and Motor Vehicle and Highway Automation Spark New Microwave Applications, *Microwave Journal, 34*, pp. 26–40, Aug. 1991.
123. K. E. Ehwald, D. Knoll, B. Heinemann, K. Chang, J. Kirchgessner, R. Mauntel, I. S. Lim, J. Steele, P. Schley, B. Tillack, A. Wolff, K. Blum, W. Winkler, M. Pierschel, U. Jagdhold, R. Barth, T. Grabolla, H. J. Erzgräber, B. Hunger, and H. J. Osten, Modular Integration of High-Performance SiGe:C HBTs in a Deep Submicron, Epi-Free CMOS Process, *IEDM Tech. Dig.*, pp. 561–564, 1999.
124. G. Freeman, D. Ahlgreen, D. R. Greenberg, R. Groves, F. Huang, G. Hugo, B. Jagannathan, S. J. Jeng, J. Johnson, K. Schonenberg, K. Stein, R. Volant, and S. Subbanna, A 0.18μm 90 GHz f_T SiGe HBT BiCMOS, ASIC-Compatible, Copper Interconnect Technology for RF and Microwave Applications, *IEDM Tech. Dig.*, pp. 569–572, 1999.

125. J. Browne, InGaP/GaAs Provides High-Linearity HBTs, *Microwaves & RF, 39,* pp. 121–129, Feb. 2000.
126. P. Chavarkar and U. Mishra, High Electron Mobility Transistors, in: M. Golio (ed.), *The RF and Microwave Handbook,* CRC Press, Boca Raton, 2001.
127. B.-U. Klepser and W. Klein, Ramp-up of First SiGe Circuits for Mobile Communications: Postioning of SiGe vs. GaAs and Silicon, *Proc. GaAs IC Symp.,* pp. 37–40, 1999.
128. W. Mickanin, Don't Rule out the MESFET!, *Microwave J., 40,* pp. 112–120, July 1998.
129. E. I. Sobolewski, U.S. Applications of HEMT Technology in Military and Commercial Systems, in: R. L. Ross, S. P. Svensson, and P. Lugli (eds.), *Pseudomorphic HEMT Technology and Applications,* Kluwer, Dordrecht, 1996.
130. G. Bechtel, The 1999 Outlook for GaAs IC Markets and Technology, *Proc. GaAs IC Symp.,* pp. 7–9, 1999.
131. M. Telford, Larger Wafers Boosting GaAs and InP Electronics, *III-Vs Review, 13,* pp. 32–37, Issue 5 2000.
132. -, Bei HF-Anwendungen unschlagbar, *Elektronik,* pp. 88-91, Issue 16, 2000.
133. R. Lachner, SiGe HBT Technology–Survey and Challenges, *GMM Workshop Integrated Silicon Hetero Devices,* Munich April 2002.
134. -, 10 Years of RF-CMOS - But How Many Products Today?, *Dig. ISSCC,* pp. 104–105, 2001.
135. A. Hellemans, Bonding Gallium Arsenide to Silicon Substrate, *IEEE Spectrum 38,* pp. 33–34, Oct. 2001.
136. L. F. Eastman and U. K. Mishra, The Toughest Transistor Yet, *IEEE Spectrum 39,* pp. 28–33, May 2002.

CHAPTER 2

BASIC SEMICONDUCTOR PHYSICS

2.1 INTRODUCTION

Since all microwave transistors are fabricated from a class of materials known as semiconductors, it is appropriate and necessary that we discuss the physics, nature, and properties of semiconducting materials prior to the coverage of microwave transistor physics. In spite of the fact that there exist many published research papers on the experimentally and theoretically determined values of semiconductor properties, no comprehensive and condensed survey of the material properties for microwave transistors is currently available. Therefore, the values of important material properties are also presented in a unified manner in this chapter. The knowledge of these material parameters is of crucial importance for device modeling, simulation, and design.

Some remarks on the background of material parameters are advisable. The values given in this book are carefully selected from a large number of recent publications. The selection is a complex procedure, because even for the apparently well-known materials such as Si, GaAs, or InP, the published experimental and theoretical values of certain material parameters can differ considerably from reference to reference. The situation becomes even worse for less matured materials (e.g., SiC or GaN), alloys and strained layers. In some cases no data are published at all. The values given in this chapter are those considered the most reliable by the authors.

We will focus on eleven semiconductor materials, which we call the basic semiconductors, and on alloys composed of these basic semiconductors. These basic semiconductors are Ge, Si, GaAs, AlAs, InAs, InP, GaP, SiC (4H and 6H), GaN, and AlN, all of which are materials commonly used for microwave transistors that are commercially available or under development. The first seven materials are of the cubic crystal type, and the last four are of the hexagonal (wurtzite) crystal type. In microwave transistors, not only the basic semiconductors, but also their binary, ternary, or even quaternary alloys like $Si_{1-x}Ge_x$, $Al_xGa_{1-x}As$, and $In_xGa_{1-x}As_yP_{1-y}$ are being used. Therefore, in the following sections composition-dependent material parameters for these alloys will be given as well.

Sometimes material parameters for an alloy are not available. In such cases, the material parameters can be approximated by a linear interpolation scheme [1–3]. Let us consider a ternary alloy of composition $A_xB_{1-x}C$ (e.g., $Al_{0.3}Ga_{0.7}As$). If the material parameters of interest are known for the two basic binary compounds AC and BC, then the composition-dependent material parameter $T_{A_xB_{1-x}C}(x)$ of the ternary compound can be estimated by

$$T_{A_xB_{1-x}C}(x) = xB_{AC} + (1-x)B_{BC} \qquad (2\text{-}1)$$

where B_{AC} and B_{BC} are the material parameters of the two basic binary compounds AC and BC, respectively, and x is the fractional content of element A. For example, in the $Al_{0.3}Ga_{0.7}As$ alloy Al corresponds to element A, Ga to element B, As to element C, and 0.3 to x. It should be noted that not the ordering of the elements in the formula (A, B, C) is of importance, but rather the fractional contents x and $(1-x)$ of the elements A and B, respectively, solely define the physical properties of the compound.

If a semiconductor is a quaternary alloy of composition $A_xB_{1-x}C_yD_{1-y}$, then it can be thought to consist of the four basic binary compounds AC, AD, BC, and BD. The relevant material parameter $Q_{A_xB_{1-x}C_yD_{1-y}}(x, y)$ of the quaternary alloy can be estimated by

$$Q(x, y) = xyB_{AC} + x(1-y)B_{AD} + (1-x)yB_{BC} + (1-x)(1-y)B_{BD} \qquad (2\text{-}2)$$

where B is the material parameter of the binary compound and the subscript designates the corresponding binary compound.

The interpolation method should be used only when reliable data for the alloy of interest are not available, because often the composition dependence of the material properties obeys relationships more complicated than the linear interpolation. This is especially true for alloys consisting of two binary compounds with one being a direct bandgap semiconductor and the other an indirect bandgap semiconductor. Examples of such materials are $Al_xGa_{1-x}As$ and $Al_xIn_{1-x}As$. GaAs and InAs are direct bandgap semiconductors, whereas AlAs is an indirect one. The x values at which the transition from direct to indirect bandgap occurs are $x \approx 0.42$ for $Al_xGa_{1-x}As$ and $x \approx 0.65$ for $Al_xIn_{1-x}As$. At the transition point, material parameters related to the bandstructure (e.g., bandgap, effective mass, and effective density of states) may be nonlinear and even abrupt functions of x, whereas other parameters (e.g., dielectric constant) can still be approximated by the interpolation formulas given above. Caution must also be taken when strained layers are considered. Certain material parameters, such as the bandgap, of strained materials may differ considerably from those of unstrained bulk material.

Collections of semiconductor material parameters can be found in the emis Datareview Series [4–14] and in the excellent reviews by Adachi [1–3], Beadle et al. [15], Madelung [16], Levinshtein et al. [17, 18], and Vurgaftman et al. [19]. Unless otherwise stated, the material parameter values are taken from [1–19] and references therein.

To theoretically investigate transistor characteristics and to predict transistor performance, device engineers can utilize either advanced numerical device simulation tools or rather simple analytical models. Although numerical device simulation accounts for more physical effects and is often more accurate, analytical approaches nonetheless enjoy great popularity. From analytical equations, the basic correlations between transistor operation and transistor design can be easily deduced. Such information is lost for the most part when numerical simulators are used. In this book, we concentrate on analytical approaches to describe the basic semiconductor and heterostructure physics as well as the physics of the different types of microwave transistors.

In the following sections, we will first describe the concept of energy band diagrams and deal with free-carrier statistics. This is followed by the discussions of free-carrier transport in semiconductors. The physics of pn and Schottky junctions, which are the heart of semiconductor devices, are then presented. Finally, the issues of impact ionization and self-heating are addressed. Whenever appropriate, relevant material parameters will be given. Only homojunctions (devices made of a single material) are considered in this chapter. The concept and physics of heterojunctions frequently used in microwave devices will be covered in Chapter 3.

2.2 FREE-CARRIER DENSITIES

2.2.1 Band Diagrams and Band Structure

There are two types of free carriers in a semiconductor: free electrons and free holes. Currents passing through semiconductor devices are directly related to the numbers of free electrons and holes. Conventionally, the free carriers are expressed in terms of carrier densities, i.e., the number of carriers per cubic centimeter. Before we proceed with the discussion, the terms thermal equilibrium and nonequilibrium need to be introduced. Under the thermal equilbrium condition, heat is the only energy applied to the semiconductor. If other types of energy, such as light or electric energy in the form of an applied voltage and the resulting field, are applied, the equilibrium may be perturbated. Such a condition is called nonequilibrium. In the following, the thermal equilibrium condition is assumed unless otherwise stated.

We first consider a pure semiconductor without any impurities, the so-called intrinsic or undoped semiconductor. Electrons in an undoped semiconductor cannot be found at any arbitrary energy E, but only at certain allowed energy ranges called the allowed energy bands. No electrons can occupy the energy levels in the forbidden bandgaps outside these allowed energy bands. For the conductivity of semiconductors, only the free electrons in the conduction band and the free holes in the valence band are of interest. Figure 2.1(a) shows the energy band diagram of an undoped semiconductor. The valence electrons, which are the electrons in the outermost orbit of the semiconductor atoms, occupy energy states in the valence band. Above the top of the valence band, E_V, we find a forbidden bandgap, followed by the conduction band occupied by free electrons. The energy separation between E_V and the bottom of the conduction band, E_C, is called the bandgap, $E_G = E_C - E_V$. It should be noted that energy states are frequently called energy levels. These two terms are used synonymously.

The transition of an valence electron from the valence band to the conduction

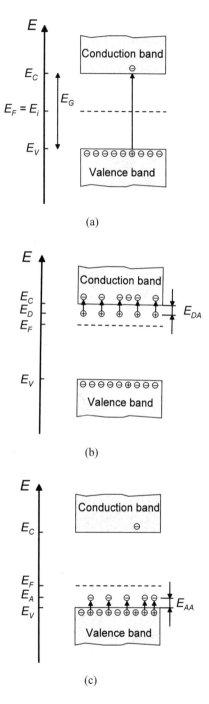

Figure 2.1. Semiconductor band diagram. (a) Intrinsic semiconductor, (b) n-type semiconductor, and (c) p-type semiconductor.

band is only possible by supplying extra energy to the electron. The extra energy must be at least equal to the bandgap energy. Then the electron is lifted to the conduction band, leaving behind an unoccupied state in the valence band, which is called a hole. Accordingly, the hole is positively charged. Such a process is called generation. The excess energy of a conduction band electron relative to the bottom of the conduction band is just the kinetic energy of the free electron. The opposite process, called recombination, occurs when a free electron falls down from the conduction band to the valence band, thereby releasing energy equal to or larger than the bandgap and filling the hole.

Since electric current is a spatial transfer of charges, free electrons and holes are necessary constituents of the conductivity in semiconductors. Actually, the motion of free holes is the motion of valence electrons jumping into unfilled energy states, thus filling the holes and creating "new" holes at their previous positions before the jump. For simplicity, the free electrons and holes are called electrons and holes hereafter.

The Fermi level E_F of an intrinsic semiconductor is located close to the middle of the bandgap. We call it the intrinsic Fermi level E_i. At absolute zero temperature ($T = 0$ K) all allowed states below E_F are filled with electrons and all states above E_F are empty. At higher temperatures, due to thermally assisted transitions a portion of the allowed energy states above E_F is occupied by electrons while a number of states below E_F are empty. The higher the temperature, the more thermal energy is fed to the semiconductor and the more electrons will be able to jump from the valence band to the conduction band. Thus, the density of free carriers (electrons and holes) increases with increasing temperature.

The density of electrons and holes can also be increased by intentionally incorporating impurities (dopants) into the semiconductor. A semiconductor containing dopants is called doped or extrinsic material. There are two types of dopants: donors and acceptors. Donors are atoms possessing one valence electron more than needed to complete their chemical bonds to the neighboring semiconductor atoms. Only a small amount of energy (typically a few tens of meV) is necessary to release this extra electron from its host donor to the conduction band, which creates one free electron and leaves behind a positively charged donor ion. This energy amount is the donor activation energy E_{DA}. In the energy band diagram, the incorporation of donors can be represented by adding an energy level E_D slightly below the conduction band edge. The donor activation energy E_{DA} is equal to the energy difference of E_C and E_D ($E_{DA} = E_C - E_D$). Simply speaking, donors deliver free electrons to the semiconductor. Since free electrons carry a negative charge, a semiconductor doped with donors is called an n-type semiconductor. Here electrons are the majority carriers and holes are the minority carriers. Figure 2.1(b) shows the band diagram of an n-type semiconductor.

The second type of dopants are acceptors. These are atoms with one valence electron less than necessary to complete the bonds to the neighboring semiconductor atoms. Thus, the acceptor tends to capture an electron from one of the neighboring lattice atoms to complete its chemical bonding. This is represented in the energy band diagram by introducing an energy level E_A slightly above the valence band edge. Only a small activation energy $E_{AA} = E_A - E_V$ is necessary to capture one valence electron from a neighboring atom. The electron being trapped at the energy level E_A leaves behind a hole in the valence band and the acceptor becomes negatively charged. Since acceptors deliver positively charged holes to the semiconductor, a material doped with acceptors is of the p-type. Here, holes are the majority carriers and electrons are minority carriers. The band diagram of a p-type semiconductor is shown in Fig. 2.1 (c).

As can be seen in Fig. 2.1, the Fermi levels are located in the upper and lower halves of the band diagram for n- and p-type semiconductors, respectively. Increasing the donor (acceptor) concentration causes a shift of the Fermi level toward the conduction (valence) band edge. For materials with very high doping levels, the Fermi level can be located either in the valence band (p-doping) or in the conduction band (n-doping). We will come back to the issue of Fermi level and explain its physical meaning in the following section.

More detailed information on the allowed carrier energies in semiconductors is provided by the energy band structure. The band structure shows the correlation between carrier energy and carrier wave vector, which in turn describes the momentum of the carrier. Figure 2.2 shows the band structures of Si, Ge, InP, and GaAs. The upper part of the band structure (above the gap) depicts the different valleys of the conduction band and the lower part (below the gap) represents the valence band. The correlation between the band diagram shown in Fig. 2.1 and band structure shown in Fig. 2.2 is as follows. The energy of the lowest valley of the conduction

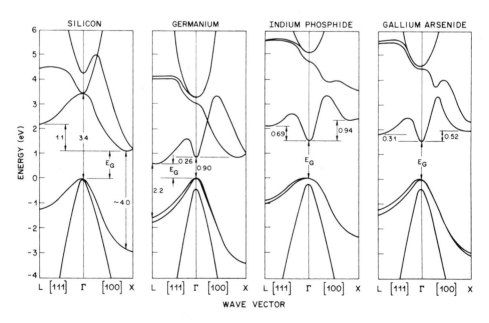

Figure 2.2. Energy band structure of Si, Ge, InP, and GaAs. Taken from [20].

band (for Si slightly above 1 eV near the X point on the wave vector axis) corresponds to the conduction band edge E_C in the band diagram. Analogously, the energy of the highest point of the valence band (for Si 0 eV at the Γ point on the wave vector axis) corresponds to the valence band edge E_V. Thus, the band diagram represents only two extreme points of the band structure.

2.2.2 Carrier Statistics

We now intuitively understand that the carrier densities depend on the dopant concentration and on the dopant activation energy, as well as on the temperature. Our primary goal now is to calculate the densities of electrons and holes, n_0 and p_0, respectively (the subscript 0 denotes the thermal equilibrium condition) in a semiconductor. For this we need the density of allowed energy states (in the following designated as density of states) and the probability that a certain state may be occupied by an electron. The density of states in the conduction band, $N(E)$, is a function of energy and can be expressed as [21]

$$N(E) = 4\pi \left(\frac{2m_{n,\text{ds}}}{h^2}\right)^{3/2} (E - E_C)^{1/2} \tag{2-3}$$

where $m_{n,\text{ds}}$ is the density-of-states effective electron mass (which may be different from the conductivity effective electron mass) and h is the Planck constant. We now introduce the effective density of states in the conduction band, N_C, defined as

$$N_C = 2\left(\frac{2\pi m_{n,\text{ds}} k_B T}{h^2}\right)^{3/2} \tag{2-4}$$

where k_B is the Boltzmann constant and T is the temperature. Putting Eq. (2-4) into Eq. (2-3) gives us

$$N(E) = \frac{2}{\sqrt{\pi}} N_C \left(\frac{E - E_C}{k_B T}\right)^{1/2} \frac{1}{k_B T} \tag{2-5}$$

The unit of $N(E)$ is cm^{-3}eV^{-1}. Thus, $N(E)$ is the number of states at a certain energy level per volume and energy unit. The probability that an energy level E is occupied by an electron is governed by the Fermi–Dirac distribution function $f(E)$ [21]:

$$f(E) = \frac{1}{1 + \exp\left(\dfrac{E - E_F}{k_B T}\right)} \tag{2-6}$$

From Eq. (2-6) it follows that at $E = E_F$ the occupation probability is equal to one half.

Thus, $f(E)$ is smaller than 1/2 (or 50%) for $E > E_F$, and is larger than 1/2 for $E < E_F$.

The density of electrons at a certain energy is thus the product of the density of states that can be occupied by electrons and the occupation probability. For device modeling, in most cases the density of all electrons in the conduction band is of interest. It is obtained by integrating the product $N(E) \times f(E)$ over all allowed energies above the conduction band edge:

$$n_0 = \int_{E_C}^{\infty} N(E)f(E)dE \tag{2-7}$$

Inserting Eqs. (2-5) and (2-6) in Eq. (2-7) we obtain

$$n_0 = \frac{2N_C}{\sqrt{\pi}} \int_{E_C}^{\infty} \left(\frac{E-E_C}{k_BT}\right)^{1/2} \times \frac{1}{1+\exp\left(\frac{E-E_F}{k_BT}\right)} \times \frac{1}{k_BT} dE \tag{2-8}$$

Changing the integration variable from E to x with

$$x = \frac{E-E_C}{k_BT} \tag{2-9}$$

yields

$$n_0 = \frac{2N_C}{\sqrt{\pi}} \int_0^{\infty} \frac{\sqrt{x}}{1+\exp\left(x-\frac{E_F-E_C}{k_BT}\right)} dx \tag{2-10}$$

The integral in Eq. (2-10) is called the Fermi integral of the order ½, which unfortunately cannot be solved analytically.

In a similar way, the hole density in the valence band is derived. The density of states in the valence band $P(E)$ is given by

$$P(E) = \frac{2}{\sqrt{\pi}} N_V \left(\frac{E_V-E}{k_BT}\right)^{1/2} \frac{1}{k_BT} \tag{2-11}$$

where N_V is the effective density of states in the valence band written as

$$N_V = 2\left(\frac{2\pi m_{h,ds} k_B T}{h^2}\right)^{3/2} \tag{2-12}$$

with $m_{h,ds}$ being the density-of-states effective hole mass. We already know that a hole is generated when an electron is activated from an energy state in the valence band to an acceptor energy level in the bandgap or to the conduction band. Therefore, to calculate the hole density, we need the probability $f_h(E)$ that the energy state is empty, i.e., *not* occupied by an electron:

$$f_h(E) = 1 - f(E) = 1 - \frac{1}{1 + \exp\dfrac{E - E_F}{k_B T}} \tag{2-13}$$

After some algebraic manipulations, the hole density in the valence band is expressed as

$$p_0 = \frac{2N_V}{\sqrt{\pi}} \int_0^\infty \frac{\sqrt{x}}{1 + \exp\left(x - \dfrac{E_V - E_F}{k_B T}\right)} \, dx \tag{2-14}$$

where

$$x = \frac{E_V - E}{k_B T} \tag{2-15}$$

2.2.3 Approximations for the Carrier Densities

A. Boltzmann Statistics. The formulas (2-10) and (2-14) give the electron and hole densities in a semiconductor based on the Fermi statistics, which is the physically rigorous description. However, under specific conditions the Fermi integrals can be well approximated. For $E - E_F > 3k_B T$, the exponential term in the denominator of Eq. (2-6) is much larger than unity, and Eq. (2-6) can be approximated by

$$f(E) \approx \exp\left(-\frac{E - E_F}{k_B T}\right) \tag{2-16}$$

Similarly, for $E_F - E > 3k_B T$, Eq. (2-13) is reduced to

$$f_h(E) \approx \exp\left(-\frac{E_F - E}{k_B T}\right) \tag{2-17}$$

The probability functions $f(E)$ and $f_h(E)$ are called the Boltzmann distribution functions. Inserting Eqs. (2-5) and (2-16) in Eq. (2-7) and carrying out the integration, we obtain for the electron density

$$n_0 = N_C \exp\left(-\frac{E_C - E_F}{k_B T}\right) \tag{2-18}$$

Analogously, for the hole density we get

$$p_0 = N_V \exp\left(-\frac{E_F - E_V}{k_B T}\right) \tag{2-19}$$

The simplified expressions Eqs. (2-18) and (2-19) can be used calculate n_0 and p_0 in n-type semiconductors with $E_C - E_F > 3k_BT$ and in p-type semiconductors with $E_F - E_V > 3k_BT$ with sufficient accuracy. Such semiconductors are called nondegenerate. For degenerate semiconductors (n-type material with $E_C - E_F < 3k_BT$ and p-type material with $E_F - E_V < 3k_BT$), however, the Fermi–Dirac statistics should be used. Nondegenerate and degenerate semiconductors correspond to lightly and heavily doped materials, respectively.

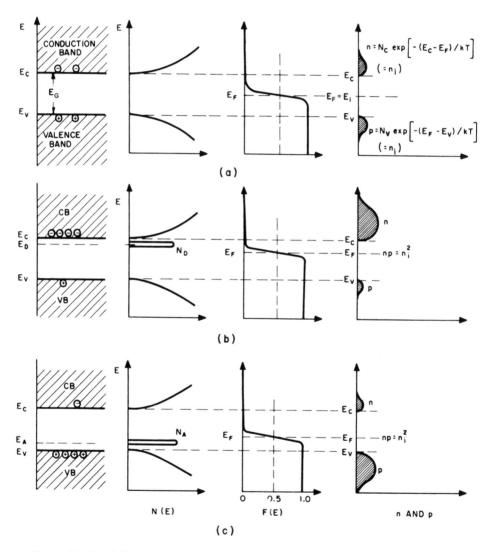

Figure 2.3. Band diagram, density of states, Fermi–Dirac distribution function and carrier densities for (a) intrinsic, (b) n-type, and (c) p-type semiconductors. Taken from [22].

Figure 2.3 summarizes our considerations regarding the free-carrier densities in the conduction and valence bands based on the Boltzmann statistics for intrinsic, p-type, and n-type semiconductors.

Frequently, it is useful to describe the electron and hole densities by alternative expressions. Equations (2-18) and (2-19) are valid for any nondegenerate n-type, p-type, or intrinsic semiconductor under the thermal equilibrium condition. The electron density is equal to the hole density in an intrinsic semiconductor, called the intrinsic carrier density n_i, and the Fermi level in such a material is the intrinsic Fermi level E_i. Thus, Eqs. (2-18) and (2-19) can be rewritten as

$$n_0 = n_i = N_C \exp\left(-\frac{E_C - E_i}{k_B T}\right) \quad (2\text{-}20)$$

and

$$p_0 = n_i = N_V \exp\left(\frac{E_i - E_V}{k_B T}\right) \quad (2\text{-}21)$$

Solving the above equations for N_C and N_V and inserting the results in Eqs. (2-18) and (2-19) gives

$$n_0 = n_i \exp\left(\frac{E_F - E_i}{k_B T}\right) \quad (2\text{-}22)$$

and

$$p_0 = n_i \exp\left(\frac{E_i - E_F}{k_B T}\right) \quad (2\text{-}23)$$

Multiplying Eqs. (2-18) and (2-19) and combining the resulting equation with the product of Eqs. (2-22) and (2-23) yields the following extremely useful expression relating the carrier densities under the thermal equilibrium condition to the intrinsic carrier concentration:

$$n_0 p_0 = n_i^2 = N_C N_V \exp\left(-\frac{E_G}{k_B T}\right) \quad (2\text{-}24)$$

Equation (2-24) holds not only for intrinsic but also for doped semiconductors and is known as the mass action law.

It should be noted that the charge neutrality is maintained in any semiconductor device regardless of the doping. This means that the sum of all positive charges (holes and ionized donors) must be equal to the sum of all negative charges (electrons and ionized acceptors):

$$p + N_D^+ = n + N_A^- \quad (2\text{-}25)$$

where N_D^+ and N_A^- are the densities of ionized donors and acceptors, respectively. Equation (2-25) is known as the charge neutrality relation. Electrons and holes are mobile charges, whereas ionized dopants are fixed.

When applying numerical device simulations, users commonly have the option of using the more accurate Fermi–Dirac distribution function [Eqs. (2-10) and (2-14)] or the approximated Boltzmann formulation [Eqs. (2-18) and (2-19)] for the calculation of the carrier densities. For analytical modeling, however, even the Boltzmann approximation is not sufficiently convenient, because both Eqs. (2-18) and (2-19) still contain two unknowns, namely the carrier densities n_0 or p_0, and the Fermi level E_F. To determine the carrier densities using these equations, one of the following approximations may be used.

B. Complete Dopant Ionization—Approximation 1. If we assume all donors are ionized (complete ionization) in an n-type material and the dopant concentration is much higher than n_i, then according to Eq. (2-25) the electron density can be approximated by

$$n_0 \approx N_D \qquad (2\text{-}26)$$

and the hole density is calculated using the mass action law by

$$p_0 = \frac{n_i^2}{N_D} \qquad (2\text{-}27)$$

Analogously, for a p-type semiconductor we have

$$p_0 \approx N_A \qquad (2\text{-}28)$$

and

$$n_0 = \frac{n_i^2}{N_A} \qquad (2\text{-}29)$$

In the above equations, N_D and N_A are the donor and acceptor concentrations.

C. Complete Dopant Ionization—Approximation 2. Consider the case where all donors are ionized but n_i is no longer negligibly small compared to the dopant density (e.g., in the case of low doping densities or a large intrincic carrier concentration). In this case, the neutrality condition in an n-type material is

$$n_0 \approx N_D + p_0 \qquad (2\text{-}30)$$

Using Eq. (2-24), the electron density becomes

$$n_0 = N_D + \frac{n_i^2}{n_0} \qquad (2\text{-}31)$$

which is a quadratic equation with the solution

$$n_0 = \frac{N_D + \sqrt{N_D^2 + 4n_i^2}}{2} \tag{2-32}$$

Combining this with Eq. (2-30), the hole density can be calculated by

$$p_0 = \frac{\sqrt{N_D^2 + 4n_i^2} - N_D}{2} \tag{2-33}$$

Similarly, for a p-type semiconductor we obtain

$$p_0 = \frac{N_A + \sqrt{N_A^2 + 4n_i^2}}{2} \tag{2-34}$$

and

$$n_0 = \frac{\sqrt{N_A^2 + 4n_i^2} - N_A}{2} \tag{2-35}$$

D. Incomplete Dopant Ionization. Here we consider the case of incomplete dopant ionization. This means that the dopant atoms are not 100% ionized. Thus, $N_D > N_D^+$ in an n-type semiconductor and $N_A > N_A^-$ in p-type material. This occurs either at low temperatures or when the dopant's activation energy is high. At room temperature, complete ionization can be assumed for most semiconductors, e.g., Si, GaAs, and InGaAs. In other semiconductors such as SiC, E_{DA} and E_{AA} are larger than those in the above-mentioned materials, and at room temperature only a portion of the dopants are ionized.

Let us first focus on the n-type semiconductor. The density of ionized donors with activation energy E_{DA} and donor level E_D is described using the Fermi-like distribution function f_D:

$$f_D = \frac{1}{1 + \frac{1}{g}\exp\left(\frac{E_D - E_F}{k_B T}\right)} \tag{2-36}$$

The difference between Eq. (2-36) and the original Fermi–Dirac distribution given in Eq. (2-6) is the addition of the ground state degeneracy factor g (equal to 2) in the denominator. This factor accounts for the mechanism whereby an electron of either spin can occupy an empty donor state, but after one electron occupies the state no other electron can enter it, even if its spin is different from the spin of the occupying electron. Following a similar approach leading to Eq. (2-13), the density of ionized donors (i.e., the density of donor states not occupied by an electron) can be expressed as

$$N_D^+ = N_D(1-f_D) = N_D\left(1 - \cfrac{1}{1+\cfrac{1}{g}\exp\left(\cfrac{E_D - E_F}{k_B T}\right)}\right) \quad (2\text{-}37)$$

Assuming that the electron density is approximately equal to the density of ionized donors, together with some straightforward algebraic manipulations, Eq. (2-37) can be transformed to

$$n_0 = N_D \cfrac{1}{1 + g\exp\left(\cfrac{E_F - E_D}{k_B T}\right)} \quad (2\text{-}38)$$

The above expression describes the electron density including the effect of incomplete donor ionization. Because of the existence of two unknowns (n_0 and E_F), however, it is not yet suited for analytical modeling. Rearranging Eq. (2-18) leads to

$$E_F = k_B T \ln \frac{n_0}{N_C} + E_C \quad (2\text{-}39)$$

This can be inserted into Eq. (2-38), thus giving a quadratic equation for n_0 with the solution

$$n_0 = \frac{N_C}{2g}\exp\left(-\frac{E_C - E_D}{k_B T}\right)\left(\sqrt{1 + 4g\frac{N_D}{N_C}\exp\frac{E_C - E_D}{k_B T}} - 1\right) \quad (2\text{-}40)$$

The only unknown quantity in Eq. (2-40) is now the electron density. Using a similar approach, an expression for the hole density in p-type semiconductors including the effect of incomplete acceptor ionization can be obtained. It should be noted, however, that the density of ionized acceptors given by

$$N_A^- = N_A f_A \quad (2\text{-}41)$$

is already the density of acceptor states occupied by an electron [compare Eq. (2-41) with Eq. (2-37) to see the difference for ionized donors and acceptors]. The probability f_A of an acceptor state occupied by an electron can be written as

$$f_A = \cfrac{1}{1 + g\exp\left(\cfrac{E_A - E_F}{k_B T}\right)} \quad (2\text{-}42)$$

In Eq. (2-42), the degeneracy factor g for acceptor levels is 4 (or 2×2). The first factor of 2 results from the fact that the acceptor can accept one hole of either spin in the same way as the acceptance of electrons by a donor level discussed above. The second factor of 2 stems from the two-fold degeneracy of the uppermost valence band maximum (see the two valence band lines at the Γ points of the four band structures in Fig. 2.2). Finally, we obtain for the hole density, including incomplete acceptor ionization

$$p_0 = \frac{N_V}{2g} \exp\left(-\frac{E_A - E_V}{k_B T}\right)\left(\sqrt{1 + 4g\frac{N_A}{N_V} \exp\frac{E_A - E_V}{k_B T}} - 1\right) \quad (2\text{-}43)$$

E. Joyce–Dixon Approximation. When a highly doped semiconductor is considered, where the Boltzmann statistics become invalid and Fermi–Dirac statistics have to be used, analytical solutions are hard to find. To alleviate this problem, the so-called Joyce–Dixon approximation can be used. It allows the calculation of the position of the Fermi level in degenerate semiconductors provided the majority carrier density is given using [23]

$$\frac{E_F - E_C}{k_B T} = \ln\left(\frac{n_0}{N_C}\right) + a_1\left(\frac{n_0}{N_C}\right) + a_2\left(\frac{n_0}{N_C}\right)^2 + a_3\left(\frac{n_0}{N_C}\right)^3 + a_4\left(\frac{n_0}{N_C}\right)^4 \quad (2\text{-}44)$$

for degenerate n-type material and

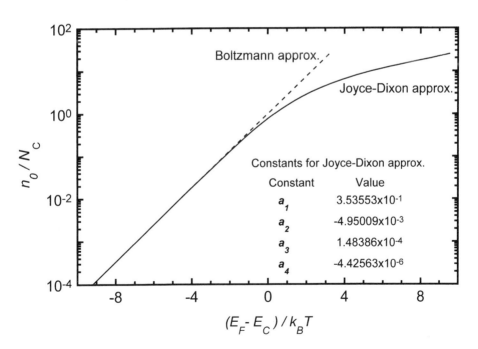

Figure 2.4. Fermi integral of an n-type semiconductor calculated using the Boltzmann approximation in Eq. (2-18) and the Joyce–Dixon approximation in Eq. (2-44).

Table 2.1. Bandgaps, effective densities of states in the conduction band, and effective density of states in the valence band for the 11 basic semiconductors, as well as for unstrained $Al_{0.3}Ga_{0.7}As$, $In_{0.53}Ga_{0.47}As$, and $In_{0.52}Al_{0.48}As$ at T = 300 K

Material	E_G, eV	N_C, cm^{-3}	N_V, cm^{-3}
Ge	0.664	2.08×10^{19}	4.03×10^{18}
Si	1.124	3.22×10^{19}	1.82×10^{19}
GaAs	1.422	3.97×10^{17}	9.40×10^{18}
AlAs (X valley)	2.164		
AlAs (Γ valley)	3.003	2.84×10^{19}	1.81×10^{19}
InAs	0.354	8.14×10^{16}	6.46×10^{18}
InP	1.353	5.57×10^{17}	1.21×10^{19}
GaP (X valley)	2.273	5.32×10^{19}	1.55×10^{19}
GaP (Γ Valley)	2.777		
4H SiC	3.200	1.35×10^{19}	1.18×10^{19}
6H SiC	2.860	4.61×10^{19}	2.54×10^{19}
GaN	3.445	2.25×10^{18}	1.80×10^{19}
AlN	6.138	4.54×10^{18}	6.06×10^{19}
$Al_{0.3}Ga_{0.7}As$	1.800	7.00×10^{17}	1.44×10^{18}
$In_{0.53}Ga_{0.47}As$	0.737	2.08×10^{17}	9.16×10^{18}
$In_{0.52}Al_{0.48}As$	1.451	5.16×10^{17}	7.34×10^{18}

$$\frac{E_V - E_F}{k_B T} = \ln\left(\frac{p_0}{N_V}\right) + a_1\left(\frac{p_0}{N_V}\right) + a_2\left(\frac{p_0}{N_V}\right)^2 + a_3\left(\frac{p_0}{N_V}\right)^3 + a_4\left(\frac{p_0}{N_V}\right)^4 \quad (2\text{-}45)$$

for degenerate p-type material. The parameters a_1, a_2, a_3, and a_4 have the same values in both equations. Figure 2.4 compares the normalized electron density in an n-type material calculated using the Boltzmann approximation and the Joyce–Dixon approximation. The two approximations start to deviate from each other when $(E_F - E_C)/k_B T$ is larger than –2 while the Joyce–Dixon approximation results in nearly the same values for the n_0/N_C versus $(E_F - E_C)/k_B T$ as the Fermi–Dirac statistics over the entire range shown in Fig. 2.4. Given in the figure are also the values of the parameters a_1–a_4.

The material properties of the basic semiconductors required for the calculation of carrier densities are presented in Table 2.1, which lists the bandgaps and the effective densities of states in the conduction and valence bands for the basic semiconductors and three frequently used alloys at room temperature.

The bandgap of a semiconductor is temperature-dependent. It decreases with rising temperature and can be described by the empirical formula

$$E_G(T) = E_G(T = 0 \text{ K}) - \frac{aT^2}{b + T} \quad (2\text{-}46)$$

The parameters $E_G(T = 0 \text{ K})$, a, and b for different semiconductors are given in Table 2.2.

Table 2.2. Parameters for the calculation of the temperature-dependent bandgap

Material	$E_G(T = 0K)$, eV	a, 10^{-4}eV/K	b, K
Ge*	0.744	4.77	235
Si	1.17	4.73	625
GaAs	1.519	5.405	204
AlAs (X valley)	2.24	7.0	530
AlAs (Γ Valley)	3.099	8.85	530
InAs	0.417	2.76	93
InP	1.424	3.63	162
GaP (X valley)	2.35	5.77	372
GaN	3.507	9.09	1020
AlN	6.23	17.99	1462
$Al_{0.3}Ga_{0.7}As$	1.899	5.41	204
$In_{0.53}Ga_{0.47}As$	0.81	4.91	305

*Data for Ge are taken from [22].

2.3 CARRIER TRANSPORT

2.3.1 Introduction

The are two driving forces for carrier transport in semiconductors: electric field and spatial variation of the carrier concentration. The former gives rise to the drift currents and the latter provides the mechanism for the diffusion currents. Both driving forces lead to a directional motion of carriers superimposed on the random thermal motion. To calculate the directional carrier motion and thus the currents in a semiconductor, classical as well as nonclassical models can be used. The classical models are based on the assumption that variation of the electric field in time is sufficiently slow so that the transport properties of carriers, such as mobility or diffusivity, can follow the changes of the field immediately. This condition is known as the stationary state. If, however, carriers are exposed to a fast-varying field, they may not be able to adjust their transport properties instantaneously to variations of the field, and carrier mobility and diffusivity may be different from their steady-state values. This kind of transport is called nonstationary.

It should be noted that nonstationary carrier transport can occur in electron devices under both dc and ac bias conditions. The type of the voltages applied (dc or ac) is not important but rather the question whether the field acting on the carriers is varying in time or not. Assume the case of a dc bias applied to a device. Then the resulting field will cause a directional motion of carriers. Consider an electron moving from a low-field region of the device to a high-field region in the course of a time interval Δt. Then during Δt, the electric field acting on the electron changes and nonstationary carrier transport takes place.

In general, nonstationary transport occurs in all microwave transistors. Its effect on transistor operation, however, depends on the semiconductor material and on the critical device dimensions, e.g., the gate length of FETs. In III–V transistors, nonstationary transport plays an important role, but it is much less important in Si transistors. In SiC transistors, the nonstationary transport effects can be entirely neglected.

2.3.2 Classical Description of Carrier Transport

A. Carrier Drift. First, let us consider a semiconductor sample at a temperature $T > 0$ having a spatially homogeneous carrier concentration with no applied electric field (i.e., thermal equilibrium condition). Thus, no driving force exists in the sample for directional carrier motion. The carriers, however, are not in standstill condition but are in continuous motion because they possess kinetic energy. For electrons in the conduction band, the kinetic energy E_{kin} is given by

$$E_{kin} = \frac{3}{2}k_B T = \frac{m_n^*}{2} v_{th}^2 \qquad (2\text{-}47)$$

where v_{th} is the thermal velocity and m_n^* is the conductivity effective electron mass. At room temperature, v_{th} is of the order of 10^7 cm/s, and the motion of a single electron is a sequence of different scattering events (collisions with scattering centers such as lattice atoms or impurities), each of which changes the direction of electron motion. The average time between two scattering events is the mean free time and the average distance a carrier travels between two collisions is the mean free path. The electron motion is random, as shown in Fig. 2.5(a), and does not result in a net displacement of the carriers.

Now we apply a voltage V to the sample, which creates an electric field $\mathscr{E}$. The field adds a directional component to the random motion of the electron, as shown in Fig. 2.5 (b). If the field is directed from the right to the left, the velocity and the net displacement of the electron are from the left to the right. The distance the elec-

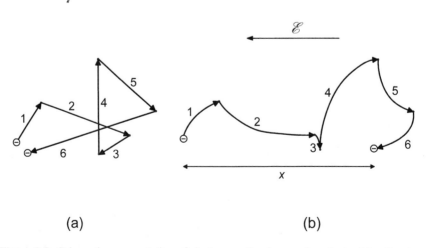

(a)   (b)

Figure 2.5. Schematic representation of electron motion in a semiconductor (a) without applied field and (b) with electric field applied.

tron is displaced by the influence of the field during the time of observation t is x. The mean electron velocity is determined by the value of the electric field and can be expressed by

$$v_n = -\mu_n \mathscr{E} \tag{2-48}$$

where μ_n is the electron mobility. The directed unilateral motion of carriers caused by the electric field is called drift and the velocity v_n is the electron drift velocity. In a similar manner, an applied electric field also influences the motion of holes:

$$v_p = \mu_p \mathscr{E} \tag{2-49}$$

where μ_p is the hole mobility. Two differences can be observed when comparing the drift of electrons and holes. First, holes move in the same direction as the field because of their positive charge (i.e., opposite to the direction of electrons), and second, the value of the hole drift velocity v_p is lower than v_n for a given applied field. In other words, the hole mobility is lower than the electron mobility. Equations (2-48) and (2-49) contain the basic assumption of the classical transport theory that the drift velocity is a direct function of electric field. In other words, a

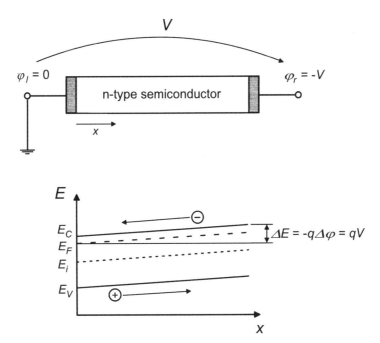

Figure 2.6. Band diagram (with band bending) of an n-type semiconductor with applied voltage V.

change in the electric field instantaneously results in a change of the drift velocity.

The effect of an applied voltage on the band diagram can be seen in Fig. 2.6. If we connect the left-hand terminal to ground (potential $\varphi_l = 0$) and the right-hand terminal to a negative potential $\varphi_r = -V$, then the energy bands at the right terminal are lifted upward by the amount $\Delta E = qV$. Thus the relationship among energy, potential and voltage is given by

$$\Delta E = -q\Delta\varphi = qV \tag{2-50}$$

The bending of the conduction and valence bands, as well as that of the intrinsic and actual Fermi levels, is the same, i.e., $qV = \Delta E_C = \Delta E_V = \Delta E_i = \Delta E_F$. Under the bias conditions shown in Fig. 2.6, the electric field is in the positive x-direction, and electrons move from the right to the left.

B. Low-Field Carrier Drift. At low fields, the drift velocity shows a linear dependence on the field, and the mobility is a constant called the low-field mobility. It is μ_{0n} and μ_{0p} for electrons and holes, respectively. For higher fields, on the other hand, the mobility becomes field-dependent and decreases with increasing field.

The low-field mobilities μ_{0n} and μ_{0p} depend on both the doping concentration and on the temperature. Conventionally, the doping dependence of the low-field mobility is modeled by the empirical relation

$$\mu_0 = \mu_{\min} + \frac{\mu_{\max} - \mu_{\min}}{1 + \left(\dfrac{N}{N_{\text{ref}}}\right)^\alpha} \tag{2-51}$$

proposed by Caughey and Thomas [24]. Here, N is the total doping concentration and $\mu_{\max}$, $\mu_{\min}$, α, and N_{ref} are parameters describing the shape of the μ_0 versus N characteristics. These parameters are determined by fitting mobility data obtained experimentally. The parameter $\mu_{\max}$ represents the low-field mobility in an undoped material, where lattice scattering is the main scattering mechanism, whereas $\mu_{\min}$ is the mobility in a highly doped sample, where impurity scattering is dominant. N_{ref} is the doping concentration at which the mobility is half way be-

Table 2.3. Parameters for the low-field mobility model in Eq. (2-51)

Material	$\mu_{\min}$, cm²/Vs	$\mu_{\max}$, cm²/Vs	N_{ref}, cm⁻³	α	Ref.
Si, n	74.5	1430	8.6×10^{16}	0.77	[25]
Si, p	49.7	468	1.6×10^{17}	0.70	[28]
GaAs, n	1750	8625	6.0×10^{16}	0.55	Fitted from Fig. 2.7(b)
4H SiC, n	40.0	950	2.0×10^{17}	0.76	[29]
4H SiC, p	15.9	124	1.7×10^{19}	0.34	[30]
6H SiC, n	30.0	420	6.0×10^{17}	0.80	[29]
6H SiC, p	6.8	99	2.1×10^{19}	0.31	[30]

tween μ_{min} and μ_{max}, and α is a measure of how quickly the mobility changes from μ_{max} to μ_{min}. Table 2.3 summarizes these fitting parameters for electron and hole low-field mobilities of different semiconductors at room temperature. It should be noted that the mobilities of majority carriers differ from those of minority carriers. The mobility of minority carriers can be considerably higher than that of majority carriers. This issue is important for the modeling of bipolar transistors and will be treated in detail in Sections 7.2 and 8.2. The parameters given in Table 2.3 are related to the majority carrier mobilities. Unfortunately, reliable values for μ_{0n} and μ_{0p} are not available for some of the basic semiconductors considered. The electron low-field mobilities in Si and GaAs versus doping concentration at room temperature are shown in Figs. 2.7(a) and (b), respectively. Clearly, a much higher electron mobility is found in GaAs over a wide range of doping densities.

The low-field hole mobility in GaAs can be calculated using the following alternative expression [23]:

$$\mu_0 = \frac{380}{(1 + 3.17 \times 10^{-17} \text{ cm}^3 \, N)^{0.266}} \text{ cm}^2/\text{Vs} \qquad (2\text{-}52)$$

Figures 2.8(a) and (b) show the low-field electron and hole mobilities, respectively, of different undoped semiconductors at room temperature.

The low-field electron and hole mobilities are temperature-dependent. The temperature range of interest for microwave transistor operation is around and above room temperature. In this temperature range, the $\mu_0(T)$ dependence can be accounted for by making the four parameters in Eq. (2-51) temperature-dependent, e.g.,

$$\alpha(T) = \alpha(300 \text{ K}) \times \left(\frac{T}{300 \text{ K}}\right)^{\eta_\alpha} \qquad (2\text{-}53)$$

or simply by setting

$$\mu_0(T) = \mu_0(300 \text{ K}) \times \left(\frac{T}{300 \text{ K}}\right)^{\eta_\mu} \qquad (2\text{-}54)$$

where η_α and η_μ are fitting parameters. Figure 2.9 shows the temperature-dependent electron low-field mobilities for some of the basic undoped semiconductors. The linear shape of μ_0 in this log–log plot follows the temperature dependence described by Eq. (2-54). In general, a high low-field mobility is preferable for microwave transistors.

C. High-Field Carrier Drift. When larger fields are applied, the drift velocity is no longer linearly proportional to the field. Consider, for example, an n-type Si sample. A low voltage applied to the sample creates a low field that causes the electrons to drift with a velocity $v = \mu_0|\mathcal{E}|$. At higher fields, however, the drift velocity has a sublinear field dependence. In this case, the mobility is field-dependent and its

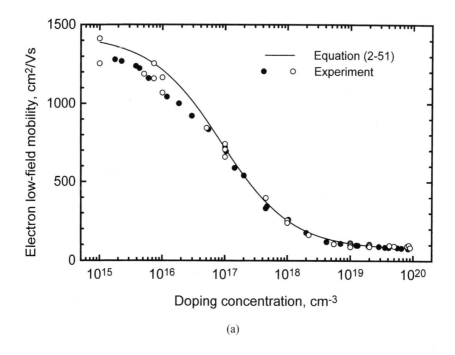

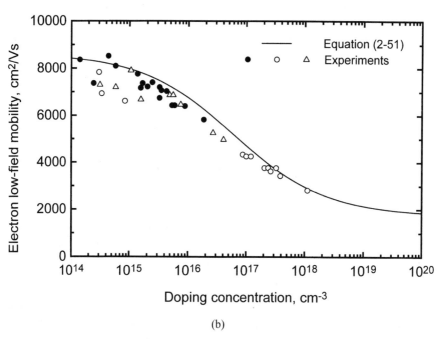

Figure 2.7. Doping-dependent low-field mobility at room temperature. (a) Si and (b) GaAs. The symbols are measured data [25–27] and lines are calculated by Eq. (2-51).

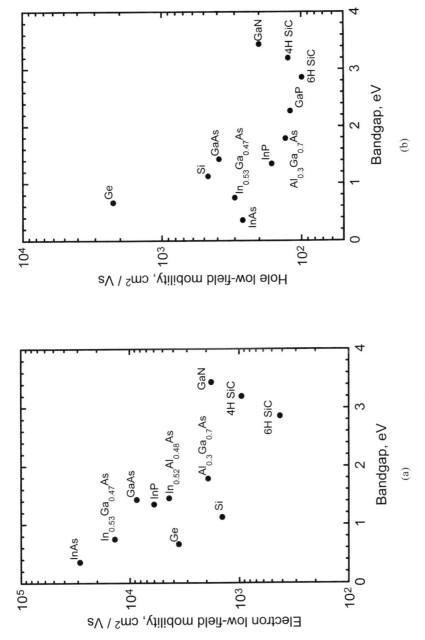

Figure 2.8. Low-field (a) electron and (b) hole mobilities versus bandgap for different undoped semiconductors.

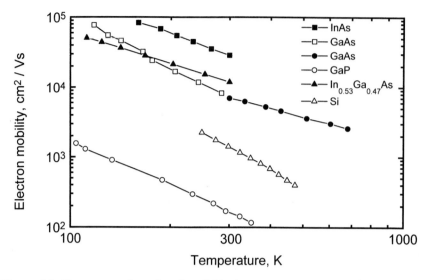

Figure 2.9. Temperature-dependent low-field electron mobilities for different undoped semiconductors. Full squares: InAs. Open squares and full circles: GaAs. Full triangles: In$_{0.53}$Ga$_{0.47}$As. Open triangles: Si. Open Circles: GaP. Data taken from [3, 15], and references therein.

value is lower than μ_0, because the energy picked up by the electrons only partially translates into a higher drift velocity. Part of the energy is transferred to the lattice through the scattering events whose efficiencies rise with increasing carrier energy. Increasing the field further will finally lead to a saturation of the drift velocity. Such a behavior can be observed for both types of carriers. The drift velocity versus field (v–$\mathscr{E}$) characteristics for Si and for semiconductors with similar v–$\mathscr{E}$ characteristics can be modeled by the empirical expression

$$v(\mathscr{E}) = \mu_0 |\mathscr{E}| \left(\frac{1}{1 + \left(\frac{\mu_0 |\mathscr{E}|}{v_{sat}} \right)^\beta} \right)^{1/\beta} \qquad (2\text{-}55)$$

proposed by Caughey and Thomas [24]. Note that the direction of the electron motion is opposite to the direction of the field. Thus, the electron velocity and the field have opposite signs [see Eq. (2-48)]. The equation is valid for both electrons and holes when appropriate values for the parameters μ_0, v_{sat}, and β are used. Like the low-field mobility, the saturation velocity v_{sat} is temperature-dependent and decreases with increasing temperature. For electrons and holes in Si, v_{sat} can be described by [22]

$$v_{sat} = \frac{2.4 \times 10^7}{1 + 0.8 \exp\left(\dfrac{T}{600 \text{ K}} \right)} \text{ cm/s} \qquad (2\text{-}56)$$

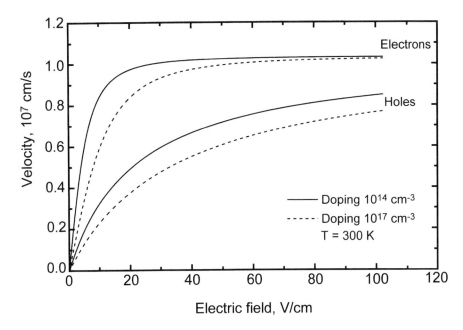

Figure 2.10. Drift velocity versus electric field characteristics for electrons and holes in Si calculated from Eqs. (2-51), (2-55), and (2-56) at room temperature.

Figure 2.10 shows the velocity-field characteristics for electrons and holes in silicon calculated using Eqs. (2-51), (2-55), and (2-56).

In materials such as GaAs or InP, another high-field effect also exists. In these materials, not only can a sublinear slope of the v–$\mathscr{E}$ characteristics be observed, but the velocity actually decreases after reaching a peak value at a certain critical field and approaches asymptotically to a saturation value. This phenomenon can be explained using the band structure. The conduction band of any semiconductor normally consists of several valleys at different energy levels. The band structure of GaAs shown in Fig. 2.2 reveals a narrow lower valley at the Γ point and a much broader upper valley (0.31 eV higher) at the L point on the wave vector axis. The curvature of the valleys is directly related to the electron mobility. Electrons occupying the lower, large-curvature valley possess a relatively large mobility μ_l. On the contrary, electrons in the upper, small-curvature valley possess a much smaller mobility μ_u and therefore can only achieve a lower velocity for a given field.

Consider an n-type GaAs sample with a total electron concentration n. When a low field is applied to the sample, all electrons reside in the lower valley and enjoy the high mobility μ_l. At higher fields, part of the electrons gain enough energy to transfer to the upper valley. Thus, we find electrons of concentration n_l in the lower valley possessing a high mobility and traveling with a high speed, and electrons of concentration n_u in the upper valley possessing a low mobility and moving much slower. The average mobility of all electrons can be calculated by

$$\mu_{av}(\mathscr{E}) = \frac{\mu_l n_l(\mathscr{E}) + \mu_u n_u(\mathscr{E})}{n} \qquad (2\text{-}57)$$

This results in an average electron drift velocity

$$v_{av} = \mu_{av}|\mathscr{E}| \qquad (2\text{-}58)$$

The concentration n_u rises and n_l reduces with increasing field. Thus, when the field exceeds a critical field $\mathscr{E}_c$, electrons transferring from the lower to the upper valley become significant, and the average velocity declines. Figure 2.11 illustrates the electron population in the lower and upper valleys and the stationary v–$\mathscr{E}$ characteristics for GaAs.

The term carrier velocity always refers to the average velocity of all conduction band electrons or valence band holes, even if it is not mentioned explicitly. The same applies to the velocity in Figs. 2.10 and 2.11, and in Eq. (2-55). Figures 2.12 and 2.13 show the stationary velocity-field characteristics for electrons of further semiconductors. The slope of the v–$\mathscr{E}$ characteristics near the origin corresponds to the low-field mobility μ_0. In addition to a large μ_0, high peak and/or saturation velocities are always desirable for microwave transistors.

D. Drift Current Densities. The drift current density J_{Dr} is described by the product of the density, the charge, and the drift velocity of the drifting carriers. Thus, the electron and hole drift current densities $J_{Dr,n}$ and $J_{Dr,p}$ are

$$J_{Dr,n} = -qnv_n = qn\mu_n(\mathscr{E})\mathscr{E} \qquad (2\text{-}59)$$

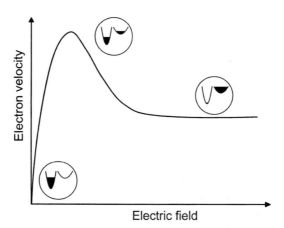

Figure 2.11. Schematic drift velocity versus electric field characteristics for electrons in GaAs. The inserts show the electron populations in the lower high-mobility and the upper low-mobility conduction band valleys.

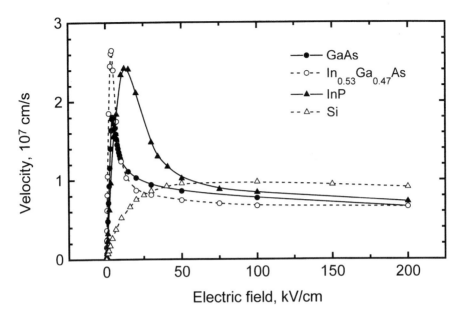

Figure 2.12. Electron drift velocity versus electric field characteristics for GaAs, $In_{0.53}Ga_{0.47}As$, InP, and Si. The doping level is 10^{17} cm^{-3} and the temperature is 300 K for all cases. Data obtained from Monte Carlo simulations [31–33].

and

$$J_{Dr,p} = qpv_p = qp\mu_p(\mathcal{E})\mathcal{E} \quad (2\text{-}60)$$

Under low-field conditions where $v = \mu_0 \mathcal{E}$ holds, one can write

$$J_{Dr,n} = qn\mu_{0n}\mathcal{E} \quad (2\text{-}61)$$

and

$$J_{Dr,p} = qp\mu_{0p}\mathcal{E} \quad (2\text{-}62)$$

The signs of the first and second terms of Eq. (2-59) are different, as mandated by the fact that the electron velocity v_n is in a direction opposite to that of the field $\mathcal{E}$. The total drift current density J_{Dr} is the sum of electron and hole drift current densities $J_{Dr,n}$ and $J_{Dr,p}$: $J_{Dr} = J_{Dr,n} + J_{Dr,p}$.

E. Carrier Diffusion and Diffusion Current Densities. In a semiconductor having a nonuniform carrier distribution, carriers tend to move from one location with higher carrier concentration to another location with lower carrier concentration. The carrier motion caused by this mechanism is called diffusion.

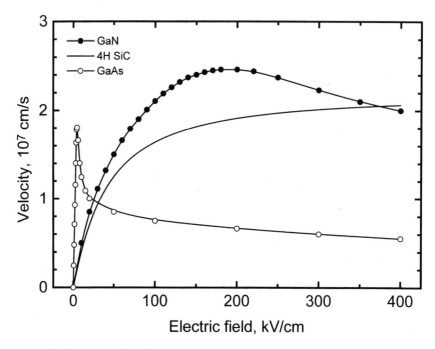

Figure 2.13. Electron drift velocity versus electric field characteristics for 4H SiC, GaN, and GaAs. The doping level is 10^{17} cm^{-3} and the temperature is 300 K for all cases. Data obtained from Monte Carlo simulations [31, 32, 34]. The SiC curve was obtained from Eqs. (2-51), (2-55), and (2-56) with the parameters taken from [29].

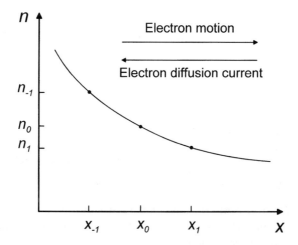

Figure 2.14. Electron concentration as a function of distance in a semiconductor sample.

Consider an n-type semiconductor sample without an applied field and under uniform temperature, but with a spatially dependent electron density as shown in Fig. 2.14. The electron concentrations are n_{-1}, n_0, and n_1 at the locations x_{-1}, x_0, and x_1, respectively. The distance Δx between x_{-1} and x_0 is less than the mean free path and equal to the distance between x_0 and x_1. Because of the uniform temperature and the absence of an external electric field, all electrons move randomly with the thermal velocity v_{th}. At a certain location x, e.g., at $x = x_0$, the probability that an electron moves to the left (in the $-x$ direction) is the same as the probability that it moves to the right. Therefore, the average rate P_1 of electrons flowing from x_1 toward x_0 (per unit area) and crossing the plane $x = x_0$ is given by

$$P_1 = \frac{1}{2} n_1 v_{\text{th}} \tag{2-63}$$

If we approximate the electron concentration at x_1 by the first two terms of a Taylor series, P_1 becomes

$$P_1 \approx \frac{1}{2} v_{\text{th}} \left(n_0 + \frac{dn}{dx} \Delta x \right) \tag{2-64}$$

Analogously, the average rate of electrons flowing from x_{-1} toward x_0 can be obtained as

$$P_{-1} \approx \frac{1}{2} v_{\text{th}} \left(n_0 - \frac{dn}{dx} \Delta x \right) \tag{2-65}$$

The net rate P (per unit area) of electrons crossing the plane at x_0 in the positive x direction is the difference between P_{-1} and P_1 and given as

$$P = v_{\text{th}} \left(-\frac{dn}{dx} \Delta x \right) \tag{2-66}$$

Introducing the electron diffusion coefficient $D_n = v_{\text{th}} \Delta x$, Eq. (2-66) becomes

$$P = -D_n \frac{dn}{dx} \tag{2-67}$$

The electron diffusion current density $J_{\text{Di},n}$ can be obtained from the net rate (per unit area) of electron diffusion by multiplying P with the charge of an electron, i.e., by $-q$, as

$$J_{\text{Di},n} = qD_n \frac{dn}{dx} \tag{2-68}$$

In this particular example, electrons diffuse from left to right, and by conventional current definition, the electron diffusion current flows from right to left. Similarly, the equation

$$J_{\text{Di},p} = -qD_p \frac{dp}{dx} \tag{2-69}$$

for the hole diffusion current density $J_{\text{Di},p}$, with D_p being the hole diffusion coefficient, can be derived. The total diffusion current density J_{Di} is the sum of the electron and hole diffusion current densities $J_{\text{Di},n}$ and $J_{\text{Di},p}$. As can be seen from Eqs. (2-68) and (2-69), the driving force for a diffusion current is the gradient of the carrier concentration, whereas the driving force for a drift current [see Eqs. (2-61) and (2-62)] is the electric field. The main difference between these driving forces is that the field acts directly on the carriers and gives rise to a directional motion of every carrier in the sample, whereas an individual carrier does not feel any force from a carrier concentration gradient.

In the drift and diffusion current density equations, the mobility (μ_n and μ_p) and the diffusion coefficient (D_n and D_p) are important parameters. They describe whether it is easy or difficult for one of the driving forces to create a net carrier displacement and thus a current. Therefore, it is not surprising that a relationship between mobility and diffusion coefficient does exist. This relationship is known as the Einstein relation, which has the form

$$D_n = \frac{k_B T}{q} \mu_n \tag{2-70}$$

for electrons and

$$D_p = \frac{k_B T}{q} \mu_p \tag{2-71}$$

for holes.

F. Alternative Expressions for the Current Density. Based on the physics given in the preceding sections, it is clear that the total current density J in a semiconductor consists of four components: the electron drift and diffusion current densities and the hole drift and diffusion current densities. Thus

$$J = J_{\text{Dr},n} + J_{\text{Di},n} + J_{\text{Dr},p} + J_{\text{Di},p} \tag{2-72}$$

The total electron current density, J_n, is given by Eqs. (2-59) and (2-68) as

$$J_n = J_{\text{Dr},n} + J_{\text{Di},n} = qn\mu_n \mathcal{E} + qD_n \frac{dn}{dx} \tag{2-73}$$

The relation between the electric field $\mathcal{E}$ and and the electrostatic potential φ is given by

$$\mathcal{E} = -\frac{d\varphi}{dx} \tag{2-74}$$

Using Eq. (2-74) for the field, Eq. (2-22) for the electron density, the Einstein relation to express the diffusion coefficient in terms of the mobility, and applying Eq. (2-50) to relate the potential and energy, Eq. (2-73) reduces to

$$J_n = n\mu_n \frac{dE_F}{dx} \quad (2\text{-}75)$$

Similarly,

$$J_p = p\mu_p \frac{dE_F}{dx} \quad (2\text{-}76)$$

It is clearly indicated in Eqs. (2-75) and (2-76) that the general driving force for a current is the gradient of the Fermi level. It is impossible, however, to identify from Eqs. (2-75) and (2-76) whether this driving force is caused by a field or a gradient of the carrier concentration.

2.3.3 Nonclassical Description of Carrier Transport

As mentioned earlier, the basic assumption of the classical description of carrier drift is that the carrier velocity is a direct function of the electric field. This means that a sudden change of the field results in an immediate change of carrier drift velocity. This assumption, however, is not correct. According to classical mechanics, a particle with a certain mass cannot change its velocity instantaneously because of inertia, even if the driving force for the motion does. Carriers possess a certain mass (the effective mass) and need a certain transition time to change their kinetic energy and their velocity after a sudden variation of the field. The behavior of carriers during the transition period is called carrier dynamics or nonstationary carrier transport. Such a transport was first investigated using Monte Carlo simulations for Si and GaAs [35]. It can also be described easily by the so-called relaxation time approximation (RTA). The heart of the RTA is the energy and the momentum balance equations, which for a homogeneous semiconductor sample can be written as [31, 36, 37]

$$\frac{dE}{dt} = q\mathcal{E}v - \frac{E - E_0}{\tau_E} \quad (2\text{-}77)$$

$$\frac{d(m^*v)}{dt} = q\mathcal{E} - \frac{m^*v}{\tau_P} \quad (2\text{-}78)$$

where τ_E and τ_P are the energy-dependent effective relaxation times for carrier energy and momentum, respectively, E_0 is the carrier energy at thermal equilibrium (i.e., $\mathcal{E} = 0$), m^* is the conductivity effective electron mass (i.e., the effective mass related to transport phenomena), and $m^* \times v$ is the momentum P of the carrier. Equations (2-77) and (2-78) describe the balance between the amount of energy and momentum the carriers gained from the applied field (first terms on the right-hand side), and the energy and momentum loss caused by scattering (second terms on the right-hand side). The effects of all different scattering events are combined in the relaxation times τ_E and τ_P.

In the following, we briefly describe how Eqs. (2-77) and (2-78) can be employed to simulate nonstationary transport when a carrier is subjected to a field change. To this end, the stationary v–$\mathcal{E}$, E–$\mathcal{E}$, and m^*–$\mathcal{E}$ dependences are necessary, and they have been calculated by Monte Carlo simulations for several different semiconductors [31–33]. For illustrations, Figs. 2.15(a)–(c) show these stationary characteristics for electrons in GaAs.

Let us consider an n-type GaAs sample and assume that between $t_{-\infty}$ and t a constant field $\mathcal{E} = \mathcal{E}(t_{-\infty}) = \mathcal{E}(t)$ was applied to the sample. Thus, stationary conditions exist in the sample. At time t, the electrons possess the energy $E(t)$, the momentum $P(t)$, and the effective mass $m^*(t)$, and travel with the velocity $v(t)$. The values for $v(t)$, $E(t)$, and $m^*(t)$ corresponding to $\mathcal{E}(t)$ can be deduced from the stationary v–$\mathcal{E}$, E–$\mathcal{E}$, and m^*–$\mathcal{E}$ characteristics, and the momentum is given by $P(t) = m^*(t) \times v(t)$. We introduce the stationary field, velocity, energy, and effective mass given by $\mathcal{E}_{st} = \mathcal{E}(t)$, $v_{st} = v(t)$, $E_{st} = E(t)$, and $m^*_{st} = m^*(t)$. The subscript st designates the stationary value of the corresponding quantity. Let us now assume that during the small time interval Δt after t the applied field changes from $\mathcal{E}(t)$ to $\mathcal{E}(t + \Delta t)$. The evolution of the electron velocity caused by the change of the electric field can be calculated based on the RTA by subsequently processing the following steps.

1. Calculate the relaxation times $\tau_E(t + \Delta t)$ and $\tau_P(t + \Delta t)$ by rearranging Eqs. (2-77) and (2-78) and assuming stationary conditions, i.e., d/dt is zero:

$$\tau_E(t + \Delta t) = \frac{E_{st}(t) - E_0}{q\mathcal{E}_{st}(t)v_{st}(t)} \tag{2-79}$$

$$\tau_P(t + \Delta t) = \frac{m^*_{st}(t)v_{st}(t)}{q\mathcal{E}_{st}(t)} \tag{2-80}$$

2. Calculate the electron energy and momentum at $t + \Delta t$. To this end, the balance equations are discretized in a way such that for the calculation of the carrier energy and momentum at a time $t + \Delta t$ only the knowledge of energy and momentum at time t are necessary. This leads to

$$E(t + \Delta t) = E(t) + \Delta t \left(q\mathcal{E}(t + \Delta t)v(t) - \frac{E(t) - E_0}{\tau_E(t + \Delta t)} \right) \tag{2-81}$$

$$P(t + \Delta t) = P(t) + \Delta t \left(q\mathcal{E}(t + \Delta t) - \frac{P(t)}{\tau_P(t + \Delta t)} \right) \tag{2-82}$$

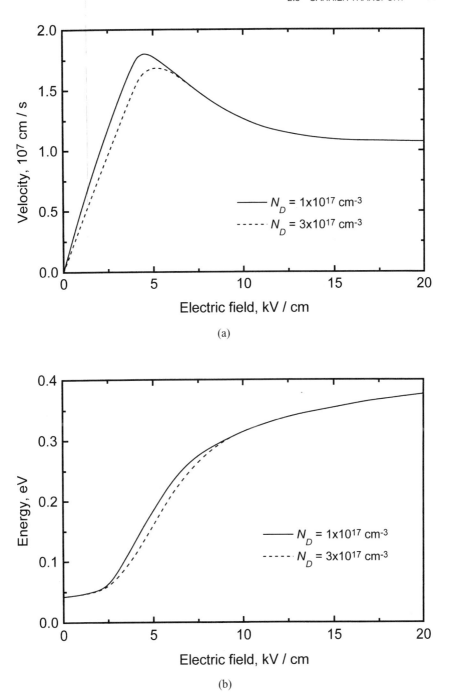

Figure 2.15. Stationary transport characteristics for electrons in GaAs. (a) velocity-field characteristics, (b) energy-field characteristics. Data taken from [31].

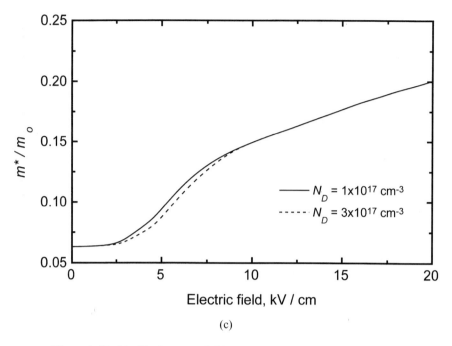

Figure 2.15. (c) effective mass-field characteristics. Data taken from [31].

3. Set $E_{st}(t + \Delta t) = E(t + \Delta t)$ and deduce the new stationary values $\mathscr{E}_{st}(t + \Delta t)$, $v_{st}(t + \Delta t)$, and $m^*_{st}(t + \Delta t)$ from the stationary v–$\mathscr{E}$, E–$\mathscr{E}$, and m^*–$\mathscr{E}$ characteristics in Fig. 2.15.
4. Calculate electron velocity $v(t + \Delta t)$ from

$$v(t + \Delta t) = \frac{P(t + \Delta t)}{m^*_{st}(t + \Delta t)} \tag{2-83}$$

5. Increase the time by another Δt and proceed with step 1.

Note that in Eqs. (2-79) and (2-80) the stationary quantities, in Eqs. (2-81) and (2-82) the dynamic (nonstationary) quantities, and in Eq. (2-83) the dynamic momentum and the stationary effective mass have to be used.

Figures 2.16 and 2.17 show the evolution of the electron velocity as a function of time after applying a field to n-type Si and GaAs samples, respectively, calculated using the algorithm described above. When a relatively large field is applied, the electron velocity in both materials reaches a much higher peak value than that predicted by the stationary v–$\mathscr{E}$ characteristics. This effect is called the velocity overshoot. The peak velocities and the duration of the transition period are much larger for GaAs than for Si. In other words, nonstationary carrier transport plays a more important role in GaAs devices compared to its Si counterparts.

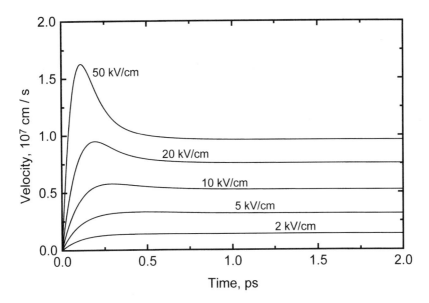

Figure 2.16. Evolution of the electron drift velocity in Si (doping concentration 10^{17} cm^{-3}, temperature 300 K) after applying an electric field, calculated by the RTA method. For $t < 0$ the electric field is zero. The stationary transport characteristics are taken from [33].

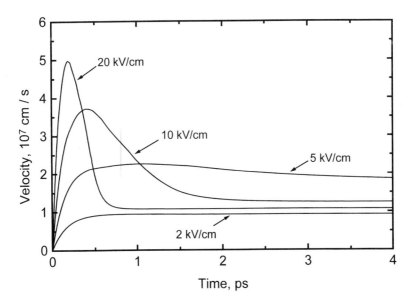

Figure 2.17. Evolution of the electron drift velocity in GaAs (doping concentration 10^{17} cm^{-3}, temperature 300 K) after applying an electric field, calculated by the RTA method. The stationary transport characteristics are taken from Fig. 2.15.

96 BASIC SEMICONDUCTOR PHYSICS

For other semiconductors (e.g., SiC and GaN), not all three characteristics necessary for RTA calculations are known, but only the stationary v–$\mathscr{E}$ and E–$\mathscr{E}$ characteristics are available in the literature. An approximation described in [38] can be used to estimate the effects of nonstationary carrier transport in these materials as well.

2.4 PN JUNCTIONS

A pn junction is formed when an n-type and a p-type semiconductor are brought in contact with each other. The pn junction is an important semiconductor structure. It is the heart of semiconductor diodes, and finds applications in many microwave transistors as well. Examples are the emitter–base and base–collector junctions in bipolar transistors and the source–bulk and bulk–drain junctions in MOSFETs. We will confine the discussions of pn junctions to the issues directly relevant to the understanding of microwave transistors. In particular, we will focus on the properties of the space–charge region of a pn junction, on the minority carrier concentrations at the edges of the space–charge region when a voltage is applied to the junction, and on the current–voltage characteristics of a pn junction.

We start with a qualitative discussion of pn junctions. On the left of Fig.

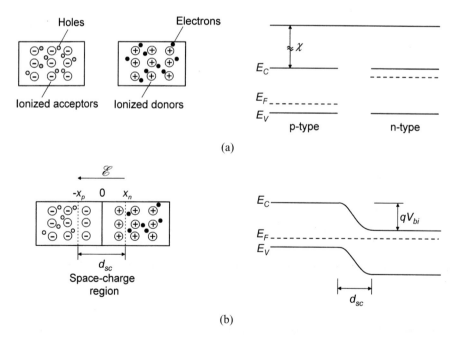

Figure 2.18. Schematic cross section and band diagram of a pn junction for (a) separated n-type and p-type materials and (b) n-type and p-type materials in contact.

2.18(a), two separate semiconductors, one of the p-type and the other of the n-type, are shown. The doping concentrations are N_A and N_D. In the p-type semiconductor, holes are majority carriers and electrons are minority carriers, and if complete ionization is assumed, the hole density is approximately equal to N_A. Similarly, in the n-type semiconductor, the electron density is approximately equal to N_D. If we bring the two materials in contact with each other as shown in Fig. 2.18(b), a large gradient of the carrier concentration occurs at the boundary, i.e., at the metallurgical junction. This gradient is the driving force for carrier diffusion (see Section 2.3.2). Holes diffuse from the p-type to the n-type region where they become the minority carriers, and electrons diffuse in the opposite direction. The minority holes survive a certain time in the n-type region before they are annihilated by recombination. The same holds for the electrons diffused from the n-type to the p-type region.

Each hole diffusing to the n-type region leaves behind an uncompensated negatively charged acceptor ion. Similarly, uncompensated positively charged donor ions are left when electrons diffuse to the p-type region. Thus, near the pn junction, a space–charge region is created, and an electric field arises due to the ionized acceptors and donors. The thickness of the space–charge region is d_{sc}. In the example of Fig. 2.18, the field is in the negative x-direction and is the driving force for a drift current. This drift tendency is in the opposite direction of the diffusion tendency. With no bias applied to the pn junction, the diffusion and drift tendencies compensate each other, and no net carrier flow occurs.

The band diagrams of the pn junction are given on the right of Fig. 2.18. When the semiconductors are separated, the vacuum level E_{vac} is chosen as the reference energy. The electron affinity χ is the energy difference between the vacuum level and the conduction band edge. When the materials are brought into contact with each other, under the thermal equilibrium condition (i.e., no voltage applied, no net current flows through the structure), the Fermi levels in the two materials must be aligned at the same horizontal position throughout the entire structure. To obtain this condition, the band diagram of the n-type region has to be pulled down and that of the p-type region has to be lifted upward while pinning the conduction and valence bands at the junction. The resulting band diagram with band bendings near the junction is shown on the right of Fig. 2.18(b).

Away from the pn junction, i.e., in the bulk, the bands are flat, and the separation between the band edges and the Fermi level is the same as that shown in Fig. 2.18(a). Near the pn junction, however, the bands are bent. The band bending in the n-type region near the junction leads to an increased separation between conduction band edge and Fermi level ($E_C - E_F$) compared to the bulk region. According to Eq. (2-18), a larger value of ($E_C - E_F$) translates into a lower electron density. In other words, the actual electron density in the n-type region near the junction is now lowered compared to the electron density in the bulk region outside the space–charge region. The positively charged donor ions are fixed and their density is not influenced. Thus positive net charges arise on the n-type side near the pn junction, and similarly negative net charges appear on the p-type side. The uncompensated acceptors and donors thus constitute the space charge in the space–charge region. Outside the space–charge region are the neutral bulk regions. An electric field is present in the space–charge region but is absent in the neutral bulk regions. The energy barrier between the p- and n-type regions is the same for the conduction and valence bands and corresponds to the so-called built-in voltage V_{bi} of the pn junction.

The built-in voltage can be derived as follows. Consider the abrupt pn junction given in Fig. 2.18. The origin of the coordinate system is located at the metallurgical pn junction and the coordinates of the space–charge region edges are $-x_p$ and x_n on the p- and n-type layers, respectively. When no external voltage is applied, neither a net electron current nor a net hole current flows across the junction. Thus, for the electron current density

$$J_n = J_{\text{Dr},n} + J_{\text{Di},n} = q\mathscr{E}n\mu_n + qD_n\frac{dn}{dx} = 0 \qquad (2\text{-}84)$$

hold. Rearranging this expression, using Eq. (2-74) to express the electric field, and applying the Einstein relation (2-70) lead to

$$d\varphi = \frac{k_BT}{q}\frac{dn}{n} \qquad (2\text{-}85)$$

where φ is the electrostatic potential. Integrating the above equation across the space–charge region (i.e., from $x = -x_p$ to $x = x_n$), and applying the mass action law yields

$$V_{bi} = \frac{k_BT}{q}\ln\frac{N_DN_A}{n_i^2} \qquad (2\text{-}86)$$

To find the thickness of the space–charge region, we start with the Poisson equation that relates potential and electric field to the space–charge density:

$$\frac{d^2\varphi}{dx^2} = -\frac{d\mathscr{E}}{dx} = -\frac{q}{\varepsilon}(p - n + N_D - N_A) \qquad (2\text{-}87)$$

Table 2.4. Relative dielectric constant ε_r of the basic semiconductors

Semiconductor	ε_r	Semiconductor	ε_r
Ge	16.2	GaP	11.1
Si	11.9	4H SiC	9.7
GaAs	12.85	6H SiC	9.7
AlAs	10.1	GaN	9.5
InAs	15.15	AlN	9.1
InP	12.56		

where $\varepsilon = \varepsilon_0 \varepsilon_r$ is the dielectric constant (ε_0 is the dielectric constant of vacuum and ε_r is the relative dielectric constant of the semiconductor). Table 2.4 gives the relative dielectric constant ε_r of the basic semiconductors. For alloys, ε_r can be calculated using the linear approximation scheme discussed in Section 2.1.

In the space–charge region the so-called depletion approximation is applicable, i.e., both the majority and minority carrier densities in the space–charge region are negligible compared to the doping concentration. The bulk regions of the pn structure, which are the regions outside the space–charge region, are electrically neutral, and thus have zero electric field under the equilibrium condition. Note that these regions are assumed to be charge neutral even if a voltage is applied to the junction (nonequilibrium condition), and therefore the regions are also called the quasineutral regions.

It is convenient to derive separately the expressions for the p-type and n-type regions. In the space–charge region on the p-type side, the Poisson equation simplifies to

$$\frac{d^2\varphi}{dx^2} = \frac{q}{\varepsilon} N_A \qquad (2\text{-}88)$$

Carrying out the indefinite integration and applying the boundary condition such that at the edge of the space–charge region at $x = -x_p$ the electric field becomes zero, we obtain the field inside the space–charge region on the p-type side of the junction as

$$\mathscr{E} = -\frac{q}{\varepsilon} N_A (x + x_p) \qquad (2\text{-}89)$$

In a similar manner, the field inside the space–charge region on the n-type side can be found as

$$\mathscr{E} = \frac{q}{\varepsilon} N_D (x - x_n) \qquad (2\text{-}90)$$

To get the potential distribution inside the space–charge region, Eqs. (2-89) and (2-90) have to be integrated. Again we use indefinite integration and the following boundary conditions: at the edge of the space–charge region on the p-type side ($x = -x_p$) the potential is zero, and at the space–charge region edge on the n-type side ($x = x_n$) the potential is equal to the built-in voltage. Thus we get

$$\varphi(x) = \frac{q}{2\varepsilon} N_A (x + x_p)^2 \qquad (2\text{-}91)$$

for the potential inside the space–charge region on the p-type side, and

$$\varphi(x) = V_{bi} - \frac{q}{2\varepsilon} N_D (x - x_n)^2 \qquad (2\text{-}92)$$

for the potential inside the space–charge region on the n-type side. Because the potential in a pn junction is continuous, both Eqs. (2-91) and (2-92) must result in the same value for $\varphi(x)$ at the metallurgical junction ($x = 0$). Using this condition and rearranging Eqs. (2-91) and (2-92), expressions for the thickness of the space–charge region are obtained:

$$x_p = \sqrt{\frac{2\varepsilon V_{bi}}{q} \frac{N_D}{N_A(N_D + N_A)}} \qquad (2\text{-}93)$$

and

$$x_n = \sqrt{\frac{2\varepsilon V_{bi}}{q} \frac{N_A}{N_D(N_D + N_A)}} \qquad (2\text{-}94)$$

The total thickness of the space–charge region, d_{sc}, is the sum of x_p and x_n:

$$d_{sc} = \sqrt{\frac{2\varepsilon V_{bi}}{q} \frac{N_D + N_A}{N_D N_A}} \qquad (2\text{-}95)$$

Equation (2-95) describes the thickness of the space–charge region d_{sc} under the thermal equilibrium condition, i.e., no voltage is applied. It is evident that d_{sc} depends on the potential barrier between the p-type and n-type regions, which is the built-in voltage V_{bi} at the thermal equilibrium. When a voltage V_{pn} is applied across the junction, the potential barrier at the junction is either increased when reverse biased ($V_{pn} < 0$) or decreased when forward biased ($V_{pn} > 0$). Equations (2-93) to (2-95) can be used to describe x_n, x_p, and d_{sc} for the case of an applied voltage, provided V_{bi} is replaced with ($V_{bi} - V_{pn}$). Thus, for the case of a voltage applied, Eqs. (2-93)–(2-95) become

$$x_p = \sqrt{\frac{2\varepsilon(V_{bi} - V_{pn})}{q} \frac{N_D}{N_A(N_D + N_A)}} \qquad (2\text{-}96)$$

$$x_n = \sqrt{\frac{2\varepsilon(V_{bi} - V_{pn})}{q} \frac{N_A}{N_D(N_D + N_A)}} \qquad (2\text{-}97)$$

and

$$d_{sc} = \sqrt{\frac{2\varepsilon(V_{bi} - V_{pn})}{q} \frac{N_D + N_A}{N_D N_A}} \qquad (2\text{-}98)$$

We next discuss the minority carrier concentrations at the edges of the space–charge region ($x = -x_p$ and $x = x_n$). If no external voltage is applied, the minority carrier concentrations are equal to the equilibrium concentrations given by Eqs. (2-22) and (2-23): $n(-x_p) = n_{0p}$ and $p(x_n) = p_{0n}$. Here the subscript 0 stands for

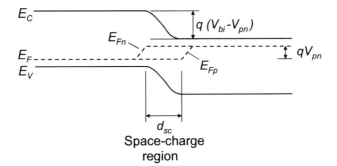

Figure 2.19. Band diagram of a pn junction under the forward bias condition ($V_{pn} > 0$).

the equilibrium case, and the subscripts p and n denote the p-type and n-type regions, respectively, of the pn junction. In the following we explain the procedure for finding $n(-x_p)$ under the forward-bias condition, i.e., $V_{pn} > 0$.

Figure 2.19 shows the band diagram of the pn junction for $V_{pn} > 0$. If the p-type region is grounded, a forward bias means the potential at the contact of the n-type region is negative. Under this condition, the bands at the contact of n-type region are lifted upward and the potential barrier is decreased (see Fig. 2.6). Now the drift and diffusion currents no longer compensate each other. The extrinsic field caused by the applied voltage is in the opposite direction compared to the intrinsic field originated from the uncompensated acceptor and donor ions in the space–charge region. Thus, the resulting field is smaller compared to that of the equilibrium case, the drift current is decreased, and the diffusion current becomes the dominant tendency. This leads to an injection of electrons from the n-type region across the pn junction into the p-type region, and holes are injected into the n-type region. Thus the minority carrier concentrations in the space–charge region and at the edges of the space–charge region are increased. Furthermore, we see that the Fermi level is no longer a straight horizontal line across the pn junction. The concept of a single Fermi level no longer applies and Eqs. (2-22) and (2-23) are not valid under these nonequilibrium conditions. To keep the convenient form of the equations for the calculation of the carrier densities, at this stage the so-called quasi-Fermi levels are introduced. At any position in the junction where the carrier densities deviate from the equilibrium densities given by Eqs. (2-22) and (2-23), the Fermi level is split into two separate quasi-Fermi levels for electrons E_{Fn} and for holes E_{Fp}, as shown in Fig. 2.19. In the regions of the pn junction where $E_{Fn} = E_{Fp} = E_F$ holds, Eqs. (2-22) and (2-23) can be used, whereas in the regions with separate quasi-Fermi levels the following equations have to be applied to calculate the nonequilibrium carrier densities:

$$n = n_i \exp \frac{E_{Fn} - E_i}{k_B T} \qquad (2\text{-}99)$$

$$p = n_i \exp\frac{E_i - E_{Fp}}{k_B T} \tag{2-100}$$

Multiplying Eq. (2-99) with Eq. (2-100) and considering the fact that the difference between E_{Fn} and E_{Fp} is equal to qV_{pn} in the space–charge region (see Fig. 2.19), we have

$$np = n_i^2 \exp\frac{E_{Fn} - E_{Fp}}{k_B T} = n_i^2 \exp\frac{qV_{pn}}{k_B T} = n_i^2 \exp\frac{V_{pn}}{V_T} \tag{2-101}$$

where V_T is the thermal voltage equal to $k_B T/q$. For a forward bias applied to the pn junction ($V_{pn} > 0$), Eqs. (2-99)–(2-101) result in increased minority carrier densities in the entire space–charge region and in the bulk regions adjacent to the space–charge region, as expected from the preceding discussion.

As in the case of zero applied bias, the entire regions of the pn structure outside the space–charge region are electrically neutral. Therefore, in the p-type bulk region, any increase in the minority carrier concentration Δn is compensated by an increase of the majority carrier concentration Δp of the same amount. This holds as well for the minority carrier density at the edge of the space–charge region. Thus we obtain for $x = -x_p$

$$p(-x_p) - p_{0p} = n(-x_p) - n_{0p} \tag{2-102}$$

In most cases, the equilibrium electron concentration n_{0p} is much smaller compared to the other charge components in Eq. (2-102) and can be neglected. Combining Eqs. (2-102) and (2-101), $n(-x_p)$ is found to be

$$n(-x_p) = \frac{p_{0p}}{2}\left(\sqrt{1 + \frac{4n_{0p}}{p_{0p}} \exp\frac{V_{pn}}{V_T}} - 1\right) \tag{2-103}$$

Similarly, the following expression for the minority carrier density at $x = x_n$, i.e., for the hole density at the edge of the space–charge region in the n-type region, can be derived:

$$p(x_n) = \frac{n_{0n}}{2}\left(\sqrt{1 + \frac{4p_{0n}}{n_{0n}} \exp\frac{V_{pn}}{V_T}} - 1\right) \tag{2-104}$$

For the case of low-level injection, i.e., for small V_{pn} and low injected minority carrier densities, the second addend in the square root in Eq. (2-103) is much smaller than unity, and the square root can be expressed as a binomial series taking only the first two terms: $(1 + x)^n \approx 1 + nx$ with $n = \frac{1}{2}$. This gives

$$n(-x_p) = n_{0p} \exp\frac{V_{pn}}{V_T} = \frac{n_i^2}{N_A} \exp\frac{V_{pn}}{V_T} \tag{2-105}$$

for the electron density at $x = -x_p$, and

$$p(x_n) = p_{0n} \exp\frac{V_{pn}}{V_T} = \frac{n_i^2}{N_D}\exp\frac{V_{pn}}{V_T} \qquad (2\text{-}106)$$

for the hole density at $x = x_n$. When high-level injection prevails (large V_{pn}), the second addend in the square root in Eq. (2-103) becomes much larger than unity, and the minority electron and hole concentrations at the edges of the space–charge region are given by

$$n(-x_p) = n_i \exp\frac{V_{pn}}{2V_T} \qquad (2\text{-}107)$$

and

$$p(x_n) = n_i \exp\frac{V_{pn}}{2V_T} \qquad (2\text{-}108)$$

Now an expression for the current density across the pn junction can be derived. For simplicity, we consider the case of low-level injection and use Eqs. (2-105) and (2-106) to express $n(-x_p)$ and $p(x_n)$. Furthermore, recombination in the space–charge region is neglected. Thus, the number of electrons (holes) entering the space–charge region at x_n ($-x_p$) is equal to the number of electrons (holes) leaving the space–charge region at $-x_p$ (x_n). As discussed in Section 2.3, the current passing through a semiconductor is the sum of four components: the electron drift and diffusion currents, and the hole drift and diffusion currents. Consider the edge of the space–charge region on the p-side ($x = -x_p$). Here, only the two diffusion components remain because the electric field vanishes at the edges of the space–charge region. Thus, the current density becomes

$$J = J_{\text{Di},n}(-x_p) + J_{\text{Di},p}(-x_p) \qquad (2\text{-}109)$$

Because we have assumed no electron–hole recombination taking place inside the space–charge region, the hole diffusion current densities at x_n and at $-x_p$ must be the same. Consequently, an equivalent to Eq. (2-109) is

$$J = J_{\text{Di},n}(-x_p) + J_{\text{Di},p}(x_n) \qquad (2\text{-}110)$$

Using Eqs. (2-68) and (2-69) the current density is expressed as

$$J = qD_n\frac{dn}{dx}\bigg|_{-x_p} - qD_p\frac{dp}{dx}\bigg|_{x_n} \qquad (2\text{-}111)$$

At this point, the gradients of p at x_n and of n at $-x_p$ have to be determined. As shown in Fig. 2.20, due to the electron–hole recombination in the quasineutral re-

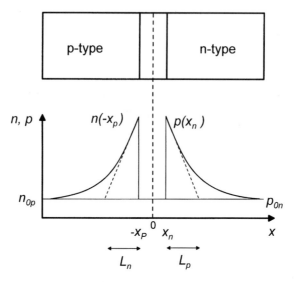

Figure 2.20. Minority carrier distributions in a forward biased pn junction (long-base junction).

gions outside the space–charge region, the nonequilibrium minority carrier densities in these regions decline exponentionally from their values at the edges of the space–charge region $n(-x_p)$ and $p(x_n)$ to the equilibrium values n_0 and p_0. The characteristic lengths L_n and L_p in Fig. 2.20 are the so-called minority carrier diffusion lengths for electrons and holes, respectively. They can be interpreted as the average distance an excess minority carriers can move by diffusion in a sea of majority carries before they disappear by recombination and are defined as

$$L_n = \sqrt{D_n \tau_n} \tag{2-112}$$

and

$$L_p = \sqrt{D_p \tau_p} \tag{2-113}$$

where D_n and D_p are the minority electron and hole diffusion coefficients, and τ_n and τ_p are the minority electron and hole lifetimes.

To determine the gradients in Eq. (2-111), we assume as a first-order approximation that the minority carrier densities approach their equilibrium values within a distance L_n (L_p) from the space–charge region edge. Thus the current density can be expressed as

$$J = qD_n \frac{n(-x_p) - n_{0p}}{L_n} - qD_p \frac{p_{0n} - p(x_n)}{L_p} \tag{2-114}$$

Assuming complete ionization, the equilibrium carrier densities can be expressed by Eqs. (2-27) and (2-29), and the nonequilibrium minority carrier densities at $-x_p$

and x_n can be modeled by Eqs. (2-105) and (2-106). This yields the final expression for the current density across a pn junction:

$$J = q\left(D_n \frac{n_i^2}{N_A L_n} + D_p \frac{n_i^2}{N_D L_p}\right)\left(\exp\frac{V_{pn}}{V_T} - 1\right) \quad (2\text{-}115)$$

This equation is valid for the so-called long-base junction, that is, a junction in which the contacts are located far away from the edges of the space–charge region. This implies significant electron–hole recombination takes place in the quasineutral regions, and all excess minority carriers vanish before reaching the contacts. In real devices, however, the quasineutral regions frequently are much shorter. Junctions with quasineutral regions shorter than the minority carrier diffusion lengths are so-called short-base junctions. In such a junction, the distribution of the nonequilibrium minority carrier densities in the quasineutral region is almost linear, as shown in Fig. 2.21. At the contacts, the equilibrium minority carrier densities are preserved, and the distance in which the densities decline from the values at the space–charge region edges to the equilibrium values are $(x_{Bn} - x_n)$ for holes on the n-type side, and $(x_{Bp} - x_p)$ for electrons on the p-type side. Thus, the current density in a short-base junction is

$$J = q\left(D_n \frac{n_i^2}{N_A(x_{Bp} - x_p)} + D_p \frac{n_i^2}{N_D(x_{Bn} - x_n)}\right) \times \left(\exp\frac{V_{pn}}{V_T} - 1\right) \quad (2\text{-}116)$$

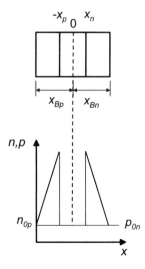

Figure 2.21. Minority carrier distributions in a forward biased pn junction (short-base junction).

Equations (2-115) and (2-116) can be written in a more compact form with all voltage-independent parameters merged into the so-called saturation current density J_S:

$$J = J_S \left(\exp \frac{V_{pn}}{V_T} - 1 \right) \qquad (2\text{-}117)$$

The approach described above to calculate the current density is called the drift-diffusion model.

2.5 SCHOTTKY JUNCTIONS

A Schottky junction is a rectifying junction formed by a metal and a semiconductor. For microwave transistors, the most important Schottky junction is the junction between a metal and an n-type semiconductor, as shown in Fig. 2.22(a), which serves as a gate electrode in MESFETs and HEMTs. According to the theory of ideal Schottky junctions, the difference between the work functions of the metal and the semiconductor governs the electrical properties of the junction. In real Schottky junctions, however, there are charged interface states located at the metal–semiconductor interface, and the effect of the interface states on the behavior of the Schottky junction is typically more significant compared to that of the work function difference itself.

As shown in Fig. 2.22(a), the negatively charged interface states repel the electrons in the n-type semiconductor from the interface. Thus, a space–charge region is present even if no bias is applied to the junction. Although many theories for the Schottky junction have been proposed, commonly the built-in voltage V_{bi} is not calculated using an expression derived from a physical model, but rather is taken from experimental data. A quantity closely related to V_{bi} is the Schottky barrier height

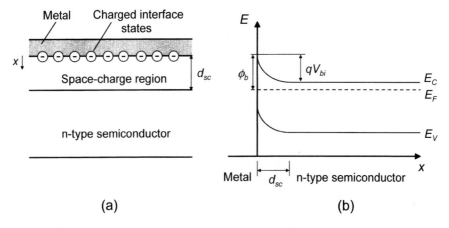

Figure 2.22. (a) Schematic cross section and (b) band diagram of a Schottky junction.

Table 2.5. Schottky barrier height of different Schottky junctions

Semiconductor	Metal	ϕ_b
		eV
(n-type)		
GaAs	Al, Ag, Au, Pt, W	0.8 ... 0.9
$Al_xGa_{1-x}As$ ($0 \leq x \leq 0.35$)	Au	$0.9 + 0.65 \times x$
	Al	$0.8 + 0.83 \times x$
	Mo	$0.78 + 0.69 \times x$
InP	Al	0.3 ... 0.4
	Au	0.42 ... 0.44
	Cu	0.42 ... 0.46
	Ag	0.51 ... 0.54
$In_{0.53}Ga_{0.47}As$		0.2
$In_{0.52}Al_{0.48}As$	Ti	0.59
	Ti/Pt/Au	0.82
SiC		1.1
GaN	Pt	1 ... 1.15
	Ni	0.55 ... 1.15
	Au	0.84 ... 0.94
$Al_xGa_{1-x}N$*	Ni	$0.84 + 1.3 \times x$

*The data for AlGaN is taken from [39].

ϕ_b. As shown in Fig. 2.22(b), it is defined as the energy barrier between the Fermi level and the tip of the conduction band at the metal–semiconductor interface. The relation between ϕ_b and the built-in voltage is given by

$$qV_{bi} = \phi_b - (E_C - E_F) \tag{2-118}$$

Table 2.5 lists values of ϕ_b for different Schottky junctions on n-type semiconductors.

The thickness of the space–charge region can be calculated as follows. The starting point is again the Poisson equation. As in the derivation in the preceding section, we employ the depletion approximation, i.e., the free carrier concentration in the space–charge region is assumed to be zero. Then, the Poisson equation in the n-type semiconductor underneath a Schottky contact is given by

$$\frac{d^2\varphi(x)}{dx^2} = -\frac{qN_D}{\varepsilon} \tag{2-119}$$

First we consider the case of zero applied voltage. Carrying out an indefinite integration twice and applying the boundary conditions that at the space–charge region edge (i.e., at $x = d_{sc}$) the electric field vanishes and the potential is equal to V_{bi} (the metal is grounded), we get

$$\varphi(x) = V_{bi} - \frac{qN_D}{2\varepsilon}(x - d_{sc})^2 \tag{2-120}$$

Setting $x = 0$ and rearranging the above equation, one obtains for the thickness of the space–charge region

$$d_{sc} = \sqrt{\frac{2\varepsilon V_{bi}}{qN_D}} \tag{2-121}$$

The thickness of the space–charge region in a Schottky junction with an applied voltage V_{app} between the gate and the semiconductor is given by

$$d_{sc} = \sqrt{\frac{2\varepsilon(V_{bi} - V_{app})}{qN_D}} \tag{2-122}$$

2.6 IMPACT IONIZATION

As discussed in Section 2.3, carriers in a semiconductor are accelerated and pick up energy when an electric field is applied. During the scattering process, they transfer part of the energy to the lattice. If the field is sufficiently high, a carrier can gain so much energy that during a collision with the lattice it can break a bond and create a free electron. This means that due to the energy transfer, a valence electron is moved to the conduction band, thereby creating a free electron and a free hole. As a result, an electron–hole pair is generated and we have three carriers after the scattering event compared to only one carrier before the process. This process is called impact ionization.

Conservation of energy requires that the sum of the energies of the three carriers after scattering must be equal to the energy of the original carrier before scattering. The same holds for the momentum. After impact ionization, all three carriers are accelerated by the field. If the field remains high enough, they can create further electron–hole pairs, which are again accelerated and create more electron–hole pairs. Such an enormous increase in the total number of carriers is called avalanche multiplication. Avalanche can lead to device breakdown, and, if the currents are not limited, may destroy the device. The voltage at which avalanche becomes noticeable is the breakdown voltage BV. Obviously the minimum kinetic energy a carrier must gain to cause impact ionization is slightly larger than the bandgap energy.

The increase in carrier concentration by impact ionization is described by the electron–hole pair generation rate G_{II} defined by

$$G_{II} = \alpha_n n v_n + \alpha_p p v_p \tag{2-123}$$

where α_n and α_p are the electron and hole ionization rates, i.e., the number of electron–hole pairs generated by an electron (hole) per unit distance traveled. Classically, the so-called local electric field models are used to calculate α_n and α_p. A frequently used local electric field model has been proposed by Selberherr [40]:

$$\alpha_n = a_n \exp\left[-\left(\frac{b_n}{\mathscr{E}}\right)^{\beta_n}\right] \tag{2-124}$$

$$\alpha_p = a_p \exp\left[-\left(\frac{b_p}{\mathscr{E}}\right)^{\beta_p}\right] \tag{2-125}$$

In Figs. 2.23–2.25, the dependence of α_n and α_p on the inverse of the field for various semiconductors is shown. From these curves, the values of the parameters a_n, b_n, β_n, and a_p, b_p, β_p can be obtained by fitting schemes. Two conclusions can be drawn from the data in Figs. 2.23–2.25. First, the ionization coefficients increase considerably with increasing field. This is because at higher fields, the electrons and holes gain more energy, and the probability of impact ionization increases. Second, semiconductors with larger bandgaps show lower ionization coefficients for a given field. This is expected, because a larger bandgap means that a greater amount of energy is necessary to move an electron from the valence to the conduction band.

Unfortunately, local electric field models normally overestimate the

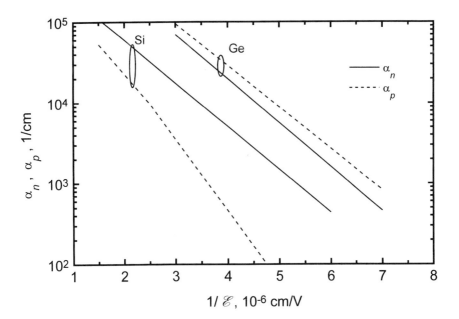

Figure 2.23. Electron and hole impact ionization coefficients for Si and Ge plotted as a function of inverse electric field. Data for Ge taken from [41], data for Si taken from [42] and [43].

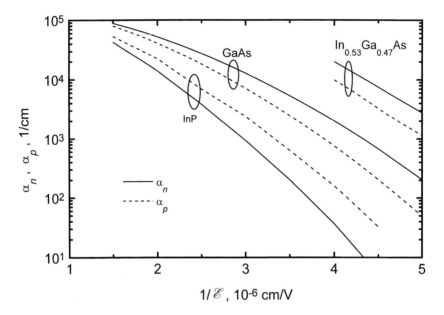

Figure 2.24. Electron and hole impact ionization coefficients for InP, GaAs, and $In_{0.53}Ga_{0.47}As$ plotted as a function of inverse electric field. Data taken from [5].

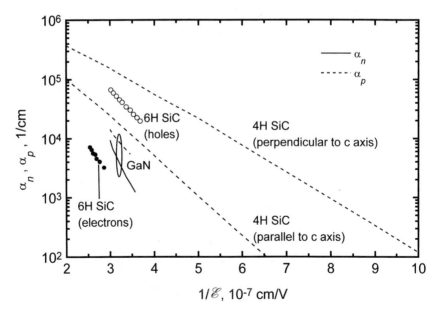

Figure 2.25. Electron and hole impact ionization coefficients for GaN and SiC plotted as a function of inverse electric field. Data for GaN taken from [44], data for 4H SiC taken from [45], and data for 6H SiC taken from [46].

electron–hole pair generation rate, leading to an underestimated breakdown voltage. Strictly speaking, not the electric field but the kinetic energy of traveling electrons and holes is the main reason for impact ionization. In other words, an appropriate impact ionization model should not be a function of the local field but rather the carrier energy.

It is extremely difficult to obtain realistic field and carrier energy distributions by analytical means. Therefore, to analytically describe avalanche multiplication and breakdown behavior, neither the local electric field nor the carrier-energy-dependent impact ionization models are suitable. A useful first order estimation of microwave transistor breakdown voltage, however, is feasible. The largest electric fields and the highest probabilities for impact ionization in microwave transistors occur in the space–charge regions of pn junctions or Schottky junctions. In bipolar transistors, the critical region is the collector–base space–charge region, in MOSFETs it is the drain–bulk pn junction, and in MESFETs and HEMTs it is the Schottky junction of the gate near the drain region. Concerning breakdown, all these junctions behave like a one-sided abrupt pn junction.

Figure 2.26 shows the critical field $\mathscr{E}_C$ for the onset of avalanche breakdown versus the doping density in one-sided abrupt Ge, Si, GaAs, GaP, and SiC pn junctions [47–49]. From the critical field $\mathscr{E}_C$, the breakdown voltage BV can be estimated according to [22]

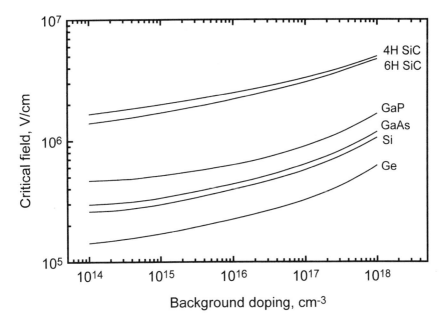

Figure 2.26. Critical field for the onset of avalanche multiplication in one-sided abrupt pn junctions as a function of doping concentration at the lower doped side.

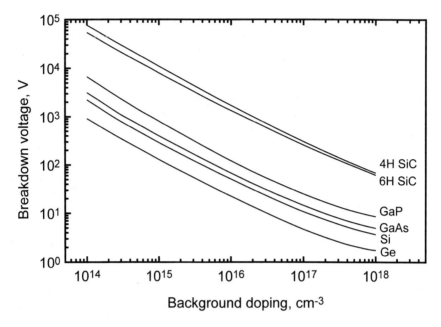

Figure 2.27. Breakdown voltage of abrupt one-sided pn junctions as a function of doping concentration calculated using Eq. (2-126).

$$BV = \frac{\varepsilon \mathscr{E}_C^2}{2q} \frac{1}{N} \qquad (2\text{-}126)$$

where N is the ionized background concentration in the lightly doped side of the pn junction or in the semiconductor side of the Schottky junction. The resulting breakdown voltages as a function of doping concentration for the six semiconductors considered in Fig. 2.26 are given in Fig. 2.27.

To get information on the critical field for semiconductors other than those given in Fig. 2.26, a fitting scheme can be devised. As discussed before, the probability of impact ionization is related to the bandgap. From the values of the critical fields in Fig. 2.26 and the bandgaps in Table 2.1, a polynomial fit of the form $\mathscr{E}_C = f(E_G)$ can be made. Using fitting, one can calculate the approximate critical fields for a semiconductor not given in Fig. 2.26. The approximated critical field and Eq. (2-126) then lead to the breakdown voltage for any semiconductor device of interest. In general, a large bandgap and a large breakdown voltage are desirable for microwave transistors.

2.7 SELF-HEATING

When a voltage is applied to a semiconductor device and thereby a current passes through it, electric power is dissipated into heat, or in other words, electric energy is

2.7 SELF-HEATING

transformed into thermal energy. In the case of dc operation, the power dissipated, P_{th}, is given by

$$P_{th} = VI \qquad (2\text{-}127)$$

where V is the dc voltage drop across the device and I is the dc current flowing through it. The heat spreads throughout the semiconductor and finally leaves the semiconductor chip, i.e., it is transferred to the surroundings. Furthermore, the heat leads to an increase in the temperature inside the device. This effect is called self-heating. Because all the material parameters of semiconductors are temperature dependent, the knowledge of the temperature inside the device is critical for the accurate modeling of device behavior. This is especially true for power transistors, in which a large amount of heat is generated. For the analysis of self-heating and for the thermal design of power transistors, the temperature rise in the transistor, the temperature distribution, and the thermal resistance of the transistor are of primary concern.

In general, there are three mechanisms by which heat can leave the device and the chip: convection, radiation, and conduction. Of these, heat conduction is by far the most important one. Thus, we will confine our discussions to heat conduction, frequently called heat flow. Furthermore, we consider only the stationary condition, in which the heat flow in the semiconductor device does not vary with time. This assumption is reasonable because the operating frequencies of microwave transistors are much higher than the frequencies the heat flow could follow. The following equation, which is called the heat flow equation, describes the heat flow

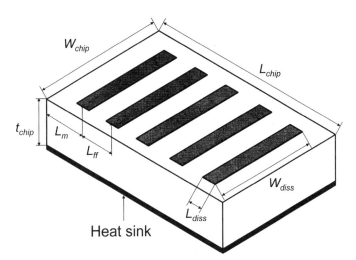

Figure 2.28. Schematic of a microwave power transistor chip. The shaded areas are the transistor fingers.

and the temperature distribution in a semiconductor device under stationary conditions:

$$\frac{\partial^2 T}{\partial x^2} + \frac{\partial^2 T}{\partial y^2} + \frac{\partial^2 T}{\partial z^2} = 0 \qquad (2\text{-}128)$$

where T is the temperature. Typical microwave power transistors are designed using the so-called multifinger structure. As will be shown later, these fingers can be gate fingers in field-effect transistors or emitter fingers in bipolar transistors. Figure 2.28 shows the basic structure and dimensions of a power transistor chip. The shaded areas are the heat sources that correspond to the transistor fingers. The dimensions of the heat sources are L_{diss} and W_{diss}, L_{ff} is the distance between the centers of two neighboring heat sources, and L_{m} is the separation between the center of the outermost heat source and the edge of the semiconductor chip. The x-axis is directed along L_{chip}, the y-axis along W_{chip}, and the z-axis along t_{chip}.

Applying appropriate boundary conditions, an analytical solution of Eq. (2-128) can be found. The conventionally used boundary conditions are:

- Power dissipation is assumed to occur only at the chip surface and the depth of the transistor structure is much smaller than the chip thickness.
- The backside of the chip is a heat sink with a constant temperature equal to the ambient temperature.
- All other surfaces (top and four edge surfaces) are assumed to be adiabatic, i.e., no heat flows across these surfaces.

With these boundary conditions, the solution of Eq. (2-128) is [50, 51]:

$$T(x, y, z) = \frac{4}{\kappa_0 W_{\text{chip}} L_{\text{chip}} W_{\text{diss}} L_{\text{diss}}}$$

$$\sum_{i=1}^{N} \left\{ \frac{L_{\text{diss}}}{2} \sum_{m=1}^{\infty} \frac{1}{\lambda_m^2} F_{zm} F_{yi} \cos(\lambda_m y) + \frac{W_{\text{diss}}}{2} \sum_{n=1}^{\infty} \frac{1}{\mu_n^2} F_{zn} F_{xi} \cos(\mu_n x) + \right.$$

$$\sum_{m=1}^{\infty} \sum_{n=1}^{\infty} \frac{1}{\lambda_m \mu_n \gamma_{mn}} F_z F_{xi} F_{yi} \cos(\lambda_m y) \cos(\mu_n x) + \qquad (2\text{-}129)$$

$$\left. \frac{W_{\text{diss}} L_{\text{diss}}(t_{\text{chip}} - z)}{4} \right\} \times P_{\text{diss},i} + T_0$$

where

$$\lambda_m = \frac{m\pi}{W_{\text{chip}}}, \quad \mu_n = \frac{n\pi}{L_{\text{chip}}}, \quad \gamma_{mn} = \sqrt{\lambda_m^2 + \mu_n^2}$$

2.7 SELF-HEATING

$$F_{xi} = \sin\left[\mu_n\left(x_i + \frac{L_{\text{diss}}}{2}\right)\right] - \sin\left[\mu_n\left(x_i - \frac{L_{\text{diss}}}{2}\right)\right]$$

$$F_{yi} = \sin\left[\lambda_m\left(y_i + \frac{W_{\text{diss}}}{2}\right)\right] - \sin\left[\lambda_m\left(y_i - \frac{W_{\text{diss}}}{2}\right)\right]$$

$$F_z = \tanh(\gamma_{mn}t_{\text{chip}})\cosh(\gamma_{mn}z) - \sinh(\gamma_{mn}z)$$

F_{zm} and F_{zn} have the form of F_z, but γ_{mn} has to be replaced by λ_m and μ_n, respectively, N is the number of separate heat sources, $P_{\text{diss},i}$ is the power dissipated in the ith heat source, x_i and y_i are the center locations of the ith heat source, T_0 is the temperature of the heat sink, κ_0 is the thermal conductivity of the chip material at T_0, and $T(x, y, z)$ is the temperature somewhere in the chip. This approach works well for heat sources having uniform temperature, but is questionable for large-area heat sources with nonuniform temperature distribution within the sources. Thus, if the temperature distribution across the gate or emitter fingers is of interest, the fingers have to be divided into subfingers, and each subfinger has to be treated as a separate, constant-temperature heat source in Eq. (2-129). In this case, the power dissipated in each subfinger is $P_{\text{diss},i}$.

In Eq. (2-129), the thermal conductivity is assumed to have a constant value κ_0. Actually, however, the thermal conductivity is temperature-dependent as well. In the temperature range of interest to microwave transistors, i.e., around and above 300 K, the thermal conductivity decreases when the temperature rises and can be approximated by

$$\kappa(T)\left[\frac{W}{\text{cm K}}\right] = aT^b \qquad (2\text{-}130)$$

where a and b are fitting parameters.

Applying Eq. (2-129), one neglects the temperature dependence of κ and calculates a fictitious temperature that is lower than the actual temperature. From the knowledge of the fictitious temperature from Eq. (2-129) and the temperature-dependent κ given in Eq. (2-130), however, the temperature in a semiconductor device can be calculated easily. We designate the fictitious temperature calculated using Eq. (2-129) as T_{fic} and the actual temperature as T_{act}. Joyce [52] has shown that between these temperatures the relation

$$T_{\text{fic}} = T_0 + \frac{1}{\kappa_0}\int_{T_0}^{T_{\text{act}}} \kappa(T)dT \qquad (2\text{-}131)$$

holds. Putting Eq. (2-130) into Eq. (2-131) for $\kappa(T)$, carrying out the integration, and rearranging the resulting equation yield

$$T_{act} = \left[\frac{\kappa_0(b+1)}{a}(T_{fic} - T_0) + T_0^{b+1} \right]^{1/b+1} \qquad (2\text{-}132)$$

Table 2.6 gives the values of the fitting parameters a and b in Eq. (2-130) for several semiconductors. Because sapphire is frequently used as substrate material for GaN transistors, the data for sapphire is also included in Table 2.6. The fitting parameters are calculated from the thermal conductivity data given in [6, 16, 53, 54].

Although Eq. (2-129) looks complicated, it is extremely useful and easy to transform into a computer program. Actually, it has been used frequently to calculate the temperature distribution in microwave power transistors and to optimize transistor structures for reduced self-heating [51, 55–57]. A computational consideration is the appropriate choice of the upper limit of the summation indices for the several sums in Eq. (2-129). In general, the upper limits should be large enough to ensure sufficient accuracy of the calculated temperature. Experience showed that 500 is a reasonable upper limit for m and n. The appropriate limits depend, however, on the actual chip design, e.g., on the number and dimensions of the individual heat sources. For safety, trail calculations are recommended.

In the following, a second method to estimate the temperature rise due to self-heating is presented [58]. It leads to a much simpler solution than the approach discussed above. The drawback of the second method is that it provides only the average temperature rise at the chip surface and gives no information on the temperature distribution across the chip surface and inside the chip. In other words, the temperature of the N fingers (heat sources) is assumed to be equal. Furthermore, heat flow in the y-direction (the direction along W_{diss}) is neglected. Nevertheless, such a method has proven to be useful as an approximation tool and

Table 2.6. Fitting parameters of the thermal conductivity for several semiconductors and sapphire[a]

Semiconductor	a	b
Ge	338.3	−1.113
Si	617.6	−1.079
GaAs	802.5	−1.295
InP	2205	−1.41
InAs	508	−1.319
GaP	103.6	−0.8045
GaN	16.94	−0.4594
AlN	22058	−1.57
6H SiC	4517	−1.29
Sapphire	276.7	−1.1356

[a] Note that the thermal conductivity of alloys cannot be interpolated using the linear interpolation scheme given in Eqs. (2-1) and (2-2).

gives the same trends as Eq. (2-129) for the thermal design, e.g., the dependence of self-heating on the chip thickness, the number of fingers, the distance between the fingers, etc.

The method is based on the calculation of the thermal resistance of a mutlifinger structure in analogy to the electrostatic capacitance of multiple, coupled transmission lines [56, 58]. The final result after elaborate derivations is the following equation for the thermal resistance R_{th} [58]:

$$R_{th} = \frac{1}{W_{diss}\kappa_0} \frac{N}{\frac{2\pi(N-1)}{\ln M} - \frac{\pi(N-2)}{\ln P}} \qquad (2\text{-}133)$$

where

$$M = \frac{2\left[\frac{\cosh\left(\pi\frac{L_{ff}+L_{diss}}{4t_{chip}}\right)}{\cosh\left(\pi\frac{L_{ff}-L_{diss}}{4t_{chip}}\right)}\right]^{1/2} + 1}{\left[\frac{\cosh\left(\pi\frac{L_{ff}+L_{diss}}{4t_{chip}}\right)}{\cosh\left(\pi\frac{L_{ff}-L_{diss}}{4t_{chip}}\right)}\right]^{1/2} - 1}, \qquad P = 2\left(\frac{1+\operatorname{sech}\frac{\pi L_{diss}}{4t_{chip}}}{1-\operatorname{sech}\frac{\pi L_{diss}}{4t_{chip}}}\right)^{1/2}$$

Using the thermal resistance from Eq. (2-133), the temperature at the chip surface is calculated by

$$T = R_{th}P_{diss} + T_0 \qquad (2\text{-}134)$$

The temperature given by Eq. (2-134) is again the fictitious temperature at the surface if the thermal conductivity has a temperature-independent value κ_0. The actual temperature taking into account the temperature dependence of κ can be obtained using equation (2-132).

REFERENCES

1. S. Adachi, GaAs, AlAs, and $Al_xGa_{1-x}As$: Material Parameters for Use in Research and Device Applications, *J. Appl. Phys., 58,* pp. R1-R29, 1985.
2. S. Adachi, Band Gaps and Refractive Indices of AlGaAsSb, GaInAsSb, and InPAsSb: Key Properties for a Variety of the 2–4-μm Optoelectronic Device Applications, *J. Appl. Phys., 61,* pp. 4869–4876, 1987.
3. S. Adachi, *Physical Properties of III–V Semiconductor Compounds,* Wiley, New York, 1992.

4. S. Adachi (ed.), *Properties of Aluminium Gallium Arsenide,* emis Datareview Series vol. 7, INSPEC, London, 1993.
5. P. Bhattacharya (ed.), *Properties of Lattice-Matched and Strained Indium Gallium Arsenide,* emis Datareview Series vol. 8, INSPEC, London, 1993.
6. J. H. Edgar (ed.), *Properties of Group III Nitrides,* emis Datareview Series vol. 11, INSPEC, London, 1994.
7. E. Kasper (ed.), *Properties of Strained and Relaxed Silicon Germanium,* emis Datareview Series vol. 12, INSPEC, London, 1995.
8. G. L. Harris (ed.), *Properties of Silicon Carbide,* emis Datareview Series vol. 13, INSPEC, London, 1995.
9. P. Bhattacharya (ed.), *Properties of III–V Quantum Wells and Superlattices,* emis Datareview Series vol. 15, INSPEC, London, 1996.
10. M. R. Brozel and and G. E. Stillman (eds.), *Properties of Gallium Arsenide,* 3rd Edition, emis Datareview Series vol. 16, INSPEC, London, 1996.
11. R. Hull (ed.), *Properties of Crystalline Silicon,* emis Datareview Series vol. 20, INSPEC, London, 1999.
12. T. P. Pearsall (ed.), *Properties, Processing and Applications of Indium Phosphide,* emis Datareview Series vol. 21, INSPEC, London, 2000.
13. J. H. Edgar, S. Strite, I. Akasaki, H. Amano, and C. Wetzel (eds.), *Properties, Processing and Applications of Gallium Nitride and Related Semiconductors,* emis Datareviews Series vol. 23, INSPEC, London, 1999.
14. E. Kasper and K. Lyutovich (eds.), *Properties of Silicon Germanium and SiGe:Carbon,* emis Datareview Series vol. 24, INSPEC, London, 2000.
15. W. E. Beadle, J. C. C. Tsai, and R. D. Plummer, *Quick Reference Manual for Silicon Integrated Circuit Technology,* Wiley, New York, 1985.
16. O. Madelung, *Semiconductors Group IV Elements and III–V Compounds,* Springer-Verlag, Berlin, 1991.
17. M. Levinshtein, S. Rumyantsev, and M. Shur (eds.), *Handbook Series on Semiconductor Parameters, vol. 1: Si, Ge, C (Diamond), GaAs, GaP, GaSb, InAs, InP, InSb,* World Scientific, Singapore, 1996.
18. M. Levinshtein, S. Rumyantsev, and M. Shur (eds.), *Handbook Series on Semiconductor Parameters, vol. 2: Ternary and Quarternary III–V Compounds,* World Scientific, Singapore, 1996.
19. I. Vurgaftman, J. R. Meyer, and L. R. Ram-Mohan, Band Parameters for III–V Compound Semiconductors and Their Alloys, *J. Appl. Phys., 89,* pp. 5815–5875, 2001.
20. J. C. Bean, Materials, Technologies, and Device Building Blocks, in: S. M. Sze (ed.), *High-Speed Semiconductor Devices,* Wiley, New York, 1990.
21. C. Kittel, *Introduction to Solid-State Physics,* Wiley, New York, 1986.
22. S. M. Sze, *Physics of Semiconductor Devices,* Wiley, New York, 1981.
23. W. Liu, *Fundamentals of III–V Devices,* Wiley, New York, 1999.
24. D. M. Caughey and R. E. Thomas, Carrier Mobilities in Silicon Empirically Related to Doping and Field, *Proc. IEEE, 52,* pp. 2192–2193, 1967.
25. W. R. Thurber, R. L. Mattis, and Y. M. Liu, Resistivity-Dopant Density Relationship for Phosphorus-Doped Silicon, *J. Electrochem. Soc., 127,* pp. 1807–1812, 1980.
26. T. I. Tosic, D. A. Tjapkin, and M. M. Jevtic, Mobility of Majority Carriers in Doped Noncompensated Silicon, *Solid-State Electron., 24,* pp. 577–582, 1981.

27. W. R. Wisseman and W. R. Frensley, GaAs Technology Perspective, in N. G. Einspruch and W. R. Wisseman (eds.), *VLSI Electronics Microstructure Science, vol. 11: GaAs Microelectronics,* Academic Press, Orlando, 1985.
28. D. A. Antoniadis, A. G. Gonzales, and R. W. Dutton, Boron in Near Intrinsic <100> and <111> Silicon under Inert and Oxidizing Ambients—Diffusion and Segregation, *J. Electrochem. Soc., 125,* pp. 813–819, 1978.
29. M. Roschke and F. Schwierz, Electron Mobility Models for 4H, 6H, and 3C SiC, *IEEE Trans. Electron Dev., 48,* pp. 1442–1447, 2001.
30. W. J. Schaffer, G. H. Negley, K. G. Irvine and J. W. Palmour, Conductivity Anisothropy in Epitaxial 6H and 4H SiC, *Mat. Res. Soc. Symp. Proc., 339,* pp. 595–600, 1994.
31. B. Carnez, A. Cappy, A. Kaszynski, E. Constant, and G. Salmer, Modeling of a Submicrometer Gate Field-Effect Transistor Including Effects of Nonstationary Electron Dynamics, *J. Appl. Phys., 51,* pp. 784–790, 1980.
32. M. V. Fischetti, Monte Carlo Simulation of Transport in Technologically Significant Semiconductors of the Diamond and Zinc-Blende Structures – Part I: Homogeneous Transport, *IEEE Trans. Electron Devices, 38,* pp. 634–649, 1991.
33. B. Meinerzhagen, Private Communication.
34. J. Kolnik, I. H. Oguzman, K. Brennan, R. Wang, P. P. Ruden, and Y. Wang, Electronic Transport Studies of Bulk Zincblende and Wurtzite Phases of GaN Based on an Ensemble Monte Carlo Calculation Including a Full Zone Band Structure, *J. Appl. Phys., 78,* pp. 1033–1038, 1995.
35. J. G. Ruch, Electron Dynamics in Short Channel Field-Effect Transistors, *IEEE Trans. Electron Devices, 19,* pp. 652–654, 1972.
36. M. Shur, Influence of Nonuniform Field Distribution on Frequency Limits of GaAs Field-Effect Transistors, *Electron. Lett., 12,* pp. 615–616, 1976.
37. J. P. Nougier, J. C. Vaissiere, D. Gasquet, J. Zimmermann, and E. Constant, Determination of Transient Regime of Hot Carriers in Semiconductors Using the Relaxation Time Approximation, *J. Appl. Phys., 52,* pp. 825–832, 1981.
38. F. Schwierz and G. Gobsch, The Potential of III-Nitrides for Use in High-Speed Field-Effect Transistors, *Ext. Abstr. SSDM,* pp. 214–215, 1999.
39. O. Ambacher, B. Foutz, J. Smart, J. R. Shealy, N. G. Weimann, K. Chu, M. Murphy, A. J. Sierakowski, W. J. Schaff, L. F. Eastman, R. Dimitrov, A. Mitchell, and M. Stutzmann, Two Dimensional Electron Gases Induced by Spontaneous and Piezoelectric Polarization in Undoped and Doped AlGaN/GaN Heterostructures, *J. Appl. Phys., 87,* pp. 334–344, 2000.
40. S. Selberherr, *Analysis and Simulation of Semiconductor Devices,* Springer-Verlag, Wien, 1984.
41. S. Tiwari, *Compound Semiconductor Device Physics,* Academic Press, San Diego, 1992.
42. R. van Overstraeten and H. de Man, Measurement of Ionization Rates in Diffused Silicon p-n Junctions, *Solid-State Electron., 13,* pp. 583–608, 1970.
43. Y. Taur and T. H. Ning, *Fundamentals of Modern VLSI Devices,* Cambridge University Press, Cambridge, 1998.
44. K. F. Brennan, E. Bellotti, M. Farahmand, J. Haralson II, P. P. Ruden, J. D. Albrecht, and A. Sutandi, Materials Theory Based Modeling of Wide Band Gap Semiconductors: From Basic Properties to Devices, *Solid-State Electron., 44,* pp. 195–204, 2000.
45. E. Bellotti, H.-E. Nilsson, K. F. Brennan, P. P. Ruden, and R. Trew, Monte Carlo Calcu-

lation of Hole Initiated Impact Ionization in 4H Phase SiC, *J. Appl. Phys., 87,* pp. 3864–3871, 2000.
46. P. A. Ivanov and V. E. Chelnokov, Recent Developments in SiC Single-Crystal Electronics, *Semicond. Sci. Technol., 7,* pp. 863–880, 1992.
47. S. M. Sze and G. Gibbons, Avalanche Breakdown Voltages of Abrupt and Linearly Graded p-n Junctions in Ge, Si, GaAs, and GaP. *Appl. Phys. Lett., 8,* pp. 111–113, 1966.
48. A. O. Konstantinov, Q. Wahab, N. Nordell, and U. Lindefeldt, Study of Avalanche Breakdown and Impact Ionization in 4H Silicon Carbide, *J. Electronic Mat., 27,* pp. 335–341, 1998.
49. E. Stefanov, L. Bailon, and J. Barbolla, Numerical Study of Avalanche Breakdown of 6H-SiC Planar PN-Junctions, *Diamond and Related Materials, 6,* pp. 1500–1503, 1997.
50. R. D. Lindsted and R. J. Surty, Steady-State Junction Temperature of Semiconductor Chips, *IEEE Trans. Electron Devices, 19,* pp. 41–44, 1972.
51. C-W. Kim, N. Goto, and K. Honjo, Thermal Behavior Dependending on Emitter Finger and Substrate Configurations in Power Heterojunction Bipolar Transistors, *IEEE Trans. Electron Devices, 45,* pp. 1190–1195, 1998.
52. W. B. Joyce, Thermal Resistance of Heat Sinks with Temperature-Dependent Conductivity, *Solid-State Electron., 18,* pp. 321–322, 1975.
53. St. G. Müller, R. Eckstein, J. Fricke, D. Hofmann, R. Horn, H. Mehling, and O. Nilsson, Experimental and Theoretical Analysis of the High Temperature Thermal Conductivity of Monocrystalline SiC, *Mat. Sci. Forum, 264-268,* pp. 623–626, 1998.
54. Y.-F. Wu, B. P. Keller, S. Keller, D. Kapolnek, S. P. Denbaars, and U. K. Mishra, Measured Microwave Power Performance of AlGaN/GaN MODFET, *IEEE Electron Device Lett., 17,* pp. 455–457, 1996.
55. G.-B. Gao, M.-Z. Wang, X. Gui, and H. Morkoc, Thermal Design Studies of High-Power Heterojunction Bipolar Transistors, *IEEE Trans. Electron Devices, 36,* pp. 854–863, 1989.
56. H. F. Cooke, Thermal Effects and Reliability, in J. L. B. Walker (ed.), *High-Power GaAs FET Amplifiers,* Artech House, Norwood, 1993.
57. W. Liu and B. Bayraktaroglu, Theoretical Calculations of Temperature and Current Profiles in Multi-Finger Heterojunction Bipolar Transistors, *Solid-State Electron., 36,* pp. 125–132, 1993.
58. H. F. Cooke, Precise Technique Finds FET Thermal Resistance, *Microwaves & RF, 25,* pp. 85–87, Aug. 1986.

CHAPTER 3

HETEROSTRUCTURE PHYSICS

3.1 INTRODUCTION

A heterostructure is a structure consisting of at least two layers of different semiconducting materials. The interface between two of these layers is called a heterojunction or heterointerface. Since the layers consist of materials with different bandgaps, discontinuities of the conduction and valence bands occur at the heterojunction. These discontinuities are called the conduction and valence band offsets ΔE_C and ΔE_V and are related to the bandgap difference ΔE_G by

$$\Delta E_G = \Delta E_C + \Delta E_V \qquad (3\text{-}1)$$

The bandgap difference and the band offsets are extremely important factors for the performance and operation of heterostructure devices such as HBTs and HEMTs.

As mentioned in Section 1.3.2, there are lattice-matched and pseudomorphic (or strained) heterostructures. Lattice-matched heterostructures consist of semiconductors having the same lattice constant. From Fig. 1.13 we see that $Al_xGa_{1-x}As$/GaAs, $In_{0.51}Ga_{0.49}P$/GaAs, and $In_{0.52}Al_{0.48}As$/$In_{0.53}Ga_{0.47}As$/InP are of the lattice matched type. Pseudomorphic heterostructures, on the other hand, consists of materials with different bulk lattice constants. Examples of pseudomorphic heterostructures are $Al_xGa_{1-x}As$/$In_yGa_{1-y}As$/GaAs, $In_{0.52}Al_{0.48}As$/$In_xGa_{1-x}As$/InP with $x > 0.53$, $Al_xGa_{1-x}N$/GaN, and $Si_{1-x}Ge_x$/Si.

Table 3.1 shows the lattice constants of the eleven basic semiconductors. Materials with a cubic lattice structure, such as Ge, Si, or GaAs, are characterized by the lattice constant a. Wurtzite crystals, such as SiC, GaN, and AlN, consisting of hexagonal prisms are characterized by the length a of the basal hexagon as well as the height c of the hexagonal prism.

The lattice constants of unstrained bulk binary, ternary, or quaternary alloys of the basic semiconductors can be estimated using the linear interpolation scheme described in Section 2.1.

Another feature distinguishing the heterojunctions is the position of the conduction and valence band edges of the materials with respect to the vacuum level before they are brought into contact with each other. For most heterojunctions, the bandgap of the narrow-gap material lies within the bandgap of the wide-gap material as shown in Fig. 3.1(a)–(c). This is called the type I heterojunction. In a lattice-

Table 3.1. Lattice constants of the eleven basic semiconductors for unstrained bulk material

Semiconductor	Lattice constant a, Å	Lattice constant c, Å
Ge	$5.65791 + 4.11 \times 10^{-5} \times (T - 300)$*	
Si	$5.43102 + 1.76 \times 10^{-5} \times (T - 300)$	
GaAs	$5.65325 + 3.88 \times 10^{-5} \times (T - 300)$	
AlAs	$5.6611 + 2.90 \times 10^{-5} \times (T - 300)$	
InAs	$6.0583 + 2.74 \times 10^{-5} \times (T - 300)$	
InP	$5.8697 + 2.79 \times 10^{-5} \times (T - 300)$	
GaP	$5.4505 + 2.92 \times 10^{-5} \times (T - 300)$	
4H SiC	3.073	10.053
6H SiC	3.0806	15.1173
GaN	3.189	5.185
AlN	3.112	4.982

*Whenever available, the temperature (in Kelvin) dependence is given as well. Note that the wurtzite semiconductors are characterized by the two lattice constants a and c.

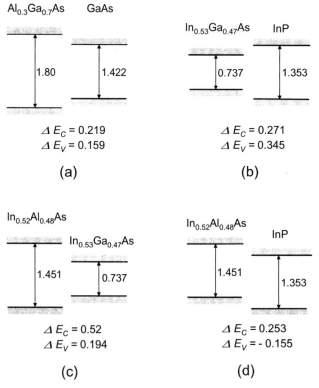

Figure 3.1. Band diagrams for different heterostructure pairs. Type I heterojunctions are shown in (a)–(c) and type II heterojunction is shown in (d). The values of the bandgaps and band offsets are in eV.

matched $In_{0.52}Al_{0.48}As/InP$ heterojunction, the situation is different because the valence band of the narrow-gap material (InP, $E_G = 1.353$ eV) is located below the valence band of the wide-gap material ($In_{0.52}Al_{0.48}As$, $E_G = 1.451$ eV), as shown in Fig. 3.1(d). Such a heterojunction is called type II. The common notation is that in a type I heterojunction, both the conduction and valence band offsets are positive quantities. In a type II heterojunction such as $In_{0.52}Al_{0.48}As/InP$, however, the valence band offset is negative. Moreover, if the conduction band of the narrow-gap material lies above the conduction band of the wide-gap material, the conduction band offset is also negative. Thus, the type of heterojunction decides the signs of ΔE_C and ΔE_V.

An important quantity for pseudomorphic heterostructures is the critical layer thickness. Consider an $AlGaAs/In_xGa_{1-x}As/GaAs$ structure grown on GaAs substrate as shown in Fig. 1.14. Both the AlGaAs layer and the GaAs substrate have a lattice constant of 5.65 Å, but bulk $In_xGa_{1-x}As$ has a larger lattice constant (e.g., $a = 5.714$ Å for $x = 0.15$). If the InGaAs layer is not too thick, the atoms of this layer adjust themselves according to the lattice structure of the GaAs substrate. In other words, the lattice of the grown InGaAs layer can accommodate that of the GaAs substrate. This, however, leads to more densely packed atoms in the grown InGaAs layer than those in the bulk InGaAs, and the grown InGaAs layer is under compressive strain. The amount of strain in the InGaAs layer depends on the In content and thus on the difference between the bulk lattice constants of InGaAs and GaAs as well as on the thickness of the InGaAs layer. When the strain becomes too strong, the grown layer relaxes, thus rearranging itself to its original bulk lattice structure and consequently causing a large number of dislocations near the heterointerface. The maximum thickness at which the InGaAs layer with a given In content remains strained is the critical layer thickness t_c.

There are several theories to calculate the critical layer thickness [1, 2]. Experience has shown, however, that the predicted critical thicknesses can deviate considerably from those obtained experimentally. In many cases, strained layers much thicker than the theoretically predicted ones have been successfully grown. This is because the critical thickness not only depends on the strain built up in the grown layer but also on the growth conditions, particularly on the growth temperature. Thus the prediction of t_c is rather difficult, and one should use measured data as a more reliable source.

The critical layer thicknesses for different strained layer systems are shown in Figs. 3.2(a) and (b). The role of the growth conditions on the critical thickness is obvious from the results in Fig. 3.2(a), which show t_c versus Ge content x for $Si_{1-x}Ge_x/Si$ heterostructures. Below the lower curve we find the region of stable strain. SiGe layers in this region are unconditionally stable. The area between both curves indicates the region of metastable strain. The symbols within this region designate experimental data. The squares correspond to SiGe layers grown at temperatures of around 550°C [3], the circles are results obtained at growth temperature of 750°C [4], and the triangles are data of growth temperatures between 900°C and 950°C [5, 6]. Obviously, lower growth temperatures result in larger critical thicknesses. The term "metastable" indicates that the layers are stable in principle after

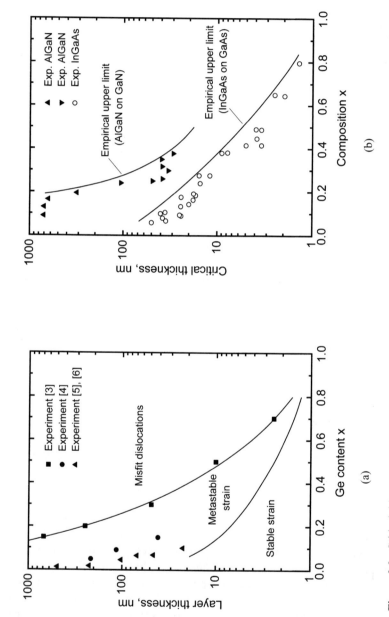

Figure 3.2. Critical thickness as a function of layer composition for pseudomorphic heterostructures of (a) $Si_{1-x}Ge_x$ grown on Si [3-6], and (b) $In_xGa_{1-x}As$ grown on GaAs [7] and $Al_xGa_{1-x}N$ grown on GaN (data taken from [8] and references therein). The upper limit curves are obtained from empirical fitting.

growth, but that they can relax under thermal stress, e.g., when the layers are exposed to temperatures exceeding the growth temperature during subsequent processing steps. Finally, above the upper curve is the region of misfit dislocations where the layers are relaxed. The two curves in Fig. 3.2(b) represent the upper limits of the critical thicknesses for InGaAs on GaAs and AlGaN on GaN. Typically, in microwave transistors only unrelaxed, i.e., strained, InGaAs/GaAs, AlGaN/GaN, and SiGe/Si heterostructures are of interest. Thus, Fig. 3.2 provides the design criteria in regard to the thickness of strained InGaAs, SiGe, and AlGaN layers.

3.2 BAND DIAGRAMS

3.2.1 The Anderson Model

To gain insight into the physics of heterostructures and to construct their energy band diagrams, we first discuss the electron affinity model proposed by Anderson [9]. Although the Anderson model considers an ideal heterojunction and often fails to predict the band offsets accurately, it is instructive for the understanding of the properties of heterojunctions. Let us consider a heterojunction consisting of n-type AlGaAs and p-type GaAs. This type of junction is called an Np junction. The capital letter N indicates that the bandgap of the n-type material is larger compared to that of the p-type material. Figure 3.3(a) shows the band diagrams of the two separate semiconductors. The parameters associated with the AlGaAs side on the left of Fig. 3.3(a) are designated by subscript 1 and those associated with the GaAs side on the right of Fig. 3.3(a) by subscript 2. The vacuum level E_{vac} is chosen as the reference. The different values of the electron affinity χ in AlGaAs and GaAs (theoretically, χ is 3.86 eV for $Al_{0.3}Ga_{0.7}As$ and 4.07 eV for GaAs) lead to different positions of the conduction and valence band edges in both materials. The positions of the Fermi levels relative to the band edges in the two materials are determined by the doping concentrations and can be calculated using the formulas given in Section 2.2.

Next the materials are brought into contact with each other. Under the condition that no voltage is applied (thermal equilibrium condition), the Fermi levels in the two materials must be aligned at the same horizontal position throughout the entire structure. To obtain this condition, the band diagram of AlGaAs has to be pulled down and that of GaAs has to be lifted upward while pinning the conduction and valence band edges at the heterojunction. The resulting band diagram with band bendings near the heterojunction is shown in Fig. 3.3(b). Because the two conduction bands in Fig. 3.3(a) do not coincide, there is a conduction band offset ΔE_C at the heterojunction. The same holds for the valence bands, where a valence band offset ΔE_V occurs. According to the Anderson model, the band offsets are given by $\Delta E_C = \chi_2 - \chi_1$ and $\Delta E_V = \Delta E_G - \Delta E_C$.

In the regions where the bands are flat, i.e., in the bulk, the separation between the band edges and the Fermi level is the same as that shown in Fig. 3.3(a). The electron and hole densities in the bulk are the equilibrium densities n_{01} and p_{01} on

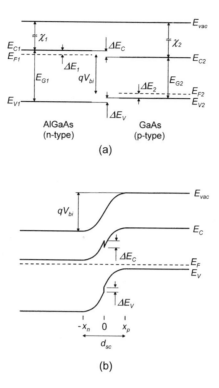

Figure 3.3. Band diagram of an abrupt Np heterojunction when (a) the materials are separated, and (b) materials are in contact. The materials of the wider and narrower bandgaps are n-type and p-type, respectively.

the AlGaAs side, and n_{02} and p_{02} on the GaAs side. Near the heterointerface, large gradients of the electron and hole concentrations exist. Immediately after the two materials are brought into contact, electrons tend to diffuse from the n-type AlGaAs (where electrons are majority carriers) to the p-type GaAs (where electrons are minority carriers). Likewise, holes tend to diffuse in the opposite direction. This leads to uncompensated negatively and positively charged ions near the heterojunction, and a space–charge region arises. The electric field in the space–charge region resulted from the charged ions gives rise to a drift current opposite to the diffusion current. Thus, in a very short period of time, the system reaches its equilibrium state and the transport of free carriers in the space–charge region stops. This is the same physics as in a homo-type pn junction discussed in Section 2.4. In the heterojunction, however, there is an additional effect influencing the carrier distribution. The conduction band on the AlGaAs side lies energetically higher than that on the GaAs side. Thus, the energy of the electrons in the conduction band of the AlGaAs layer is higher than that of the GaAs layer. Because electrons try to occupy the lowest allowed energy state, they are encouraged to move from AlGaAs to GaAs. In other words, at the N-AlGaAs/p-GaAs heterojunction, there are two driving forces for the electron motion from AlGaAs to GaAs. The first driving force is the gradient of the electron concentration and the second one results from the different energetic levels of the conduction bands in the two materials.

3.2.2 Built-In Voltage and Thickness of the Space–Charge Region

In the following, formulas will be derived to quantitatively describe the heterojunction properties under the thermal equilibrium condition. We set the origin of the coordinate system at the AlGaAs/GaAs metallurgical junction (i.e., heterointerface), with the N-AlGaAs layer residing in the $-x$ direction and the p-GaAs layer residing in the $+x$ direction, as shown in Fig. 3.3(b). Thus, the space–charge region has a thickness d_{sc} extending from $-x_n$ to x_p. As in the case of the homojunction, the depletion approximation can be employed, and we can assume that the free carrier densities are negligible compared to the doping densities in the space–charge region. Thus, on the AlGaAs side $N_{D1} \gg n_{01}$ between $-x_n$ and 0, and on the GaAs side $N_{D2} \gg p_{02}$ between 0 and x_p. Applying these conditions, the Poisson equation in the space–charge region can be simplified to

$$\frac{d\mathscr{E}(x)}{dx} = \frac{q}{\varepsilon_1} N_{D1} \quad \text{for } -x_n < x \leq 0 \text{ (AlGaAs side)} \quad (3\text{-}2)$$

and

$$\frac{d\mathscr{E}(x)}{dx} = -\frac{q}{\varepsilon_2} N_{A2} \quad \text{for } 0 \leq x < x_p \text{ (GaAs side)} \quad (3\text{-}3)$$

where $\mathscr{E}$ is the electric field. Integration of these equations yields

$$\mathscr{E}(x) = \frac{q}{\varepsilon_1} N_{D1}(x + x_n) \quad \text{on the AlGaAs side} \quad (3\text{-}4)$$

and

$$\mathscr{E}(x) = \frac{q}{\varepsilon_2} N_{A2}(x_p - x) \quad \text{on the GaAs side} \quad (3\text{-}5)$$

In deriving Eqs. (3-4) and (3-5), the boundary condition of zero field at the edges of the space–charge region [$\mathscr{E}(-x_n) = \mathscr{E}(x_p) = 0$] has been used. To obtain the potential distribution in the space–charge region, a second integration is necessary. Choosing the reference that the AlGaAs bulk is grounded, then the potential $\varphi(-x_n)$ at the edge of the space–charge region on the AlGaAs side is zero. Integrating Eq. (3-4) yields

$$\varphi(x) = -\frac{q}{2\varepsilon_1} N_{D1}(x + x_n)^2 \quad \text{on the AlGaAs side} \quad (3\text{-}6)$$

The potential at the heterojunction is obtained from Eq. (3-6) setting $x = 0$ as

$$\varphi(0) = -\frac{q}{2\varepsilon_1} N_{D1} x_n^2 \qquad (3\text{-}7)$$

Note that $\varphi(0)$ is negative. This is correct because we assumed the bulk of the AlGaAs to be grounded and the bands at the heterojunction are bent upward [see also Eq. (2-50) and Fig. 2.6)]. Now we integrate Eq. (3-5) using Eq. (3-7) as the boundary condition and obtain

$$\varphi(x) = \frac{q}{\varepsilon_2} N_{A2} \left(\frac{x^2}{2} - xx_p \right) - \frac{q}{2\varepsilon_1} N_{D1} x_n^2 \quad \text{on the GaAs side} \qquad (3\text{-}8)$$

The potential difference across the space–charge region under zero applied voltage is the built-in voltage V_{bi} of the heterojunction, which can also be regarded as the potential difference between the GaAs and AlGaAs bulks. The potential difference V_{bi1} between $-x_n$ and the heterojunction at $x = 0$ is the portion of the built-in potential on the AlGaAs side and is equal to $|\varphi(0)|$ from Eq. (3-7):

$$V_{bi1} = \frac{q}{2\varepsilon_1} N_{D1} x_n^2 \qquad (3\text{-}9)$$

From Eq. (3-8), we obtain V_{bi2}, which is the portion of the built-in voltage on the GaAs side:

$$V_{bi2} = \frac{q}{2\varepsilon_2} N_{A2} x_p^2 \qquad (3\text{-}10)$$

The total built-in voltage V_{bi} is simply the sum of V_{bi1} and V_{bi2}:

$$V_{bi} = V_{bi1} + V_{bi2} = \frac{q}{2\varepsilon_1} N_{D1} x_n^2 + \frac{q}{2\varepsilon_2} N_{A2} x_p^2 \qquad (3\text{-}11)$$

From Fig. 3.3(a), an alternative expression for the built-in voltage can be obtained:

$$V_{bi} = \frac{E_{G2} - \Delta E_1 - \Delta E_2 + \Delta E_C}{q} \qquad (3\text{-}12)$$

where ΔE_1 is the difference between conduction band edge and the Fermi level in the bulk of AlGaAs and ΔE_2 is the difference between the Fermi level and valence band edge in the bulk of GaAs.

At this point, the thickness of the space–charge region, $d_{sc} = x_n + x_p$, is still not known, but equations for x_n and x_p can be easily derived. At the heterointerface (at $x = 0$), in the absence of interface charges, the displacement $\mathscr{D} = \varepsilon \times \mathscr{E}$ must be continuous:

$$\varepsilon_1 \mathscr{E}(0^-) = \varepsilon_2 \mathscr{E}(0^+) \tag{3-13}$$

Inserting Eqs. (3-4) and (3-5) into Eq. (3-13) results in

$$N_{D1}x_n = N_{A2}x_p \tag{3-14}$$

Solving this equation first for x_p and then for x_n, and inserting the resulting equations into Eq. (3-11), we obtain

$$x_n = \left(\frac{2\varepsilon_1\varepsilon_2 N_{A2} V_{bi}}{q N_{D1}(\varepsilon_1 N_{D1} + \varepsilon_2 N_{A2})}\right)^{1/2} \tag{3-15}$$

and

$$x_p = \left(\frac{2\varepsilon_1\varepsilon_2 N_{D1} V_{bi}}{q N_{A2}(\varepsilon_1 N_{D1} + \varepsilon_2 N_{A2})}\right)^{1/2} \tag{3-16}$$

The total thickness of the space–charge layer, d_{sc}, is the sum of x_n and x_p. When an external voltage V_{pN} is applied to the junction, the thickness of the space–charge region changes. The space–charge region becomes thinner when a forward bias (negative potential on the n-type side and positive potential on the p-type side resulting in a positive V_{pN}) is applied, while a reverse bias (negative V_{pN}) results in a wider space–charge region. For the case of an applied external voltage, x_n and x_p are calculated by

$$x_n = \left(\frac{2\varepsilon_1\varepsilon_2 N_{A2}(V_{bi} - V_{pN})}{q N_{D1}(\varepsilon_1 N_{D1} + \varepsilon_2 N_{A2})}\right)^{1/2} \tag{3-17}$$

and

$$x_p = \left(\frac{2\varepsilon_1\varepsilon_2 N_{D1}(V_{bi} - V_{pN})}{q N_{A2}(\varepsilon_1 N_{D1} + \varepsilon_2 N_{A2})}\right)^{1/2} \tag{3-18}$$

Combining Eq. (3-14) with Eq. (3-9) to Eq. (3-11) yields the following useful expression

$$\frac{V_{bi1}}{V_{bi2}} = \frac{\varepsilon_2 N_{A2}}{\varepsilon_1 N_{D1}} \tag{3-19}$$

The method described above is in general applicable not only to an Np AlGaAs/GaAs junction but to any other abrupt Np and pN heterojunction as well. Although we stated earlier that the Anderson model may not be sufficiently accurate to describe the properties of the heterojunction, the formulas developed above are correct. The reason is that the electron affinities do not appear in these expressions. The only condition is that the conduction band offset must be known precisely.

To overcome the drawbacks of the Anderson model and to calculate the band offsets, several other theories had been developed for heterojunctions. Unfortunately, none of these delivers accurate values for the band offsets for the large variety of different heterostructures used in microwave transistors. Therefore, measured band offsets are commonly used. Detailed information for the bandgaps of alloys and for band offsets in frequently used heterostructures is given below.

3.2.3 Bandgaps and Band Offsets

The relation between the band offsets and the bandgap difference has been given in Eq. (3-1). If the bandgaps of the two materials of the heterojunction and at least one of the two band offsets are known, the missing band offset can be calculated by Eq. (3-1). The bandgaps of the basic semiconductors were given in Table 2.1. An important factor concerning the bandgaps of layers in heterostructures is that strain can significantly shift the position of the band edges and thus alter the bandgap of a semiconductor. Because both unstrained bulk and strained layers are used in microwave transistors, information on the bandgaps and band offsets in unstrained and strained heterostructures is required.

A. Bandgap of Alloys. We start with the bandgaps of unstrained alloys. An excellent review on this topic is given in [10]. It includes data from about 1000 references, delivers consistent data sets, makes remarks on the reliability of the cited data, and recommends which values really should be used. For the unstrained alloys of interest to microwave transistors, the composition dependence of the bandgap is not linear. Therefore the interpolation scheme presented in Section 2.1 is not applicable. Instead, a quadratic interpolation is commonly used. Consider a ternary alloy $A_xB_{1-x}C$ (e.g., $In_{0.53}Ga_{0.47}As$). From the knowledge of the bandgaps of the two basic materials AC and BC (InAs and GaAs), the bandgap of the alloy can be calculated by

$$E_{G,A_xB_{1-x}C} = xE_{G,AC} + (1-x)E_{G,BC} - x(1-x)c \qquad (3\text{-}20)$$

where c is the so-called bowing parameter. Table 3.2 lists the bowing parameters for alloys important for microwave transistors. Equation (3-20), together with the

Table 3.2. Bowing parameters for the calculation of the bandgap of unstrained (bulk) alloys

Alloy (unstrained)	c, eV	Comment
$Al_xGa_{1-x}As$	0.458	$x < 0.42$, $E_{G,AlAs} = 3.003$ eV (related to Γ valley)
$In_xGa_{1-x}As$	0.477	
$In_xAl_{1-x}As$	0.7	$x < 0.65$, $E_{G,AlAs} = 3.003$ eV (related to Γ valley)
$In_xGa_{1-x}P$	0.877	$x < 0.65$, $E_{G,GaP} = 2.777$ eV (related to Γ valley)
$Al_xGa_{1-x}N$	1	
$Si_{1-x}Ge_x$	$0.206x/(x-1)$	$x < 0.85$

Note: the bowing parameter for SiGe is not a constant but composition-dependent.

bandgaps of the basic semiconductors given in Table 2.1 and the bowing parameters from Table 3.2, allows one to calculate the bandgap of an alloy of the basic semiconductors.

A remark on the bandgap calculation for the $Al_xGa_{1-x}As$, $In_xAl_{1-x}As$, and $In_xGa_{1-x}P$ alloys is necessary. These alloys are direct semiconductors for small x values and become indirect materials when x exceeds a certain value. The transition from direct to indirect occurs at $x \approx 0.42$ for $Al_xGa_{1-x}As$ and $x \approx 0.65$ for both $In_xAl_{1-x}As$ and $In_xGa_{1-x}P$. The composition range of interest to microwave transistors is below 0.4 for $Al_xGa_{1-x}As$. $In_xAl_{1-x}As$ is lattice matched to InP for $x = 0.52$, and $In_xGa_{1-x}P$ is lattice matched to GaAs for $x = 0.49$. Thus the three alloys used in microwave transistors are direct semiconductors, and bandgaps associated with the Γ valley have to be used for the calculation of the bandgaps of the alloys. Figures 3.4(a) and (b) show the calculated composition-dependent bandgaps associated with the Γ, X, and L valleys in $Al_xGa_{1-x}As$ and $In_xAl_{1-x}As$, respectively. The crossover points for the transition from direct to indirect can be seen clearly.

Next let us consider the bandgaps of strained materials. Strained semiconductor layers possess bandgaps different from their bulk counterparts of the same composition. Thus, the interpolation scheme in Eq. (3-20) and the bowing parameters given in Table 3.2 do not apply to the strained alloys. To estimate the bandgaps of strained alloys of interest to microwave devices, the following quadratic polynomial fit can be used:

$$E_G(x) = a + bx + cx^2 \qquad (3-21)$$

where a, b, and c are fitting parameters and are given in Table 3.3.

B. Band Offsets. A great deal of information about the bandgaps of alloys can be found in the literature. Although the published bandgap values are somewhat scattered, the differences are quite small and the values given in Tables 3.2 and 3.3 are considered reliable. The same cannot be said for the band offsets, however. The ratio $\Delta E_C/\Delta E_V$ for certain heterojunctions can vary significantly from one study to another. Therefore, the band offsets given below should serve only as a guided reference.

We first deal with the band offsets of lattice-matched heterojunctions. For AlGaAs/GaAs heterojunctions, the valence band offset can be assumed to vary linearly with x from $\Delta E_V = 0.53$ eV for $x = 1$ to $\Delta E_V = 0$ for $x = 0$, whereas the effect of bowing influences only the conduction band offset. The ΔE_V value of 0.53 eV for an AlAs/GaAs junction corresponds to 34% of the bandgap difference (Γ valley) between AlAs and GaAs.

In Table 3.4, the bandgap difference and the band offsets ΔE_V and ΔE_C for the widely used lattice-matched heterojunctions are listed. Note that $In_{0.52}Al_{0.48}As/InP$ heterojunction is of type II and has a negative ΔE_V.

We now turn our attention to the band offsets of strained heterojunctions. For both $In_xGa_{1-x}As/GaAs$ and $In_xGa_{1-x}As/AlGaAs$ heterojunctions with x between 0 and 0.2, the conduction and valence band offsets are equal to around 70% and

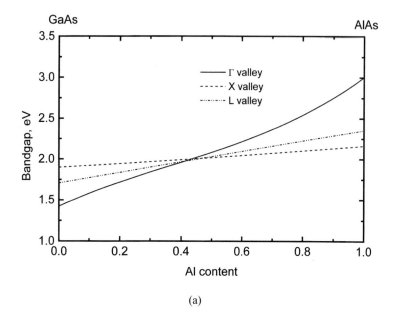

(a)

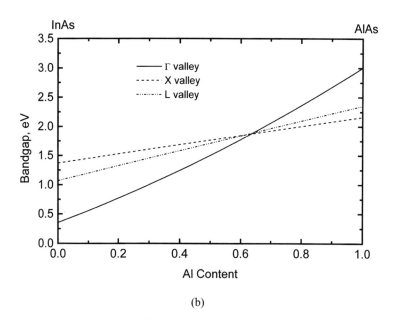

(b)

Figure 3.4. Γ, X, and L valley bandgaps of (a) $Al_xGa_{1-x}As$ and (b) $In_xAl_{1-x}As$.

Table 3.3. Fitting parameters for the bandgap calculation of different strained semiconductor alloys according to Eq. (3-21)

Alloy (strained)	a, eV	b, eV	c, eV	Comment
$In_xGa_{1-x}As$	$E_{G,GaAs}$	−1.11	0.45	small x, grown on GaAs
$In_xGa_{1-x}As$	1.211	−1.107	0.45	x around 0.53, grown on InP
$Si_{1-x}Ge_x$	$E_{G,Si}$	−0.896	0.396	$x < 0.5$, grown on Si
Si	$E_{G,Si}$	−0.643	−0.174	$x < 0.5$, grown on relaxed $Si_{1-x}Ge_x$
$Al_xGa_{1-x}N$	$E_{G,GaN}$	1.71	1.0	

Note: parameters for strained Si and $Al_xGa_{1-x}N$ are taken from [11] and [12], respectively.

30% of the bandgap difference, respectively. Conduction band offsets of 0.315 eV for $x = 0.69$ and 0.175 eV for $x = 0.37$ have been reported for $In_xGa_{1-x}As/InP$ junctions.

The composition dependence of ΔE_C in $In_xGa_{1-x}As/In_{0.52}Al_{0.48}As$ junctions can be described by

$$\Delta E_C [eV] = 0.384 + (0.254 \times x) \quad \text{for } x < 0.54 \quad (3\text{-}22)$$

followed by a quite abrupt shift around $x = 0.56$, and

$$\Delta E_C [eV] = 0.344 + (0.487 \times x) \quad \text{for } x > 0.58 \quad (3\text{-}23)$$

It is well accepted that in heterojunctions consisting of strained $Si_{1-x}Ge_x$ layers grown on Si substrate, a majority of ΔE_G transforms to a positive valence band offset. The value of the conduction band offset, on the other hand, and even the sign of ΔE_C, of such heterojunctions are still controversial. Measurements by King et al. [13] suggested a positive ΔE_C (i.e., type I heterojunction), which amounts to 23 meV for a Ge content x of 0.15 and increases up to about 40 meV for x in the range between 0.25 and 0.35. The offset ratio was estimated to be 15:85. The band offsets for the system of strained Si on relaxed $Si_{1-x}Ge_x$ substrates are [11]

$$\Delta E_C [eV] = -0.53x - 0.2x^2 \quad (3\text{-}24)$$

Table 3.4. Bandgap difference and band offsets of lattice-matched heterojunctions

Heterojunction	ΔE_G, eV	ΔE_C, eV	ΔE_V, eV
$Al_{0.3}Ga_{0.7}As/GaAs$	0.378	0.219	0.159
$In_{0.53}Ga_{0.47}As/InP$	0.616	0.271	0.345
$In_{0.53}Ga_{0.47}As/In_{0.52}Al_{0.48}As$	0.714	0.52	0.194
$In_{0.52}Al_{0.48}As/InP$	0.098	0.253	−0.155
$In_{0.49}Ga_{0.51}P/GaAs$ (ordered)	0.438	0.034	0.404
$In_{0.49}Ga_{0.51}P/GaAs$ (disordered)	0.46	0.22	0.24

and

$$\Delta E_V[\text{eV}] = -0.47x + 0.06x^2 \qquad (3\text{-}25)$$

The conduction band offset in $Al_xGa_{1-x}N/GaN$ is around 70% of the bandgap difference between $Al_xGa_{1-x}N$ and GaN.

Care must be taken when inserting the composition parameter x in Eqs. (3-20)–(3-25), because two different notations have been used in the literature. For example, to describe the composition of an InGaAs alloy, the term $In_xGa_{1-x}As$ is frequently used, whereas in some references the designation $In_{1-x}Ga_xAs$ is preferred. Thus for $In_{0.15}Ga_{0.85}As$, x is 0.15 in the first notation but x is 0.85 in the other.

3.3 CARRIER TRANSPORT ACROSS HETEROJUNCTIONS

The carrier flow in HBTs passes through at least one heterojunction (emitter–base heterojunction). Thus, the transport of free carriers across a heterojunction is a key issue in the understanding of HBT behavior. Before discussing such a mechanism, we need to further classify the types of heterojunctions into spike heterojunctions and smooth heterojunctions. The abrupt Np AlGaAs/GaAs heterojunction shown in Fig. 3.3(a) possesses a spike in the conduction band and therefore is designated as a spike heterojunction. In this structure, the Al content in the AlGaAs layer is constant and changes abruptly to zero at the heterointerface. The doping in the p-type GaAs base of an AlGaAs/GaAs HBT is normally very high and the base material is degenerate. Thus, the Fermi level can be located below the valence band edge. In this case, the conduction band spike at the abrupt emitter–base heterojunction becomes even more prominent and the tip of the spike is considerably higher than the conduction band in the base. On the other hand, in a heterojunction where the Al content does not change abruptly at the heterointerface, but rather changes gradually within a thin layer (called the graded layer) located adjacent to the left side of the heterointerface, the spike is reduced or eliminated all together. Such a heterojunction is called a smooth junction. Figures 3.5(a) and (b) compare the band diagrams of spike and smooth heterojunctions, respectively, under the equilibrium condition. Because most HBTs are Npn transistors having a wide bandgap emitter and a narrow bandgap base, we will concentrate below on the carrier transport across an Np heterojunction.

3.3.1 Currents Across Spike Heterojunctions

In this section we deal with the current across a spike Np heterojunction. In an HBT, the wide bandgap n-type region constitutes the emitter and the narrow bandgap p-type region acts as the base. The base of HBTs is much shorter than the minority electron diffusion length L_n. Thus the following discussions are made for a short-base Np junction. The band diagram of such a junction under the thermal equilibrium condition (no voltage applied) has been shown in Fig. 3.5(a). To

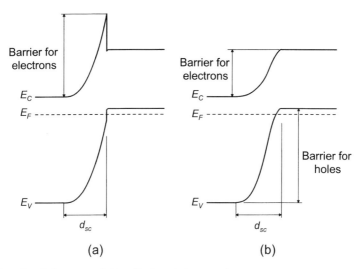

Figure 3.5. Band diagrams of (a) spiked, and (b) smooth Np heterojunctions under thermal equilibrium condition.

drive a current across the Np spike heterojunction, a voltage has to be applied. If the HBT is to be operated as an amplifier, the emitter–base junction has to be forward biased. In a homo-type pn junction, a forward bias causes the injection of electrons from the n-type region across the junction to the p-type region, and holes are injected in the opposite direction. This effect has been discussed in Section 2.4. In general, the same tendency exists in a spike Np junction as well. The existence of the conduction band and valence band offsets, however, leads to modified conditions.

As can be seen from Fig. 3.5(a), the energy barrier for holes at the junction is considerably larger than that for electrons. This is the case not only for the thermal equilibrium condition but for the case of an applied forward bias as well. Thus, it is much more difficult for holes to move from the p-type to the N-type region than for electrons to move from the N-type to the p-type region. The hole injection is effectively suppressed and for the calculation of the forward current across an Np junction only the electron injection needs to be considered.

The band diagram of an Np heterojunction with a forward bias applied is shown in Fig. 3.6(a). The applied voltage is designated as V_{pN} and is positive, i.e., the potential of the p-type region is more positive than that of the n-type region. The spike in the conduction band represents a potential barrier, which hinders the electron motion from the N-type region to the p-type region. Nevertheless, electron injection clearly prevails and hole injection can be neglected as discussed above. The transport of electrons across the spike is governed by two mechanisms. The first is called the thermionic emission, through which electrons with energies higher than the tip of the spike can surmount the barrier. The second mechanism is called the field

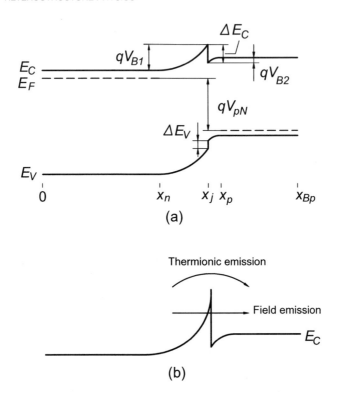

Figure 3.6. Band diagram of a spike Np heterojunction. (a) With a forward bias V_{pN} applied. (b) Current mechanisms across the junction.

emission or tunneling. It describes the ability of electrons with energies lower than the tip of the spike to tunnel through the energy barrier. Both mechanisms can occur concurrently, and the resulting mechanism is referred to as the thermionic field emission. Figure 3.6(b) illustrates the current mechanisms.

Deriving the equations enabling one to calculate the current in a spike heterojunction is fairly lengthy. Nevertheless, the model development provides useful insights into the physics of a spike heterojunction and can serve as a tool for the modeling of spike HBTs. We will discuss the basic ideas of the modeling, present the important equations, and refer the details to appropriate references.

Thermionic-emission current across a spike barrier will first be treated. The basic theory of such a current was described in [14] and later applied to HBTs in [15]. The electron current density $J_n(x_j)$ at the heterointerface can be considered as the difference between the two opposite electron fluxes at that position:

$$J_n(x_j) = -qv_n \left[n(x_j^-) - n(x_j^+) \exp - \frac{\Delta E_C}{k_B T} \right] \qquad (3\text{-}26)$$

where $n(x_j^-)$ is the electron density on the N-type side of the heterointerface, $n(x_j^+)$ is the electron density on the p-type side of the heterointerface, and the velocity v_n is given by

$$v_n = \sqrt{\frac{k_B T}{2m_n^*}} \quad (3\text{-}27)$$

where m_n^* is the conductivity effective electron mass. For simplicity, we have assumed the same effective electron mass in the N-type and p-type regions. In analogy to Eq. (2-105), the electron densities on both sides of the heterointerface can be expressed as

$$n(x_j^-) = n(x_n)\exp\left(-\frac{V_{B1}}{V_T}\right) \quad (3\text{-}28)$$

and

$$n(x_j^+) = n(x_p)\exp\frac{V_{B2}}{V_T} \quad (3\text{-}29)$$

The locations x_j, x_n, and x_p are denoted in Fig. 3.6(a), and the barrier voltages V_{B1} and V_{B2} corresponding to the conduction band bending on the N-type and p-type sides are

$$V_{B1} = (V_{bi} - V_{pN})\frac{N_{D1}\varepsilon_1}{N_{A2}\varepsilon_2 + N_{D1}\varepsilon_1} \quad (3\text{-}30)$$

and

$$V_{B2} = (V_{bi} - V_{pN})\left(1 - \frac{N_{D1}\varepsilon_1}{N_{A2}\varepsilon_2 + N_{D1}\varepsilon_1}\right) \quad (3\text{-}31)$$

The parameters associated with the N-type region are designated by subscript 1 and those associated with p-type region by subscript 2.

Since electrons are the majority carriers in the N-type region, $n(x_n)$ can be approximated by the equilibrium electron concentration, which is equal to N_{D1} if complete ionization is assumed. There are still two unknowns in Eqs. (3-26)–(3-29), namely $J_n(x_j)$ and $n(x_p)$. Neglecting recombination in the space–charge region, the current density at x_j should be equal to the current density at x_p:

$$J_n(x_p) = qD_{n2}\frac{n(x_{Bp}) - n(x_p)}{x_{Bp} - x_p} \quad (3\text{-}32)$$

where $x_{Bp} - x_p$ is the distance between the edge of the space–charge region and the contact of the p-type region. Note that we assume the case of a short-base junction. Equating $J_n(x_j)$ in Eq. (3-26) and $J_n(x_p)$ in Eq. (3-32) and solving the resulting equation for $n(x_p)$ yields

$$n(x_p) = \frac{v_n(x_{Bp} - x_p)N_{D1}\exp\left(-\dfrac{V_{B1}}{V_T}\right) + D_{n2}n(x_{Bp})}{D_{n2} + v_n(x_{Bp} - x_p)\exp\dfrac{V_{B2} - \Delta E_C/q}{V_T}} \quad (3\text{-}33)$$

where V_T is the thermal voltage equal to $k_B T/q$. Because the minority electron concentration at the contact (at x_{Bp}) is equal to the electron equilibrium concentration in the p-type region, we set $n(x_{Bp}) = n_{02}$. Equations (3-32) and (3-33) form the basic model for the thermionic-emission current density across a spike Np heterojunction.

Next we consider the mechanism of field emission. A physically rigorous treatment of the field emission requires a quantum mechanical derivation. An approximated and simplified approach will be presented here, assuming the shape of the spike is triangular and the transparency of this triangular spike can be expressed based on an approach suggested in [15–17]. Then the thermionic-field-emission current density including field emission can be obtained by multiplying the thermionic-emission current density given in Eq. (3-26) with the electron tunneling coefficient γ_n:

$$J_n(x_j) = -qv_n\left[n(x_j^-) - n(x_j^+)\exp\left(-\frac{\Delta E_C}{k_B T}\right)\right]\gamma_n \quad (3\text{-}34)$$

The tunneling coefficient is expressed as

$$\gamma_n = 1 + \frac{1}{V_T}\exp\frac{V_{B1}}{V_T} \times \int_{V^*}^{V_{B1}} D(v)\exp\left(-\frac{V}{V_T}\right)dV \quad (3\text{-}35)$$

where $D(v)$ is the barrier transparency and $v = V/V_{B1}$. The lower limit of the integral in Eq. (3-35), V^*, is

$$V^* = \begin{cases} V_{B1} - \dfrac{\Delta E_C}{q} & \text{for } V_{bi} > \dfrac{\Delta E_C}{q} \\ 0 & \text{otherwise} \end{cases} \quad (3\text{-}36)$$

For a triangular barrier, the transparency can be calculated by

$$D(v) = \exp\left[-\frac{qV_{B1}}{E_{00}}\left(\sqrt{1-v} + \frac{1}{2}v\ln v - v\ln(1 + \sqrt{1-v})\right)\right] \quad (3\text{-}37)$$

where

$$E_{00} = \frac{hq}{4\pi} \frac{N_{D1}}{m_n^* \varepsilon_1} \tag{3-38}$$

with h being the Planck constant.

3.3.2 Currents Across Smooth Heterojunctions

Since there is no abrupt change in the free-carrier energy in a smooth heterojunction, the drift–diffusion theory applies, and the description of carrier transport across a smooth Np heterojunction follows closely to the treatment of a homojunction discussed in Section 2.4. Comparing the band diagrams of a homojunction (Fig. 2.18) and of a smooth heterojunction [Fig. 3.5(b)], we see very similar characteristics. This means that we can simply repeat the derivations given in Section 2.4 for the minority carrier concentrations at the edges of the space–charge region and for the current density. The only difference is that we have to consider the fact that the N- and p-type regions of the heterojunction consist of different materials having different material parameters, such as bandgap, dielectric constant, intrinsic carrier concentrations, etc. The band diagram in Fig. 3.5(b) reveals that the energy barrier electrons have to surmount from the N-type to the p-type region is much lower compared to the energy barrier for holes moving from the p-type to the N-type region. Thus, we can expect the electron current being much larger than the hole current in a forward biased Np smooth heterojunction.

In analogy to Eq. (2-115), an expression for the current density across a smooth long-base heterojunction junction is given by

$$J_{hj} = q\left(D_{n2}\frac{n_{i2}^2}{N_{A2}L_{n2}} + D_{p1}\frac{n_{i1}^2}{N_{D1}L_{p1}}\right)\left(\exp\frac{V_{pN}}{V_T} - 1\right) \tag{3-39}$$

and for the current density across a smooth short-base heterojunction,

$$J_{hj} = q\left(D_{n2}\frac{n_{i2}^2}{N_{A2}(x_{Bp}-x_p)} + D_{p1}\frac{n_{i1}^2}{N_{D1}x_n}\right)\left(\exp\frac{V_{pN}}{V_T} - 1\right) \tag{3-40}$$

where V_{pN} is the voltage across the heterojunction. The meaning of x_{Bp}, x_p, and x_n is the same as in Fig. 3.6, and L_{n2} and L_{p1} are the minority carrier diffusion lengths in the p-type and N-type regions, respectively. The two addends inside the left parentheses on the right-hand side of Eqs. (3-39) and (3-40) correspond to the electron and hole currents across the heterojunction, respectively. Because the intrinsic carrier concentration in region 1 (N-AlGaAs), n_{i1}, is several orders of magnitude less than n_{i2}, the electron current clearly dominates the current across a forward biased Np heterojunction. This is exactly what we expected above.

As in the case of the current passing through a homojunction, we can transform Eqs. (3-39) and (3-40) into a more compact form by merging all voltage-independent factors into a prefactor J_S:

$$J_{hj} = J_S\left(\exp\frac{V_{pN}}{V_T} - 1\right) \tag{3-41}$$

3.4 CARRIER TRANSPORT PARALLEL TO HETEROJUNCTIONS AND TWO-DIMENSIONAL ELECTRON GAS

3.4.1 Band Diagram

Free-carrier transport parallel to a heterojunction is the main transport mechanism governing the operation of HEMTs. In principle, the concepts presented in Section 2.3 can be used to aid the understanding of such a transport. As in a bulk material, the current parallel to a heterojunction can be a drift current, a diffusion current, or a combination of both. There are, however, some essential differences in the behavior of carriers moving parallel to a heterojunction in HEMTs compared to carrier motion in a bulk material. To explain these, consider an N^+-AlGaAs/p^--GaAs heterojunction. The + and − signs indicate that the AlGaAs and GaAs layers are heavily and lightly doped, respectively. This type of heterojunction is the heart of the conventional AlGaAs/GaAs HEMT.

First we discuss the band diagram of such a heterojunction under the thermal equilibrium condition. The basic principle described in Section 3.2 still applies, but emphasis will be slightly different. Figure 3.7(a) shows the band diagram of the two materials still separated from each other. The vacuum energy is used as the reference level. Thus, the conduction band edge of the N^+-AlGaAs is located energetically higher than that of the p^--GaAs. Therefore, the energy of the electrons in the AlGaAs is higher compared to that in the GaAs. Typically, the heavily doped layer with the higher conduction band is called the barrier layer, and the lightly doped layer with the lower conduction band is called the channel layer. Immediately after both materials are brought in contact to each other, as shown in Fig. 3.7(b), electrons move from the AlGaAs to the GaAs because of the gradient of the electron concentration at the junction and the tendency of electrons to occupy the lowest allowed energy state. Thus, in the vicinity of the heterointerface, the electron concentration is reduced on the AlGaAs side and is increased on the GaAs side. At the same time holes also diffuse from the p-type channel layer to the barrier layer, although this effect is much less significant due to the presence of the valence band offset. In a very short period of time, a counterbalancing drift tendency arises and the net transport of electrons and holes stops. The result of such an initial free-carrier transport across the heterojunction is a band-bending and a space–charge region with thickness $d_{sc} = d_{sc1} + d_{sc2}$. An electric field is present in the space–charge region mainly due to the uncompensated donor ions on the AlGaAs side and due to the accumulated electrons on the GaAs side (which originally stem from the now uncompensated donor ions in the AlGaAs), and to a much lesser extent due to the uncompensated acceptor ions on the GaAs side. The electrons near the heterointerface on the GaAs side are confined in a very narrow

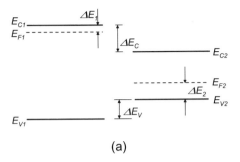

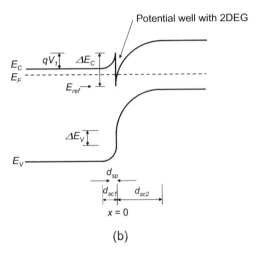

Figure 3.7. Band diagram of an N^+-AlGaAs/p^--GaAs heterojunction as used in AlGaAs/GaAs HEMTs under thermal equilibrium condition when (a) materials are separated and (b) materials are in contact.

potential well with a thickness of only a few nanometers. They can move only in a two-dimensional plane parallel to the heterointerface but not perpendicular to it because of the two potential barriers confining the potential well. Thus, these electrons constitute the so-called two-dimensional electron gas (2DEG). In a similar manner, a two-dimensional hole gas (2DHG) can be found at a P^+-AlGaAs/n^--GaAs heterointerface. Because of their higher mobility, 2DEGs are far more popular and practical than 2DHGs in microwave transistors. Therefore, the following discussions are focused on 2DEGs.

Frequently in HEMTs, a so-called spacer layer with a thickness d_{sp} of a few nanometers is inserted between the heavily doped barrier layer and the lightly doped channel layer. The spacer consists of the same material as the barrier layer, i.e., AlGaAs in our example, and is undoped. It is used to further separate the 2DEG and ions in the AlGaAs layer, thereby enhancing the 2DEG mobility.

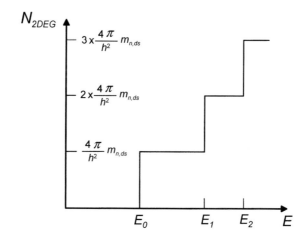

Figure 3.8. Density of states versus energy for the 2DEG of an AlGaAs/GaAs heterostructure.

3.4.2 2DEG Sheet Density

The calculation of the electron density in a 2DEG differs from that in bulk material. Unlike the electrons in bulk materials, the density of states $N_{2DEG}(E)$ of a 2DEG in the conduction band does not possess a continuous function of energy, but rather a staircase-like function. Figure 3.8 shows the 2DEG density of states versus energy in GaAs of an AlGaAs/GaAs heterojunction. The 2DEG in the Γ valley consists of n subbands (i.e., $0, 1, \ldots, n-1$), and the energies E_0, E_1, and E_2 indicated in Fig. 3.8 are the minima of the first three subbands, each with its own characteristic density of states. The energy levels E_0, E_1, and E_2 are called the subband energies. In Fig. 3.8, the origin of the E-axis corresponds to the conduction band edge in the GaAs channel layer. Thus, in the lowest part of the conduction band between $E = E_C$ and $E = E_C + E_0$, there are no allowed energy states in the channel layer adjacent to the heterojunction, and the channel layer has an "effective" bandgap larger than the bulk GaAs.

To describe the electron density in a 2DEG, the 2DEG sheet density n_s typically is used. Note that n_s is in cm^{-2}, whereas the electron density n in bulk material is in cm^{-3}. The knowledge of the electron density in the 2DEG is essential for the description of the currents in a HEMT. To calculate n_s correctly, the Schrödinger wave equation and the Poisson equation would have to be solved self-consistently [18]. This is a complicated process and requires the use of quantum mechanical descriptions. For analytical modeling, the self-consistent solution of the Schrödinger and Poisson equations is not practical, and we will present simplified approximations to estimate n_s. First, a simplified approach for the calculation of the sheet density in an ungated AlGaAs/GaAs structure containing a single heterojunction will be described. Afterward, the discussion will be extended to include the effect of a Schottky gate.

These simplified approaches are not restricted to AlGaAs/GaAs structures but are applicable to all heterostructures consisting of a single heterojunction. For more complicated heterostructures having two heterojunctions, such as AlGaAs/InGaAs/GaAs (used in pHEMTs on GaAs) or InAlAs/InGaAs/InP (used in InP HEMTs), however, the Schrödinger–Poisson exact solutions are recommended.

Consider an N^+-AlGaAs/p^--GaAs heterojunction as used in AlGaAs/GaAs HEMTs. The doping concentration in the N^+-AlGaAs is N_{D1} and that in the p^--GaAs is N_{A2}. We focus on the equilibrium condition and set the origin of the x-axis at the heterointerface. From Fig. 3.7(b), the shape of the conduction band on the GaAs side can be approximated by a triangular potential well. The subband energies E_n in such a well are given by [19]

$$E_n = \left(\frac{h^2}{8\pi^2 m_{n,l}}\right)^{1/3} \left(\frac{3}{2}\pi q\right)^{2/3} \left(n + \frac{3}{4}\right)^{2/3} \mathscr{E}_{i2}^{2/3} \quad (3\text{-}42)$$

where $m_{n,l}$ is the longitudinal effective mass of the 2DEG electrons, $\mathscr{E}_{i2}$ is the electric field at the heterointerface on the GaAs side, and n is the number of the subband. In our AlGaAs/GaAs heterostructure, the 2DEG resides in the GaAs layer. In GaAs, the longitudinal and transverse effective masses of electrons are equal: $m_{n,l} = m_{n,t} = 0.063 \times m_0$ where m_0 is the electron rest mass. Thus we obtain $E_0 = 4.227 \times 10^{-5}$ eV · $\mathscr{E}_{i2}^{2/3}$, $E_1 = 7.437 \times 10^{-5}$ eV · $\mathscr{E}_{i2}^{2/3}$, and $E_2 = 1.005 \times 10^{-4}$ eV · $\mathscr{E}_{i2}^{2/3}$.

Next we try to find an expression for the electric field $\mathscr{E}_{i2}$ at the heterointerface. The Poisson equation in the triangular potential well on the GaAs side is given by

$$\frac{d\mathscr{E}_2}{dx} = -\frac{q}{\varepsilon_2}[n_2(x) + N_{A2}] \quad (3\text{-}43)$$

where the subscript 2 again designates the quantities on the GaAs side of the heterojunction. The integration of Eq. (3-43) with the limits of $x = 0$ (location of the heterointerface) and $x = d_{sc2}$ (edge of the space–charge region on the GaAs side where the electric field becomes zero) leads to the field at the heterointerface

$$\mathscr{E}_{i2} = \frac{q}{\varepsilon_2}(n_s + N_{A2}d_{sc2}) \quad (3\text{-}44)$$

Since the GaAs side is lightly doped, $n_s \gg N_{A2}d_{sc2}$, Eq. (3-44) becomes

$$\mathscr{E}_{i2} = \frac{q}{\varepsilon_2}n_s \quad (3\text{-}45)$$

The same result is obtained in the case of a lightly doped n-type GaAs layer, and for an undoped GaAs layer. Combining Eqs. (3-42) and (3-45) we obtain $E_0 = 1.143 \times 10^{-9}$ eV · $n_s^{2/3}$, $E_1 = 2.011 \times 10^{-9}$ eV · $n_s^{2/3}$, and $E_2 = 2.717 \times 10^{-9}$ eV · $n_s^{2/3}$.

The next step is the derivation of an expression for n_s. From Eq. (2-7), it is clear that the electron density in a semiconductor is described by integrating the product of the density of states and the Fermi–Dirac distribution function over the entire conduction band. The density of states in a 2DEG, $N_{2DEG}(E)$, is given by [20]

$$N_{2DEG}(E) = \begin{cases} 0 & \text{for } 0 < E < E_0 \\ \dfrac{4\pi}{h^2} m_{n,ds} = D & \text{for } E_0 \leq E < E_1 \\ 2\dfrac{4\pi}{h^2} m_{n,ds} = 2D & \text{for } E_1 \leq E < E_2 \end{cases} \quad (3\text{-}46)$$

where $m_{n,ds}$ is the density of states effective mass which is equal to $0.063 \times m_0$ in GaAs. Note that the unit of $N(E)$ in Eq. (2-7) is cm^{-3}eV^{-1}, whereas $N_{2DEG}(E)$ in Eq. (3-46) has the unit cm^{-2}eV^{-1}. The occupation probability in the 2DEG is the same as that in the bulk material, i.e., $f(E)$ given in Eq. (2-6) still applies. Typically it is sufficient to take only the two lowest subbands into consideration, because almost all 2DEG electrons reside in these subbands. Doing this, the sheet density of the 2DEG becomes

$$n_s = D \int_{E_0}^{E_1} \frac{dE}{1 + \exp\dfrac{E - E_F}{k_B T}} + 2D \int_{E_1}^{\infty} \frac{dE}{1 + \exp\dfrac{E - E_F}{k_B T}} \quad (3\text{-}47)$$

Carrying out the integration we finally arrive at

$$n_s = Dk_B T \ln\left(1 + \exp\frac{E_F - E_0}{k_B T}\right) + Dk_B T \ln\left(1 + \exp\frac{E_F - E_1}{k_B T}\right) \quad (3\text{-}48)$$

Equation (3-48) still contains two unknowns, namely the 2DEG sheet density n_s and the Fermi level E_F. Therefore we need a second expression. As will be shown later, the expression for the electric field $\mathscr{E}_{i1}$ at the heterointerface in the AlGaAs layer serves this need.

The Poisson equation in the AlGaAs layer, including the spacer, is

$$\frac{d\mathscr{E}_1}{dx} = \frac{q}{\varepsilon_1} N_{D1}(x) \quad (3\text{-}49)$$

where subscript 1 denotes the quantities on the AlGaAs side, i.e., $N_{D1}(x)$ is the doping density in the AlGaAs layer, which is 0 for $-d_{sp} < x < 0$ (in the spacer layer) and is N_{D1} for $-d_{sc1} < x < -d_{sp}$. The integration of Eq. (3-49) yields the electric field in the space–charge region on the AlGaAs side:

$$\mathscr{E}_1 = \frac{qN_{D1}}{\varepsilon_1}(x - d_{sp}) \quad \text{for } -d_{sc1} < x < -d_{sp} \quad (3\text{-}50)$$

3.4 CARRIER TRANSPORT PARALLEL TO HETEROJUNCTIONS

and

$$\mathcal{E}_1 = \frac{qN_{D1}}{\varepsilon_1}(d_{sc2} - d_{sp}) \quad \text{for } -d_{sp} < x < 0 \tag{3-51}$$

Because the spacer is undoped, the electric field inside the spacer is constant and the field at the heterojunction (i.e., at $x = 0$), $\mathcal{E}_{i1}$, on the AlGaAs side is given by Eq. (3-51). The band bending in the AlGaAs layer, qV_1, which corresponds to V_{bi1} given in Eq. (3-9), is obtained by integrating the Poisson equation one more time over the space–charge region in the AlGaAs layer, i.e., from $x = -d_{sc1}$ to $x = 0$:

$$V_1 = -\int_{-d_{sc1}}^{0} \mathcal{E}_1(x)dx = \int_{0}^{-d_{sc1}} \mathcal{E}_1(x)dx = \int_{0}^{-d_{sp}} \mathcal{E}_{i1}dx + \int_{-d_{sp}}^{-d_{sc1}} \mathcal{E}_1(x)dx \tag{3-52}$$

The result of the integration is

$$V_1 = \frac{qN_{D1}}{2\varepsilon_1}(d_{sc1}^2 - d_{sp}^2) \tag{3-53}$$

Rearranging Eq. (3-53) we obtain an expression for d_{sc1}. Putting this into Eq. (3-51) leads to

$$\mathcal{E}_{i1} = \frac{qN_{D1}}{\varepsilon_1}\left(\sqrt{\frac{2\varepsilon_1 V_1}{qN_{D1}} + d_{sp}^2} - d_{sp}\right) \tag{3-54}$$

At the heterointerface, the electric displacement $\mathcal{D} = \varepsilon \times \mathcal{E}$ must be continuous:

$$\varepsilon_1 \mathcal{E}_{i1} = \varepsilon_2 \mathcal{E}_{i2} = qn_s \tag{3-55}$$

Thus we have two equations for the 2DEG density, namely Eqs. (3-48) and (3-55). Combining these equations gives

$$n_s = \frac{1}{q}\sqrt{2\varepsilon_1 V_1 qN_{D1} + q^2 N_{D1}^2 d_{sp}^2} - N_{D1}d_{sp} \tag{3-56}$$

From the band diagram in Fig. 3.7, we get another expression for V_1:

$$qV_1 = \Delta E_C - \Delta E_1 - E_F \tag{3-57}$$

The reference energy is the position of the conduction band on the GaAs side of the heterojunction. The 2DEG sheet density can be calculate numerically in the following way. We start with a reasonable value for E_F (experience shows that a not too negative value is appropriate) and calculate the potential V_1 using Eq. (3-57). Then the field at the heterointerface, $\mathcal{E}_{i1}$, is deduced from Eq. (3-54) and used to calculate

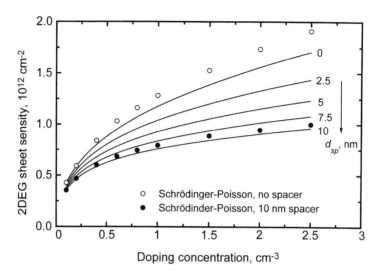

Figure 3.9. Calculated 2DEG sheet density versus the barrier layer doping density in $Al_{0.3}Ga_{0.7}As/GaAs$ heterojunctions having difference spacer thicknesses. Solid lines: analytic approach based on Eqs. (3-42)–(3-57). Circles: results of self consistent Schrödinger–Poisson solutions.

E_0 and E_1 in Eq. (3-42). Finally, n_s is obtained from Eq. (3-48). This value is compared with n_s calculated using Eq. (3-56). If the difference between the two n_s values is smaller than a predefined error Δn_s, the calculation is finished and n_s is found. Otherwise E_F has to be changed and the procedure repeated.

Figure 3.9 shows the 2DEG sheet densities as a function of the doping concentration in $Al_{0.3}Ga_{0.7}As/GaAs$ heterojunctions with different spacer thicknesses d_{sp} calculated using the analytic approach described above. For comparison, the results of the self-consistent Schrödinger–Poisson solution for d_{sp} of zero and 10 nm are also included. It can be seen that the analytic results are in good agreement with the Schrödinger–Poisson simulation. The sheet densities obtained from the analytic approach are slightly lower than the exact solutions. This is due to the use of the triangular potential well approximation in the analytic approach. As can be seen from Fig. 3.7, the well is of triangular shape only in the immediate vicinity to the heterointerface. Farther away from the heterojunction, the triangular approximation becomes questionable. Nevertheless, the analytical approach does provide valuable physical insight into heterojunctions and can be used to estimate the influence of the layer design (material combination, doping density, and spacer thickness) on the 2DEG sheet density. Obviously, the sheet density decreases with decreasing doping density in the barrier layer and/or with increasing spacer thickness.

The drain current in a HEMT is constituted by the flow of 2DEG electrons. The 2DEG sheet density n_s, and thus the drain current, is controlled by the voltage applied to a Schottky gate on top of the barrier layer. We now consider a gated hetero-

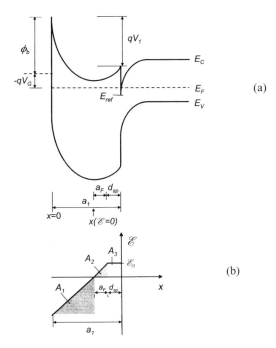

Figure 3.10. (a) Energy band diagram and (b) field distribution in a gated AlGaAs/GaAs heterostructure.

junction and derive an expression for the 2DEG sheet density, including the effect of gate bias condition (i.e., 2DEG sheet density under nonequilibrium condition). Figures 3.10(a) and (b) show the band diagram and the field distribution, respectively, in a gated AlGaAs/GaAs heterojunction. The AlGaAs layer has a total thickness a_1 and consists of an undoped spacer (thickness d_{sp}) and a homogeneously doped part (thickness $a_1 - d_{sp}$). The GaAs substrate is grounded and the origin of the x-axis is now located at the interface between the gate metal and the AlGaAs barrier layer. In this structure, the band bending in the barrier layer is influenced by two factors. The first factor is the electron transfer from the AlGaAs barrier layer to the GaAs layer as discussed above, and the second is the effect of the Schottky gate. The latter is a function of the Schottky barrier height, ϕ_b, and the voltage applied to the Schottky gate, V_G.

From Fig. 3.10(a), it can be seen that the slope of the conduction band edge and thus the electric field change their signs somewhere in the AlGaAs layer. The AlGaAs layer is fully depleted by the combination of two space–charge regions. The first space–charge region is associated with the AlGaAs/GaAs heterojunction and the other is associated with the gate/AlGaAs Schottky junction.

The electric field $\mathcal{E}_{i1}$ at the heterointerface on AlGaAs side, including the gate effect, is obtained by integrating the Poisson equation in the limits of $x = x(\mathcal{E} = 0)$ and $x = x(\mathcal{E} = 0) + a_F$. The integration results in

$$\mathcal{E}_{i1} = \frac{q}{\varepsilon_1} N_{D1} a_F \tag{3-58}$$

where a_F is the distance between the spacer and the point in the doped AlGaAs layer where the slope of energy band is zero [see Fig. 3.10(a)]. Note that the electric field in the spacer is constant because of zero doping. The voltage V_1 is defined as the difference of the potentials at the metal/AlGaAs interface and the heterointerface, i.e., by the difference between the AlGaAs conduction band edge at the metal/AlGaAs interface and the tip of the conduction band at the heterointerface. It can be derived from the integration of Poisson equation over the AlGaAs layer in the limits of $x = 0$ and $x = a_1$:

$$V_1 = -\int_0^{a_1} \mathcal{E}(x)dx \tag{3-59}$$

where a_1 is the thickness of the barrier layer. The integration can be easily carried out by graphically determining the areas A_1, A_2, and A_3 shown in Fig. 3.10(b). These areas are given by

$$\left. \begin{array}{l} A_1 = -\dfrac{\mathcal{E}_{i1}}{2a_F}[(a_1 - d_{sp})^2 + a_F(-2a_1 + a_F + 2d_{sp})] \\[2mm] A_2 = \dfrac{a_F \mathcal{E}_{i1}}{2} \\[2mm] A_3 = d_{sp} \mathcal{E}_{i1} \end{array} \right\} \tag{3-60}$$

Putting Eq. (3-58) into Eq. (3-60) and rearranging the resulting equation yields

$$V_1 = -(A_1 + A_2 + A_3) = \frac{qN_{D1}}{2\varepsilon_1}(a_1 - d_{sp})^2 - \mathcal{E}_{i1} a_1 = V_{po} - \mathcal{E}_{i1} a_1 \tag{3-61}$$

Here, V_{po} is the pinch-off voltage. It is the gate voltage needed to deplete the entire AlGaAs layer under the following conditions:

- the built-in voltage of the Schottky junction is zero
- no heterojunction is located underneath the AlGaAs layer
- the bottom of the AlGaAs layer is grounded

Equation (3-61) can be converted to

$$\varepsilon_1 \mathcal{E}_{i1} = \frac{\varepsilon_1}{a_1}(V_{po} - V_1) \tag{3-62}$$

We go back to the band diagram in Fig. 3.10(a). Choosing the position of the conduction band on the GaAs side of the heterojunction as the reference energy [E_{ref} in Fig. 3.10(a)], the following alternative expression for V_1 can be deduced:

$$V_1 = \frac{\phi_b}{q} - V_G + \frac{E_F}{q} - \frac{\Delta E_C}{q} \tag{3-63}$$

Inserting this into Eq. (3-62) and combining with Eq. (3-55), we get

$$qn_s = \frac{\varepsilon_1}{a_1}\left(V_{\text{po}} - \frac{\phi_b}{q} + V_G - \frac{E_F}{q} + \frac{\Delta E_C}{q}\right) \tag{3-64}$$

Equation (3-64) relates the 2DEG sheet density to the applied gate voltage V_G. Unfortunately, it still contains two unknowns, namely n_s and E_F, but these two unknowns can be correlated. Applying Eq. (3-42) to Eq. (3-57) and the method described above to calculate the sheet density in an ungated heterojunction, we can calculate the dependence of the Fermi level on the 2DEG sheet density. This dependence is universally valid and does not depend on whether the structure is gated or not. Figure 3.11 shows the calculated E_F versus n_s characteristics for an $Al_{0.24}Ga_{0.76}As/GaAs$ heterojunction with a 3 nm spacer layer. The variation of the sheet density and thus of the Fermi level is obtained by changing the doping concentration of the AlGaAs layer. As a first order approximation, the nonlinear E_F versus n_s curve can be described by a linear fit according to

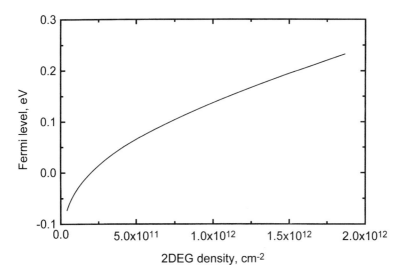

Figure 3.11. Fermi level as a function of 2DEG sheet density calculated using Eqs. (3-42)–(3-57). The variation of E_F and n_s is obtained by varying the doping concentration of the barrier layer.

$$E_F = E_{F0} + a n_s \tag{3-65}$$

where E_{F0} and a are fitting parameters. Inserting Eq. (3-65) into Eq. (3-64) yields

$$q n_s = \frac{\varepsilon_1}{a_1 + \Delta a_1} \left(V_{po} - \frac{\phi_b}{q} + V_G - \frac{E_{F0}}{q} + \frac{\Delta E_C}{q} \right) \tag{3-66}$$

where

$$\Delta a_1 = \frac{\varepsilon_1 a}{q^2} \tag{3-67}$$

It should be noted that up to now we assumed that the 2DEG electrons are located immediately adjacent to the heterojunction. In reality, however, the 2DEG electrons are distributed across a certain distance near the heterojunction, with the peak electron concentration being located a few nanometers from the heterointerface. This can be seen clearly from the Schrödinger–Poisson solutions given in Fig. 3.12. The correction term Δa_1 in Eq. (3-66) accounts for the dependence of E_F on V_G, but the effect that the 2DEG electrons are not located precisely at the heterojunction has not been accounted for in Eq. (3-66). In Fig. 3.12(a), the conduction band edge and the electron concentration in an AlGaAs/GaAs heterostructure are shown. The peak electron concentration is located about 3 nm away from the heterointerface. To take this into account, the denominator of the prefactor in Eq. (3-66) has to be replaced by $a_1 + \Delta a_1 + \Delta$, where Δ is the distance between the heterointerface and the position of the peak 2DEG electron density, i.e., $\Delta = 3$ nm for the $Al_{0.26}Ga_{0.76}As$/GaAs structure in Fig. 3.12(a).

To give a more comprehensive picture of the conditions in heterostructures used in HEMTs, Figs. 3.12(b)–(d) show the electron concentration and the conduction band edge in three more advanced heterostructures. The makeup of these heterostructures is shown in Table 3.5. Clearly, heterostructures with larger conduction band offsets show larger sheet densities. The sheet density can be further increased by introducing a second heterojunction at the bottom of the channel layer [see structures 2 and 3 in Fig. 3.12(b,c)] and by introducing a second doped barrier layer [see structure 4 in Fig. 3.12(d)].

The linear fitting for the E_F versus n_s characteristics mentioned in Eq. (3-65) is a little tricky. In [20], such a fitting for n_s has been made with n_s ranging between 1.5×10^{12} cm^{-2} and 4×10^{12} cm^{-2}. The fitting parameters a and E_{F0} have been determined to be 0.09427×10^{-12} eV and 0.0518 eV, respectively, at 300 K. This n_s range is too large, as the typical sheet density in AlGaAs/GaAs HEMTs is around or below 10^{12} cm^{-2}. For microwave devices, the n_s range typical for HEMTs under normal operating conditions should be selected for the linear fitting. Figure 3.13 shows the calculated 2DEG sheet density as a function of the gate voltage in a gated AlGaAs/GaAs heterojunction. The solid line with circles was obtained from the Schrödinger–Poisson solution, the dot-dash line (fit 2) represent the linear fit based on the fitting parameters given in [20], and the dashed line (fit 1) rep-

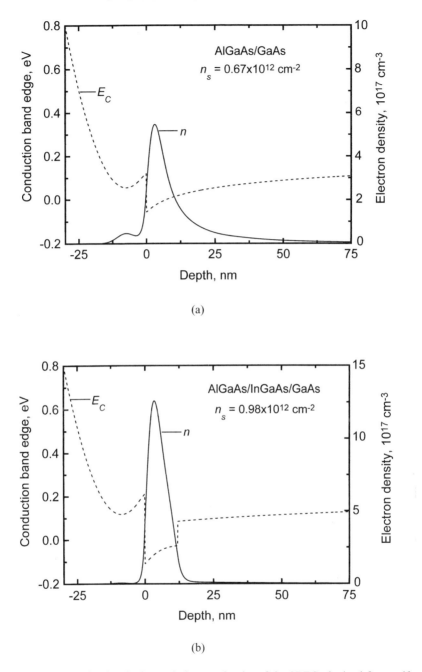

Figure 3.12. Conduction band edge and electron density of the 2DEG obtained from self-consistent Schrödinger–Poisson simulations for (a) sturcture 1—AlGaAs/GaAs, (b) structure 2—AlGaAs/InGaAs/ GaAs.

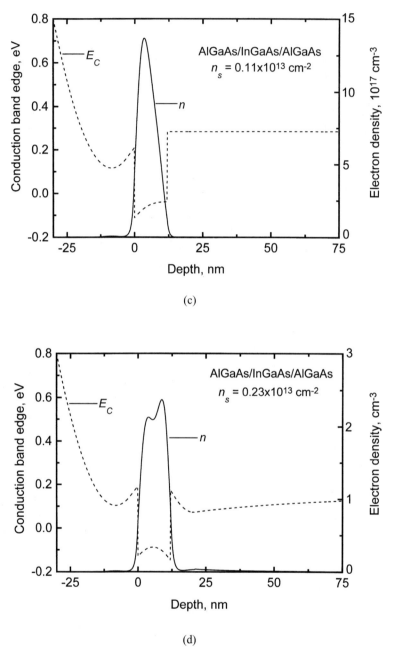

Figure 3.12. (c)–(d) structures 3 and 4—AlGaAs/InGaAs/AlGaAs/GaAs.

3.4 CARRIER TRANSPORT PARALLEL TO HETEROJUNCTIONS

Table 3.5. Four heterostructures considered in Figs. 3.12(a)–(d)

Structures	Layer sequence	Thickness, doping	2DEG sheet density
1 [see Fig. 3.12(a)]	$Al_{0.24}Ga_{0.76}$, barrier	27 nm, 2×10^{18} cm^{-3}	
	$Al_{0.24}Ga_{0.76}$, spacer	3 nm, undoped	
	GaAs, channel	undoped	6.7×10^{11} cm^{-2}
	GaAs, bulk	undoped	
2 [see Fig. 3.12(b)]	$Al_{0.24}Ga_{0.76}$, barrier	27 nm, 2×10^{18} cm^{-3}	
	$Al_{0.24}Ga_{0.76}$, spacer	3 nm, undoped	
	$In_{0.15}Ga_{0.85}As$, channel	undoped	9.8×10^{11} cm^{-2}
	GaAs, bulk	undoped	
3 [see Fig. 3.12(c)]	$Al_{0.24}Ga_{0.76}$, barrier	27 nm, 2×10^{18} cm^{-3}	
	$Al_{0.24}Ga_{0.76}$, spacer	3 nm, undoped	
	$In_{0.15}Ga_{0.85}As$, channel	undoped	1.1×10^{12} cm^{-2}
	$Al_{0.24}Ga_{0.76}$, bulk	undoped	
4 [see Fig. 3.12(d)]	$Al_{0.24}Ga_{0.76}$, barrier	27 nm, 2×10^{18} cm^{-3}	
	$Al_{0.24}Ga_{0.76}$, spacer	3 nm, undoped	
	$In_{0.15}Ga_{0.85}As$, channel	undoped	2.3×10^{12} cm^{-2}
	$Al_{0.24}Ga_{0.76}$, spacer	1.5 nm, undoped	
	$Al_{0.24}Ga_{0.76}$, second barrier	7.5 nm, 2×10^{18} cm^{-3}	
	$Al_{0.24}Ga_{0.76}$, bulk	undoped	

Note: the gate voltage (applied voltage plus built-in voltage) is 0.8 V for all cases.

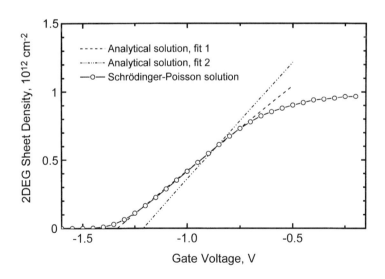

Figure 3.13. 2DEG sheet density n_s as a function of gate voltage V_G in an $Al_{0.24}Ga_{0.76}As/GaAs$ heterojunction obtained from Schrödinger–Poisson solutions (solid line) and from two different fitting schemes (dashed and dot-dash lines). The spacer thickness is 3 nm, and the thickness of the doped AlGaAs layer is 27 nm ($N_D = 2 \times 10^{18}$cm^{-3}). The fitting constants for linear fit 1 are $E_{F0} = -0.072$ eV and $a = 3.083 \times 10^{-13}$ eVcm^2.

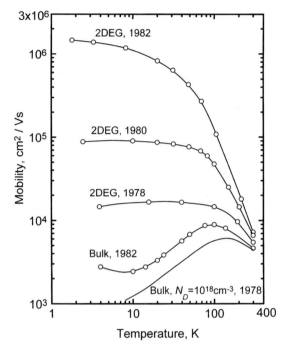

Figure 3.14. Electron mobility versus temperature in several AlGaAs/GaAs heterostructures and in undoped bulk GaAs. Data taken from [21, 22]. The increase of the mobility between 1978 and 1982 indicates the progress in epitaxial growth and the improvement of epitaxial layer quality during that time.

resents the linear fit based on the more realistic n_s range of 0.5×10^{11} cm^{-2} to 5×10^{11} cm^{-2}. Clearly, the results of fit 1 compare more favorably with the exact solutions than that of fit 2.

3.4.3 2DEG Mobility

In the late 1970s, a series of experiments on the electron transport in AlGaAs/GaAs multiquantum wells (several GaAs thin layers interleaved with AlGaAs layers) was conducted. A major finding was that in structures consisting of doped AlGaAs and undoped GaAs layers, the electrons parallel to the heterointerfaces can have a much higher mobility compared to that in the doped bulk GaAs [21]. Figure 3.14 compares the electron mobilities versus temperature in AlGaAs/GaAs structures and in bulk GaAs reported between 1978 and 1982. It is evident that the 2DEG mobility is higher than the bulk mobility over a wide range of temperature. At low temperatures, the 2DEG and the bulk mobilities differ by one or more orders of magnitude, whereas at room temperature the difference is reduced to a factor below 2. The effect of the enhanced mobility in heterostructures is exploited in HEMTs.

3.4 CARRIER TRANSPORT PARALLEL TO HETEROJUNCTIONS

The mobility is influenced by several different scattering mechanisms (impurity, optical phonon, acoustic phonon, and surface roughness scattering). At room temperature, the main scattering mechanism limiting the mobility is optical phonon scattering found in both the 2DEG and the bulk material. Impurity scattering also plays a certain role, as discussed in Section 2.3. At low temperatures, on the other hand, ionized impurity and acoustic phonon scattering are the dominant mechanisms. The reason for the distinct mobility advantage of the AlGaAs/GaAs system at low temperatures is the fact that the 2DEG electrons located on the GaAs side are spatially isolated from the donor ions situated on the AlGaAs side of the heterojunction. Thus ionized impurity scattering (Coulomb scattering) is reduced significantly in such a heterostructure. In a doped bulk GaAs, on the other hand, the ionized donors and the free electrons are not separated, and ionized impurity scattering is much more active.

When a spacer is introduced between the AlGaAs and GaAs layers, the separation between ionized donors and 2DEG electrons is increased, and the effect of Coulomb scattering can be further reduced. There is, however, a tradeoff between spacer thickness and 2DEG sheet density. A thicker spacer layer leads to a higher mobility but a lower sheet density (see Fig. 3.9).

It should be mentioned that microwave transistors are commonly operated at room temperature (or above when self heating is nonnegligible). Therefore the benefit of reduced Coulomb scattering in these devices is less than that suggested in Fig. 3.14 on first sight.

It is no surprise that the 2DEG mobility and sheet density in a heterostructure depend strongly on the semiconductors used. The mobility in III–V heterostructures can be increased considerably by replacing GaAs with $In_xGa_{1-x}As$. Bulk $In_xGa_{1-x}As$ possesses a higher electron mobility than bulk GaAs (i.e., zero In content), which translates into a higher 2DEG mobility, as shown in Fig. 3.15(a). Interestingly, the 2DEG mobilities in pseudomorphic GaAs HEMTs (the typical In content x is in the range of 0.15–0.25) are not necessarily higher than those in conventional AlGaAs HEMTs. This suggests that the strain in the InGaAs layer of these transistors has a detrimental effect on the mobility. The choice of InGaAs instead of GaAs leads to a larger conduction band offset as well, and a larger ΔE_C results in a higher sheet density [see Eq. (3-64)]. Figure 3.15(b) shows a compilation of reported sheet densities versus In content for III–V HEMTs. The experimental data confirm the trend predicted by the results of Schrödinger–Poisson solutions given in Fig. 3.12 and Table 3.5.

A high mobility and a high 2DEG sheet density are design targets for HEMTs, which is why InGaAs layers with high In contents are widely used in high-performance HEMTs. Moreover, it is desirable that these quantities be both large, thus giving a large product $\mu_0 \times n_s$ for enhancing HEMT performance. Figure 3.16 shows the product $\mu_0 \times n_s$ versus In content for different HEMTs. The advantage of high In content in the channel layer is clearly demonstrated. The incorporation of In, however, brings the drawback of a lower bandgap and thus a lower breakdown field in the InGaAs layer.

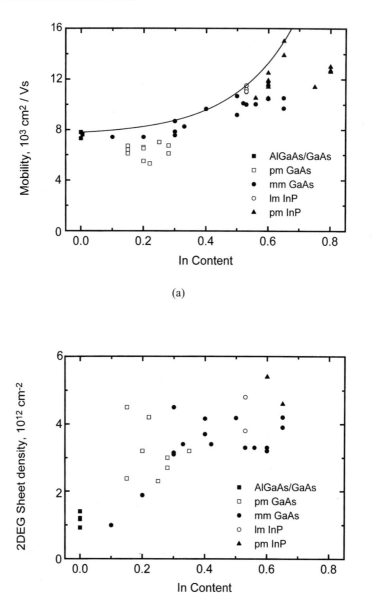

Figure 3.15. Influence of the In content in the InGaAs channel layer of a HEMT structure on (a) the 2DEG mobility and (b) the 2DEG sheet density. Data for the various types of InGaAs layers with the following abbreviations are included: pm—pseudomorphic, mm—metamorphic, and lm—lattice-matched.

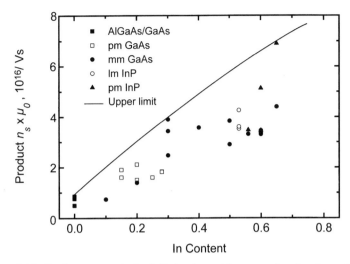

Figure 3.16. Product $\mu_0 \times n_s$ in the InGaAs channel layer as a function In content.

3.4.4 Spontaneous and Piezoelectric Polarization Effects

In Section 3.4.2 the electric field perpendicular to a heterojunction, caused by the doping conditions and the band offsets, and the resulting 2DEG sheet density have been discussed. The methods presented are applicable for calculating the field and sheet density in III–V heterostructures of interest for microwave HEMTs. The results are in most cases of sufficient accuracy to describe the electrical behavior of HEMTs. When heterostructures of wurtzite semiconductors such as AlGaN/GaN are considered, however, the methods given in Section 3.4.2 can underestimate 2DEG sheet densities. This is because additional mechanisms exist in these heterostructures that can further influence carrier confinement at the heterointerface.

These additional mechanisms are polarizations. Both AlGaN and GaN show a strong polarization, with the polarization in AlGaN being larger than that in GaN [8, 12, 23–25]. There are two types of polarization contributing to the total polarization in the AlGaN/GaN system, namely spontaneuos and piezoelectric polarizations. The former is due to the different polarizations in bulk AlGaN and GaN, whereas the latter is induced by the strain in the pseudomorphic AlGaN layer grown on the relaxed GaN layer. The change of the polarization at the AlGaN/GaN interface induces an interface charge. For AlGaN/GaN HEMTs, the so-called Ga-face heterostructures are used [12, 26], and the effect of combined spontaneous and piezoelectric polarizations results in a large positive interface charge in such devices. Electrons tend to compensate this positive interface charge and consequently build up an additional 2DEG component in the GaN layer adjacent to the heterointerface.

The sheet density of the additional polarization-induced 2DEG component and the electric field at the interface generated by the positive interface charge can be

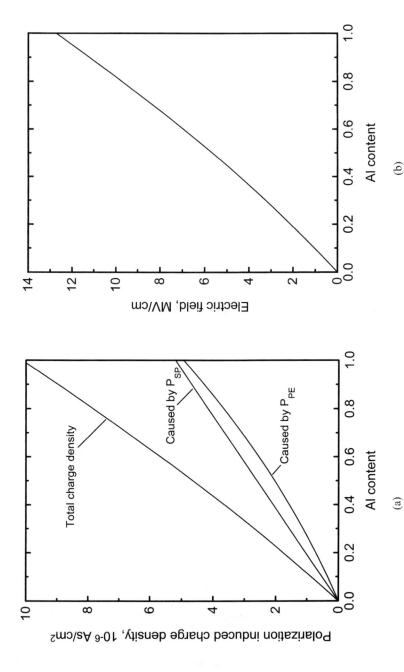

Figure 3.17. Effect of polarizations as a function of the Al content in AlGaN/GaN heterostructures on (a) calculated spontaneous, piezoelectric, and total polarizations and (b) electric field at the interface caused by polarizations.

quite large and must be taken into account to properly describe the overall 2DEG in the AlGaN/GaN heterojunction. The spontaneous and piezoelectric polarizations, P_{SP} and P_{PE}, have been discussed and modeled in [8, 12, 24]. Figure 3.17(a) shows the polarization-induced charge in AlGaN/GaN heterojunctions calculated based on the dependences of the material properties on the Al content given in [8, 12, 24], where a linear interpolation scheme has been used. More recent information on a nonlinear interpolation can be found in [27]. The unit of the polarization is charge per unit area, i.e., As/cm². It should be noted that strained, nonrelaxed AlGaN layers have been assumed in Fig. 3.17(a). In other words, the results are valid only for AlGaN layers with a thickness below the critical thickness [see Fig. 3.2(b)]. The curve designated by the total charge density is the positive interface charge. It is simply the sum of P_{SP} and P_{PE}. It can be seen that for a given Al content, the values for P_{SP} and P_{PE} are nearly the same. The electric field at the interface on the AlGaN side, $\mathscr{E}_{i1,pol}$, caused by the interface charge is

$$\mathscr{E}_{i1,pol} = \frac{P_{SP} + P_{PE}}{\varepsilon} \tag{3-68}$$

and is shown in Fig. 3.17(b). Depending on the Al content of the AlGaN layer, the field can amount to several MV/cm. It should be pointed out that $\mathscr{E}_{i1,pol}$ from Eq. (3-68) is the field component originated exclusively from the polarizations. The total field is therefore the sum of this field and the field associated with the electron

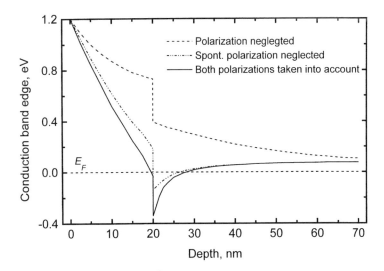

Figure 3.18. Conduction band edges for a gated $Al_xGa_{1-x}N/GaN$ heterostructure obtained from self-consistent solutions of Schrödinger and Poisson equations with and without polarizations. Data taken from [28], in which the following parameters were used in the calculation: $a_1 = 20$ nm, $d_{sp} = 5$ nm, $N_{D1} = 10^{18}$ cm⁻³, $\phi_b = 1.2$ V, zero gate voltage.

transfer from the n-AlGaN layer to the GaN layer, as explained in Section 3.3.2. When self-consistently solving the Schrödinger and Poisson equations for the electron distribution and the 2DEG sheet density in the AlGaN/GaN heterostructure, the total polarization charge $P = P_{SP} + P_{PE}$ can be modeled by placing a positive sheet charge layer of same quantity, i.e., P, on the AlGaN side immediately to the heterointerface.

To simplify the analysis of AlGaN/GaN heterojunctions, the following fits can be used to calculate the different polarizations and the interface field induced by the polarization charge:

$$P_{SP} = 5.2 \times 10^{-6} x \tag{3-69}$$

$$P_{PE} = 3.204 \times 10^{-6} x + 1.725 \times 10^{-6} x^2 \tag{3-70}$$

$$P = P_{SP} + P_{PE} = 8.404 \times 10^{-6} x + 1.725 x^2 \tag{3-71}$$

and

$$\mathcal{E}_i = 9.929 x + 2.772 x^2 \tag{3-72}$$

where x is the Al content in the AlGaN layer, and the units for P and $\mathcal{E}$ are As/cm² and 10^6V/cm, respectively.

The effect of the polarizations on the behavior of AlGaN/GaN heterostructures becomes evident when comparing the conduction band edge calculated with and without taking into account the polarizations. Figure 3.18 shows the conduction band edge of an $Al_{0.2}Ga_{0.8}N$/GaN heterostructure. The conduction band edge at the heterointerface is located 0.3 eV below the Fermi level when polarizations are taken into account, which gives rise to a very high 2DEG sheet density. When polarizations are neglected, however, the conduction band edge at the heterointerface is located 0.4 eV above the Fermi level (an energy of about 0.7 eV higher than the case of including polarizations) resulting in an electron density at the heterointerface of zero, i.e., no 2DEG would exist. Consequently, the effect of polarizations can lead to the formation of high-density 2DEG in AlGaN/GaN heterostructures even when the AlGaN layer is not doped.

REFERENCES

1. J. H. van der Merwe, Crystal Interfaces. Part I. Semi-Infinite Crystals, *J. Appl. Phys., 34,* pp. 117–122, 1963.
2. J. M. Matthews and A. E. Blakeslee, Defects in Epitaxial Multilayers. Misfit Dislocations in Layers, *J. Crystal Growth, 27,* pp. 118–125, 1974.
3. J. C. Bean, L. C. Feldman, A. T. Fiory, S. Nakahara, and I. K. Robinson, Ge(x)Si(1-x)/Si Strained-Layer Superlattice Grown by Molecular Beam Epitaxy, *J. Vac. Sci. Technol., A2,* pp. 436–440, 1984.

4. E. Kasper, H.-J. Herzog, and H. Kibbel, A One-Dimensional Superlattice Grown by UHV Epitaxy, *Appl. Phys.*, *8,* pp. 199–205, 1975.
5. M. L. Green, B. E. Weir, D. Bransen, Y. F. Hsieh, G. Higashi, A. Feygenson, L. C. Feldman, and R. L. Headrick, Mechanically and Thermally Stable Si-Ge Films and Heterojunction Bipolar Transistors Grown by Rapid Thermal Chemical Vapor Deposition at 900°C, *J. Appl. Phys.*, *69,* pp. 745–751, 1991.
6. D. C. Houghton, C. J. Gibbings, C. G. Tuppen, M. H. Lyons, and M. A. G. Halliwell, Equilibrium Critical Thickness for $Si_{1-x}Ge_x$ Strained Layers on (100) Si, *Appl. Phys. Lett.*, *56,* pp. 460–462, 1990.
7. L. D. Nguyen, L. E. Larson, and U. K. Mishra, Ultra-High-Speed Modulation-Doped Field-Effect Transistors: A Tutorial Review, *Proc. IEEE*, *80,* pp. 494–518, 1992.
8. O. Ambacher, B. Foutz, J. Smart, J. R. Shealy, N. G. Weimann, K. Chu, M. Murphy, A. J. Sierakowski, W. J. Schaff, L. F. Eastman, R. Dimitrov, A. Mitchell, and M. Stutzmann, Two Dimensional Electron Gases Induced by Spontaneous and Piezoelectric Polarization in Undoped and Doped AlGaN/GaN Heterostructures, *J. Appl. Phys.*, *87,* pp. 334–344, 2000.
9. R. L. Anderson, Experiments on Ge-GaAs Heterojunctions, *Solid-State Electron.*, *5,* pp. 341–351, 1962.
10. I. Vurgaftman, J. R. Meyer, and L. R. Ram-Mohan, Band Parameters for III–V Compound Semiconductors and Their Alloys, *J. Appl. Phys.*, *89,* pp. 5815–5875, 2001.
11. D. Nuernbergk, *Simulation des Transportverhaltens in Si/$Si_{1-x}Ge_x$/Si-Heterobipolartransistoren*, H. Utz Verlag, Munich, 1999.
12. O. Ambacher, J. Smart, J. R. Shealy, N. G. Weimann, K. Chu, M. Murphy, W. J. Schaff, L. F. Eastman, R. Dimitrov, L. Wittmer, M. Stutzmann, W. Rieger, and J. Hilsenbenck, Two-Dimensional Electron Gases Induced by Spontaneous and Piezoelectric Polarization Charges in N- and Ga-Face AlGaN/GaN Heterostructures, *J. Appl. Phys.*, *85,* pp. 3222–3233, 1999.
13. C. A. King, J. L. Hoyt, and J. F. Gibbons, Bandgap and Transport Properties of $Si_{1-x}Ge_x$ by Analysis of Nearly Ideal Si/$Si_{1-x}Ge_x$/Si Heterojunction Bipolar Transistors, *IEEE Trans. Electron Devices*, *36,* pp. 2093–2104, 1989.
14. S. S. Pearlman and D. L. Feucht, p-n Heterojunctions, *Solid-State Electronics*, *7,* pp. 911–923, 1964.
15. A. A. Grinberg, M. S. Shur, R. J. Fischer, and H. Morkoc, An Investigation of the Effect of Graded Layers and Tunneling on the Performance of AlGaAs/GaAs Heterojunction Bipolar Transistors, *IEEE Trans. Electron Devices*, *31,* pp. 1758–1765, 1984.
16. F. A. Padovani and R. Stratton, Field and Thermionic-Field Emission in Schottky Barriers, *Solid-State Electronics*, *9,* pp. 695–707, 1966.
17. R. Stratton, Theory of Field Emission From Semiconductors, *Phys. Rev.*, *125,* pp. 67–82, 1969.
18. T. Ando, A. B. Fowler, and F. Stern, Electronic Properties of Two-Dimensional Systems, *Rev. Mod. Phys.*, *54,* pp. 437–672, 1982.
19. D. Delagebeaudeuf and N. T. Linh, Metal-(n) AlGaAs-GaAs Two-Dimensional Electron Gas FET, *IEEE Trans. Electron Devices*, *29,* pp. 955–960, 1982.
20. W. Liu, *Fundamentals of III–V Devices*, Wiley, New York, 1999.
21. R. Dingle, H. L. Störmer, A. C. Gossard, and W. Wiegmann, Electron Mobilities in

Modulation-Doped Semiconductor Heterojunction Superlattices, *Appl. Phys. Lett., 33,* pp. 665–667, 1978.
22. R. Dingle, New High-Speed III–V Devices for Integrated Circuits, *IEEE Trans. Electron Devices, 31,* pp. 1662–1667, 1984.
23. R. Dimitrov, A. Mitchell, L. Wittmer, O. Ambacher, M. Stutzmann, J. Hilsenbeck, and W. Rieger, Comparison of N-Face and Ga-Face AlGaN/GaN-Based High Electron Mobility Transistors Grown by Plasma-Induced Molecular Beam Epitaxy, *Jpn. J. Appl. Phys., 38,* pp. 4962–4968, 1999.
24. O. Ambacher, Growth and Applications of Group III-Nitrides, *J. Phys. D: Appl. Phys., 31,* pp. 2653–2710, 1998.
25. M. Singh, Y. Zhang, J. Singh, and U. Mishra, Examination of Tunnel Junctions in the AlGaN/GaN System: Consequences of Polarization Charge, *Appl. Phys. Lett., 77,* pp. 1867–1869, 2000.
26. L. F. Eastman, V. Tilak, J. Smart, B. M. Green, E. M. Chumbes, R. Dimitrov, H. Kim, O. S. Ambacher, N. Weimann, T. Prunty, M. Murphy, W. J. Schaff, and J. R. Shealy, Undoped AlGaN/GaN HEMTs for Microwave Power Amplification, *IEEE Trans. Electron Devices, 48,* pp. 479–485, 2001.
27. O. Ambacher, J. Majewski, C. Miskys, A. Link, M. Eickhoff, F. Bernadini, V. Fiorentini, V. Tilak, B. Schaff, and L. T. Eastman, Pyroelectric Properties of Al(In)GaN/GaN Hetero- and Quantum Well Structures, *J. Phys: Condens. Matter, 14,* pp. 3399–3434, 2002.
28. F. Sacconi, A. Di Carlo, P. Lugli, and H. Morkoc, Spontaneous and Piezoelectric Polarization Effects on the Output Characteristics of AlGaN/GaN Heterojunction Modulation Doped FETs, *IEEE Trans. Electron Devices, 48,* pp. 450–457, 2001.

CHAPTER 4

MESFETs

4.1 INTRODUCTION

The term MESFET stands for *Me*tal–*S*emiconductor *F*ield-*E*ffect *T*ransistor. Thanks to its simplicity and good high-speed performance, this type of transistor has been one of the most important microwave devices for more than thirty years. The development of the MESFET started when a semiconductor device superior to the Si bipolar transistor for high-speed applications was needed. GaAs has been a material of interest because of its high electron mobility. Unfortunately, experiments on GaAs bipolar transistors in the late 1950s and early 1960s were not successful, and GaAs field-effect transistors gained prominence. A difficulty arose in the realization of GaAs FETs, however. Unlike the SiO_2 layer for Si MOSFETs, high-quality and stable insulators cannot be grown on GaAs. Thus, the concept of MOSFET fabricated on GaAs was ruled out and an alternative concept for the gate electrode had to be found.

As early as 1952, W. Shockley had theoretically described a FET using a reverse-biased pn junction as the gate electrode [1]. This type of FET is called the junction field-effect transistor (JFET) and its basic structure is shown in Fig. 4.1(a). In 1966, C. Mead used a similar concept to realize a GaAs FET, called MESFET, built with a Schottky junction instead of a pn junction as the gate [2]. Although this first MESFET was not suitable for microwave operations, only one year later a GaAs MESFET with a maximum frequency of oscillation f_{max} of 3 GHz had been reported [3]. In 1973, 80 to 100 GHz f_{max} were reported from 0.5 μm gate length GaAs MESFETs [4], and the suitability of the MESFET for microwave applications was established. In the mid 1970s, both low-noise and power GaAs MESFETs were commercially available and became the workhorses of GHz electronics. Although the market share of the GaAs MESFET is currently declining, it is still an important and popular microwave device. MESFETs fabricated on other materials have also been developed. For example, in the 1990s, MESFETs made from wide bandgap materials, such as SiC and GaN, had been studied for power amplification at GHz frequencies [5, 6], and in 1999 the first commercial SiC MESFET arrived at the market [7].

The basic structure of a MESFET is shown in Fig. 4.1(b). The transistor consists of a doped active layer (channel layer) located on a semi-insulating buffer or substrate. Commonly, the active layer is n-type (n-channel MESFET), although p-channel MESFETs have occasionally been used. The reason for the dominance of

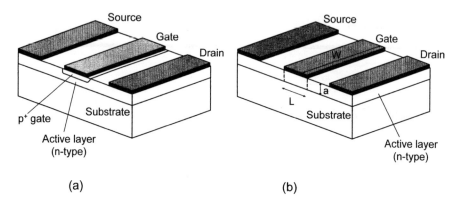

Figure 4.1. Basic structures of (a) JFET and (b) MESFET.

n-channel MESFETs in microwave applications stems from the fact that the electron mobility is much higher compared to the hole mobility in any semiconductor. For this reason, we will concentrate on n-channel MESFETs. On top of the active layer, there are three metal–semiconductor contacts: the source and drain contacts, which are ohmic electrodes, and the gate contact, which is a Schottky electrode. The gate length of the MESFET is L and the gate width is W. The active layer has a thickness a and a doping concentration N_D.

The basic operation principle of the MESFET is as follows. First, consider the case of zero drain–source applied voltage V_{DS}. Normally, the source is grounded (source potential equals zero) and the Schottky contact is reverse-biased. For some applications, the gate may be slightly forward-biased, but in any case a space–charge region with a thickness d_{sc} is located underneath the gate and extends into the channel layer, and only a negligible current flows to the gate contact. Thus, the top portion of the active layer is depleted, and a conducting channel with a height b exists between the edge of the space–charge region and the bottom of the active layer. This is shown in Fig. 4.2(a). As discussed in Section 2.5, the thickness of the space–charge region depends on the built-in voltage of the Schottky gate and on the applied gate–source voltage V_{GS}. A more negative V_{GS} gives rise to a wider space–charge region and thus a narrower conducting channel, as shown in Fig. 4.2(b).

The relationship among d_{sc}, the built-in voltage, and the doping density of the semiconductor layer in a Schottky junction has been presented in Section 2.5 [see Eq. (2-122)]. Consider a GaAs MESFET with an active layer doping of 10^{17} cm^{-3} and a gate built-in voltage of 0.9 V. Then, at zero applied V_{GS}, d_{sc} is about 113 nm. If the thickness of the active layer is 200 nm, then the height of the conducting channel is 87 nm, and the device can conduct current when a voltage is applied to the drain terminal. A gate voltage of -1.91 V would extend the space–charge region all the way down to the bottom of the active layer and thus close the channel. This type of MESFET for which a negative gate voltage has to be applied to fully deplete

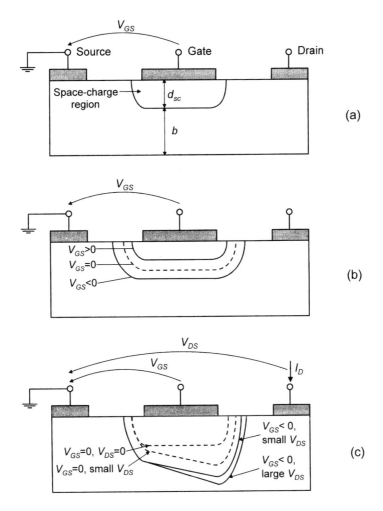

Figure 4.2. Conditions in a MESFET with applied voltages. (a) Space–charge region with thickness d_{sc}, and channel with height b. (b) Space–charge region for different gate voltages and zero drain voltage. (c) Space–charge region for different gate voltages and nonzero drain voltages.

the channel layer is called the depletion-mode MESFET and is frequently designated as the normally-on transistor.

For the case of an active layer thickness of only 100 nm at zero V_{GS}, the space–charge region extends beyond the bottom of the active layer and into the substrate. A positive gate voltage (i.e., the gate is forward biased) has to be applied to open up the channel. For the device considered, a gate voltage of 0.4 V results in a d_{sc} of around 84 nm and a conducting channel height of 16 nm. A MESFET for which a positive gate voltage has to be applied to open the channel is called the enhancement-mode MESFET and is frequently designated as the normally-off transistor. Caution is required to avoid a considerable gate current when applying a positive gate voltage. Typically, the gate–source voltage of GaAs MESFETs is kept below 0.5 V.

Now let us consider the case in which a positive voltage V_{DS} is applied between the source and drain terminals, which gives rise to an electric field in the $-x$ direction in the channel. The x-axis is directed from the source toward the drain parallel to the semiconductor surface. Driven by such a field, electrons move from the source toward the drain, which constitutes the flow of the drain current I_D. This drain current is defined as positive when electrons flow in the x direction and leave the transistor at the drain, or, in other words, when positive charges flow into the drain. A change in the gate voltage results in a change in d_{sc}, and thus a change in the drain current. The drain current increases with increasing gate voltage, and decreases when the gate voltage becomes more negative. Because the potential difference between the channel and gate becomes larger if we move from source (source potential $V_S = 0$) to drain (drain potential $V_D = V_{DS}$), d_{sc} increases and b decreases towards the drain, as shown in Fig. 4.2(c). Thus, the height of the conducting channel decreases from the source to drain.

The region between the dotted lines underneath the gate in Fig. 4.1(b) is the intrinsic transistor. It governs the main function of MESFET, i.e., the ability to control currents and to amplify signals. The remainder is the extrinsic region necessary to connect the intrinsic transistor to the outside world. The extrinsic region does not enhance the transistor performance but rather deteriorates device behavior due to its parasitic resistances and capacitances.

In the following sections, we will present in detail two physically based models to describe the dc and small signal-behavior of MESFETs. These models will also be extended to treat the high-frequency noise of MESFETs. This is followed by the analysis of the power performance of MESFETs. Finally, the design and performance issues of GaAs MESFETs and wide bandgap MESFETs are discussed.

4.2 DC ANALYSIS

Two dc MESFET models have been widely used and will be described in this section. The first one is the so-called Pucel–Haus–Statz (PHS) model [8, 9], named after the three scientists who originally developed the model. It is probably the most widely used MESFET model. The model is quite simple, allows an almost fully analytical treatment of the transistor, and can easily be extended to treat the small-signal and noise behavior of MESFETs. In spite of its popularity, however, the PHS model has some shortcomings. First, it does not take into account nonstationary transport effects such as velocity overshoot, and its validity is questionable for short-channel MESFETs. Second, it is difficult to apply the PHS model to MESFETs based on materials having velocity–field (v–$\mathscr{E}$) characteristics with pronounced nonlinearities over large ranges of the electric field, such as GaN.

To overcome these drawbacks, another dc model, called the Cappy model, was introduced [10]. It is a numerical model allowing for arbitrary v–$\mathscr{E}$ characteristics. Furthermore, it is well suited to include nonstationary electron transport and can also be extended to describe the small-signal and noise behavior of MESFETs [11].

4.2.1 The PHS Model

The electrical behavior of a MESFET is in principle very similar to that of a JFET. The main difference between the two devices is their gate electrodes: a pn junction in the JFET and a Schottky junction in the MESFET. When comparing the n-channel JFET and MESFET shown in Fig. 4.1, we see that the p^+-gate of the JFET serves the same functionality as the metal gate of the MESFET.

The operation principle of the JFET was first analyzed by W. Shockley [1]. The basic assumptions of his approach, some of which have also been used in the PHS MESFET model, are:

1. The space–charge region underneath the gate is fully depleted and the space–charge density is equal to the doping concentration, i.e., the conventional depletion approximation is applicable.
2. The electron density in the conducting channel is equal to the doping concentration; in the case of incomplete donor ionization as described in Section 2.2, the electron density is equal to the ionized donor concentration.
3. The field distribution underneath the gate can be treated as a superposition of two separate fields. The first is the longitudinal field between drain and source in the direction of current flow, and the second is the transverse field in the direction perpendicular to the channel. The longitudinal field prevails and the transverse field is negligible in the conducting channel, and vise versa in the space–charge region.
4. The channel height decreases gradually from source to drain.
5. The mobility in channel does not depend on the electric field and is equal to the low-field mobility μ_0.
6. The drain current passing through the transistor is entirely drift current, and generation/recombination in the channel is neglected.

Shockley's approach is known as the gradual channel approximation (GCA). Strictly speaking, only assumption (4) is related to the gradually changing conditions in the channel, but commonly the term GCA refers to a transistor model applying all assumptions mentioned above.

With regard to microwave transistors, the most critical issue of the GCA is assumption (5). Consider a typical low-noise GaAs MESFET with a gate length of 0.3 µm and a source-to-drain separation L_{SD} of about 1 µm. Suppose a typical drain–source voltage of 3 V is applied, then the average longitudinal field in the channel, $\mathscr{E}_{av} = V_{DS}/L_{SD}$, is in the neighborhood of 30 kV/cm. This is clearly in a region of the v–$\mathscr{E}$ characteristics where the electron drift velocity is saturated and far beyond the field range in which the mobility can be considered as field-independent (see Figs. 2.11 and 2.15). The effect of velocity saturation has long been recognized and taken into account in FET modeling. The common approach [8, 9, 12], which is used in the PHS model, is to divide the FET channel into two regions and utilize a piecewise linear v–$\mathscr{E}$ characteristics, as shown in Fig. 4.3. In region 1 of the channel, the field $\mathscr{E}$ is relatively low. Thus the GCA holds, the mobility is approximately constant, and the electron drift velocity v is given by

$$v = -\mu_0 \mathscr{E} \quad (4\text{-}1)$$

This is lower than the saturation velocity v_S. The boundary between regions 1 and 2 is defined as the point where the magnitude of the electric field reaches the critical value $\mathscr{E}_S$, at which the electron velocity is equal to the saturation velocity v_S:

$$v_S = \mu_0 \mathscr{E}_S \quad (4\text{-}2)$$

Note that $\mathscr{E}_S$ is always a positive number because it is defined as the magnitude of the electric field. Across the entire region 2 the electrons travel with their saturation velocity.

The electron density in regions 1 and 2 of the channel is assumed to be equal to the doping concentration [assumption (2) of the GCA]. As described above, the conducting channel height b in region 1 decreases toward the drain. To maintain a constant drain current in the MESFET, the electron velocity increases to the same degree. In region 2, on the other hand, the channel height must not decrease further because of the constant electron density and drift velocity (v_S).

At low V_{DS}, the electric field in the entire channel is lower than $\mathscr{E}_S$, and the GCA holds everywhere. The field increases with increasing V_{DS}, and at a certain V_{DS} the electric field at the drain end of the gate eventually reaches $\mathscr{E}_S$ and the carrier drift

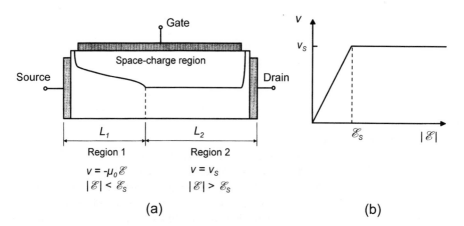

Figure 4.3. Two-region MESFET model. (a) MESFET with two channel regions. (b) Piecewise linear velocity-field characteristics.

velocity becomes v_S. This drain–source voltage is called the drain–source saturation voltage V_{DSS}. If V_{DS} is increased further, the saturation field $\mathscr{E}_S$ is reached before the drain end of the gate, and the MESFET has to be divided into region 1 and region 2 as shown in Fig. 4.3.

The equations describing the dc behavior of MESFETs are presented below. The intrinsic transistor will first be considered, and the influence of the extrinsic part will be added later. Figure 4.4 shows the intrinsic MESFET depicting the coordinate system used in the following as well as the dimensions and voltages. The lengths of regions 1 and 2 are L_1 and L_2, respectively. The drain current at any x-position in the channel is obtained by multiplying the current density with the channel area perpendicular to current flow:

$$I_D = qN_D v(x) W b(x) = qN_D v(x) W[a - d_{sc}(x)] \tag{4-3}$$

As shown in Section 2.5, the thickness of the space–charge region underneath the Schottky contact depends on the voltage drop across the space–charge region. The voltage between a position x in the channel and the gate is the sum of the built-in voltage of the Schottky junction, V_{bi}, the applied voltage between source and gate, i.e., $-V_{GS}$, and the potential difference between x and source, $V(x)$. Thus, the thickness of the space–charge region becomes

$$d_{sc} = \sqrt{\frac{2\varepsilon(-V_{GS} + V(x) + V_{bi})}{qN_D}} \tag{4-4}$$

where ε is the dielectric constant of the active layer. The voltage necessary to deplete the entire active layer, i.e., to create a space–charge region under the gate with a thickness a, assuming V_{bi} to be zero, is the pinch-off voltage V_{po}:

$$V_{po} = \frac{qN_D a^2}{2\varepsilon} \tag{4-5}$$

Introducing the normalized potential $w(x)$

$$w(x) = \sqrt{\frac{-V_{GS} + V_{bi} + V(x)}{V_{po}}} \tag{4-6}$$

and putting this into Eq. (4-4), together with Eq. (4-5) we obtain

$$d_{sc}(x) = aw(x) \tag{4-7}$$

The combination of Eqs. (4-3) and (4-7) results in the basic equation for the drain current of MESFETs:

$$I_D = qN_D v(x) Wa[1 - w(x)] \tag{4-8}$$

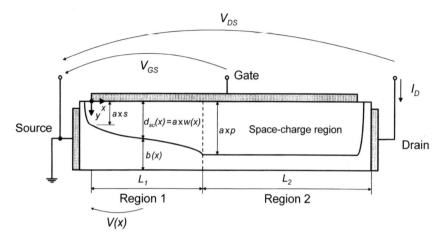

Figure 4.4. Schematic showing the details of the MESFET two-region model.

Based on Eq. (4-8), we will now derive an expression for the drain current valid in region 1. Putting $v(x)$ from Eq. (4-1) into Eq. (4-8), we get

$$I_D = -qN_D W a \mu_0 \mathcal{E}(x)[1 - w(x)] \tag{4-9}$$

Using Eq. (4-6), the electric field in the channel can be expressed as

$$\mathcal{E} = -\frac{dV(x)}{dx} = -2w(x)\frac{dw(x)}{dx}V_{po} \tag{4-10}$$

To make the expressions more compact, the x-dependence of the different quantities (i.e., the x in parantheses) is omitted in the following. Inserting Eq. (4-10) in Eq. (4-9) and rearranging the result yields the following differential equation with separated variables for the drain current:

$$I_D dx = qN_D W a \mu_0 V_{po} 2(w - w^2) dw \tag{4-11}$$

This equation has to be integrated over region 1 of the transistor: on the left-hand side from $x = 0$ to $x = L_1$, and on the right-hand side from $w = s$ to $w = p$, where s and p are the normalized potentials at the source and at the boundary between regions 1 and 2, respectively. From Eq. (4-6), we obtain for s and p:

$$s = \sqrt{\frac{-V_{GS} + V_{bi}}{V_{po}}} \tag{4-12a}$$

$$p = \sqrt{\frac{-V_{GS} + V(L_1) + V_{bi}}{V_{po}}} \tag{4-12b}$$

The result of the integration is

$$I_D = \frac{qN_D W a \mu_0 V_{po}}{L_1}\left[p^2 - s^2 - \frac{2}{3}(p^3 - s^3)\right] = \frac{qN_D W a \mu_0 V_{po}}{L_1} f_1 \quad (4\text{-}13)$$

At the boundary between regions 1 and 2, and beyond the boundary toward the drain, the electrons travel with their saturation velocity, and using Eq. (4-8) the drain current can be expressed as

$$I_D = qN_D v_S W a (1 - p) \quad (4\text{-}14)$$

Because the drain current at any position along the channel must be constant, we can equate Eq. (4-13) with Eq. (4-14) and obtain an expression for the length of region 1:

$$L_1 = \frac{\mu_0 V_{po}}{v_S} \frac{f_1}{1 - p} \quad (4\text{-}15)$$

The drain–source voltage is the sum of the voltage drops across region 1 and region 2. Using Eqs. (4-12a) and (4-12b), one obtains the voltage drop V_1 in region 1:

$$V_1 = V_{po}(p^2 - s^2) \quad (4\text{-}16)$$

Because in region 2 the carriers travel with saturation velocity and the channel height is constant, our approach would result in a zero potential drop across region 2. We can avoid this difficulty by considering a two-dimensional potential and field distribution in region 2. This is modeled by placing positive charges at the drain and their negative counterparts at the portion of the gate above region 2. A rather lengthy derivation is given in detail in [12], which leads to the following expression for the voltage drop V_2 in region 2:

$$V_2 = \frac{2a v_S}{\pi \mu_0} \sinh \frac{\pi L_2}{2a} \quad (4\text{-}17)$$

The sum of V_1 and V_2 is equal to the drain–source voltage:

$$V_{DS} = V_1 + V_2 = V_{po}(p^2 - s^2) + \frac{2a v_S}{\pi \mu_0} \sinh \frac{\pi L_2}{2a} \quad (4\text{-}18)$$

In Eq. (4-18), L_2 is equal to $L - L_1$ and can be expressed with the help of Eq. (4-15). For given voltages V_{GS} and V_{DS}, Eqs. (4-14) and (4-18) make up a system of two equations with two unknowns (I_D and p) that can be solved numerically.

When simulating the dc behavior of a MESFET under specific bias conditions, one has to determine first whether the applied drain–source voltage is above or below the drain saturation voltage V_{DSS}. At $V_{DS} = V_{DSS}$ the condition

$$L = \frac{\mu_0 V_{po}}{v_S} \frac{d^2 - s^2 - \frac{2}{3}(d^3 - s^3)}{(1-d)} \qquad (4\text{-}19)$$

holds, with d being the normalized potential at the drain:

$$d = \sqrt{\frac{-V_{GS} + V_{DSS} + V_{bi}}{V_{po}}} \qquad (4\text{-}20)$$

For $V_{DS} < V_{DSS}$, the drain current can be calculated from Eq. (4-13), provided p and L_1 be replaced by d and L, respectively. For $V_{DS} > V_{DSS}$, the drain current can be calculated from Eqs. (4-14) and (4-18).

Now that the model for the intrinsic MESFET is in place, the extrinsic components need to be included. The important extrinsic components for dc operations are the parasitic source and drain resistances, R_S and R_D. Because of these resistances, the voltage drop in the intrinsic MESFET is smaller than the external voltage applied to the MESFET, as shown in Fig. 4.5. The relations between the external voltages (subscript ext) and the internal voltages (subscript int) are given by

$$V_{DS,ext} = V_{DS,int} + I_D(R_S + R_D) \qquad (4\text{-}21)$$

and

$$V_{GS,ext} = V_{GS,int} + I_D R_S \qquad (4\text{-}22)$$

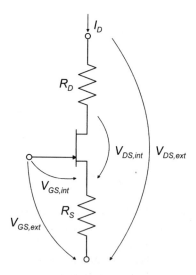

Figure 4.5. Intrinsic MESFET with the parasitic resistances R_S and R_D.

Note that in Eqs. (4-4)–(4-20) the intrinsic drain–source and gate–source voltages have to be used.

Two questions still have to be answered before the PHS model can be used to simulate MESFETs: How can R_S and R_D be estimated, and what values should be used for the low-field mobility μ_0 and for the saturation velocity v_S?

We start with the parasitic resistances and discuss the approach for the GaAs MESFET. Similary, MESFETs based on different materials can be treated as well. The best way is to use measured R_S and R_D for a given technology. If no measured data is available, these resistances have to be estimated. However, simple and reliable methods to estimate R_S and R_D are hard to find in the literature. Both R_S and R_D consist of several components: the transition resistance of the metal–semiconductor contact, the spreading resistance, and the resistance of the active layer between source/drain and the intrinsic transistor. To decrease the parasitic resistances, the active layer underneath the source and drain contacts is typically made thicker and more heavily doped than the active layer directly underneath the gate. Fukui in 1979 developed empirical formulas to estimate the parasitic resistances based on measured dc data [13]. These formulas have been widely used in MESFET modeling. It should be noted, however, that they are based on experimental results from the late 1970s.

In the following an alternative approach to estimate R_S and R_D is described. Consider the source-to-gate region of a GaAs MESFET with a recessed gate as shown in Fig. 4.6. The source resistance R_S is composed of the contact resistance R_{co}, which includes the transition resistance of the metal–semiconductor contact and the spreading resistance, and of the resistances of the different parts of the epitaxial layer structure R_{S11}, R_{S12}, and R_{S2} (see Fig. 4.6). The uncovered GaAs surface is depleted due to the effect of negatively charged surface states. In analogy to the space–charge region of a Schottky junction, the thickness of the surface depletion region, $d_{sc,surf}$, can be modeled by

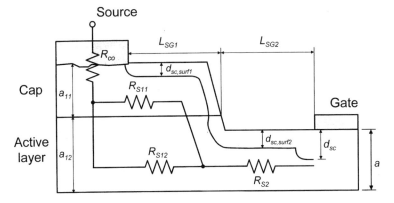

Figure 4.6. Schematic showing the various resistance components in the source–gate region of an epitaxial recessed-gate GaAs MESFET.

$$d_{sc,surf} = \sqrt{\frac{2\varepsilon V_{surf}}{qN_D}} \qquad (4\text{-}23)$$

where V_{surf} is the potential describing the effect of the surface states. For GaAs, V_{surf} is around 0.5 to 0.7 V. From their corresponding geometries, the resistances R_{S11}, R_{S12}, and R_{S2} are given by

$$R_{S11} = \frac{1}{q\mu_{0,cap}N_{D,cap}} \frac{L_{SG1}}{(a_{11} - d_{sc,surf1})W}$$

$$R_{S12} = \frac{1}{q\mu_0 N_D} \frac{L_{SG1}}{a_{12}W} \qquad (4\text{-}24)$$

$$R_{S2} = \frac{1}{q\mu_0 N_D} \frac{L_{SG2}}{(a - d_{sc,surf2})W}$$

L_{SG1}, L_{SG2}, a_{11}, a_{12}, and a are the lengths and thicknesses of the different parts of the layer structure as shown in Fig. 4.6, $N_{D,cap}$ and N_D are the doping concentrations of the cap and the active layer, $d_{sc,surf1}$ and $d_{sc,surf2}$ are the widths of the surface depletion regions for the cap and active layer, and $\mu_{0,cap}$ and μ_0 are the low-field mobilities in the cap and active layer described by the mobility model given in Section 2.3. The total source resistance is then given by

$$R_S = R_{co} + \frac{R_{S11}R_{S12}}{R_{S11} + R_{S12}} + R_{S2} \qquad (4\text{-}25)$$

A similar expression can be developed for R_D. Figure 4.7 shows a compilation of the measured source resistance R_S of reported experimental GaAs MESFETs as a function of the gate length. The solid line in Fig. 4.7 represents the source resistance calculated from Eqs. (4-23)–(4-25) based on the following conditions: (a) the gate–source separation is equal to the gate length and $L_{SG1} = L_{SG2} = L_{SG}/2$, (b) the thickness of the active layer underneath the gate is scaled in a way that for any transistor a constant $L/a = 3$ ratio is maintained, (c) the doping density in the active layer is chosen so that the pinch-off voltage is a linear function of the gate length (i.e., $V_{po,L=0.1\mu m} = 2$ V, $V_{po,L=1\mu m} = 3$ V), and (d) the value of the contact resistance R_{co} is 0.1 Ωmm, which is typical for the state of the art GaAs technology. Clearly, the source resistance model in Eqs. (4-23)–(4-25) gives reasonable predictions. Note that R_S in Fig. 4.7 is given in Ωmm, and to obtain R_S in Ω the values need to be divided by the gate width W (in mm).

The choice of proper values for μ_0 and v_S is another critical issue, as μ_0 and v_S are two important parameters describing the piecewise linear v–$\mathscr{E}$ characteristics used in the PHS model. It should be noted that the two-piece linear v–$\mathscr{E}$ characteristics used in the PHS model can only coarsely reproduce the real v–$\mathscr{E}$ characteristics. For GaAs and other III–V materials, the use of either measured low-field mobility or a doping dependent low-field mobility calculated using the mobility model de-

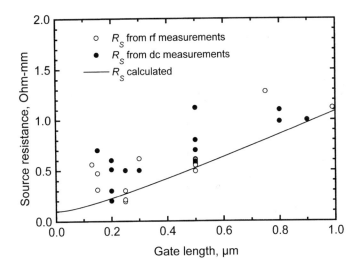

Figure 4.7. Measured and simulated parasitic source resistance versus the gate length of GaAs MESFETs.

scribed in Section 2.3 is recommended. In addition, a saturation velocity v_S between the peak velocity v_p and the saturation velocity v_{sat} obtained from the stationary v–$\mathscr{E}$ characteristics at high fields. A reasonable approximation for v_S is

$$v_S \approx \frac{v_p + v_{sat}}{2} \tag{4-26}$$

Figure 4.8 shows the dc current–voltage characteristics obtained from the PHS model for a 1-μm gate GaAs MESFET. The transistor dimensions, the values for R_S, R_D, and μ_0, as well as the measured I–V data (symbols) were taken from [9], and the built-in voltage is set to 0.8 V. Two different regions of operation can be clearly distinguished. At low drain–source voltages ($V_{DS} < V_{DSS}$), the drain current follows a nearly linear V_{DS} dependence (in fact it is a $V_{DS}^{3/2}$ dependence), while at large V_{DS} ($V_{DS} > V_{DSS}$), the drain current is almost independent of the drain–source voltage. The former region is called the active region, and latter is the saturation region.

To demonstrate the influence of velocity saturation, Fig. 4.9 compares the saturation drain currents (I_D at $V_{DS} = 3.5$ V and $V_{GS} = 0$ V), calculated from the GCA and PHS models for GaAs MESFETs having different gate lengths. Evidently, the GCA model considerably overestimates the current. This is due to the fact that the effect of velocity saturation is neglected in the GCA model.

To model the v–$\mathscr{E}$ characteristics of wide-bandgap semiconductors, such as SiC, where velocity saturation is achieved only at much higher fields compared to GaAs, the piecewise linear v–$\mathscr{E}$ characteristics need to be modified. The linear v–$\mathscr{E}$ dependence in region 1 of the MESFET has to be changed to

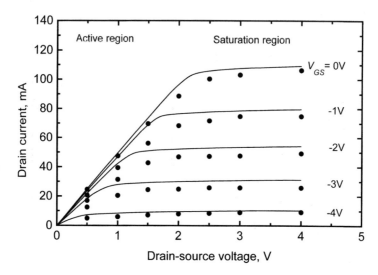

Figure 4.8. Calculated (full lines) and measured (closed circles) output characteristics of a GaAs MESFET with $L = 1$ μm, $W = 500$ μm, $a = 0.34$ μm, $N_D = 6.5 \times 10^{16}$ cm^{-3}, $\mu_0 = 4500$ cm^2/Vs, $R_S = 6.5$ Ω, $R_D = 11.3$ Ω. Transistor dimensions and experimental data points taken from [9].

$$v = -\frac{\mu_0 \mathcal{E}}{1 - \dfrac{\mu_0 \mathcal{E}}{\gamma v_S}} \tag{4-27}$$

Here, the factor γ accounts for the bowing of the v–$\mathcal{E}$ characteristics and allows the modeled characteristics be fitted to the real data. When γ approaching infinity, the denominator in Eq. (4-27) becomes unity and, according to Eq. (4-1), the velocity is a linear function with respect to electric field. For $\gamma \leq 1$, at the other extreme, the velocity asymptotically approaches v_S but true saturation is never reached. Thus, Eq. (4-27) allows for a smooth transition between $\mu = \text{const.} = \mu_0$ and $\mu = \mu(\mathcal{E})$ at $\mathcal{E} = \mathcal{E}_S$.

Putting the modified v–$\mathcal{E}$ characteristics in Eq. (4-27) into Eq. (4-13), the following expression for the drain current valid in region 1 is obtained

$$I_D = \frac{q N_D W a \mu_0 V_{po}}{L_1 + \dfrac{V_{po}\mu_0}{\gamma v_S}(p^2 - s^2)} f_1 \tag{4-28}$$

4.2.2 The Cappy Model

The Cappy model allows to take into account nonstationary carrier transport and to consider v–$\mathcal{E}$ characteristics of arbitrary shape. Thus it overcomes two major draw-

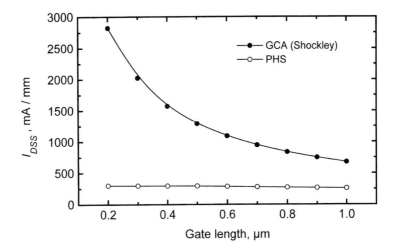

Figure 4.9. Saturation drain current versus gate length calculated using the GCA model (closed circles) and PHS model (open circles).

backs of the PHS model. This is achieved, however, at the expense of increased model complexity.

In the framework of the Cappy model, the MESFET structure is divided (discretized) into small channel segments, each with a length Δx, in the direction of current flow. The bias conditions are given by applying a gate–source voltage V_{GS} and feeding a certain drain current I_D into the transistor. The sum of the voltage drops across the channel segments caused by I_D is the drain–source voltage for the given bias conditions. This is different from the PHS model where the bias conditions were defined by the applied voltages V_{DS} and V_{GS}. Figure 4.10 illustrates the model MESFET. The basic assumptions of the Cappy model are:

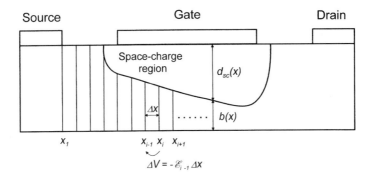

Figure 4.10. Schematic of the model MESFET used in the Cappy model.

1. Assumptions (1), (3), (4), and (6) in the gradual channel approximation (see Section 4.2.1)
2. Carrier transport is described by the relaxation time approximation (see Section 2.3).

It should be noted that the electron density in the channel, n, is not required to be constant as in the GCA and PHS models, but rather it can deviate from the donor concentration N_D. For the simulation of MESFETs based on the Cappy model, the knowledge of stationary v–$\mathscr{E}$ (velocity–field), E–$\mathscr{E}$ (energy–field), and m^*–$\mathscr{E}$ (conductivity effective mass–field) characteristics of the semiconductor of interest are required. For GaAs, these characteristics are given in Fig. 2.15. The following four equations form the basis of the Cappy model:

$$\frac{dE}{dx} = -q\mathscr{E} - \frac{E - E_0}{v\tau_E(E)} \qquad \text{energy balance equation} \qquad (4\text{-}29)$$

$$v\frac{d[m^*(E)v]}{dx} = -q\mathscr{E} - \frac{m^*(E)v}{\tau_P(E)} \qquad \text{momentum balance equation} \qquad (4\text{-}30)$$

$$\frac{d\mathscr{E}}{dx} = \frac{q}{\varepsilon}(N_D - n) \qquad \text{Poisson equation} \qquad (4\text{-}31)$$

$$I_D = qnvbW \qquad \text{drift current equation} \qquad (4\text{-}32)$$

where τ_E and τ_P are the energy and momentum relaxation times, respectively, m^* is the conductivity effective mass, and E_0 is the carrier energy at thermal equilibrium. Equations (4-29) and (4-30) are obtained from Eqs. (2-77) and (2-78) using $v = dx/dt$. Note that $\mathscr{E}$ is negative in n-channel MESFETs. This results in a positive value of the term $-q\mathscr{E}$ in Eqs. (4-29) and (4-30). It is possible to discretize the above equations in a way that the quantities at a position x_i can be determined only from the known quantities at the previous point x_{i-1}. Thus iterations are avoided, and the simulation reduces to a simple recursion. Because the discretization is not straightforward and is discussed in the original papers only briefly, the details of the approach are given below.

First the bias conditions are defined chosing V_{GS} and I_D. From Eq. (4-32), the velocity at source, i.e., at x_1, is obtained by setting $n_1 = N_D$ and $b_1 = a$. From the stationary v–$\mathscr{E}$ and E–$\mathscr{E}$ characteristics, the corresponding stationary values $\mathscr{E}_{st,1} = \mathscr{E}_1$ and $E_{st,1} = E_1$ can be deduced. Then, starting at x_2 and proceeding sequentially towards the drain, in all segments the following steps have to be run through subsequently.

1. Calculate the energy relaxation time τ_{Ei} according to Eq. (2-79) as

$$\tau_{Ei} = \frac{E_{st,i-1} - E_0}{q\mathscr{E}_{st,i-1}v_{st,i-1}} \qquad (4\text{-}33)$$

2. Calculate the electron energy at position x_i from the discretized energy balance equation given by

$$E_i = E_{i-1} + \Delta x \left(-q\mathcal{E}_{i-1} - \frac{E_{i-1} - E_0}{v_{i-1}\tau_{Ei}} \right) \tag{4-34}$$

3. Set $E_{st,i} = E_i$ from Eq. (4-34) and deduce the new stationary values $\mathcal{E}_{st,i}$, $v_{st,i}$, and $m^*_{st,i}$ from the stationary v–$\mathcal{E}$, E–$\mathcal{E}$, and m^*–$\mathcal{E}$ characteristics in Fig. 2.15.

4. Calculate the momentum relaxation time τ_{Pi} according to Eq. (2-80) as

$$\tau_{Pi} = \frac{m^*_{st,i} v_{st,i}}{q\mathcal{E}_{st,i}} \tag{4-35}$$

5. Calculate the electron velocity at position x_i, v_i, using the rearranged momentum balance equation. To this end, we need expressions for the electric field $\mathcal{E}_i$ and the electron density n_i at x_i, which can be obtained from the discretized Poisson equation in Eq. (4-31):

$$\mathcal{E}_i = \mathcal{E}_{i-1} + \Delta x \frac{q}{\varepsilon}(N_D - n_i) \tag{4-36}$$

and from the rearranged current equation

$$n_i = \frac{I_D}{qv_i b_i W} \tag{4-37}$$

The discretized momentum balance equation is given by

$$v_i = -\frac{q\tau_{P,i}}{m^*_{st,i}}\mathcal{E}_i - v_{i-1}\tau_{P,i}\frac{v_i - v_{i-1}}{\Delta x} \tag{4-38}$$

Combining Eqs. (4-36)–(4-38) leads to a quadratic equation for the carrier velocity v_i at x_i:

$$D_1 v_i^2 + D_2 v_i + D_3 = 0 \tag{4-39}$$

with

$$D_1 = 1 + \frac{v_{i-1}\tau_{P,i}}{\Delta x}$$

$$D_2 = -\frac{q\tau_{P,i}}{m^*_{st,i}}\left(-\mathcal{E}_{i-1} - \frac{\Delta x q}{\varepsilon} N_D + \frac{m^*_{st,i}}{q}\frac{v_{i-1}^2}{\Delta x} \right) \tag{4-40}$$

$$D_3 = -\frac{q\tau_{\text{P},i}\Delta x I_\text{D}}{\varepsilon m^*_{\text{st},i} b_i W}$$

6. Calculate the electron density n_i from Eq. (4-37) and the electric field $\mathscr{E}_i$ using Eq. (4-36).
7. Move to postion x_{i+1} and proceed with step 1.

It is not clear at this point as to how the channel height b_i at x_i is determined. If charged surface states are neglected, the channel height near the source and drain is equal to the thickness of the active layer. Directly underneath the gate and in the gate fringing regions (these are the parts of the space–charge region extending beyond the gate edges towards the source and drain, see Fig. 4.10), the channel height b_i is $a - d_{\text{sc},i}$, with $d_{\text{sc},i}$ being the x-dependent thickness of the space–charge region at position x_i given by

$$d_{\text{sc},i} = d_{\text{sc},i-1} + \frac{\varepsilon|\mathscr{E}_{i-1}|\Delta x}{qd_{\text{sc},i-1}n_{i-1}} \tag{4-41}$$

According to the results of two-dimensional simulations [14], the shape of the space–charge region edge in the fringing regions can be described by ellipses.

A simplified version of the Cappy model neglecting nonstationary carrier transport can be developed using only Eqs. (4-36), (4-37), and (4-41). After calculating the electric field at x_i using Eq. (4-36) using the electron density from the previous point, i.e., n_{i-1}, the velocity v_i is deduced from the stationary v–$\mathscr{E}$ characteristics. Then the electron density is obtained from Eq. (4-37).

Figs. 4.11(a) and (b) show the I–V characteristics simulated using the Cappy model for 0.25 μm gate and 1 μm gate GaAs MESFETs, respectively. The influence of nonstationary carrier transport is evident from the difference between the solid lines (complete Cappy model) and the dotted lines (simplified Cappy model). Obviously, the effects of nonstationary carrier transport become more prominent when the gate length is decreased. Interestingly, the currents calculated neglecting and including nonstationary carrier transport are the same at low drain–source voltages. This suggests that at low voltages, i.e., low fields, nonstationary transport plays a negligible role.

Comparing Figs. 4.8 and 4.11(b), we see that the PHS and Cappy models result in similar descriptions for the dc behavior of GaAs MESFETs. The PHS model is certainly easier to be implemented into a computer program. Nonetheless, it has the drawbacks of employing the less accurate piecewise linear v–$\mathscr{E}$ characteristics and excluding nonstationary carrier transport.

4.3 SMALL-SIGNAL ANALYSIS

4.3.1 Small-Signal Equivalent Circuit

The small-signal condition for a transistor can be described as follows. A dc operating point is specified by a dc bias condition (i.e., V_{DS} and V_{GS}), and small-signals

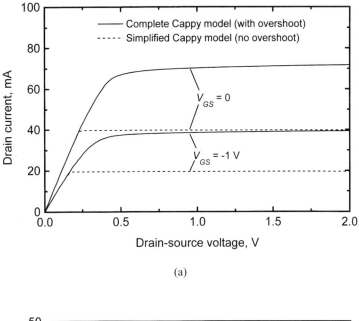

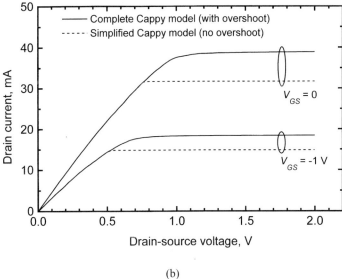

Figure 4.11. Simulated output characteristics using the complete and the simplified Cappy model for MESFETs with gate lengths of (a) 0.25 μm and (b) 1.0 μm. The following parameters have been used in the calculations: $V_{bi} = 0.8$ V, $a = 0.15$ μm, $N_D = 2 \times 10^{17}$ cm^{-3}, $L_{SG} = L$, $L_{GD} = 2 \times L$, $W = 100$ μm, and $R_{co} = 0$.

(ac signals) are superimposed to such a dc condition. For example, the ac signal can be a sinusoidal voltage v_{GS} applied to the gate and to be amplified by the transistor. The term small-signal means that the magnitude of the signal voltage is on the order of $k_B T/q$. Under such a condition, the nonlinear characteristics of semiconductor devices can be linearized. This linearization is used as a main tool to describe the small-signal behavior of microwave transistors.

Under the small-signal condition, a field-effect transistor can be described by a small-signal equivalent circuit with lumped elements, as shown in Fig. 4.12. In principle, this equivalent circuit is applicable for MESFETs, HEMTs, and MOSFETs. Each of the equivalent circuit elements is related to a physical effect in the transistor. We will first focus on the intrinsic MESFET inside the dashed lines in Fig. 4.12.

The drain current of a MESFET is dependent of both the gate–source and the drain–source voltages. The transconductance g_m describes the change of the drain current with respect to small variations of the gate–source voltage V_{GS} when V_{DS} is fixed:

$$g_m = \left. \frac{dI_D}{dV_{GS}} \right|_{V_{DS}=\text{const.}} \tag{4-42}$$

At very high frequencies, the drain current cannot follow immediately the gate–source voltage variation. The effect of this delay is commonly described by a frequency-dependent transconductance y_m given by

$$y_m = g_m \exp(-j\omega\tau) \approx \frac{g_m}{1+j\omega\tau} \approx g_m(1-j\omega\tau) \tag{4-43}$$

where $\omega = 2\pi f$ is the angular frequency, τ is a time constant, f is the operating fre-

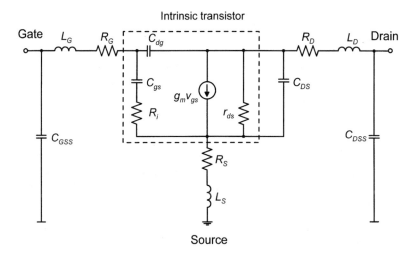

Figure 4.12. Small-signal equivalent circuit of a MESFET.

4.3 SMALL-SIGNAL ANALYSIS

quency, and j is the imaginary unit. τ is on the order of 1 ps and can be related to the carrier transit time, i.e., the time carriers needed to travel across the channel from the source end of the gate to the drain end of the gate, but no reliable physically based expression exists for this parameter.

The variation of the drain current caused by a variation of V_{DS} at fixed V_{GS} is modeled by the drain conductance g_{ds} given by

$$g_{ds} = \frac{dI_D}{dV_{DS}}\bigg|_{V_{GS}=\text{const.}} \quad (4\text{-}44)$$

The corresponding drain resistance r_{ds} is $1/g_{ds}$.

As discussed in Section 4.1, the thickness of the space–charge region, and thus the depletion charge stored in this region, varies with the applied voltages. Because under any bias conditions, charge neutrality must exist in the transistor, a change in the charge stored in the space–charge region causes a change in the gate charge of same magnitude but opposite sign. In the small-signal equivalent circuit, the gate–source capacitance C_{gs} and the gate–drain capacitance C_{gd} describe this charge modulation and are defined as

$$C_{gs} = \frac{dQ_{\text{gate}}}{dV_{GS}}\bigg|_{V_{DS}=\text{const.}} = -\frac{dQ_{sc}}{dV_{GS}}\bigg|_{V_{DS}=\text{const.}} \quad (4\text{-}45)$$

$$C_{gd} = \frac{dQ_{\text{gate}}}{dV_{DS}}\bigg|_{V_{GS}=\text{const.}} = -\frac{dQ_{sc}}{dV_{DS}}\bigg|_{V_{GS}=\text{const.}} \quad (4\text{-}46)$$

where Q_{gate} and Q_{sc} are the charges at the gate and in the space–charge region, respectively.

Note that the equivalent circuit elements given in Eqs. (4-42) to (4-46) are for the intrinsic MESFET, and therefore the intrinsic drain–source and gate–source voltages have to be used for calculating these elements.

The only remaining element of the intrinsic MESFET to be dealt with is the resistance R_i. It acts as a charging resistance for the gate–source capacitance. Even though the physical effect of this resistance is clear, there is no commonly accepted model for R_i. An estimate of R_i is possible, however, in the following way. The time τ_T needed for electrons to travel from the source end to the drain end of the gate can be calculated by

$$\tau_T = \int_0^L \frac{1}{v(x)} dx \quad (4\text{-}47)$$

The time to charge or discharge C_{gs} is certainly related to τ_T. Thus, R_i can be estimated by

$$R_i = \gamma \frac{\tau_T}{C_{gs}} \quad (4\text{-}48)$$

where γ is an empirical factor with a typical value of 5.

Next we turn to the extrinsic portion of the small-signal equivalent circuit. The parasitic source and drain resistances R_S and R_D have already been discussed in section 4.2, but the gate resistance R_G requires explanation. When the voltages at the MESFET terminals vary, Q_{gate} and Q_{sc} vary as well. In other words, charges enter or leave the gate electrode under small-signal conditions, and the gate resistance R_G hinders such a charge transport.

The capacitances C_{DS}, C_{DSS}, and C_{GSS} are external parasitic capacitances (stray capacitances between the electrodes and pad capacitances). In principle it would be possible to account for the effects of both C_{DS} and C_{DSS} using a single parasitic capacitance. To be in accordance with the commonly used equivalent circuits in the literature, however, we keep both C_{DS} and C_{DSS} in our circuit. Finally, L_S, L_D, and L_G are the bond and lead inductances.

4.3.2 Modeling the Equivalent Circuit Elements Based on the PHS Model

Analytical expressions for g_m, r_{ds}, C_{gs}, and C_{gd}, can be derived from the PHS dc model discussed earlier. We consider the more important case that the MESFET is operated in the saturation region, i.e., $V_{DS} > V_{DSS}$. The expressions of the equivalent circuit elements for the active region can be derived in a similar manner. The transconductance g_m given in Eq. (4-42) can be rewritten using the chain rule

$$g_m = \frac{dI_D}{dp} \times \frac{dp}{ds} \times \frac{ds}{dV_{GS}}\bigg|_{V_{DS}=\text{const.}} \quad (4\text{-}49)$$

The first term on the right-hand side, dI_D/dp, is obtained from Eq. (4-14):

$$\frac{dI_D}{dp} = -qN_DWav_S \quad (4\text{-}50)$$

The second term, dp/ds, is found by taking the derivative of Eq. (4-18) with respect to s. We use $L_2 = L - L_1$ with L_1 from Eq. (4-15), fix V_{DS} ($dV_{DS}/ds = 0$), and get

$$\frac{dp}{ds} = \frac{\left(\cosh\frac{\pi L_2}{2a}\right)2s(1-s) - 2s(1-p)}{\left(\cosh\frac{\pi L_2}{2a}\right)\left(2p(1-p) + \frac{f_1}{1-p}\right) - 2p(1-p)} \quad (4\text{-}51)$$

Note that f_1 is defined in Eq. (4-13). The third term on the right-hand side of (4-49) can be deduced from Eq. (4-12a) and is given by

$$\frac{ds}{dV_{GS}} = -\frac{1}{2sV_{po}} \quad (4\text{-}52)$$

Combining Eqs. (4-49)–(4-52) we finally arrive at

$$g_m = \frac{qN_D W a v_S}{V_{po}} f_g \tag{4-53}$$

with

$$f_g = \frac{\left(\cosh\frac{\pi L_2}{2a}\right)(1-s) - (1-p)}{\left(\cosh\frac{\pi L_2}{2a}\right)\left(2p(1-p) + \frac{f_1}{1-p}\right) - 2p(1-p)} \tag{4-54}$$

Equations (4-53) and (4-54) describe the intrinsic transconductance. The relationship between the intrinsic transconductance and the external transconductance $g_{m,ext}$, which can be easily measured at the transistor terminals, is

$$g_{m,ext} = \frac{g_m}{1 + g_m R_S} \tag{4-55}$$

In a similar manner, the drain resistance r_{ds} can be derived as

$$r_{ds} = \frac{1}{g_{ds}} = \frac{V_{po}}{qN_D W a v_S} \left\{ \left(\cosh\frac{\pi L_2}{2a}\right)\left[2p + \frac{f_1}{(1-p)^2}\right] - 2p \right\} \tag{4-56}$$

To model the two capacitances C_{gs} and C_{gd}, we first have to find an expression for the gate charge, or the corresponding charge stored in the space–charge region. The charge Q_{sc1} in the space–charge region in region 1 of the MESFET is given by

$$Q_{sc1} = -Q_{gate,1} = qN_D W \int_0^{L_1} d_{sc}(x)dx = qN_D W a \int_0^{L_1} w(x)dx \tag{4-57}$$

Combining Eqs. (4-11) and (4-13) and carrying out the integration yield

$$Q_{sc1} = -Q_{gate,1} = qN_D W a L_1 \frac{f_2}{f_1} \tag{4-58}$$

with

$$f_2 = \left[\frac{2}{3}(p^3 - s^3) - \frac{1}{2}(p^4 - s^4)\right] \tag{4-59}$$

The charge in the space–charge region in region 2 is

$$Q_{sc2} = -Q_{gate,2} = qN_D W a L_2 p \tag{4-60}$$

In the earlier discussion of the voltage drop across region 2 (see Section 4.2.1), a third gate charge component was introduced, which is given by [9]

$$Q_{\text{gate},3} = -qN_DWa\frac{a^2v_S}{\pi\mu_0 V_{po}}\left[\cosh\left(\frac{\pi L_2}{2a}\right) - 1\right] \quad (4\text{-}61)$$

The last charge component to be considered is the charge stored in the fringing portions of the space–charge region near the gate edges, which can be approximated by

$$Q_{\text{fringe}} = 2 \times \frac{0.5\pi a^2 s^2}{4} \times qN_DW \quad (4\text{-}62)$$

The first factor in Eq. (4-62), i.e., 2, takes into account the fact that there are two fringing regions: one at the source side and the other at the drain side. The second factor is the area of the fringing regions, which is approximated by a quarter of an ellipse. The ratio of the axes of the ellipse is taken from the results of [14].

The total gate charge is the sum of the four charge components given in Eqs. (4-57)–(4-62). The gate–source capacitance is the sum of the derivatives of these charge components with respect to the gate–source voltage:

$$C_{gs} = C_{gs1} + C_{gs2} + C_{gs3} + C_{\text{fringe}} \quad (4\text{-}63)$$

with

$$C_{gs1} = \frac{qN_D\mu_0 aW}{v_S}\left[f_g\frac{2p^2(1-p)^2 + f_2}{(1-p)^2} - \frac{s(1-s)}{1-p}\right] \quad (4\text{-}63\text{a})$$

$$C_{gs2} = \frac{qaWN_D}{V_{po}}\left[L_2 f_g + V_{po}\frac{v_S}{\mu_0}\cosh\left(\frac{\pi L_2}{2a}\right)p(1 - 2pf_g)\right] \quad (4\text{-}63\text{b})$$

$$C_{gs3} = 2\varepsilon W\left(\frac{1}{2} - pf_g\right)\tanh\left(\frac{\pi L_2}{2a}\right) \quad (4\text{-}63\text{c})$$

and

$$C_{\text{fringe}} = \frac{\pi}{2}\varepsilon W \quad (4\text{-}63\text{d})$$

In a similar manner, the gate–drain capacitance C_{gd} can be modeled, and the expression for C_{gd} (without taking into account the fringing capacitance) is

$$C_{gd} = \varepsilon W\frac{2}{af_3}\left[\frac{V_{po}\mu_0}{v_S}\frac{f_2 - f_1 p}{(1-p)^2} + L_2 + \frac{a(f_3 - 2p)}{2}\tanh\left(\frac{\pi L_2}{2a}\right)\right] \quad (4\text{-}64)$$

where

$$f_3 = 2p - \left[2p + \frac{f_1}{(1-p)^2}\right]\cosh\left(\frac{\pi L_2}{2a}\right) \qquad (4\text{-}65)$$

An expression for the carrier transit time τ_T necessary for estimating R_i is obtained by carrying out the integral in (4-47) leading to

$$\tau_T = \frac{L_1}{v_S f_1(1-p)}\left[p^2 - s^2 - \frac{4}{3}(p^3 - s^3) + \frac{1}{2}(p^4 - s^4)\right] + \frac{L_2}{v_S} \qquad (4\text{-}66)$$

Now having determined the elements of the intrisic transistor, we turn to the elements of the part of the equivalent circuit are treated. The parasitic resistances R_S and R_D are modeled using Eq. (4-25). To estimate R_G, Fukui proposed the following empirical expression for MESFETs with a rectangular-shaped gate [13]:

$$R_G = \frac{17 W_f^2}{WhL} \qquad (4\text{-}67)$$

where h and L are the gate height and length (in µm), and W and W_f are the gate width and the gate finger width (in mm). Multiple gate fingers are commonly used in MESFETs, the details of which will be discussed later in Section 4.6.

Finding simple models to estimate the extrinsic capacitances and inductances is problematic. The approach given in [9] allows at least a rough estimate of these capacitances. Concerning the inductances, we recommend using typical values published in the literature (see Table 4.1 in Section 4.3.4). These values are extracted from the measured frequency-dependent S parameters of MESFETs.

All elements of the intrinsic equivalent circuit are bias dependent. As an example, Fig. 4.13 shows the transconductance and the gate–source capacitance of a 0.2 µm gate GaAs MESFET calculated using the small-signal models presented above. It can be seen that both g_m and C_{gs} increase with increasing gate–source voltage.

4.3.3 Modeling the Equivalent Circuit Elements Based on the Cappy Model

Analytic expressions for the equivalent circuit elements cannot be derived from the Cappy model. It is possible, however, to determine the values of elements numerically as follows. First we have to simulate the transistor for the bias condition of interest. We call this condition the operating point 1, which is characterized by applied the gate–source voltage V_{GS1}, the drain current I_{D1} fed into the transistor, and the resulting drain–source voltage V_{DS1}. For this operating point, the individual gate charge component of a channel segment i can be expressed by

$$Q_{\text{gate},i} = -Q_{\text{sc},i} = -qW d_{\text{sc},i} \Delta x N_D \qquad (4\text{-}68)$$

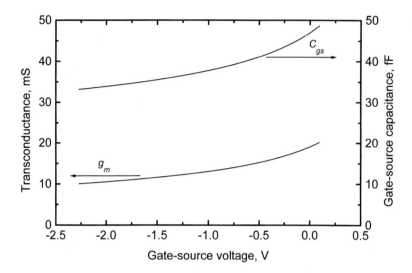

Figure 4.13. Transconductance and gate–source capacitance calculated using the PHS model for a 0.2 μm gate GaAs MESFET using the PHS model with $V_{bi} = 0.8$ V, $a = 0.15$ μm, $N_D = 2 \times 10^{17}$ cm^{-3}, $L_{SG} = L$, $L_{GD} = 2 \times L$, $W = 100$ μm, and $R_{co} = 0.1$ Ωmm.

The total gate charge for operating point 1, Q_{gate1}, is obtained by adding up $Q_{gate,i}$ in all the segments in the channel from source to drain.

The transistor is then simulated for operating point 2, which is characterized by $V_{GS2} = V_{GS1} + \Delta V_{GS}$, I_{D2}, and $V_{DS2} = V_{DS1}$ (i.e., fixed V_{DS} but a variable V_{GS} added leading to a change of I_D), and the total gate charge Q_{gate2} is calculated. Following this, a third operating point with bias conditions $V_{GS3} = V_{GS1}$, I_{D3}, and $V_{DS3} = V_{DS1} + \Delta V_{DS}$ (i.e., fixed V_{GS} but a variable V_{DS} added leading to a change of I_D) is used and the total gate charge Q_{gate3} is determined. Finally, following the definitions given in Eqs. (4-42)–(4-46), the equivalent circuit elements can be obtained as follows. The transconductance is calculated by

$$g_m = \frac{I_{D2} - I_{D1}}{V_{GS2} - V_{GS1}} \qquad (4\text{-}69)$$

and for the drain resistance we get

$$r_{ds} = \frac{1}{g_{ds}} = \frac{V_{DS3} - V_{DS1}}{I_{D3} - I_{D1}} \qquad (4\text{-}70)$$

Similarly, expressions for C_{gs} and C_{gd} according to Eqs. (4-45) and (4-46) can be found. The electron transit time τ_T is the sum of the individual times τ_i the carriers needed to travel across the individual channel segments from the source end to the drain end of the gate, taking into account the fringing regions. The transit time τ_i in each segment is

$$\tau_i = \frac{\Delta x}{v_i} \qquad (4\text{-}71)$$

It should be pointed out that in the Cappy model the voltages V_{DS} and V_{GS} are the external voltages because the effect of the parasitic resistances R_S and R_D (except for the contact resistance R_{co}) is automatically taken into account. Thus, the circuit elements calculated as described above include the effects of R_S and R_D as well. If the intrinsic circuit elements are to be calculated, first the bias conditions have to be chosen in a way that the intrinsic drain–source voltage $V_{DS2,int}$ and the intrinsic gate–source voltage $V_{GS3,int}$ are equal to the intrinsic voltages $V_{DS1,int}$ and $V_{GS1,int}$, respectively. Second, in Eqs. (4-69) and (4-70), and in the corresponding expressions for C_{gs} and C_{gd}, the intrinsic voltages have to be used.

The parasitic resistances R_S and R_D can be calculated using

$$R_S = R_{co} + \frac{V_{GS,ext} - V_{GS,int}}{I_D} \qquad (4\text{-}72)$$

and

$$R_D = R_{co} + \frac{V_{GD,ext} - V_{GD,int}}{I_D} \qquad (4\text{-}73)$$

where $V_{GD,ext}$ and $V_{GD,int}$ are the external and internal gate–drain voltages, respectively.

4.3.4 Small-Signal Parameters and Gains

To describe the frequency-dependent small-signal power and current gains of the MESFET, the small-signal parameters of the equivalent circuit have to be determined. This is commonly done using Y parameters, as defined in Appendix 4. The following Y parameters describe the intrinsic MESFET (i.e., all extrinsic elements are assumed to be zero):

$$y_{11} = \frac{\omega^2 C_{gs}^2 R_i}{1 + \omega^2 R_i^2 C_{gs}^2} + j\omega \left(\frac{C_{gs}}{1 + \omega^2 R_i^2 C_{gs}^2} + C_{gd} \right) \qquad (4\text{-}74a)$$

$$y_{12} = -j\omega C_{gd} \qquad (4\text{-}74b)$$

$$y_{21} = \frac{g_m}{1 + \omega^2 \tau^2} - j\omega \left(\frac{g_m \tau}{1 + \omega^2 \tau^2} + C_{gd} \right) \qquad (4\text{-}74c)$$

$$y_{22} = g_{ds} + j\omega C_{gd} \qquad (4\text{-}74d)$$

Based on these intrinsic Y parameters and using the two-port network theory, the complete Y parameters of the entire transistor are obtained by successively hooking up the extrinsic elements and calculating the corresponding Y parameters. This can be done analytically, but the resulting expressions for the complete Y parameters of the transistor are so complicated that a computer program has to be used to calculate them. Then, from the complete Y parameters, the different small-signal current and power gains (h_{21}, MSG, MAG, and U) as well as the characteristic frequencies f_T and f_{max} can be found using the definitions given in Section 1.2.

Approximated Y parameters are frequently utilized to derive simple and analytical expressions for the small-signal current and power gains, and the characteristic frequencies f_T and f_{max}. The major benefit of such expressions is that they provide insights into the effects of the circuit elements on these figures of merit. This advantage, however, comes at the expense of accuracy. Many different approximations for the gains and the characteristic frequencies of MESFETs have been reported in the literature. After careful evaluation, we recommend a set of approximated expressions based mainly on the work of Tasker and Hughes [15] for the current gain and the cutoff frequency, and of Ohkawa et al. [16] and Wolf [17] for the power gains and the maximum frequency of oscillation. The basic assumptions involved are (a) the expression $\omega^2 C_{gs}^2 R_i^2$ in the denominator of Eq. (4-74a) is much smaller than unity and thus is negligible, (b) the influence of the frequency dependence of g_m described by the time constant τ in Eq. (4-43) can be neglected (i.e., $\tau = 0$), and (c) the only extrinsic elements important and taken into account are R_S and R_G. Because the derivation is lengthy we give only the final results below.

The small-signal current gain can be approximated by

$$|h_{21}| = \left|\frac{i_2}{i_1}\right| = \left|\frac{y_{21}}{y_{11}}\right| \approx \frac{g_m}{2\pi f} \frac{1}{(C_{gs}+C_{gd})(1+g_{ds}R_S) + C_{gd}g_m R_S} \quad (4\text{-}75)$$

Frequently, the gains are not given as dimensionless ratios, but in decibels (dB). The current gain in dB is

$$|h_{21}|[dB] = 20 \log_{10}|h_{21}| \quad (4\text{-}76)$$

The cutoff frequency f_T is obtained by setting $|h_{21}|$ equal to unity:

$$f_T \approx \frac{g_m}{2\pi} \frac{1}{(C_{gs}+C_{gd})(1+g_{ds}R_S) + C_{gd}g_m R_S} \quad (4\text{-}77)$$

Frequently, the cutoff frequency of MESFETs (and of HEMTs and MOSFETs as well) is approximated by either

$$f_T \approx \frac{g_m}{2\pi C_{gs}} \quad (4\text{-}78)$$

or

$$f_T \approx \frac{v_S}{2\pi L} \quad (4\text{-}79)$$

We will present the derivation of these expressions and discuss their limitations. Consider the simplified equivalent circuit of a MESFET shown in Fig. 4.14. It consists only of the gate–source capacitance C_{gs} and the transconductance g_m. Any other circuit elements of the intrinsic transistor and the entire extrinsic transistor are neglected. Then, the ac current at the input, i_1, is given by

$$i_1 = j\omega C_{gs} v_1 \quad (4\text{-}80)$$

where v_1 is the ac input voltage. The current at the output, i_2, is

$$i_2 = g_m v_1 \quad (4\text{-}81)$$

Thus, for the magnitude of the current gain we obtain

$$|h_{21}| = \left|\frac{i_2}{i_1}\right| = \frac{g_m}{\omega C_{gs}} = \frac{g_m}{2\pi f C_{gs}} \quad (4\text{-}82)$$

Applying the condition that at $f = f_T$ the magnitude of h_{21} becomes unity, the cutoff frequency is equal to Eq. (4-78).

If we now assume that the electron velocity in the entire channel underneath the gate is equal to v_S, the drain current obtained using the PHS model is

$$I_D = qN_D v_s a(1 - s) \quad (4\text{-}83)$$

and the charge stored in the space–charge region is

$$Q_{sc} = qN_D as WL \quad (4\text{-}84)$$

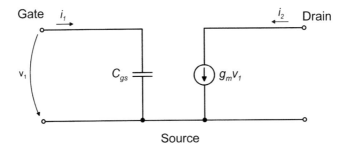

Figure 4.14. Simplified equivalent circuit.

Taking the derivative of I_D and $-Q_{sc}$ with respect to the gate–source voltage, the transconductance and the gate–source capacitance become

$$g_m = \frac{qN_D v_s a W}{2sV_{po}} \tag{4-85}$$

and

$$C_{gs} = \frac{qN_D a W L}{2sV_{po}} \tag{4-86}$$

Now we put Eqs. (4-85) and (4-86) into Eq. (4-78) and obtain the cutoff frequency from Eq. (4-79). It should be noted that Eqs. (4-78) and (4-79) are only first-order approximations for f_T.

According to [16, 17], the power gains MSG, MAG, and U can be approximated by

$$MSG \approx \frac{1}{f}\frac{g_m}{2\pi(C_{gs}+C_{gd})}\frac{C_{gs}}{C_{gd}} \tag{4-87}$$

$$MAG \approx \frac{1}{f^2}\left[\frac{g_m}{2\pi(C_{gs}+C_{gd})}\right]^2 \frac{1}{4g_{ds}(R_i+R_S+R_G)+\dfrac{2g_m C_{gd}}{C_{gs}+C_{gd}}(R_i+R_S+2R_G)} \tag{4-88}$$

and

$$U \approx \frac{1}{f^2}\left[\frac{g_m}{2\pi(C_{gs}+C_{gd})}\right]^2 \frac{1}{4g_{ds}(R_i+R_S+R_G)+4g_m R_G \dfrac{C_{gd}}{C_{gs}+C_{gd}}} \tag{4-89}$$

These power gains in dB are given by

$$\text{gain}[dB] = 10\log_{10}\text{gain} \tag{4-90}$$

The maximum frequency of oscillation f_{max} is originally defined as the frequency at which the unilateral power gain U rolls off to unity. Sometimes in the literature, however, the frequency at which the maximum available gain MAG becomes unity is designated as f_{max} as well. Therefore we present expressions for the two different definitions used for f_{max}: $f_{max,MAG}$ and $f_{max,U}$:

$$f_{max,MAG} \approx \frac{g_m}{2\pi(C_{gs}+C_{gd})}\frac{1}{\left[4g_{ds}(R_i+R_S+R_G)+\dfrac{2g_m C_{gd}}{C_{gs}+C_{gd}}(R_i+R_S+2R_G)\right]^{1/2}} \tag{4-91}$$

4.3 SMALL-SIGNAL ANALYSIS

$$f_{max,U} \approx \frac{g_m}{2\pi(C_{gs} + C_{gd})} \left[4g_{ds}(R_i + R_S + R_G) + 4g_m R_G \frac{C_{gd}}{C_{gs} + C_{gd}} \right]^{1/2} \tag{4-92}$$

If the Y parameters and gains for the complete transistor are to be calculated accurately and not from the approximated expressions in Eqs. (4-75)–(4-92), one also needs the values of the extrinsic capacitances and inductances. This can be a problem, as for these circuit elements no reliable expressions exist.

Frequently the values of the equivalent circuit elements, including the extrinsic elements, are not calculated from model equations, but are extracted from measured S parameters through optimization techniques. Table 4.1 lists the equivalent circuit elements of three different GaAs MESFETs extracted from measured S parameters [18-20].

Figure 4.15 shows the short circuit current gain and the different power gains calculated using the equivalent circuit elements for transistor 1 given in Table 4.1. The solid lines are the gains calculated using the complete equivalent circuit and the lines with symbols are the gains obtained from the approximated expressions (4-75) and (4-87)–(4-89). Reasonable accuracy is found in the approximated expressions, and only in the frequency range near f_T and f_{max} do discrepancies become notable. When the complete equivalent circuit is used, MAG and U cross 0 dB at the same frequency, i.e., $f_{max,U}$ and $f_{max,MAG}$ coincide, whereas these values differ when the approximated expressions are used. If published gains, cutoff frequencies or maximum frequencies of oscillation of different transistors from different references are to be compared, one has to make sure that they have been obtained according to the

Table 4.1. Values of small-signal equivalent circuit elements for three different GaAs MESFETs (data taken from [18–20])

Circuit elements	Transistor 1	Transistor 2	Transistor 3
$L \times W$, μm²	0.15 × 200	0.25 × 300	0.5 × 100
g_m, mS	88.9	85.8	26
r_{ds}, Ω	98.2	110	440
C_{gs}, fF	104.7	236	104
C_{gd}, fF	24	63.5	16
C_{DS}, fF	49.6	81.9	20
R_i, Ω	4.45	2.78	2.13
R_S, Ω	1.55	0.64	6.05
R_D, Ω	3.6	1.59	6.55
R_G, Ω	1.72	0.5	2.98
L_S, pH	8.11	2.32	21
L_D, pH	11.1	—	40
L_G, pH	45.9	—	51
C_{GSS}, fF	—	—	—
C_{DSS}, fF	—	—	—
τ, ps	—	0.8	—

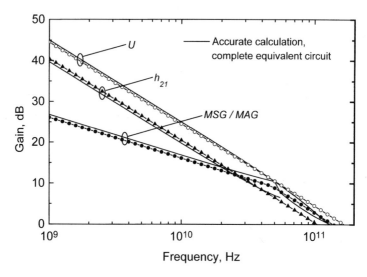

Figure 4.15. Calculated gains of a 0.15 μm gate GaAs MESFET. The elements of the small-signal equivalent circuit are taken from [18].

same definition and by the same method. Otherwise the result of the comparison can be misleading. It should be noted that because the intrinsic circuit elements are bias-dependent, the gains and the characteristic frequencies depend on the operation conditions as well.

The analyses covered in this section are related to the small-signal operation of MESFETs. Strictly speaking, they are not suited to characterizing power transistors, which typically are operated in the large-signal mode. Nonetheless, because small-signal and large-signal gains are correlated, and because theoretical large-signal considerations are much more elaborate compared to the small-signal case, the small-signal approaches are frequently used to characterize power transistors as well.

4.4 NOISE ANALYSIS

4.4.1 Noise Mechanisms

The knowledge of the sources and magnitude of noise generated in a transistor is of crucial importance to the design of low-noise transistors. We will concentrate on the analysis of the high-frequency noise, i.e., noise generated at GHz frequencies. It should be noted, however, that for specific applications, such as oscillators, the low-frequency $1/f$ noise is also important. A general overview of the $1/f$ noise is given in [21], and coverage of practical aspects of $1/f$ noise related to microwave transistors is available in [22]. For the design of power FETs, on the other hand, the issue of noise is secondary.

Basically, the physical origin of high-frequency noise in MESFETs is fluctuations in the carrier density and/or the carrier drift velocity in the conducting channel and in the parasitic resistances. These fluctuations cause random variations of the current passing through the transistor, or, equivalently, of the voltage drop across it. The following thought experiment will give a feeling of the physical mechanisms of noise.

Consider a small segment in the channel extending from x_{i-1} to x_i as shown in Fig. 4.16. The transistor is biased with the voltages V_{DS} and V_{GS}, which give rise to a dc drain current I_D. We call this current the expected current I. The expected current passing through the channel segment with length Δx is represented by the product of the number of carriers N_{i-1} entering the segment at x_{i-1} and their average velocity v_{i-1}, as well as by the product of the number of carriers N_i leaving it at x_i and their average velocity v_i. We observe the conditions in the channel segment during a certain time interval Δt and assume that the applied voltages are constant during Δt. The carriers representing I experience scattering events in the segment, and the number and types of the events are random and can be different from time to time. Thus, it might happen that during Δt the carriers are less scattered than typically. Then the carrier motion across the segment is less hampered, the average carrier velocity during Δt is higher, and the current leaving the segment at x_i is slightly higher than the expected current I by an amount of i. If we repeat the observation a bit later, the conditions may change and more scattering events could occur, thus resulting in a current at x_i slightly smaller than I. These random current fluctuations represent the noise current generated in a particular channel segment. Because of the amplification effect inherent to the MESFET, the noise current i generated in the channel segment entails an amplified noise current i_d at the drain. The current fluctuations i_d generated in all segments of the FET channel add up to be the total drain current fluctuation called the drain noise current i_D. Since the small signal output resistance of the transistor is r_{ds}, the drain noise current i_D can be transformed to the drain noise voltage v_D given by

$$v_D = i_D r_{ds} \tag{4-93}$$

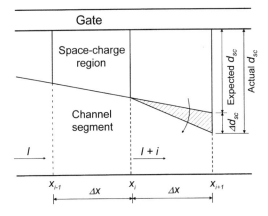

Figure 4.16. Noise generation in a channel segment of a MESFET.

The origins of the current fluctuations in different channel segments are fully independent of each other. Therefore the noise currents produced in different channel segments are independent from each other as well. In other words, they are uncorrelated.

Besides the drain noise current, there is a second mechanism in the MESFET related to the aforementioned current fluctuations. We go back to our thought experiment. The low number of scattering events during Δt led to a current leaving the segment and flowing through the following segment extending from x_i to x_{i+1}, that is slightly larger than the expected current. This automatically gives rise to an increased voltage drop across the channel region between $x = x_i$ and $x = L$. Thus, the potential difference between the gate and the channel at $x > x_i$ is larger than expected, the space–charge region extends deeper into the active layer, and the thickness of the space–charge region d_{sc} is larger than the expected thickness by Δd_{sc} at $x = x_{i+1}$. The hatched area in Fig. 4.16 corresponds to the additional charge in the space–charge region resulting from the increased d_{sc}. The charge neutrality condition requires an additional charge with same amount and opposite sign to appear at the gate. This gate charge variation during Δt is nothing but a small additional gate current i_g. If we add up the additional gate currents caused by the current fluctuations in all individual channel segments, we get the gate noise current i_G. The gate current fluctuation is a direct consequence of the drain current fluctuation, i.e., between both current fluctuations a correlation exists. The degree to which the two quantities are correlated is described by the correlation coefficient C. A correlation coefficient of unity means that the two quantities are fully correlated.

The increase of the thickness of the space–charge region by Δd_{sc} caused by the noise current i results, with a small delay, in a smaller channel height and in a smaller drain current. Thus, the increase of drain current during Δt (by the less than average number of scattering events) will be compensated somewhat by the subsequent decrease of the drain current due to the smaller channel height. This compensation is beneficial and a factor contributing to the excellent noise performance of FETs.

Because of the random nature of noise, it is impossible to accurately predict the noise, e.g., the noise current i, at a certain time. It would be obvious to deal instead with its mean value, e.g., $\langle i \rangle$. This, however, would not solve the problem because the mean value of a noise current is zero, as it fluctuates randomly around (above and below) the expected current. Therefore, noise currents are described by the mean square values $\langle i^2 \rangle$. The same holds for noise voltages.

Modeling the noise behavior of MESFETs is complicated and demanding. The derivations are quite extensive and require numerous steps. Therefore we do not present the full derivations here but refer the interested reader to the original papers for more details. We will, however, explain the basic ideas and main procedures of MESFET noise modeling based on the PHS and Cappy models.

FET noise modeling can in general be divided into two major parts. The first one is the derivation of expressions for the drain and gate noise currents i_D and i_G (to be more accurate, their mean square values $\langle i_D^2 \rangle$ and $\langle i_G^2 \rangle$) or, equivalently, the drain and gate noise voltages v_D and v_G. Based on the determined noise currents i_D and i_G, the second part of noise modeling consists of the derivation of expressions of various noise parameters which lead to noise related figures of merit, such as the minimum noise figure NF_{min}.

4.4.2 Noise Modeling Using the PHS Noise Model

The PHS noise model [8, 9] was developed based on the dc and small-signal PHS models described in the preceding sections. It is an extension of van der Ziels classical noise theory of FETs [23, 24]. The PHS noise model was the first comprehensive and theoretical treatment of FET noise including the effect of velocity saturation and is widely used in the investigation of MESFET noise.

First, the channel noise in region 1 is considered. The starting point is the calculation of the noise voltage v generated in a channel segment of region 1. Because the conditions in region 1 are ohmic, this can be done using the Nyquist formula given by

$$\langle v^2 \rangle = 4k_B T R \Delta f \quad (4\text{-}94)$$

where k_B is the Boltzmann constant, T is the temperature, R is the resistance of the channel segment, and Δf is the bandwidth. If, for example, the noise behavior of the MESFET at a frequency of 12 GHz is to be investigated, the bandwidth is 12 GHz.

This noise voltage v, although small in magnitude, changes not only the potential in the segment itself but also the potential in all channel segments through which the noise current (that corresponds to v) flows. The result is a small shift of the boundary between regions 1 and 2, a small variation of the length of region 1, and a small variation of the channel height in region 2. This in turn leads to an enhanced or amplified noise voltage v_{d1} at the drain. The subscript 1 indicates that the origin of the noise voltage is in region 1 of the transistor. The addition (integration) of all these drain voltage fluctuations caused by noise generated in the channel segments of region 1 gives the drain noise voltage v_{D1}. Note the difference between v_{d1} and v_{D1}: The quantity v_{d1} is the drain noise voltage drain caused by *one* small channel segment of region 1, whereas v_{D1} is the drain noise voltage caused by *all* channel segments of region 1, i.e., by the entire region 1.

The channel noise generated in region 2 is considered next. Because the carriers in region 2 travel with a constant saturation velocity and the conditions are nonohmic, the van der Ziel's approach does not apply here. Instead, Pucel et al. [8, 9] proposed the generation of dipole layers, as shown in Fig. 4.17, as the process responsible for the noise originated in region 2. Consider an electron in region 2 moving with the saturation velocity in the x direction. At the location designated with a cross in Fig. 4.17, scattering takes place. Following this event, in a short period of time, the electron velocity also has a y component, the x component of its velocity is smaller than the saturation velocity, and the electron takes a path (solid line) different from that it would take without the scattering event (dashed line). At the actual position of the electron after scattering, there is an "excess" electron charge, where-

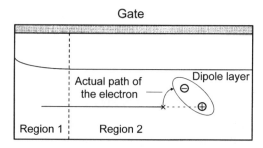

Figure 4.17. Illustration of the dipole layer generation in region 2 of the MESFET channel.

as at the position the electron supposed to arrive without scattering there is a "deficiency" electron charge. In other words, a dipole layer traveling with the saturation velocity toward the drain has been generated due to the scattering process. Because scattering is random, the dipole layer generation is random as well. After calculating the noise voltage at the drain caused by one dipole layer, v_{d2}, the corresponding noise voltages of all dipole layers generated in region 2 are summarized (i.e., are integrated), resulting in the drain noise voltage v_{D2}. This type of noise is called the diffusion noise. Finally, based on the drain noise voltages v_{D1} and v_{D2} generated by noise sources in regions 1 and 2, respectively, the total mean square drain noise current $\langle i_D^2 \rangle$ is obtained by

$$\langle i_D^2 \rangle = \frac{\langle v_{D1}^2 \rangle + \langle v_{D2}^2 \rangle}{r_{ds}^2} \tag{4-95}$$

Because the noise voltages generated in regions 1 and 2 are independent from each other, no correlation between them exists.

As mentioned above, a noise current, or equivalently a noise voltage, generated in the FET channel also leads to a variation of the gate charge, which creates a gate noise current. From the drain noise voltages v_{d1} and v_{d2}, the gate charges q_{g1} and q_{g2} are obtained. These charges are induced by the individual noise voltages generated in the channel segments of region 1 and by the noise voltages generated by the dipole layers in region 2. By adding (i.e., integrating) q_{g1} and q_{g2} over the entire transistor, the total induced gate charges q_{G1} and q_{G2} are obtained. Then we get the gate noise currents i_{G1} and i_{G2} by multiplying q_{G1} and q_{G2} by $\omega = 2\pi f$. The total mean square gate noise current $\langle i_G^2 \rangle$ is

$$\langle i_G^2 \rangle = \langle i_{G1}^2 \rangle + \langle i_{G2}^2 \rangle \tag{4-96}$$

Finally, the correlation coefficient that describes the correlation between the drain and gate noise currents is derived [8, 9]. It should be noted that the PHS noise model provides analytical expressions of all MESFET noises mentioned above.

4.4.3 Noise Modeling Using the Cappy Noise Model

In the Cappy noise model [11], the noise current i_i generated in the ith segment in the channel is be described by

$$\langle i_i^2 \rangle = \frac{4q^2 D_i n_i b_i W}{\Delta x} \Delta f \tag{4-97}$$

where D_i is the energy- or field-dependent diffusion coefficient in the direction of current flow in the ith segment and Δf is the bandwidth. The diffusion coefficient can be obtained from the mobility using the Einstein relation [see Eq. (2-70)], or from Monte Carlo calculations [25]. As already mentioned in the PHS noise model, a noise component generated somewhere in the FET channel appears amplified at the drain. The drain voltage fluctuation caused by the noise current i_i can be obtained from

$$\langle v_d^2 \rangle = \frac{4q^2 D_i n_i b_i W}{\Delta x} \Delta f \left(\frac{\Delta V_{DS}}{\Delta I} \right)_i^2 \tag{4-98}$$

The term $(\Delta V_{DS}/\Delta I)_i$ describes the variation of the drain–source voltage caused by a current fluctuation in the ith segment, and it should not be confused with the drain resistance $r_{ds} = \Delta V_{DS}/\Delta I_D$. When the drain current changes by ΔI_D, the voltage drops across all channel segments change accordingly. The noise current i_i, however, changes only the voltage drops across the segments between x_i and the drain. This issue has already been discussed in the previous section. The total drain noise voltage v_D is obtained by adding v_d produced in all channel segments:

$$\langle v_D^2 \rangle = \sum_i \frac{4q^2 D_i n_i b_i W}{\Delta x} \Delta f \left(\frac{\Delta V_{DS}}{\Delta I} \right)_i^2 \tag{4-99}$$

Using Eq. (4-93), this can be easily transformed to the total drain noise current i_D.

In a similar manner, the total gate noise current i_G can be modeled as

$$\langle i_G^2 \rangle = \omega^2 \sum_i \frac{4q^2 D_i n_i b_i W}{\Delta x} \Delta f \left(\frac{\Delta Q_G}{\Delta I} \right)_i^2 \tag{4-100}$$

The term $(\Delta Q_G/\Delta I)_i$ describes the variation of the gate charge resulting from a small drain current fluctuation in the ith channel segment. From i_D and i_G, the correlation coefficient C can be calculated [11].

4.4.4 Minimum Noise Figure

Now that the noise currents and the correlation coefficients have been determined either by the PHS or by the Cappy model, we can continue on to the second part of noise modeling. The aim here is to come up with expressions for the minimum noise figure $NF_{\min}$, a commonly used figure of merit for noise performance. The derivations leading to $NF_{\min}$ are quite extensive, and we again present only the basic ideas and the final results for the purpose of the design and optimization of low-noise MESFETs. The basis of the derivation of $NF_{\min}$ is the noise equivalent circuit shown in Fig. 4.18.

The intrinsic transistor in the dashed-line box is considered to be noise-free, and the noise sources i_G and i_D are connected to its input and output. Compared to the small-signal equivalent circuit in Fig. 4.12, fewer elements are used here, and the intrinsic transistor is described only by the gate–source capacitance C_{gs}, the charging resistance R_i, and the transconductance g_m. In principle, it is possible to take into account the other intrinsic elements, but doing so would make the noise figure calculation even more complicated. For the extrinsic elements, only the parasitic gate and source resistances R_G and R_S are included in the noise consideration, because they influence the noise behavior of the transistor considerably. The thermal noise generated in R_G and R_S is taken into account by the noise voltage sources e_G and e_S, respectively. The signal source described by the source impedance Z_{source} and its noise voltage source e_{source} is connected to the transistor input.

First, the two dimensionless noise coefficients P and R are defined as

$$P = \frac{\langle i_D^2 \rangle}{4k_B T \Delta f g_m} \tag{4-101}$$

$$R = \frac{\langle i_G^2 \rangle}{4k_B T \Delta f \omega^2 C_{gs}^2 / g_m} \tag{4-102}$$

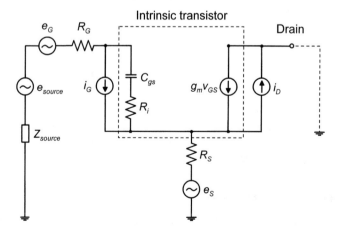

Figure 4.18. Simplified noise equivalent circuit used for the derivation of the minimum noise figure of MESFETs.

From R, P, and the correlation coefficient C, a set of three noise coefficients K_g, K_c, and K_r can be obtained. These coefficients are called the fundamental noise coefficients and are given by [9]

$$K_g = P\{[1 - C(R/P)^{1/2}]^2 + (1 - C^2)R/P\} \tag{4-103}$$

$$K_c = \frac{1 - C(R/P)^{1/2}}{[1 - C(R/P)^{1/2}]^2 + (1 - C^2)R/P} \tag{4-104}$$

and

$$K_r = \frac{R(1 - C^2)}{[1 - C(R/P)^{1/2}]^2 + (1 - C^2)R/P} \tag{4-105}$$

Using these fundamental noise coefficients, the minimum noise figure can be expressed in a power series in ω given by

$$NF_{min} = 1 + 2\frac{\omega C_{gs}}{g_m}\{K_g[K_r + g_m(R_S + R_G)]\}^{1/2}$$

$$+ 2\left(\frac{\omega C_{gs}}{g_m}\right)^2 [K_g g_m(R_S + R_G + K_c R_i)] + [\text{higher order terms}] \tag{4-106}$$

Clearly, the influence of ω on NF_{min} decreases with increasing power of ω: the linear ω term has the largest effect, the role of the ω^2 term is much smaller, and the higher-order terms can be omitted provided the frequency is not too high.

As can be seen from Fig. 4.18, the signal source influences the noise behavior of the transistor through its impedance Z_{source} and noise source e_{source}. The noise figure reaches its minimum value for a certain optimal source impedance $Z_{source,opt}$. The minimum noise figure NF_{min} given in Eq. (4-106) is the noise figure of the MESFET having the optimal source impedance.

Figure 4.19(a) shows the calculated (solid line, based on the PHS noise model) and measured (symbols) minimum noise figures for a 2 μm gate GaAs MESFET. The drain current variation (at the abscissa) is obtained by fixing the drain–source voltage and changing the gate–source voltage. The abscissa is not directly related to V_{GS} but to the current ratio I_D/I_{DSS}, where I_D is the drain current corresponding to the actual V_{GS}, and I_{DSS} is the drain current for zero V_{GS}. This is a frequently used representation of the bias dependence of NF_{min}. The measured data and the device structure are those reported in [9, 26]. Good agreement between the model calculations and measurements is observed. Note that the general bias dependence, the minimum of NF_{min} at I_D/I_{DSS} between 0.1 and 0.2, and the U-shape of the NF_{min} versus I_D/I_{DSS} curve are reproduced correctly by the model. Experience shows, however, that the NF_{min} versus I_D/I_{DSS} curve of short-gate GaAs MESFETs with small L/a ratios modeled using the PHS model shows a less pronounced U-shape.

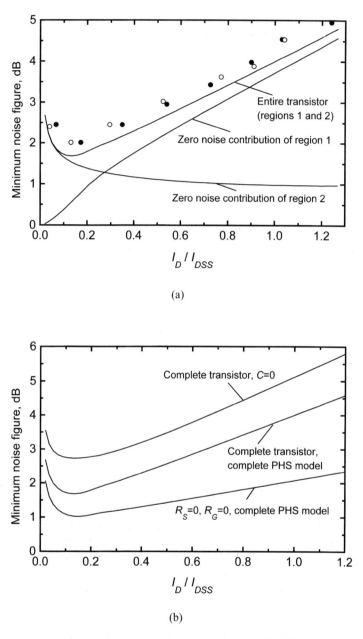

Figure 4.19. (a) Minimum noise figures calculated using the PHS noise model for a 2 μm gate GaAs MESFET with the contributions of regions 1 and 2 to the overall minimum noise figure. Also included are measured data (closed and open circles) [8, 26]. (b) Demonstration of the influence of the parasitic resistances and the correlation between drain and gate noises on the minimum noise figure.

The reason is that the PHS model underestimates the noise contribution of region 1 in such cases. The contributions of the noises generated in regions 1 and 2 of the MESFET to the total minimum noise figure can be seen from the two additional curves in Fig. 4.19(a). At low drain currents (i.e., at more negative V_{GS}), the noise is due mainly to the noise sources in region 1, whereas at high drain currents the noise generated in region 2 dominates the total noise.

The definition for the noise figure is somewhat ambiguous and confusing. Sometimes, the noise figure under arbitrary bias conditions and optimal source impedance is called the minimum noise figure NF_{min}. Other times, the noise figure under the optimal I_D/I_{DSS} condition (i.e., at the bottom of the U-shape) is also designated as NF_{min}. Therefore, the definition of NF_{min} needs to be clarified before it is used to assess the noise performance.

Figure 4.19(b) demonstrates the influence of the parasitic resistances and the correlation between the drain and gate noises on NF_{min}. When the correlation coefficient is set to zero, the calculated minimum noise figure is considerably larger than in the case when correlation is taken into account. Clearly, the correlation between i_D and i_G acts as a noise reduction mechanism and is beneficial. On the other hand, the resistances R_S and R_G increase the noise figure in two ways. First, as can be seen from the noise equivalent circuit in Fig. 4.18, these resistances introduce the thermal noise represented by the noise voltage sources e_S and e_G to the MESFET. Second, because of the presence of R_S and R_G, the noise compensation caused by the correlation between drain and gate noise currents is decreased.

Figure 4.20 shows the minimum noise figures calculated based on the Cappy model [11] for GaAs MESFETs having different gate lengths. The Cappy noise

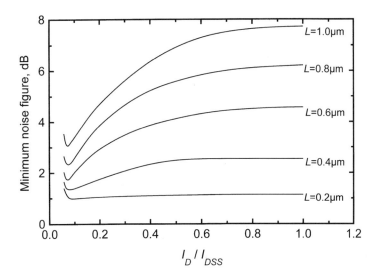

Figure 4.20. Minimum noise figures calculated using the Cappy model of GaAs MESFETs having different gate lengths. Data taken from [11].

model also reproduces the frequently seen U-shape characteristics. The bowing of the NF_{min} curve, however, is different from the PHS noise model in Fig. 4.19. Nevertheless, for transistors with not too short gates, both models describe the general trends of the noise figure and its dependence on transistor design and bias conditions essentially correctly.

In a series of papers, Fukui developed simple empirical expressions that enable an estimate of the minimum noise figure (the minimum value of the NF_{min} versus I_D/I_{DSS} curve) for GaAs MESFETs, thereby avoiding the complicated procedure described above. Fukui's expressions have the basic form of [13, 27–29]

$$NF_{min} = 1 + 2\pi f k C_{gs} \sqrt{\frac{R_S + R_G}{g_m}} \qquad (4\text{-}107)$$

where k is a fitting factor representing the quality of the channel material. Care must be taken for the units involved in Eq. (4-107). In the version of Eq. (4-107) given in [28], for example, g_m and the resistances (R_S and R_G) have to be in S and Ω, respectively, C_{gs} in pF, f in GHz, and a k value of 2.5×10^{-3} must be used. Moreover, Fukui emphasized that the bias-dependent circuit elements in Eq. (4-107) have to be considered for a bias condition of $V_{GS} = 0$.

It is instructive to see that Eq. (4-107) is closely related to the PHS noise model. It has been shown that the noise coefficient K_r in the PHS model is normally much smaller than K_g and K_c, and decreases rapidly toward zero as the gate length is decreased. Thus, for short-gate MESFETs, and considering only the linear ω term, Eq. (4-106) is reduced to

$$NF_{min} = 1 + 2\frac{\omega C_{gs}}{g_m} \sqrt{K_g g_m (R_S + R_G)} \qquad (4\text{-}108)$$

This is identical to Eq. (4-107). Thus, Fukui's formula can be considered as a specific case of the PHS noise model.

4.5 POWER ANALYSIS

When a MESFET is used in a power amplifier, it is operated in the large-signal mode. Here, the variations of the input and output voltages are much larger compared to $k_B T/q$ and can be comparable to the dc bias voltages. Amplifier circuits are divided into different classes depending on the dc operating conditions of the transistor. In the following, we concentrate on the class A amplifier, but the basic concept applies generally to class AB, B, and C amplifiers as well. Additional information on the different amplifier classes can be found in [30–34].

Figure 4.21 shows the simplified circuit diagram of a MESFET amplifier. The high-frequency ac input signal to be amplified is fed to the FET gate via the capaci-

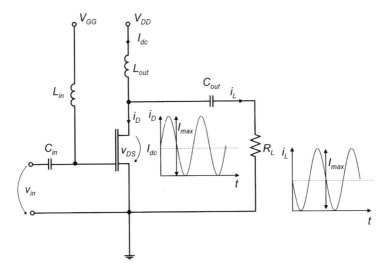

Figure 4.21. Circuit diagram of a MESFET amplifier showing the high-frequency ac currents flowing through the MESFET (i_D) and through the load (i_L).

tor C_{in} and the amplified output signal is passed to the load R_L through C_{out}. The inductors L_{in} and L_{out} act as extremely large high-frequency impedances and block the high-frequency signals from passing to the dc voltage supplies V_{GG} and V_{DD}. C_{in} and C_{out}, on the other hand, prevent the dc currents from flowing through the high-frequency path to the signal source or to the load.

Let us consider first an idealized dc output characteristics of a MESFET, having piecewise linear I_D–V_{DS} characteristics with an abrupt transition from the active to the saturation region at $V_{DS} = V_k$, as shown in Fig. 4.22. V_k is the so-called knee voltage and is approximately equal to the drain–source saturation voltage V_{DSS}. The transconductance of the MESFET is assumed to be constant, i.e., the I_D–V_{DS} curves in the saturation region are equidistant over the entire V_{GS} range. The load line of the amplifier is also shown in Fig. 4.22.

In a class A amplifier with the above-mentioned idealized I_D–V_{DS} characteristics, the dc bias is chosen in such a way that the dc operating point is located midway between points A and B, i.e., between 0 and I_{max} on the current axis. The maximum drain current I_{max} is obtained by applying either a zero V_{GS} or a small positive V_{GS} causing only a barely noticeable dc gate current to flow. In terms of voltage, the dc operating point is located midway between the knee voltage V_k and V_m. The voltage V_m is equal to $(2V_{DD} - V_k)$, where V_{DD} is the dc supply voltage. A large ac gate voltage swing causes a shift of the operating point in the limits between points A and B along the load line.

An important figure of merit for power transistors is their high-frequency output power, which is the power that can be delivered to the load R_L in the circuit in Fig.

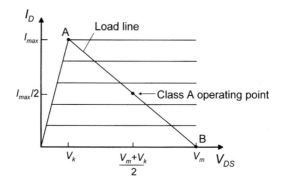

Figure 4.22. Idealized output characteristics of a MESFET and the load line of the amplifier in Fig. 4.21.

4.21. The load line for power matching, i.e., for maximum power transfer to the load, is given by

$$R_L = \frac{V_m - V_k}{I_{max}} \quad (4\text{-}109)$$

For pure sinusoidal ac input signals, the ac output power can be calculated by

$$P_{out} = \frac{1}{T}\int_0^T R_L i_L^2 \, dt = \frac{1}{T}\int_0^T R_L \left(\frac{I_{max}}{2} \sin\omega t\right)^2 dt \quad (4\text{-}110)$$

where T is the period of the high-frequency ac input signal, i.e., $1/f$, and i_L is the ac current flowing through the load. Applying $\sin^2 x = \tfrac{1}{2} \times (1 - \cos x)$ and bearing in mind that the integral of the cosine function over one full period equals zero, we get

$$P_{out} = \frac{I_{max}(V_m - V_k)}{8} \quad (4\text{-}111)$$

Obviously, the output power can be increased by increasing V_m. The upper limit of V_m is $V_{m,max}$, which is the drain–source breakdown voltage BV_{DS} at which avalanche breakdown sets in. It is given by the gate–drain breakdown voltage BV_{GD}, minus the pinch-off voltage V_{po}. The estimation of BV_{GD} is possible using the method described in Section 2.6 and will be discussed for GaAs MESFETs in more detail in Section 4.6. The dc drain–source voltage for maximum output power $P_{out,max}$ is $(V_{m,max} + V_k)/2$, and the maximum output power is given by

$$P_{out,max} = \frac{I_{max}(V_{m,max} - V_k)}{8} \quad (4\text{-}112)$$

If the dc drain–source voltage is smaller than $2 \times V_k$, the load line for power matching crosses the current axis below I_{max}, and the output power has to be calculated using the following expression:

$$P_{out} = \frac{1}{16} \frac{V_{DS}^2}{V_k} I_{max} \qquad (4\text{-}113)$$

Equation (4-112) reveals that a high output power is obtainable when the transistor possesses a high maximum drain current, a large breakdown voltage, and a small knee voltage.

Another figure of merit important for power transistors is the power-added efficiency PAE. It is of exceptional importance in applications where dc power consumption is to be minimized, such as in mobile phone handsets. The power-added efficiency is defined by

$$PAE = \frac{P_{hf}}{P_{dc}}\left(1 - \frac{1}{G}\right) \qquad (4\text{-}114)$$

where P_{hf} is the high-frequency output power whose maximum value is given in Eq. (4-112), P_{dc} is the dc power dissipated in the transistor, and G is the transistor's large signal gain. From Fig. 4.22 we can deduce P_{dc} for the case of maximum output power as

$$P_{dc} = \frac{V_{m,max} + V_k}{2} \frac{I_{max}}{2} = \frac{(V_{m,max} + V_k)I_{max}}{4} \qquad (4\text{-}115)$$

The calculation of the large-signal gain G is complicated. As a first order approximation we can assume that the large signal gain behaves similarly to its small-signal counterpart. Putting Eq. (4-115) in Eq. (4-114) and assuming $G \gg 1$, we come to

$$PAE \approx \frac{1}{2} \frac{V_{m,max} - V_k}{V_{m,max} + V_k} \qquad (4\text{-}116)$$

In the limit of infinite gain and zero knee voltage, the theoretical upper limit of PAE for a class A amplifier is 50%. From Eq. (4-116), it can be seen that a very low V_k is required for high PAE.

In the above considerations, we assumed a constant transconductance in the MESFET, which is a reasonable first-order approximation but not completely correct. The influence of a nonconstant transconductance on output power and on power added efficiency has been investigated in detail in [35].

If the maximum output power of a MESFET is to be calculated based on output characteristics simulated from the PHS model, the current I_{max} and the knee voltage can be approximated as the drain current when the intrinsic gate–source voltage is zero (which corresponds to a positive applied V_{GS}) and as the drain–source voltage equal to the saturation voltage V_{DSS}, respectively. Another option applicable to the output characteristics simulated from the PHS and the Cappy models is as follows. The output characteristics are simulated using one of the two models under the condition that the intrinsic gate–source voltage is around zero when the device is operated in the saturated region. Then the load line is rotated around V_m until the output power in Eq. (4-111) becomes maximal. During the rotation, I_{max} and V_k in Eq. (4-111) correspond to the point where the load line intersects the calculated *I–V* characteristics.

In power transistors, considerable heat is generated by the dissipation of electric energy (dc plus ac). Therefore, self heating should be taken into account in the simulation of power transistors and in the calculation of output power. This can be done by employing one of the methods described in Section 2.7. In addition, we must first know how much electric power P_{FET} is dissipated in the transistor, which is given by

$$P_{FET} = \frac{1}{T}\int_0^T i_D(t)v_{DS}(t)dt \tag{4-117}$$

where i_D and v_{DS} are the drain current and the drain–source voltage containing both the dc and ac components. In the case of MESFET operating in a class A amplifier under CW conditions, $i_D(t)$ and $v_{DS}(t)$ are given by

$$i_D(t) = \frac{I_{max}}{2} + \frac{I_{max}}{2}\sin \omega t \tag{4-118}$$

$$v_{DS}(t) = \frac{V_m + V_k}{2} - \frac{V_m - V_k}{2}\sin \omega t \tag{4-119}$$

Inserting these expressions into Eq. (4-117) and carrying out the integration yield the total power dissipated in the transistor:

$$P_{FET} = \frac{1}{8}I_{max}V_m + \frac{3}{8}I_{max}V_m \tag{4-120}$$

At this point, we have developed the tools necessary for MESFET modeling. In the following sections we will discuss the different types of MESFETs, their structures, design, and performance.

4.6 ISSUES OF GaAs MESFETs

4.6.1 Transistor Structures

Several criteria are commonly used to classify GaAs MESFETs in regard to their device technology and applications. First, the active layer of GaAs MESFETs can be realized by two different methods: epitaxy and ion implantation. Thus, regarding technology, there are epitaxial and ion-implanted GaAs MESFETs. The cross sections of an epitaxial and of an ion-implanted MESFET are shown in Figs. 4.23(a) and (b), respectively. Both types normally have an n-type active layer with a thickness a and a doping density N_D directly underneath the gate. To reduce the parasitic source and drain resistances R_S and R_D, the active layer under the source and drain contacts is more heavily doped and thicker compared to the region underneath the gate.

In epitaxial MESFETs, the doping in the active layer is carried out during the epitaxial growth process. Following epitaxial growth, the desired thickness of the active layer in the gate region is defined by an etching step. Then the source and drain contacts, and finally the gate contact, are formed. As can be seen from Fig. 4.23(a), the gate of such a transistor, which is called a recessed-gate MESFET, is located deeper than the source and drain.

In the case of ion implanted MESFETs, shown in Fig. 4.23(b), ion implantation defining the thickness and doping of the active layer underneath the gate region is first carried out. A subsequent second implantation with higher implantation dose

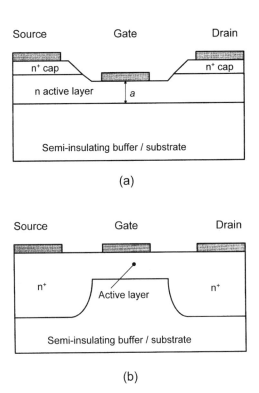

Figure 4.23. Cross section of (a) an epitaxial GaAs MESFET and (b) an ion-implanted GaAs MESFET.

and energy is performed to create the thicker and more heavily doped drain and source regions. This is followed by the contact formation.

The typical impurity dopant in GaAs is Si. An eutectic AuGe alloy is used for ohmic contacts, and either Al or Ti/Pt/Au is used for the Schottky contact. The gate length is the most important dimension of a GaAs MESFET. It determines the frequency limits, and thus the maximum operating frequency, of the transistor. Typical gate lengths are in the range between 0.1 µm and 1 µm. Doping concentration of the active layer is normally between 10^{17} and 5×10^{17} cm^{-3}. This, together with the active layer thickness underneath the gate, defines the threshold voltage V_{th} of the transistor. The threshold voltage V_{th} can be approximated by

$$V_{th} \approx -V_{po} + V_{bi} \tag{4-121}$$

Depending on the value of V_{th}, MESFETs can also be separated into the depletion- and enhancement-mode transistors. A depletion-mode MESFET has a negative threshold voltage, whereas an enhancement-mode MESFET has a positive threshold voltage. Assuming a constant doping in the active layer and applying Eq. (4-5), we find that for

$$N_D a^2 > \frac{2\varepsilon V_{bi}}{q} \tag{4-122}$$

the transistor is of the depletion type, whereas for

$$N_D a^2 < \frac{2\varepsilon V_{bi}}{q} \tag{4-123}$$

the transistor is of the enhancement type.

Concerning the applications, GaAs MESFETs can be categorized into low-noise FETs and power FETs. Low-noise transistors are designed for lowest minimum noise figure NF_{min} combined with a high associated gain G_a, whereas the issue of output power is of less importance. Power transistors, on the other hand, are optimized for high output power P_{out} and/or high power-added efficiency PAE at a given frequency, and noise is of no concern.

4.6.2 Low-Noise GaAs MESFETs

A low NF_{min} and a high G_a are the design targets for low-noise GaAs MESFETs. For the minimum noise figure, it became clear in Section 4.4 that a low-noise transistor should possess a low gate–source capacitance C_{gs} and a high transconductance g_m. The intrinsic C_{gs} is directly related to the gate length and becomes smaller when the gate length decreases. Thus, the gate length is normally made as small as the available technology permits. If the transistor is a discrete device, it needs to have a large gate bond pad, as shown in Fig. 4.24(a). It is imperative to place this bond pad on an undoped area of the chip or on an insulating layer. If the pad were

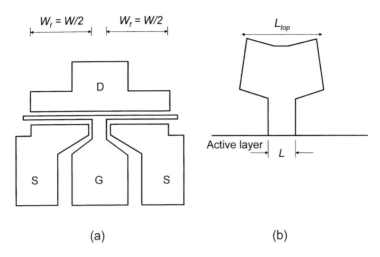

Figure 4.24. Design details of a low-noise GaAs MESFET showing (a) the layout of a two-gate-finger MESFET and (b) the mushroom gate structure.

located on the n-type active layer, a large area space–charge region would be formed under the pad, which leads to a huge parasitic capacitance in parallel to C_{gs}. Figure 4.25 illustrates this issue. The gate pad is located on the surface of the semi-insulating GaAs substrate. The gate bus connecting the pad with the gate on top of the active layer runs over the mesa step.

A short gate is desirable to achieve a high transconductance as well. There is, however, only a slight increase of g_m with shrinking gate length because the MESFET is operated in the regime of velocity saturation. A high low-field mobility and a high saturation velocity are also critical for high transconductance.

The parasitic resistances R_S and R_G should be minimized. To reduce R_S, the distance between source and gate should be small, and the sheet resistance of the active layer between source and gate should be low. To minimize R_G, there are two methods. First, a multifinger structure can be used in which the gate with a width W is divided into several parallel gate fingers. Figure 4.24(a) shows the simplest case of two parallel gate fingers, each with a finger width W_f of $W/2$. The multifinger structure has the beneficial spin-off that the distance between the gate feed point and the end of the gate is smaller than in the one-finger structure, thereby resulting in a reduction of the time needed to propagate a signal from the front to the back of the gate. We will discuss this issue in more detail in the following section. Second, the mushroom gate, frequently also called T-gate, can be used, which is illustrated in Fig. 4.24(b). Its characteristic feature is a small foot-print to ensure the required small gate length, and a much larger length and thus a much larger cross section area near the top of the gate to reduce the gate resistance. The length at the top can be several times the gate length L. In typical low-noise MESFETs, both the multifinger and mushroom shape features are used.

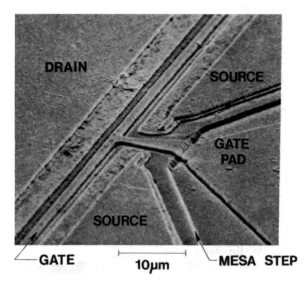

Figure 4.25. Photomicrograph showing the gate bonding pad and the gate bus running over the mesa step to the top of the active layer. Taken from [36]. © 1976 IEEE.

Figures 4.26 and 4.27 illustrate the state of the art of GaAs MESFETs in terms of f_T and f_{max}. Shown are also empirical upper limit fits for both characteristic frequencies. The record f_T obtained with a GaAs MESFET is 168 GHz (L = 60 nm) [38], and the record f_{max} is 177 GHz (L = 0.12 μm) [40]. Remarkable are also the transistors reported in [41] (f_T = 121 GHz, f_{max} = 160 GHz, L = 0.12 μm) and in [43] (f_T = 113 GHz, f_{max} = 132 GHz, L = 0.1 μm), which show simultaneously high f_T and f_{max} values.

For long-gate transistors, the upper limits for both f_T and f_{max} in Figs. 4.26 and 4.27 show a linear slope in the log–log plots, which corresponds to the well-known dependence

$$f_T, f_{max} \propto L^{-a} \tag{4-124}$$

For gates shorter than about 0.5 μm, however, the upper limits deviate from the L^{-a} dependence and the curves flatten. This behavior can be explained as follows. In a long-gate FET, the total gate–source capacitance is mainly determined by the intrinsic C_{gs}, i.e., by the components C_{gs1}, C_{gs2}, and C_{gs3} in Eq. (4-63). When the gate is shortened, the share of the fringing component of C_{gs}, the gate pad capacitance, and the stray capacitances of the total C_{gs} becomes larger. These components deteriorate the transistor behavior. Moreover, due to the short channel effect, it is difficult to maintain a small drain conductance g_{ds} in such transistors. A larger g_{ds} decreases the current and the power gains, and consequently imposes a negative influence on f_T and f_{max}. The undesirable short-channel effect can be suppressed by maintaining a certain gate length L to active layer thickness a ratio when the gate length is de-

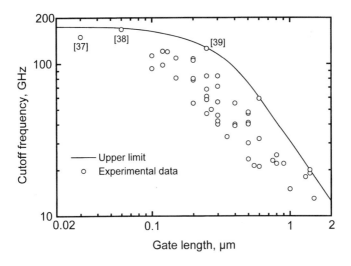

Figure 4.26. Reported cutoff frequency of GaAs MESFETs as a function of gate length.

creased. The L/a ratio is frequently called the aspect ratio. Experimental experience shows that the aspect ratio should be at least 3, preferably even larger. Thus, decreasing the gate length should be accompanied by decreasing the active layer thickness. To maintain a certain drain current and threshold voltage, however, a small active layer thickness simultaneously requires a higher doping density, which, in turn, leads to lower mobility and breakdown voltage. Evidently, there is a

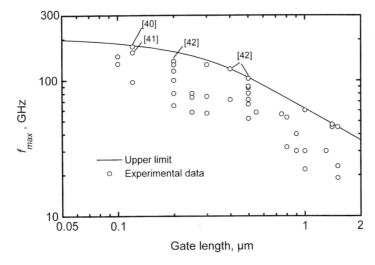

Figure 4.27. Reported maximum frequency of oscillation of GaAs MESFETs as a function of gate length.

design trade-off in regard to the gate length, aspect ratio, and active layer doping density.

To obtain a high small-signal power gain and thus a high f_{max}, the total C_{gs} and C_{gd} as well as the parasitic resistances R_S and R_G should be as small as possible and the transconductance g_m should be as high as possible. In general, the same guideline holds for f_T.

In Fig. 4.28, the high-frequency noise performance of state of the art GaAs MESFETs is shown. Minimum noise figures less than 1 dB up to 26 GHz have been obtained. Table 4.2 shows the minimum noise figures and the associated gain obtained with a 0.11 μm gate GaAs MESFET.

The impedance Z_{source} of the signal source has to be optimized in order for the MESFET to achieve its minimum noise. This optimum source impedance $Z_{source,opt}$ is different from the source impedance required for maximum gain (*MSG*, *MAG*, or *U*). Thus the associated gain G_a, i.e., the power gain of the transistor matched for minimum noise, will definitely be smaller than *MSG*, *MAG*, and *U*. One can expect a G_a being at least 3 dB lower than the *MAG* value attainable under optimum conditions. Moreover, the dc operating conditions for minimum noise and maximum gain are different.

Figures 4.29(a) and (b) show the measured NF_{min} and G_a of two different 0.5 μm gate GaAs MESFETs reported in [20] and [45], respectively. Both transistors exhibit the typical *U*-shape characteristics, and the minimum of NF_{min} is found at I_D/I_{DSS} around 0.2. It should be noted that the *U*-shape is not always as pronounced as in Fig. 4.29. GaAs MESFETs showing extremely flat NF_{min} versus I_D curves have been reported [46]. In such cases, the choice of the bias conditions is less critical because the noise figure is almost insensitive to V_{GS}.

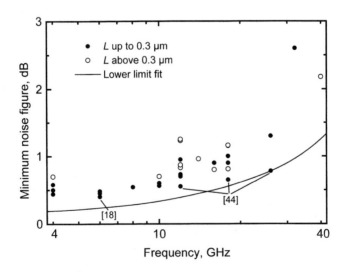

Figure 4.28. Minimum noise figure versus the frequency of experimental GaAs MESFETs. Also shown is the lower NF_{min} limit corresponding to the state of the art of GaAs MESFET in the year 2001.

Table 4.2. Minimum noise figure and associated gain of a GaAs MESFET with a gate length of 0.11 μm [44]

Frequency, GHz	12	18	26
Minimum noise figure, dB	0.55	0.65	0.78
Associated gain, dB	12.4	9.7	8.9

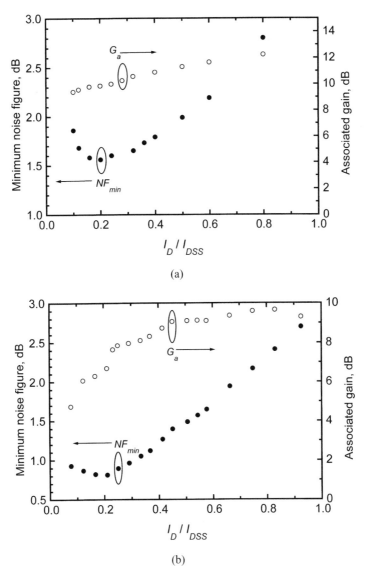

Figure 4.29. Minimum noise figure and associated gain of (a) a 0.5 μm gate GaAs MESFET at 12 GHz reported in [45] and (b) a 0.5 μm gate GaAs MESFET at 16 GHz reported in [20].

The frequency limits and the noise performance of GaAs MESFETs are connected. Simply stated, a MESFET having high f_T and f_{max} will automatically possess a low minimum noise figure.

4.6.3 Power GaAs MESFETs

The design targets for power MESFETs are a high output power P_{out} and/or high power-added efficiency PAE. The proper design of power MESFETs is more complicated compared to low-noise transistors. A high output power requires a large gate width. Depending on the operating frequency and the required output power, the gate width ranges from several 100 μm up to several 10 mm. Similar to the low-noise design, the gate is divided in parallel gate fingers. The number of gate fingers, however, is much larger in power MESFETs. Figure 4.30 shows a power MESFET layout with multifinger structure. It consists of two identically designed cells each containing one gate pad and 10 small unit cells each with one gate finger. The gate fingers are connected to the gate pads by a gate bus. The separation between the gate fingers is L_{GG}, and the maximum distance between the feed point and the endpoint of a gate finger is $L_{Gbus} + W_f$, where L_{Gbus} is the length of the gate bus and W_f is the width of a gate finger.

The power transistor design starts with the optimization of the unit cell. The layer structure has to be optimized, that is, the active layer thickness and doping density underneath the gate and in the source/drain contact regions, as well as the gate recess of the unit cell have to be designed according to the required operating and

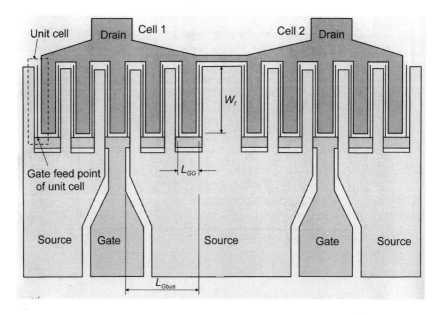

Figure 4.30. Layout of a two-cell, 20-finger GaAs power MESFET.

breakdown voltages, drain currents (on-current I_{max}), and output power density. At this point, the targeted operating frequency f_{op} has to be considered to define the gate length and, according to the $L/a > 3$ criterion, the corresponding active layer thickness, and doping density. The gate length can be estimated from the rules of thumb discussed in Section 1.2 [$f_T \geq (1 \ldots 1.5) \times f_{op}$ and $f_{max} \geq (2 \ldots 3) \times f_{op}$], and the gate length necessary to fulfill these rules can be obtained from Figs. 4.26 and 4.27, and from the experience that commonly properly designed GaAs MESFETs show a f_{max}/f_T ratio of around 2.

Typical GaAs power MESFETs have an I_{max} of 300 to 400 mA/mm, a gate–drain breakdown voltage of 20 to 30 V, and a V_k of 2 V, thus resulting in an maximum output power density of around 1 W/mm. Bearing in mind that self-heating and nonlinearities degrade the output power, 0.5 to 0.6 W/mm is a more realistic output power density. In Fig. 4.31, measured output powers (in W/mm) as a function of V_{DS} for GaAs MESFETs are compiled. Also shown are calculated output powers using Eq. (4-111), the piecewise linear I_D–V_{DS} output characteristics from Fig. 4.22, and a knee voltage of 2 V. Although a few transistors show output powers above 0.6 W/mm, most transistors confirm the rough estimate given above. Remarkable is the fact that the few reports on GaAs MESFETs with output power densities above 1 W/mm are from the late 1970s and early 1980s [47, 48].

A method to estimate the breakdown voltage of a Schottky junction has been discussed in Section 2.6. This method results in estimated breakdown voltages of about 14.6, 9.8, and 8 V for active layer doping densities of 10^{17}, 2×10^{17}, and 3×10^{17} cm^{-3}, respectively. However, GaAs MESFETs often show a higher gate–drain breakdown voltage BV_{GD}. This is because the simple method in Section 2.6 consid-

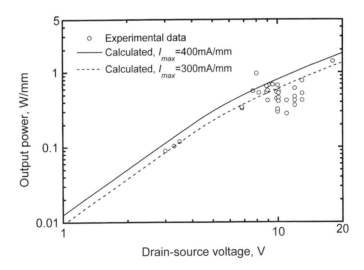

Figure 4.31. Measured output power densities of GaAs MESFETs. Also shown are the calculated output power densities assuming linearized I_D–V_{DS} characteristics and two different maximum currents I_{max}.

ered only the Schottky junction itself, i.e., the metal–semiconductor junction. In this case, the field lines are crowded near the gate edge on the drain side. The situation is different if a realistic GaAs surface is considered. At the GaAs surface, charged interface states are located. These states change the field distribution in a way that the field lines are less crowded at the gate edge, which results in a lower electric field and thus in a higher breakdown voltage. This effect has been investigated theoretically [49] and observed experimentally [50]. The following empirical relation for the gate–drain breakdown voltage has been suggested [50, 51]:

$$BV_{GD}[V] \cong \frac{9 \times 10^9 L_{\text{eff}}}{N_D a} \quad (4\text{-}125)$$

where L_{eff} is the effective gate length (in μm). In Eq. (4-125) the doping density N_D and the active layer thickness a have to be inserted in cm^{-3} and cm, respectively, and L_{eff} is 0.5 μm for transistors with gate lengths above 0.5 μm, and equal to the gate length L for transistors with $L < 0.5$ μm. Equation (4-125) holds for GaAs MESFETs with $N_D \times a$ products less than 2.3×10^{12} cm^{-2}. This condition is kept in many power FET designs. For $N_D \times a$ products higher than about 2.6×10^{12} cm^{-2}, the guidelines established in Section 2.6 should be used.

Gate–recess MESFETs show higher breakdown voltages compared to transistors with a planar gate–drain structure, and a double gate–recess structure can further enhance the breakdown behavior [52].

The next step in the design of a power MESFET is to determine the maximum allowed gate finger width. A high-frequency ac signal applied to the feed point of the gate needs a certain time to propagate to the end of the gate finger. Therefore, the gate finger portion closest to the gate feed point and the end of the finger operate out of phase. Furthermore, because the gate is a very narrow strip (short gate length), considerable signal loss occurs when signals at microwave frequencies propagate along such a strip. To minimize these effects, the finger width should not exceed a certain critical value $W_{f,\max}$.

The gate–channel structure of a FET can be considered as a microstrip line. According to the theory of microstrip lines [53], the propagation velocity of a signal along such a line can be calculated by

$$v = \frac{c}{\sqrt{\varepsilon_{\text{re}}}} \quad (4\text{-}126)$$

Here c is the velocity of light (3×10^{10} cm/s) and ε_{re} is the effective dielectric constant of the microstrip line given by the empirical equation [53]

$$\varepsilon_{\text{re}} = 0.475\varepsilon_r + 0.67 \quad (4\text{-}127)$$

where ε_r is the relative dielectric constant of the board material of the line (GaAs in our case). Based on the rule of wave propagation in vacuum ($c = \lambda \times f$), the wavelength λ in the microstrip line is

$$\lambda = \frac{v}{f} = \frac{c}{f\sqrt{\varepsilon_{re}}} \quad (4\text{-}128)$$

where f is the frequency of the propagating signal, i.e., the operation frequency of the MESFET. A frequently used condition in the design of power MESFETs is that the gate finger width should not exceed one-tenth of the length across which the phase angle of the signal shifts by $\pi/4$ rad. Because the wavelength corresponds to 2π rad, the gate finger width should not exceed $\lambda/80$ [54].

Another frequently applied criterion concerning the gate finger width requires that the gain degradation caused by signal losses along the gate should not exceed 0.5 dB [51]. The calculation of this maximum finger width, however, is more complicated compared to the calculation of $W_{f,max}$ using the $\lambda/80$ criterion. In principle, both criteria are correlated because a wide gate finger leads to a large phase shift as well as a large signal loss.

Figure 4.32 shows the gate finger widths versus the targeted operating frequency f_{op} of GaAs MESFETs. The symbols are the gate finger widths of experimental GaAs MESFETs and the dashed and solid lines are the maximum allowed gate finger widths according to the $\lambda/80$ criterion and to the -5 dB gain degradation criteria, respectively [51]. In essence, both criteria give similar maximum allowed finger widths above 5 GHz, and the gate finger widths of experimental transistors closely follow the $\lambda/80$ criterion. For operating frequencies of 10, 15, and 20 GHz, the gate finger widths should not exceed 150, 100, and 75 μm, respectively.

With the knowledge of the attainable P_{out} per mm gate width and the gate finger

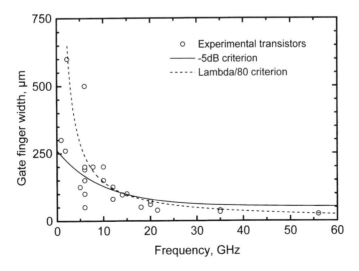

Figure 4.32. Calculated maximum gate finger width as a function of operating frequency based on the -5 dB criterion (solid line) and on the $\lambda/80$ criterion (dashed lines). The gate finger widths of reported experimental transistors are also given.

width, the total gate width and the number of gate fingers can be determined. The total gate width is simply the targeted P_{out} (in W) divided by the output power density (in W/mm) of the unit cell. The number of gate fingers, in turn, is given by the total gate width divided by the gate finger width.

Next, the gate-to-gate distance L_{GG}, the length of the gate bus L_{Gbus}, and the number of gate bond pads N_{pad} need to be considered. These three quantities are connected to each other and there are design trade-offs. From the viewpoint of thermal design, L_{GG} should be as large as possible to place the sources of heat generation, i.e., the transistor channels, as far from each other as possible. Because a gate bond pad adds a parasitic capacitance $C_{gs,pad}$ to the MESFET, the area and the number of pads should be minimized. Note that the number of gate pads is equal to the number of cells. On the other hand, the gate bus length should not exceed a certain value because along the bus a phase shift occurs, resulting in an out-of-phase operation of the different gate fingers. For L_{Gbus}, a requirement similar to the $\lambda/80$ criterion applies, and an upper limit for L_{Gbus} of $\lambda/64$ has been proposed in [54].

To get a picture on the complexity of power MESFET patterns, various GaAs power MESFET designs are listed in Table 4.3, and Fig. 4.33 shows the layout of a GaAs power MESFET with a total gate width of 15 mm.

It can be seen that the total gate width decreases with increasing operating frequency. Although transistors for f_{op} up to 12 GHz can be designed with total a gate width above 10 mm, the width reduces to a few hundred μm for transistors operating above 20 GHz. Experimental experiences show that the maximum output power of transistors possessing a gain of at least 3–4 dB at the operating frequency obeys the relation [48]

$$P_{out,max} f^2 \cong \text{const.} \qquad (4\text{-}129)$$

If the active layer design is optimized in regard to the combination of minimum V_k, maximum I_{max}, and maximum breakdown voltage, then the maximum output power is only dependent on the total gate width. Therefore Eq. (4-129) becomes

$$W_{max} \propto f^{-2} \qquad (4\text{-}130)$$

Table 4.3. Various power GaAs MESFET designs

f_{op}, GHz	L, μm	W, μm	N_{pad}	N_F*	W_f, μm	L_{GG}, μm	Ref.
6		13000	10	130	100	14	[55]
6		15000	8	75	200	25	[55]
8		15000	8	80	187.5	25	[56]
4-12	0.6	18000	12	12	125	10	[51]
18	0.3	1200	2	12	50	13	[51]
35	0.35	400		12	33	12	[57]

*N_F is the total number of gate fingers.

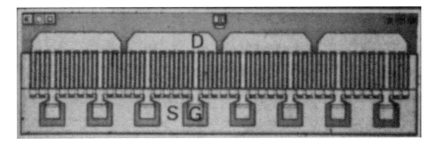

Figure 4.33. Photomicrograph showing the layout of a GaAs power MESFET. The chip size is 0.7mm × 2.2mm, the total gate width is 15 mm, and the number of gate fingers is 80. Taken from [56]. © 1980 IEEE.

In other words, the output power of a power GaAs MESFET can be maximized for a certain f_{op} by optimizing the active layer and finger pattern structure. But even if the transistor design is optimized, the attainable output power decreases rapidly when the frequency increases.

The gate finger pattern shown in Fig. 4.30 indicates that the connection of the source, gate, and drain contacts of the small unit cells to the bond pads requires crossovers. They are realized either by placing an insulator between the crossing lines, or by using air bridges. In an air bridge structure, two metal lines cross each other separated by an air gap.

Up to now, we have only dealt with the power transistor design in regard to the output power. Sometimes, the power-added efficiency is even more important than P_{out}. The power-added efficiency is scarcely influenced by the pattern layout. Most critical for achieving a high PAE is a low on-resistance of the transistor, i.e., a high I_{max} and a low V_k. This can be obtained using high-quality active layers having high low-field mobility and low parasitic resistances R_S and R_D.

The large signal gain of a power MESFET should be as large as possible, but it is normally much lower than the small signal power gains, such as MSG, MAG, and U. The reason is that the input of MESFETs in most circuits is not conjugately impedance-matched to the signal source as required by MSG, MAG, and U. Moreover, the load line is often chosen for maximum output power and/or maximum PAE, but not for maximum large-signal gain. As a rule of thumb, the large-signal gain of GaAs power MESFETs under realistic operating conditions is in the range between 5 dB and 10 dB.

The low thermal conductivity of GaAs makes it difficult to remove the heat generated in the MESFET channel to the surroundings. The choice of a large L_{GG} and placing the individual FET cells farther apart would help. The most effective way to reduce self-heating, however, is thinning the substrate and putting a plated heat sink on its back side. Typically, the substrates are thinned down to 25 to 30 μm. Furthermore, the so-called via holes etched through the substrate and filled with metal can serve as additional means to remove the heat. If the via holes are located underneath the source pads, and an electric contact is made between source pad and via hole

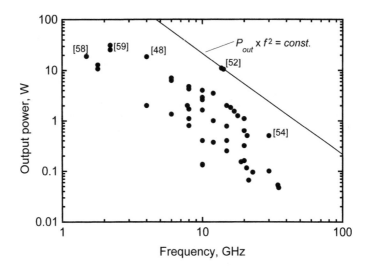

Figure 4.34. Measured output power of GaAs MESFETs. Also shown is the $P_{out} \times f^2 =$ const. line for the highest output power reported.

metal, then source bond wires are not necessary. This considerably reduces the source inductance and increases the gain.

Figure 4.34 shows the reported P_{out} values as a function of frequency for the state of the art of power GaAs MESFETs in the year 2001. The compiled data is for single transistor chips. To increase the output power, it is possible to combine several transistor chips to a power module. For example, a four-chip module with a total output power of 125 W at 2.2 GHz [59] and a two-chip module delivering 20.9 W at 14 GHz [52] have been reported.

In general, the design guidelines for GaAs MESFET (both low-noise and low-power) apply to HEMTs as well.

4.7 ISSUES OF WIDE BANDGAP MESFETs

4.7.1 Transistor Structures

Wide bandgap semiconductors, such as SiC and GaN, are promising materials for microwave power field-effect transistors. The wide bandgaps of these materials in excess of 2.5 eV lead to breakdown fields and breakdown voltages much larger than those of the traditional III–V materials. In addition, considerably higher operating temperatures are allowed for these semiconductors. These advantages have nourished high hopes of realizing wide bandgap power FETs with output power densities far above the 0.5 to 1 W/mm obtainable from GaAs MESFETs.

During the 1990s, considerable research efforts were focused on the development of SiC and GaN MESFETs, and numerous experimental transistors with promising high-frequency performance have been reported. The basic structure of wide bandgap MESFETs is very similar to that of GaAs MESFETs. A buffer layer is epitaxially grown on a substrate, followed by the n-type active layer (frequently with a gate recess). On top of this structure, the source, gate, and drain contacts are formed.

A difficulty in fabricating wide-bandgap MESFETs is the substrate. In the case of SiC, 4H and 6H substrates are commonly used. These substrates, however, are extremely expensive and can only be grown in small diameters. Two-inch (50 mm) and three-inch (75 mm) SiC wafers became commercially available only in 1997 and 2000, respectively [60]. To make things worse, the quality of the wafers is still not satisfactory because of large densities of a defect type called micropipe. Micropipes are holes several micrometers in size running through the wafers perpendicular to the surface. The micropipe densities in commercial 35 mm and 50 mm SiC wafers in the year 2000 were around 2 cm^{-2} and 50 cm^{-2}, respectively [60]. One micropipe in the active area of a power transistor can destroy the whole device. Commercial GaN substrates are not available at all, and commonly either sapphire or SiC substrates are used for GaN devices.

4.7.2 SiC MESFETs

Since 1995, both 4H and 6H SiC MESFETs have been investigated and reported. Because the low-field electron mobility in 4H SiC is about two times higher than that in 6H SiC (e.g., for a doping concentration of 10^{17}cm^{-3}, it is 610 cm^2/Vs in 4H compared to 340 cm^2/Vs in 6H), the performance of 4H SiC MESFETs is superior to their 6H counterparts. Thus, currently only 4H SiC MESFETs are extensively pursued. Both semi-insulating and conductive wafers are used, but because the parasitic capacitances of MESFETs on semi-insulating material are lower than on conductive substrates, semi-insulating wafers are normally preferred.

The low-field mobility of SiC is much lower compared to that of GaAs (610 cm^2/Vs for 4H SiC compared to 4700 cm^2/Vs for GaAs for a doping density of 10^{17}cm^{-3}), the saturation velocity is twice as much (about 2×10^7 cm/s compared to less than 10^7 cm/s) and is reached only at a much higher field than that in GaAs. Due to the lower mobility and the higher contact resistance R_{co} (several Ωmm for SiC compared to 0.1 Ωmm for GaAs), the parasitic resistances are quite high in SiC MESFETs. These factors lead to a smaller slope in the I_D–V_{DS} characteristics. Moreover, drain current saturation takes place only at relatively high V_{DS}, and consequently the knee voltage is higher than that in GaAs MESFETs.

Figure 4.35 shows the measured I_D–V_{DS} characteristics of a 1.2 μm × 100 μm gate 4H SiC MESFET [61]. A typical knee voltage of around 10 V (compared to 2 V in GaAs MESFETs) can be observed. The negative effect of high knee voltage on the output power, however, is by far compensated by the large breakdown voltage of SiC MESFETs. Both 4H and 6H SiC transistors with breakdown voltages around or above 150 V have been reported.

Figure 4.36(a) compares reported cutoff frequencies of SiC MESFETs and of GaN MESFETs (to be discussed in the next section). The aforementioned advan-

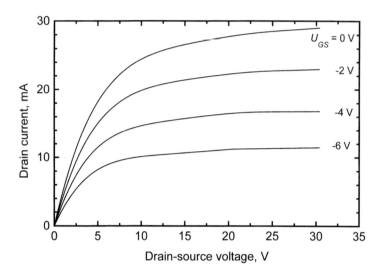

Figure 4.35. Measured output characteristics of a 1.2 μm gate 4H SiC MESFET. Data taken from [61].

tage of 4H SiC transistors is evident. In Fig. 4.36(b), published maximum frequencies of oscillation for the two MESFET types are shown. The best data are 22 GHz f_T and 50 GHz f_{max}, both reported for a 0.45 μm 4H SiC MESFET [62].

Because of the relatively low low-field mobility of SiC, SiC-MESFETs are not suited for low-noise applications. The projected application for SiC MESFETs is definitely power amplification at GHz frequencies. The design rules for SiC MESFETs are the same as those for GaAs power MESFETs discussed in the preceding section. The only difference is that, thanks to the high thermal conductivity of SiC, thermal design is less stringent and there is no need to thin the substrate.

The activation energies of nitrogen as a typical donor in 4H and 6H SiC are about 80 meV and 100 meV, respectively. This is much higher than the activation energy of donors in Si (P, E_{DA} = 45 meV) and in GaAs (Si, E_{DA} = 6 meV). Therefore, in the analysis and simulation of SiC MESFETs, the effect of incomplete dopant ionization should be taken into account (see Section 2.2).

As expected from their high breakdown voltages, SiC MESFETs have demonstrated around three times higher output power densities compared to GaAs MESFETs. Figure 4.37 shows a compilation of reported output powers versus V_{DS} for SiC MESFETs. For comparison, the calculated output power of GaAs MESFETs (solid line, from Fig. 4.31 for I_{max} = 400 mA/mm) is also included in the figure. The year 2001 state of the art for SiC MESFETs is characterized by a total output power of 120 W and an output power density of 5.6 W/mm at a frequency of 3.1 GHz [64]. The total output power of SiC MESFETs is expected to increase in the future when substrate quality improves and gate widths similar to that used in GaAs MESFETs can be realized.

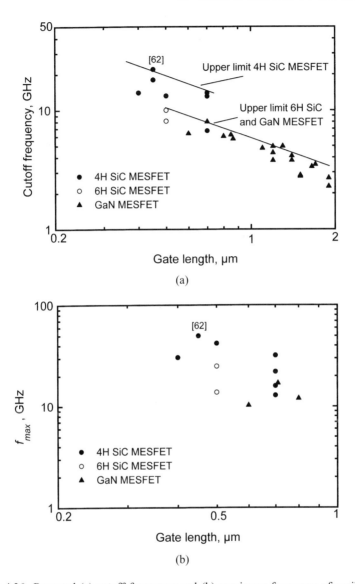

Figure 4.36. Reported (a) cutoff frequency and (b) maximum frequency of oscillation of SiC and GaN MESFETs.

4.7.3 GaN MESFETs

The first GaN MESFET was reported in the early 1990s. The motivation for the realization of these devices was to exploit the outstanding properties of GaN (large bandgap, high breakdown field, and high saturation velocity) for high-power microwave amplifiers. As shown in Figs. 4.36 (a) and (b), the frequency limits of GaN

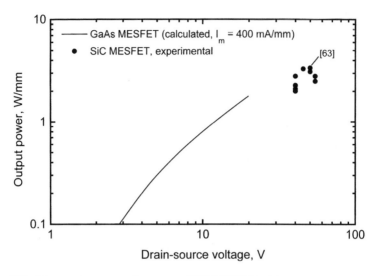

Figure 4.37. Measured power performance of SiC MESFETs. The output power of a GaAs MESFET is also included for comparison.

MESFETs follow a similar trend to those of 6H SiC MESFETs and are inferior to those of 4H SiC MESFETs. In parallel with the development of GaN MESFETs, considerable efforts have also been made to develop AlGaN/GaN HEMTs. From the beginning, GaN-based HEMTs have shown much better performance in terms of frequency limits and output power than GaN MESFETs, and consequently the interest in GaN MESFETs soon faded. The situation is not expected to change anytime soon, and GaN MESFETs will probably play only a minor role in microwave electronics in the foreseeable future.

REFERENCES

1. W. Shockley, A Unipolar "Field-Effect" Transistor, *Proc. IRE, 40,* pp. 1365–1376, 1952.
2. C. A. Mead, Schottky Barrier Gate Field Effect Transistor, *Proc. IEEE, 54,* pp. 307–308, 1966.
3. W. W. Hooper and W. I. Lehrer, An Epitaxial GaAs Field-Effect Transistor, *Proc. IEEE, 55,* pp. 1237–1238, 1967.
4. W. Baechtold, K. Daetwyler, T. Forster, T. O. Mohr, W. Walter, and P. Wolf, Half-Micron Si and GaAs Schottky-Barrier Field-Effect Transistors, *Electron. Lett., 9,* pp. 232–234, 1973.
5. C. E. Weitzel, Wide Bandgap Semiconductor Power Electronics, *IEDM Tech. Dig.,* pp. 51–54, 1998.
6. S. C. Binari, K. Doverspike, G. Kelner, H. B. Dietrich, and A. E. Wickenden, GaN FETs for Microwave and High-Temperature Applications, *Solid-State Electronics, 41,* pp. 177–180, 1997.

7. J. Browne, SiC MESFET Delivers 10-W Power at 2 GHz, *Microwaves & RF, 38,* pp. 138–139, Oct. 1999.
8. H. Statz, H. A. Haus, and R. A. Pucel, Noise Characteristics of Gallium Arsenide Field-Effect Transistors, *IEEE Trans. Electron Devices, 21,* pp. 549–562, 1974.
9. R. A. Pucel, H. A. Haus, and H. Statz, Signal and Noise Properties of Gallium Arsenide Microwave Field-Effect Transistors, *Advan. Electron. Electron Phys., 38,* pp. 195–265, 1975.
10. B. Carnez, A. Cappy, A. Kaszynski, E. Constant, and G. Salmer, Modeling a Submicrometer Gate Field-Effect Transistor Including Effects of Nonstationary Electron Dynamics, *J. Appl. Phys., 51,* pp. 784–790, 1980.
11. B. Carnez, A. Cappy, R. Fauquembergue, E. Constant, and G. Salmer, Noise Modeling in Submicrometer-Gate FET's, *IEEE Trans. Electron Dev., 28,* pp. 784–789, 1981.
12. A. Grebene and S. K. Ghandhi, General Theory for Pinched Operation of the Junction-Gate FET, *Solid-State Electron., 12,* pp. 573–589, 1969.
13. H. Fukui, Determination of the Basic Device Parameters of a GaAs MESFET, *BSTJ, 58,* pp. 771–797, 1979.
14. E. Wasserstrom and J. McKenna, The Potential Due to a Charged Metallic Strip on a Semiconductor Surface, *BSTJ, 49,* pp. 853–877, 1970.
15. P. J. Tasker and B. Hughes, Importance of Source and Drain Resistance of the Maximum f_T of Millimeter-Wave MODFET's, *IEEE Electron Device Lett., 10,* pp. 291–293, 1989.
16. S. Ohkawa, K. Suyama, and H. Ishikawa, Low Noise GaAs Field-Effect Transistors, *Fujitsu Sci. Techn. J., 11,* pp. 151–173, 1975.
17. P. Wolf, Microwave Properties of Schottky-Barrier Field-Effect Transistors, *IBM J. Res. Develop., 14,* pp. 125–141, 1970.
18. M. Feng, J. Laskar, J. Kruse, and R. Neidhard, Ultra Low-Noise Performance of 0.15-Micron Gate GaAs MESFET's Made by Direct Ion Implantation for Low-Cost MMIC's Applications, *IEEE Microwave and Guided Wave Lett., 2,* pp. 194–195, 1992.
19. M. Feng, D. R. Scherrer, P. J. Apostolakis, J. R. Middleton, M. J. McPartlin, B. D. Lauterwasser, and J. D. Oliver, Jr., Ka-Band Monolithic Low-Noise Amplifier Using Direct Ion-Implanted GaAs MESFET's, *IEEE Microwave and Guided Wave Lett., 5,* pp. 156–158, 1995.
20. C. L. Lau, M. Feng, T. R. Lepkowski, G. W. Wang, Y. Chang, and C. Ito, Half-Micrometer Gate-Length Ion-Implanted GaAs MESFET with 0.8-dB Noise Figure at 16 GHz, *IEEE Electron Device Lett., 10,* pp. 409–411, 1989.
21. A. van der Ziel, Unified Presentation of 1/f Noise in Electronic Devices: Fundamental 1/f Noise Sources, *Proc. IEEE, 76,* pp. 233–258, 1988.
22. J. Graffeuil and R. Plana, Low Frequency Noise Properties of Microwave Transistors and Their Application to Circuit Design, *Proc. Europ. Microwave Conf.,* pp. 62–75, 1994.
23. A. van der Ziel, Thermal Noise in Field-Effect Transistors, *Proc. IRE, 50,* pp. 1808–1812, 1962.
24. A. van der Ziel, Gate Noise in Field Effect Transistors at Moderately High Frequencies, *Proc. IEEE, 51,* pp. 461–467, 1963.
25. M. V. Fischetti, Monte Carlo Simulation of Transport in Technologically Significant Semiconductors of the Diamond and Zinc-Blende Structures—Part I: Homogeneous Transport, *IEEE Trans. Electron Devices, 38,* pp. 634–649, 1991.

26. G. E. Brehm and G. D. Vendelin, Biasing FET's for Optimum Performance, *Microwaves, 13,* pp. 38–44, Feb. 1974.
27. H. Fukui, Optimal Noise Figure of Microwave GaAs MESFET's, *IEEE Trans. Electron Devices, 26,* pp. 1032–1037, 1979.
28. H. Fukui, Design of Microwave GaAs MESFET's for Broad-Band Low-Noise Amplifiers, *IEEE Trans. Microwave Theory Tech., 27,* pp. 643–650, 1979.
29. H. Fukui, J. V. DiLorenzo, B. S. Hewitt, J. R. Velebir, H. M. Cox, L. C. Luther, and J. A. Seman, Optimization of Low-Noise GaAs MESFET's, *IEEE Trans. Electron Devices, 27,* pp. 1034–1037, 1980.
30. J. L. B. Walker (ed.), *High-Power GaAs FET Amplifiers,* Artech House, Norwood, MA, 1993.
31. S. C. Cripps, *RF Power Amplifiers for Wireless Communications,* Artech House, Norwood, MA, 1999.
32. P. L. D. Abrie, *Design of RF and Microwave Amplifiers and Oscillators,* Artech House, Norwood, MA, 1999.
33. A. J. Wilkinson and J. K. A. Everard, Transmission-Line Load-Network Topology for Class-E Power Amplifiers, *IEEE Trans. Microwave Theory Tech., 49,* pp. 1202–1210, 2001.
34. F. H. Raab, Maximum Efficiency and Output Power of Class-F Power Amplifiers, *IEEE Trans. Microwave Theory Tech., 49,* pp. 1162–1166, 2001.
35. L. J. Kushner, Output Performance of Idealized Microwave Power Amplifiers, *Microwave Journal, 32,* pp. 103–116, Oct. 1989.
36. C. A. Liechti, Microwave Field-Effect Transistors—1976, *IEEE Trans. Microwave Theory Tech., 24,* pp. 279–300, 1976.
37. J. A. Adams, I. G. Thayne, N. I. Cameron, M. R. S. Taylor, S. P. Beaumont, C. D. W. Wilkinson, N. P. Johnson, A. H. Kean, and C. R. Stanley, Fabrication and High Frequency Characterization of GaAs MESFETs with Gate Lengths Down to 30nm, *Microelectron. Eng., 11,* pp. 65–68, 1990.
38. M. Tokumitsu, M. Hirano, T. Otsuji, S. Yamaguchi, and K. Yamasaki, A 0.1-μm Self-Aligned-Gate GaAs MESFET with Multilayer Interconnection Structure for Ultra-High-Speed ICs, *IEDM Tech. Dig.,* pp. 211–214, 1996.
39. G. W. Wang and M. Feng, Quarter-Micrometer Gate Ion-Implanted GaAs MESFET's with an f_t of 126 GHz, *IEEE Electron Device Lett., 10,* pp. 386–388, 1989.
40. D. C. Caruth, R. L. Shimon, M. S. Heins, H. Hsia, Z. Tang, S. C. Shen, D. Becher, J. J. Huang, and M. Feng, Low-Cost 38 and 77 GHz CPW MMICs Using Ion-Implanted GaAs MESFETs, *IEEE MTT-S Dig.,* pp. 995–998, 2000.
41. H. Hsia, Z. Tang, D. Caruth, D. Becher, and M. Feng, Direct Ion-Implanted 0.12-μm GaAs MESFET with f_t of 121 GHz and f_{max} of 160 GHz, *IEEE Electron Device Lett., 20,* pp. 245–247, 1999.
42. K. Onodera, M. Tokimutsu, M. Tomizawa, and K. Asai, Effects of Neutral Buried p-Layer on High-Frequency Performance of GaAs MESFET's, *IEEE Trans. Electron Devices, 38,* pp. 429–436, 1991.
43. Y. Yamaue, K. Nishimura, K. Inoue, and M. Tokumitsu, 0.1 μm GaAs MESFET's Fabricated Using Ion-Implantion and Photo-Lithography, *GaAs IC Symp. Dig.,* pp. 185–188, 1993.
44. K. Onodera, K. Nishimura, S. Aoyama, S. Sugitani, Y. Yamane, and M. Hirano, Ex-

tremely Low-Noise Performance of GaAs MESFET's with Wide-Head T-Shaped Gate, *IEEE Trans. Electron Devices, 46,* pp. 310–319, 1999.
45. M. Feng, V. K. Eu, C. M. L. Yee, and T. Zielinski, A Low-Noise GaAs MESFET Made with Graded-Channel Doping Profiles, *IEEE Electron Dev. Lett., 5,* pp. 85–87, 1984.
46. M. Feng, D. Scherrer, P. J. Apostolakis, and J. Kruse, Temperature Dependent Study of the Microwave Performance of 0.25-μm Gate GaAs MESFET's and GaAs Pseudomorphic HEMT's, *IEEE Trans. Electron. Dev., 43,* pp. 852–860, 1996.
47. H. M. Macksey and F. H. Doerbeck, GaAs FETs Having High Output Power Per Unit Gate Width, *IEEE Electron Device Lett., 2,* pp. 147–148, 1981.
48. J. V. DiLorenzo and W. R. Wisseman, GaAs Power MESFET's: Design, Fabrication, and Performance, *IEEE Trans. Microwave Theory Tech., 24,* pp. 367–378, 1979.
49. H. Mizuta, K. Yamaguchi, and T. Takahashi, Surface Potential Effect on Gate-Drain Avalanche Breakdown In GaAs MESFETs, *IEEE Trans. Electron Dev., 34,* pp. 2027–2033, 1987.
50. S. H. Wemple, W. C. Niehaus, H. M. Cox, J. V. DiLorenzo, and W. O. Schlosser, Control of Gate-Drain Avalanche in GaAs MESFET's, *IEEE Trans. Electron Dev., 27,* pp. 1013–1018, 1980.
51. Y. Aoki and Y. Hirano, High-Power GaAs FETs, in: J. L. B. Walker (ed.), *High-Power GaAs FET Amplifiers,* Artech House, Norwood, MA, 1993.
52. Y. Saito, T. Kuzuhara, T. Ohmori, K. Kai, H. Ishimura, and H. Tokuda, Ku-Band 20 W Power GaAs FETs, *IEEE MTT-S Dig.,* pp. 343–346, 1995.
53. S. Y. Liao, *Microwave Devices and Circuits,* Prentice Hall, Englewood Cliffs, NJ, 1990.
54. Y. Hirachi, Y. Takeuchi, M. Igarashi, K. Kosemura, and S. Yamamoto, A Packaged 20-GHz 1-W GaAs MESFET with a Novel Via-Hole Plated Heat Sink Structure, *IEEE Trans. Microwave Theory Tech., 32,* pp. 309–316, 1984.
55. F. Hasegawa, Power GaAs FETs, in: J. V. DiLorenzo and D. D. Khandelwal, *GaAs FET Principles and Technology,* Artech House, Dedham, MA, 1982.
56. A. Higashisaka, Y. Takayama, and F. Hasegawa, A High-Power GaAs MESFET with an Experimentally Optimized Pattern, *IEEE Trans. Electron Devices, 27,* pp. 1025–1029, 1980.
57. T. Ho, F. Phelleps, G. Hegazi, K. Pande, H. Huang, P. Rice, and P. Pages, A Monolithics mm-Wave GaAs FET Power Amplifier for 35 GHz Seeker Applications, *Microwave J., 33,* pp. 113–122, Aug. 1990.
58. J. Morikawa, K. Asano, K. Ishikura, H. Oikawa, M. Kanamori, and M. Kuzuhara, 60W L-Band Power AlGaAs/GaAs Heterostructure FETs, *IEEE MTT-S Dig.,* pp. 1413–1416, 1997.
59. K. Ebihara, T. Takahashi, Y. Tateno, T. Igarashi, and J. Fukuya, L-Band 100-WATTs Push-Pull GaAs Power FET, *IEEE MTT-S Dig.,* pp. 703–706, 1998.
60. C. H. Carter, Jr., V. F. Tsvetskov, R. C. Glass, D. Henshall, M. Brady, St. G. Müller, O. Kordina, K. Irvive, J. A. Edmond, H.-S. Kong, R. Singh, S. T. Allen, and J. W. Palmour, Progress in SiC: From Material Growth to Commercial Device Development, *Mat. Sci. Eng., B61-62,* pp. 1–8, 1999.
61. O. Noblanc, C. Arnodo, E. Chartier, and C. Brylinski, Characterization of Power MESFETs on 4H-SiC Conductive and Semi-Insulating Wafers, *Mat. Sci. Forum, 264-268,* pp. 949–952, 1998.
62. S. T. Allen, R. A. Sadler, T. S. Alcorn, J. Sumakeris, R. C. Glass, C. H. Carter, Jr., and J.

W. Palmour, Silicon Carbide MESFET's for High-Power S-Band Applications, *Mat. Sci. Forum, 264-268,* pp. 953–956, 1998.

63. K. E. Moore, C. E. Weitzel, K. J. Nordquist, L. L. Pond, III, J. W. Palmour, S. Allen, and C. H. Carter, Jr., 4H-SiC MESFET with 65.7% Power Added Efficiency at 850 MHz, *IEEE Electron Device Lett., 18,* pp. 69–70, 1997.

64. J. C. Zolper, Wide Bandgap Semiconductor Microwave Technologies: From Promise to Practice, *IEDM Tech. Dig.,* pp. 389–392, 1999.

CHAPTER 5

HIGH ELECTRON MOBILITY TRANSISTORS

5.1 INTRODUCTION

The *H*igh *E*lectron *M*obility *T*ransistor (HEMT) is a heterostructure field-effect transistor. The development of HEMTs started in 1978, immediately after the successful experiments on modulation-doped AlGaAs/GaAs heterostructures, which revealed the formation of a two-dimensional electron gas (2DEG) with enhanced electron mobility [1]. The physics of 2DEGs has been covered in Section 3.4.

Earlier HEMTs utilized the AlGaAs/GaAs system, which was the most widely studied and best understood heterojunction system at that time, and consisted of a single heterojunction. Since then, other material systems have been used, and double heterojunction HEMTs, such as AlGaAs/InGaAs/GaAs HEMTs, have also been introduced. Figures 5.1(a) and (b) show schematically the typical layer sequences (underneath the gate) and conduction band edges of single and double heterojunction HEMTs, respectively. The heart of a HEMT is the heterojunction between the channel layer with the lower energy conduction band and the barrier layer with the higher energy conduction band. In the case of single heterojunction HEMTs, the substrate is commonly the same material as the channel layer, whereas in double heterojunction HEMTs the channel layer is sandwiched between an upper and a lower barrier layer. At the interfaces between the channel and the barrier layers a two-dimensional electron gas (2DEG) is created. The most important feature of the 2DEG is the high electron concentration and the high mobility of these electrons.

The naming of the HEMT was a little peculiar, and several different names evolved for this transistor type. The first experimental AlGaAs/GaAs HEMT was reported by Fujitsu researchers in 1980 [2], who also suggested the name HEMT. Although the initial response of the microwave device community to this device was rather restrained (see, e.g., [3]), considerable interest ignited shortly after groups from Bell Labs, University of Illinois, Cornell University, Rockwell, and Thomson-CSF reported the successful realization of AlGaAs/GaAs FETs based on the same concept. They, however, used different names for the transistor, including SDHT (*S*electively *D*oped *H*eterostructure Field-Effect *T*ransistor), MODFET (*Mo*dulation *D*oped *F*ield-*E*ffect *T*ransistor) and TEGFET (*T*wo-Dimensional *E*lectron *G*as *F*ield-*E*ffect *T*ransistor). Other names have also appeared in the literature,

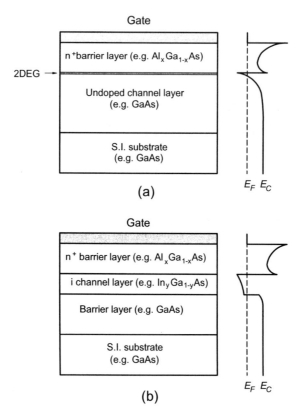

Figure 5.1. Typical layer sequence for (a) single heterojunction HEMT, e.g., AlGaAs/GaAs HEMT; and (b) double heterojunction HEMT, e.g., AlGaAs/InGaAs/GaAs HEMT (GaAs pHEMT).

such as HFET (*H*eterostructure *F*ield-*E*ffect *T*ransistor) and HJFET (*H*etero-*J*unction *F*ield-*E*ffect *T*ransistor). Since the designation HEMT is most widely used and describes the basic feature of the transistor best, it is adopted in this book.

The AlGaAs/GaAs HEMT was originally intended for low-temperature, high-speed digital applications. Worldwide efforts in HEMT research, however, soon led to transistors with impressive microwave performance as well. By the mid 1980s, AlGaAs/GaAs HEMTs outperformed the GaAs MESFET in terms of frequency limits and noise behavior. The 1985 state of the art of such devices was characterized by a 0.5 μm gate length AlGaAs/GaAs HEMT showing f_T and f_{max} values of 80 GHz and 120 GHz, respectively, and minimum noise figures of 1.2, 1.8, and 2.1 dB at 18, 30, and 40 GHz, respectively [4]. The first commercial low-noise HEMTs were introduced in 1985 [5]. Since then, AlGaAs/GaAs HEMTs have found numerous commercial low-noise and power applications.

In spite of these impressive results, however, it soon became clear that some basic properties of the AlGaAs/GaAs system set certain limits on frequency performance. One of these limitations is that the Al content x of the $Al_xGa_{1-x}As$ barrier layer should not be greater than about 0.3. The reason is that an increasing Al content will increase the donor activation energy E_{DA}, thus leading to a more significant incomplete ionization and a lower doping efficiency. Moreover, if x increases above 0.22, defects called DX centers are created and the transistor performance deteriorates. This limits the conduction band offset to about 0.22 eV and the 2DEG sheet density n_s to about 1.5×10^{12} cm^{-2}. The maximum attainable electron low-field mobility μ_0 is connected to the band structure of GaAs and is thus confined to about 8000 cm^2/Vs at room temperature. Consequently, the product $n_s \times \mu_0$, which is a key factor for the high-frequency performance of HEMTs, could not be increased above 10^{16} V^{-1}s^{-1} in experimental AlGaAs/GaAs HEMT structures (see Fig. 3.16).

Therefore, shortly after the first reports on AlGaAs/GaAs HEMTs, the search for alternative heterostructures offering the possibility of higher $n_s \times \mu_0$ products started. From the discussions in Section 2.3, it is clear that bulk $In_xGa_{1-x}As$ possesses a higher electron mobility compared to GaAs and that the mobility increases with increasing In content x. Furthermore, the conduction band offsets of AlGaAs/$In_xGa_{1-x}As$ and InAlAs/$In_xGa_{1-x}As$ heterostructures are considerably larger than that of AlGaAs/GaAs structures and increase with increasing x as well. These considerations led to the introduction of the following HEMT types in the mid 1980s [6–9]:

- GaAs pHEMT consisting of AlGaAs/$In_xGa_{1-x}As$ structures on GaAs substrate with $x \leq 0.3$, where the letter p indicates that the InGaAs channel layer is pseudomorphic
- Lattice matched InP HEMTs made from InAlAs/$In_xGa_{1-x}As$ structures on InP substrates with $x = 0.53$
- InP pHEMT consisting of InAlAs/$In_xGa_{1-x}As$ structures on InP substrates with $x > 0.53$, where the letter p stands for pseudomorphic

In addition, the concept of metamorphic layers had been introduced in the early 1990s [10], and such layers have been successfully used in GaAs metamorphic HEMTs (GaAs mHEMT) having $In_xGa_{1-x}As$ channel layers with x exceeding 0.3 on GaAs substrates [11]. Currently, GaAs pHEMTs are widely used in large-volume commercial applications. Although the technology of InP HEMTs is less matured compared to that of GaAs HEMTs, during the 1990s substantial progress was made and larger volume commercial applications were envisaged in the year 2000. In contrast, GaAs mHEMTs were still in the development phase at that time.

Three other types of HEMTs are currently under development. AlGaN/GaN HEMTs for high-power applications are intensively investigated, and their performance is quite promising (f_T and f_{max} exceeding 100 GHz [12], and output power density around 10 W/mm at 8 GHz [13]). Commercial GaN HEMTs were not yet available in 2001, however. There are also attempts being made to realize Si-based HEMTs. Such transistors having a strained Si channel layer grown on relaxed SiGe buffers showed f_T in excess of 60 GHz and f_{max} of 120 GHz [14, 15]. Finally, the so-called 6.1 Å material system consisting of AlSb, GaSb, and InAs is being investigated for low-power HEMTs. Experimental AlSb/InAs HEMTs with f_T of 160 GHz and f_{max} of 100 GHz have been reported [16, 17]. Both Si/SiGe and AlSb/InAs HEMTs are still in an early stage of development.

A mentioned above, various HEMTs based on different material systems are currently in commercial use or under development. Although the material properties and the resulting transistor performances are different, the underlying principle of HEMT operation is basically the same for all cases. Consider as an example the AlGaAs/GaAs HEMT shown in Fig. 5.2. It consists of an undoped or lightly doped GaAs channel (buffer) layer grown epitaxially on a semi-insulating GaAs substrate. The AlGaAs barrier layer is located on top of the channel layer. A thin portion of the AlGaAs barrier layer adjacent to the channel layer is undoped and serves as the spacer, whereas the upper portion of the barrier layer is n-type and acts as the electron supply layer. Finally, on top of the AlGaAs layer there is a highly doped n-type GaAs cap layer. In the middle of the HEMT structure, the GaAs cap and part of the AlGaAs layer are removed by etching, and a Schottky gate is located on the recessed surface of the n-type AlGaAs barrier layer. The gate length of the HEMT is L and the gate width (not shown in the figure) is W. The AlGaAs layer underneath the gate has the thickness a_1, which is the sum of the thickness of the doped layer, d_d, and the thickness of the spacer, d_{sp}. The doping concentration in the doped AlGaAs layer is N_D. As discussed in Section 3.4, a 2DEG is formed in the GaAs channel layer near the AlGaAs/GaAs heterojunction. The source and drain ohmic contacts are alloyed to ensure a conductive connection between the 2DEG and source and drain. The region between the dotted lines underneath the gate in Fig. 5.2 indicates the intrinsic transistor. Also shown in the figure is the spatial coordinate system we will use in the following.

Next we discuss the operation principle of the HEMT. The source terminal is commonly grounded. Let us consider first the case of zero drain–source and gate–source applied voltages V_{DS} and V_{GS}, respectively. The Schottky barrier of the

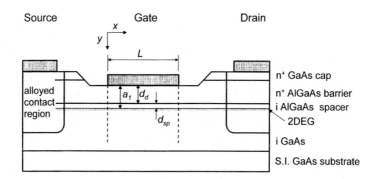

Figure 5.2. Cross section of an AlGaAs/GaAs HEMT.

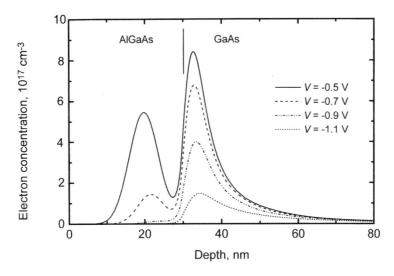

Figure 5.3. Electron concentration in an $Al_{0.24}Ga_{0.76}As/GaAs$ HEMT structure as a function of depth at different applied gate–source voltages. The HEMT consists of a 27 nm thick doped barrier layer with a doping density of 2×10^{18} cm^{-3} and a 3 nm spacer. The electron concentration has been calculated solving Schrödinger and Poisson equations self-consistently.

gate, ϕ_b, causes a space–charge region underneath the gate.* From the discussions of 2DEGs in Section 3.4, we know that the electron transfer across the heterojunction of the n-type AlGaAs and GaAs layers originates a second space–charge region in the AlGaAs layer adjacent to the heterojunction. For proper HEMT operation, the two space–charge regions should overlap so that the AlGaAs layer is fully depleted. When a negative gate–source voltage V_{GS} is applied, the transverse electric field resulting from the gate penetrates deeper into the semiconductor and starts to deplete the 2DEG. Then, the electron concentration n in the 2DEG decreases and the 2DEG sheet density, n_s, becomes smaller.

To illustrate the controlling effect of V_{GS} on the 2DEG, Fig. 5.3 shows the electron concentration n as a function of the depth y in an AlGaAs/GaAs HEMT for different gate potentials V, where V includes both V_{GS} and ϕ_b, i.e., $V = V_{GS} - \phi_b/q$. The thickness of the AlGaAs layer is 30 nm. Note that the 2DEG sheet density n_s is the integral of n within the limits $y = a_1 = 30$ nm and $y = \infty$. For the case of $V = -1.1$ V, the entire AlGaAs layer is depleted ($n \approx 0$ in the AlGaAs), all electrons reside in the GaAs layer, and both space–charge regions overlap. For $V = -0.9$ V, the electron density in the 2DEG is increased, and some electrons start to appear in the AlGaAs layer as well. When V increases to -0.7 V and finally to -0.5 V, the electron

---

*The Schottky gate of a MESFET is commonly described by the built-in voltage V_{bi}, whereas the Schottky barrier ϕ_b is typically used in the HEMT literature. V_{bi} and ϕ_b are closely related to each other [see Eq. (2-118)], and $V_{bi} \approx \phi_b/q$ for Schottky gates on heavily doped semiconductors.

density in the GaAs layer increases further, but now a considerable portion of the total number of electrons resides in the AlGaAs layer. Thus the two space–charge regions no longer overlap and we have two parallel conducting channels in the structure, namely a channel in the AlGaAs with low-mobility electrons and the GaAs channel with high-mobility electrons. The mobility of the electrons in the AlGaAs layer is low due to the following two reasons. First, the AlGaAs is doped and ionized impurity scattering reduces the mobility, and second, the mobility of AlGaAs is by nature lower than that of GaAs. This scenario is not desirable because now both the 2DEG electrons (sheet density n_s) and the electrons in the AlGaAs layer (sheet density $n_{s,AlGaAs}$) would contribute to the drain current when a drain–source voltage is applied. Thus, for HEMT turn-on operations, the thickness of the doped AlGaAs layer and the bias conditions have to be chosen such that complete depletion of the AlGaAs layer takes place, the two space–charge regions overlap, and n_s is large. When V_{GS} becomes more negative ($V < -1.1$ V in our example), the 2DEG becomes more depleted and n_s decreases further. At a certain value of V, the 2DEG is fully depleted and there are no free electrons in the heterostructure. The controlling effect of V_{GS} on n_s can also be seen from Fig. 3.13.

When comparing the controlling mechanisms of MESFETs and HEMTs, we see a key difference. In the MESFET, a variation of V_{GS} causes a change of the channel conductivity because the channel height is altered, whereas the density of carriers remains essentially constant. In the HEMT, however, the carrier density is changed, whereas the channel height, which is the thickness of the 2DEG, remains almost constant.

Let us now consider the case when a positive drain–source voltage V_{DS} is applied between the source and drain. In this case, an electric field in the $-x$ direction is present, the 2DEG electrons move from the source to the drain, and a drain current I_D flows. The magnitude of the drain current depends on the 2DEG sheet density n_s, which is controlled by the gate–source voltage as discussed above. More negative gate–source voltages result in smaller sheet densities and thus smaller drain currents. When a certain V_{DS} is applied (source grounded, i.e., source potential $V_S = 0$, drain potential V_D is positive) and a drain current flows, the channel potential $V(x)$ is increased from the source end toward the drain end of the gate. Therefore, the sheet concentration $n_s(x)$ at a location x in the channel not only depends on V_{GS}, but on $V(x)$ as well. The more negative the gate potential is compared to the channel potential, the smaller $n_s(x)$ becomes. Thus, $n_s(x)$ decreases toward the drain. This is comparable with the physics of the wedge-shaped channel in the MESFET with a decreasing channel height $b(x)$ toward the drain.

As in the case of the MESFET, there are both depletion and enhancement HEMTs. Most HEMTs are of the depletion type. For power amplifiers in mobile applications, however, enhancement power HEMTs are being developed to allow for a single-polarity power supply in the system. Depletion HEMTs, on the other hand, need a positive drain voltage and a negative gate voltage to switch the transistor off.

In the following sections, we will present details of two physically based models to describe the dc and small-signal behavior of HEMTs. These models resemble closely the MESFET models described in Sections 4.2 and 4.3, and they can also be extended to treat the high-frequency noise and the output power of HEMTs. In addition to the two models, we will present the concepts of modulation efficiency and of delay times, which are frequently used in the analysis of HEMT design and operation. Finally, design and performance issues of the different HEMT types will be addressed. Because both the HEMT and MESFET have similar design guidelines, we will omit the issues already discussed in the MESFET chapter and concentrate on the ones pertinent only to the HEMT.

5.2 DC ANALYSIS

In this section, we will describe the dc behavior of HEMTs in a manner similar to that of MESFETs covered in Section 4.2. Two dc HEMT models, which are closely related to the PHS model and the Cappy model for the MESFET, will be presented.

5.2.1 PHS-Like HEMT Model

The PHS-like HEMT model was originated from the work of Brookes [18]. His basic idea was to combine the calculation of the 2DEG sheet density (using the method of Delagebeaudeuf and Linh [19]) with a two-region transistor model very similar to the PHS MESFET model. We will closely follow Brookes's treatment and start with the derivation of expressions for the drain current of the intrinsic HEMT. The intrinsic transistor is divided into two regions as shown in Fig. 5.4(a), namely the ohmic region (region 1) with a length of L_1 and the velocity saturation

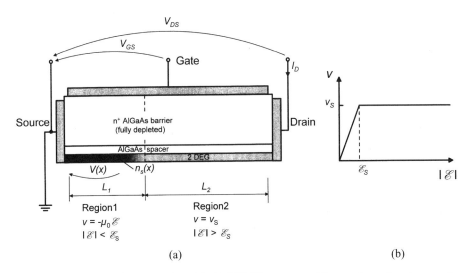

Figure 5.4. Two-region HEMT model. (a) HEMT structure with two channel regions, and (b) piecewise linear velocity-field characteristics.

region (region 2) with a length of L_2. The piecewise linear velocity-field characteristics [see Fig. 5.4(b)] is applied, with regions 1 and 2 operating in the linear and saturation velocity-field regions, respectively. We consider an AlGaAs/GaAs HEMT in the analysis, but the approaches apply generally to all types of HEMTs. The following assumptions, which are nothing but an extension of the gradual channel approximation for the MESFET, hold in region 1 of the HEMT:

1. The AlGaAs layer underneath the gate is fully depleted, i.e., $n_{s,\text{AlGaAs}}$ is zero and the two space–charge regions overlap.
2. The field distribution underneath the gate can be treated as a superposition of two separate fields. The first is the longitudinal field between drain and source in the direction of current flow, and the second is the transverse field in the direction perpendicular to the 2DEG channel. The longitudinal field prevails and the transverse field is negligible in the channel, and vice versa in the space–charge regions of the AlGaAs layer.
3. The 2DEG sheet density decreases gradually in the x direction.
4. The mobility in region 1 of the channel does not depend on the electric field and is equal to the low-field mobility μ_0. In region 2, the electrons travel with their saturation velocity, i.e., the electron velocity in region 2 is constant. Effects of nonstationary carrier transport are neglected.
5. The drain current passing through the transistor is entirely drift current, and generation/recombination in the channel is neglected.

The electric field $\mathscr{E}$ in region 1 is relatively small and, according to assumption (4), the electron drift velocity v is given by

$$v = -\mu_0 \mathscr{E} \qquad (5\text{-}1)$$

The magnitude of the field, and thus the drift velocity, increases with increasing x in region 1. At the boundary between regions 1 and 2, the magnitude of the field reaches its critical value $\mathscr{E}_S$, i.e., $|\mathscr{E}| = \mathscr{E}_S$, and the velocity becomes the saturation velocity v_S:

$$v_S = \mu_0 \mathscr{E}_S \qquad (5\text{-}2)$$

The electrons in region 2 travel with the constant velocity v_S. As mentioned above, the 2DEG sheet density n_s decreases with increasing x in region 1. To fulfill the requirement of a constant drain current in the HEMT, n_s must not decrease further in region 2 because of the constant drift velocity in the region. This condition is similar to the requirement of constant channel height in region 2 of the MESFET.

At low V_{DS}, the magnitude of the electric field everywhere in the channel is lower than $\mathscr{E}_S$, and region 1 covers the entire channel. When V_{DS} is increased, the magnitude of the field increases as well. At a certain V_{DS}, it eventually reaches $\mathscr{E}_S$ at the drain end of the gate and the carrier drift velocity becomes v_S. This particular drain–source voltage is the drain–source saturation voltage V_{DSS}. If V_{DS} is increased further, the condition $|\mathscr{E}| = \mathscr{E}_S$ is reached before the drain end of the gate, and the HEMT has to be divided into region 1 and region 2 as shown in Fig. 5.4.

The drain current at any x position in the HEMT is given by

$$I_D = qn_s(x)v(x)W \tag{5-3}$$

This equation resembles Eq. (4-3), which describes the drain current in a MESFET. Note that in Eq. (5-3), no channel height is involved because n_s is the sheet concentration given in cm^{-2}, whereas in Eq. (4-3) the carrier concentration n in cm^{-3} has been used. An expression for the 2DEG sheet density in a gated heterostructure has been derived in Section 3.4.2 [see Eq. (3-66)], and is repeated here for convenience:

$$qn_s = \frac{\varepsilon_1}{a_1 + \Delta a_1}\left(V_{po} - \frac{\phi_b}{q} + V_G - \frac{E_{F0}}{q} + \frac{\Delta E_C}{q}\right) \tag{5-4}$$

V_{po} is the pinch-off voltage and given by

$$V_{po} = \frac{qN_D}{2\varepsilon_1}(a_1 - d_{sp})^2 \tag{5-5}$$

where a_1 is the thickness of the barrier layer, ε_1 is the dielectric constant of AlGaAs, V_G is the gate voltage with respect to the GaAs substrate, ΔE_C is the conduction band offset, and Δa_1 and E_{F0} are fitting parameters. Because the source is grounded and thus the potentials of the source and substrate are equal, we can replace V_G in Eq. (5-4) by the gate–source voltage V_{GS}. An inspection of Eq. (5-4) reveals that the sheet density becomes zero when the $V_{GS} = \phi_b/q - V_{po} + E_{F0}/q - \Delta E_C/q$. Therefore, we define the threshold voltage V_{th} of the HEMT as

$$V_{th} = \frac{\phi_b}{q} - V_{po} + \frac{E_{F0}}{q} - \frac{\Delta E_C}{q} \tag{5-6}$$

This is the gate–source voltage at which the 2DEG density vanishes. Inserting Eq. (5-6) into Eq. (5-4) yields

$$qn_s = \frac{\varepsilon_1}{a_1 + \Delta a_1}(V_{GS} - V_{th}) \tag{5-7}$$

This expression gives the 2DEG sheet density at the source end of the gate. To calculate n_s at any position x in region 1, the x-dependent channel potential, i.e., the voltage drop $V(x)$ between the position x and source, must be taken into account, and Eq. (5-7) becomes

$$qn_s(x) = \frac{\varepsilon_1}{a_1 + \Delta a_1}[V_{GS} - V_{th} - V(x)] \tag{5-8}$$

The combination of Eqs. (5-8) and (5-3) leads to the following basic equation for the drain current in a HEMT:

$$I_D = \frac{\varepsilon_1}{a_1 + \Delta a_1}[V_{GS} - V_{th} - V(x)]v(x)W \quad (5\text{-}9)$$

Inserting Eq. (5-1) into Eq. (5-9) we obtain

$$I_D = -\frac{\varepsilon_1}{a_1 + \Delta a_1}[V_{GS} - V_{th} - V(x)]\mu_0 \mathcal{E}(x)W \quad (5\text{-}10)$$

At this stage we introduce the potential $w(x)$ given by

$$w(x) = V_{GS} - V_{th} - V(x) \quad (5\text{-}11\text{a})$$

and the corresponding potentials s and p at the source and at the boundary between regions 1 and 2, i.e., at $x = L_1$, are given by

$$s = V_{GS} - V_{th} \quad (5\text{-}11\text{b})$$

and

$$p = V_{GS} - V_{th} - V(L_1) \quad (5\text{-}11\text{c})$$

Note that $w(x)$, s, and p are always positive, that $s > p$, and that the expressions to be derived in the following are valid only for $V_{GS} - V_{th} > 0$, i.e., for the on-state of the transistor. The electric field $\mathcal{E}(x)$ is related to w as

$$\mathcal{E}(x) = -\frac{dV(x)}{dx} = \frac{dw(x)}{dx} \quad (5\text{-}12)$$

For simplicity, the x in parentheses indicating the x-dependence of the different quantities will be omitted in the following. Inserting Eq. (5-12) in Eq. (5-10) and rearranging the result yields the following differential equation with separated variables:

$$I_D dx = -\frac{\varepsilon_1}{a_1 + \Delta a_1} W \mu_0 w\, dw \quad (5\text{-}13)$$

Integrating Eq. (5-13) over region 1 of the HEMT, i.e., the left-hand side from $x = 0$ to $x = L_1$ and the right-hand side from $w = s$ to $w = p$, leads to

$$I_D = \frac{\varepsilon_1}{a_1 + \Delta a_1}\frac{W\mu_0}{2L_1}(s^2 - p^2) \quad (5\text{-}14)$$

5.2 DC ANALYSIS

At the boundary between regions 1 and 2, the introduced potential is p and the electron velocity is equal to the saturation velocity v_S. Thus, using Eqs. (5-3) and (5-8), the drain current can be alternatively expressed by

$$I_D = \frac{\varepsilon_1}{a_1 + \Delta a_1} p v_S W \tag{5-15}$$

Because the drain current at any position along the channel must be constant [see assumption (5)], Eqs. (5-14) and (5-15) can be equated and lead to an expression for the length of region 1:

$$L_1 = \frac{s^2 - p^2}{2p\mathscr{E}_S} \tag{5-16}$$

Next, we focus on the voltage drops across regions 1 and 2 of the HEMT. From Eqs. (5-11b) and (5-11c), the voltage drop across region 1, V_1, is obtained as

$$V_1 = s - p \tag{5-17}$$

For the voltage drop across region 2, V_2, a consideration similar to that concerning the voltage drop across region 2 of the PHS MESFET model [see Eq. (4-17)] can be made, which finally leads to

$$V_2 = \frac{2(a_1 + \Delta a_1)\mathscr{E}_S}{\pi} \sinh\left[\frac{\pi L_2}{2(a_1 + \Delta a_1)}\right] \tag{5-18}$$

The sum of V_1 and V_2 is equal to the drain–source voltage:

$$V_{DS} = V_1 + V_2 = s - p + \frac{2(a_1 + \Delta a_1)\mathscr{E}_S}{\pi} \sinh\left[\frac{\pi L_2}{2(a_1 + \Delta a_1)}\right] \tag{5-19}$$

When simulating the HEMT under specific bias conditions, one first has to determine whether the applied drain–source voltage is above or below the saturation voltage V_{DSS}. At $V_{DS} = V_{DSS}$ the condition

$$L = \frac{s^2 - d^2}{2d\mathscr{E}_S} \tag{5-20}$$

holds, with d being the potential at the drain

$$d = V_{GS} - V_{th} - V_{DS} \tag{5-21}$$

From Eqs. (5-20) and (5-21) the saturation voltage can be determined by

$$V_{DSS} = s + L\mathscr{E}_S + \sqrt{L^2\mathscr{E}_S^2 + s^2} \tag{5-22}$$

For $V_{DS} < V_{DSS}$, the drain current can be calculated from Eq. (5-14), provided p and L_1 are replaced by d and L, respectively. For $V_{DS} > V_{DSS}$, the drain current has to be calculated by numerically solving the system of Eqs. (5-14), (5-15), and (5-19).

To this point, only the intrinsic HEMT has been considered. The voltages involved in Eqs. (5-7)–(5-22) are the intrinsic voltages, which are different from the external applied voltages. The relationship between the external and internal voltages is the same as for the MESFET [see Eqs. (4-21) and (4-22)]. The determination of source and drain resistances, R_S and R_D, however, is more complicated for HEMTs.

We focus on R_S associated with the source–gate region of a HEMT, as shown in Fig. 5.5. The same approach applies to R_D as well. The source–gate region, which is located between the source contact and the intrinsic transistor, can be modeled as a network of several lumped and distributed resistive components. First, there are the contact resistances R_{co1} and R_{co2} describing the transition resistances from the source metallization to the n$^+$ cap layer, and from the source metallization to the 2DEG channel, respectively. Second, the current can flow either through the cap layer with the sheet resistance r_{sh1} or through the 2DEG having the sheet resistance r_{sh2}. The portion of the current flowing through the cap has to surmount the barrier layer somewhere between source and gate to enter the 2DEG channel. Because of the band bendings at the cap-barrier and barrier-channel interfaces, the only possibility to surmount the barrier layer is tunneling. Therefore, the electron transport from the cap to the 2DEG via the barrier layer is described by a distributed tunneling resistance R_T having a specific tunneling resistance ρ_T. The total source resistance can be obtained by treating the different resistive elements as a transmission line. This has been done by Feuer [20], who formulated a set of coupled differential equations for the current flow in the cap and in the 2DEG and presented a solution using appropriate boundary conditions. For the case of a narrow gate recess where the cap layer extends up to the gate, Feuer's final formula for the source resistance is:

$$R_S = r_{sh,eff} L_{SG} + \frac{1}{r_{sh1} + r_{sh2}} \frac{\alpha + \beta \cosh(kL_{SG}) + \gamma k \sinh(kL_{SG})}{(r_{sh1} + r_{sh2})\cosh(kL_{SG}) + (R_{co1} + R_{co2})k \sinh(kL_{SG})}$$

(5-23)

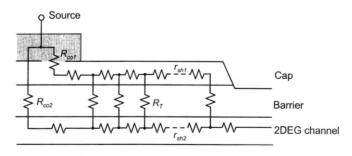

Figure 5.5. Source-to-gate region of a HEMT depicting the various resistive components contributing to the parasitic source resistance R_S.

where

$$\alpha = 2r_{sh2}(r_{sh1}R_{co2} - r_{sh2}R_{co1})$$

$$\beta = 2r_{sh2}^2 R_{co1} + (r_{sh1}^2 + r_{sh2}^2)R_{co2}$$

$$\gamma = (r_{sh1} + r_{sh2})R_{co1}R_{co2} + r_{sh2}^2 \rho_T$$

$$k = \sqrt{\frac{r_{sh1} + r_{sh2}}{\rho_T}}$$

$$r_{sh,eff} = \frac{r_{sh1}r_{sh2}}{r_{sh1} + r_{sh2}}$$

and L_{SG} is the source–drain distance. Equation (5-23) gives the source resistance in Ωmm provided that all resistance components, i.e., contact, sheet, and specific tunnel resistances, are in Ωmm as well. This is obtained by dividing the contact resistance in Ωcm taken from Eq. (5-25), the sheet resistances in $\Omega/\square$ taken from Eqs. (5-26) and (5-27), and the specific tunneling resistance in Ωcm² (shown in Fig. 5.6 for an AlGaAs/GaAs hetereostructure) by $W = 1$ mm $= 0.1$ cm. The source resistance R_S in Ωmm corresponds to R_S of a transistor with a gate width W of 1 mm. The source resistance in Ω of a HEMT with a certain gate width and the R_S value in Ωmm are related by

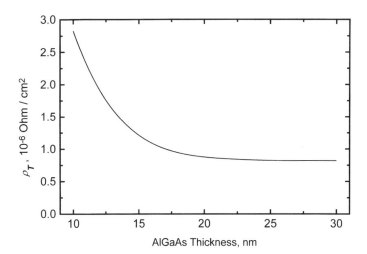

Figure 5.6. Specific tunneling resistance ρ_T of an n-GaAs/n-AlGaAs/i-GaAs heterojunction system calculated as a function of barrier layer thickness [23]. The doping concentrations of the n-GaAs and the n-AlGaAs layers are 2×10^{18} cm^{-3} and 3×10^{18} cm^{-3}, respectively.

$$R_S[\Omega] = \frac{R_S[\Omega mm]}{W[mm]} \quad (5\text{-}24)$$

To calculate R_S, we still need information on the contact resistances R_{co1} and R_{co2}, the sheet resistances r_{sh1} and r_{sh2}, and the specific tunneling resistance ρ_T. For HEMTs having a long source contact, the contact resistance is given by [21, 22]

$$R_{co1,2} = \sqrt{\rho_{co1,2} r_{sh1,2}} \quad (5\text{-}25)$$

where ρ_{co1} and ρ_{co2}, with typical values of 10^{-8} Ωcm^2 and 10^{-7} Ωcm^2 [23], are the specific transition resistances from the source contact to the cap and from the source contact to the 2DEG, respectively. The sheet resistances are

$$r_{sh1} = \frac{1}{q\mu_{0,cap} N_{D,cap} a_{cap}} \quad (5\text{-}26)$$

and

$$r_{sh2} = \frac{1}{q\mu_{0,2DEG} n_{s,2DEG}} \quad (5\text{-}27)$$

where $\mu_{0,cap}$ and $\mu_{0,2DEG}$ are the low-field mobilities in the cap layer and in the 2DEG channel, respectively, $N_{D,cap}$ is the doping concentration in the cap, and a_{cap} is the thickness of the undepleted portion of the cap. Equations (5-26) and (5-27) give the sheet resistances of a semiconductor layer whose length is equal to the width. The unit of these resistances is $\Omega/\square$.

The specific tunneling resistance depends on the thickness of the barrier layer. Figure 5.6 shows ρ_T calculated by Ando and Itoh for GaAs/AlGaAs/GaAs structures as a function of the AlGaAs barrier thickness [23].

A simple method to estimate the source resistance avoiding the quite lengthy formulas given above is to define the lower and upper limits of the source resistance, $R_{S,ll}$ and $R_{S,ul}$. The lower limit is obtained by setting the tunneling resistance to zero and is given by

$$R_{S,ll} = R_{co2} + r_{sh,eff} L_{SG} \quad (5\text{-}28)$$

The upper limit, on the other hand, is obtained by assuming an infinite tunneling resistance, suggesting a current flow only through the 2DEG:

$$R_{S,ul} = R_{co2} + r_{sh2} L_{SG} \quad (5\text{-}29)$$

The Feuer model has been extended by Ando and Itoh [23], who took into account the fact that the conditions between source and gate can vary with the position x. Figure 5.7 shows the source resistance of AlGaAs/GaAs HEMTs having different

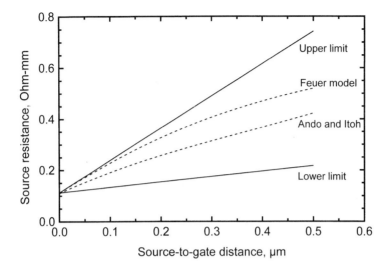

Figure 5.7. Calculated source resistance R_S of an AlGaAs/GaAs HEMT (AlGaAs layer thickness is 10 nm, r_{sh1} = 250 Ω/□, and r_{sh2} = 1260 Ω/□) using the Feuer formula. Also shown are the source resistances of the same HEMT calculated by Ando and Itoh [23], and the upper and lower limits for R_S.

source–gate separations calculated using the Feuer model, i.e., by Eq. (5-23), the lower and upper limits calculated applying Eqs. (5-28) and (5-29), and the source resistance obtained by Ando and Itoh [23]. The specific contact resistances given above and the specific tunneling resistance from Fig. 5.6 have been used to obtain the results in Fig. 5.7.

Figure 5.8 compiles reported source resistances for three different HEMT types (AlGaAs/GaAs HEMTs, GaAs pHEMTs, and InP HEMTs) as a function of gate length L. It should be noted that the gate length and the source–gate distance L_{SG} are not necessarily equal. Although L_{SG} should be as small as possible to minimize R_S, L_{SG} can be considerably larger compared to L, especially in short-gate HEMTs with L < 0.25 μm. It can be seen that on average the source resistances of GaAs pHEMTs and InP HEMTs are lower than that of AlGaAs/GaAs HEMTs. The lowest source resistances reported for the three HEMT types, however, are almost the same.

It is possible to extend the HEMT model discussed above to take into account a nonlinear v–$\mathscr{E}$ characteristics in region 1. This can be done using the same approach as for the MESFET described in Section 4.2.1. We use Eq. (4-27) to express the nonlinear dependence of the velocity on the field:

$$v = -\frac{\mu_0 \mathscr{E}}{1 - \dfrac{\mu_0 \mathscr{E}}{\gamma v_S}} \qquad (5\text{-}30)$$

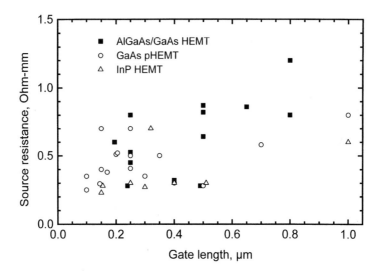

Figure 5.8. Parasitic source resistance versus gate length for different experimental HEMTs.

where γ is a factor accounting for the bowing of the v–$\mathscr{E}$ characteristics. Note that the electric field is negative because it is directed in the $-x$ direction from the positively biased drain to the source. Putting Eq. (5-30) into Eq. (5-9), the following expression for the drain current valid in region 1 is obtained:

$$I_D = \frac{\varepsilon_1}{a_1 + \Delta a_1} \frac{\mu_0 W}{L_1 + \dfrac{\mu_0}{\gamma v_s}(s-p)} \frac{s^2 - p^2}{2} \tag{5-31}$$

5.2.2 Cappy-Like HEMT Model

It is sometimes desirable to model the behavior of HEMTs taking into account both the nonlinear GaAs-like v–$\mathscr{E}$ characteristics with a pronounced peak and the effects of nonstationary carrier transport. A HEMT model of this type has been reported by Cappy et al. [24] based on the relaxation time approximation (RTA). In the following, we will develop a similar model that is based on the RTA as well. Our approach, however, is somewhat different from that in [24] and is more closely related to the MESFET model described in Section 4.2.2. Thus, once a computer program to simulate the dc MESFET behavior is written, only minor modifications are required to simulate the HEMT dc behavior. The derivation is focused on the AlGaAs/GaAs HEMT, but in principle is valid for any type of HEMT.

From the discussions in Sections 3.4 and 5.2.1, it is clear that the transfer of electrons from the AlGaAs to the GaAs creates a 2DEG and that for proper HEMT operation the AlGaAs layer underneath the gate should be depleted. The electron concentration in the AlGaAs/GaAs heterostructure is schematically shown in Fig. 5.9(a). It can be approximated by a step-like carrier profile given in Fig. 5.9(b), assuming that the 2DEG electrons are located in a narrow channel region in the GaAs layer adjacent to the heterointerface, whereas the electron concentration is zero elsewhere. The thickness of such a channel region is the effective thickness of the 2DEG, d_{2DEG}. It amounts to about 10–20 nm in single heterojunction HEMTs like AlGaAs/GaAs HEMTs, or to the thickness of the channel layer in double heterojunction HEMTs like AlGaAs/InGaAs/GaAs HEMTs. To ensure that the actual 2DEG shown in Fig. 5.9(a) and the 2DEG channel region in Fig. 5.9(b) contain the same number of electrons, the electron concentration in the 2DEG channel region, n_{ch}, is calculated by

$$n_{ch} = \frac{n_s}{d_{2DEG}} \qquad (5\text{-}32)$$

Note that n_{ch} is given in cm^{-3} and n_s in cm^{-2}.

One important idea of our approach to the Cappy-like HEMT modeling is as follows. First, for zero applied drain–source voltage and for the gate–source voltage of interest the 2DEG sheet density n_s is calculated. This can be done, for example, using the model of Delagebeaudeuf and Linh [19]. We will come back to this issue later. Under normal bias conditions, the AlGaAs layer underneath the gate is fully depleted and free electrons exist only in the 2DEG. Thus, the region under the gate consists of the AlGaAs layer with a zero electron concentration followed by the 2DEG channel region, where the electron concentration is equal to n_{ch} given by Eq. (5-32). Consider now the case where the gate voltage becomes a little more negative. Then, the electron concentration in the AlGaAs will remain zero and n_{ch} will decrease. In the following we will show that the reaction of the structure AlGaAs ($n = 0$) followed by the 2DEG channel region ($n = n_{ch}$) on a variation of the gate volt-

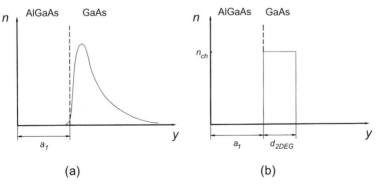

Figure 5.9. Electron concentration in an AlGaAs/GaAs HEMT. (a) Realistic electron distribution. (b) Approximated electron distribution with a step-like profile.

age is very similar to that of a structure consisting of undoped AlGaAs followed by doped GaAs with $N_D = n_{ch}$. In particular, the extension of the space–charge region into the channel region will be considered in detail. As a result we will see that the HEMT can be modeled in close analogy to a MESFET with a step-like doping profile underneath the gate.

Consider first a homogeneously doped semiconductor sample with a Schottky gate. The built-in voltage V_{bi} and the applied gate voltage V_G give rise to a space–charge region with a thickness d_{sc}. Applying Eq. (2-122), the relationship between the voltage drop V across the space–charge region and d_{sc} is obtained as

$$V = -V_G + V_{bi} = \frac{qN_D d_{sc}^2}{2\varepsilon} \tag{5-33}$$

If the doping density N_D of the sample is not homogeneous but rather position-dependent, i.e., a function of depth y, Eq. (5-33) becomes

$$V = \frac{q}{\varepsilon} \int_0^{d_{sc}} N_D(y) y \, dy \tag{5-34}$$

For the case of a step-like doping profile with two constant doping densities of N_{D1} for $0 < y < a_1$ and N_{D2} for $a_1 < y < d_{sc}$, we have

$$V = \frac{q}{\varepsilon} \int_0^{a_1} N_{D1} y \, dy + \frac{q}{\varepsilon} \int_{a_1}^{d_{sc}} N_{D2} y \, dy = \frac{qN_{D1} a_1^2}{2\varepsilon} + \frac{qN_{D2}(d_{sc}^2 - a_1^2)}{2\varepsilon} \tag{5-35}$$

If the doping concentration N_{D1} between $y = 0$ and $y = a_1$ is zero, Eq. (5-35) is reduced to

$$V = \frac{qN_{D2}(d_{sc}^2 - a_1^2)}{2\varepsilon} \tag{5-36}$$

The electric field resulting from such a voltage repels electrons away from the gate. In principle, it does not matter whether the electrons stem from the donors in the sample or come from another origin, the repelling force on the electrons is the same. Thus, the donor concentration N_D in Eqs. (5-33)–(5-36) can be replaced by the electron concentration n.

Let us now compare two gated-GaAs samples shown in Fig. 5.10. Sample I is of the n-type and homogeneously doped at 10^{18}cm^{-3}. Two different gate voltages V_{G1} and V_{G2} are applied subsequently to sample I: $-V_{G1} + V_{bi} = V_1 = 1$ V, and $-V_{G2} + V_{bi} = V_2 = 2$ V. The corresponding space–charge region thicknesses, d_{sc1} and d_{sc2}, calculated from Eq. (5-33) are 37.7 nm and 53.3 nm, respectively. The difference between d_{sc1} and d_{sc2} is caused by the difference in the applied voltages $\Delta V = V_2 - V_1 = 1$ V.

Sample II has a step-like doping profile. The region adjacent to the gate is undoped and has a thickness d_1 of 37.7 nm (the same as d_{sc1}). This is followed by a homogeneously doped region with $N_D = 10^{18} \text{ cm}^{-3}$ covering the rest of the GaAs layer.

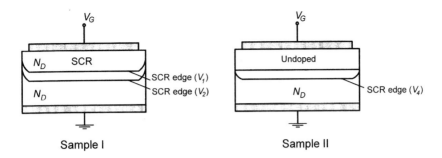

Figure 5.10. Illustration of the space–charge region (SCR) in two GaAs samples under different applied gate voltages. Sample I is homogeneously doped; sample II has a step-like doping profile.

If a gate voltage V_{G3} that exactly compensates the built-voltage (i.e., $-V_{G3} + V_{bi} = V_3 = 0$) is applied and carrier diffusion is neglected, the resulting carrier concentration profile is equal to the doping profile, i.e., step-like. Now we apply a gate voltage V_{G4} to sample II: $-V_{G4} + V_{bi} = V_4 = \Delta V = V_2 - V_1$, i.e., $V_4 = 1$ V, and the corresponding space–charge region $d_{sc4} = 53.3$ nm is obtained from Eq. (5-36), which is precisely the value of d_{sc2}. Obviously, the reactions of homogeneously and step-like doped semiconductor samples to gate voltage variations are very similar. Such device physics will be used in the following to model the dc behavior of HEMTs.

The HEMT is divided (discretized) into a number of small segments, each with a length of Δx, in the direction of current flow. The bias conditions are V_{GS} and the drain current I_D fed into the transistor. The drain–source voltage is the sum of the voltage drops ΔV across all channel segments. Figure 5.11 shows the cross section of the model HEMT structure. According to the preceding discussions, all electrons are confined to the channel region with a thickness d_{2DEG}. The basic approach to developing the Cappy-like HEMT model is the same as that for the Cappy MESFET

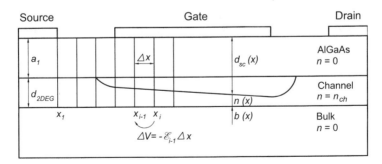

Figure 5.11. Schematic of the HEMT structure for the Cappy-like HEMT model, showing the different layers and the shape of the space–charge region in the channel. At position x_1, the extension of the space–charge region is $d_{sc}(x)$, the remaining height of the 2DEG channel is $b(x)$, and the electron concentration is $n(x)$.

model described in Section 4.2.2. Therefore, we will not repeat the details here, but rather concentrate on the steps and equations specific only to the HEMT. The basic equations of the HEMT model are the energy balance equations (4-29) and (4-30), as well as the Poisson equation and the drift current equation given by

$$\frac{d\mathscr{E}}{dx} = \frac{q}{\varepsilon}(n_{ch} - n) \qquad \text{Poisson equation} \qquad (5\text{-}37)$$

$$I_D = qnvbW = qnv(a_1 + d_{2DEG} - d_{sc}) \qquad \text{drift current equation} \qquad (5\text{-}38)$$

Here, n is the actual electron density anywhere in the channel layer, b is the thickness of the 2DEG channel, and d_{sc} is the extension of the space–charge region (underneath the surface of the barrier layer or underneath the gate). The meaning and determination of n_{ch} will be discussed below. The main difference between Eqs. (4-31)–(4-32) and Eqs. (5-37)–(5-38) is that the doping concentration N_D in Eq. (4-31) becomes n_{ch} in Eq. (5-37), and the channel height b in the HEMT is $a_1 + d_{2DEG} - d_{sc}$ in Eq. (5-38). As in the MESFET model, four equations [(4-29), (4-30), (5-37), and (5-38)] can be discretized so that the quantities at a certain point in the channel, x_i, can be calculated provided the quantities at the previous point, x_{i-1}, are known.

The calculation in the frame of the Cappy-like HEMT model begins with defining the applied gate–source voltage V_{GS}. Then, the 2DEG sheet densities in the channel between source and gate, $n_{s,2DEGsg}$, and the 2DEG sheet density in the channel under the gate, $n_{s,2DEGg}$, are determined. This can be done using the model of Delagebeaudeuf and Linh [19], solving the self-consistent Schrödinger–Possion equation, or applying one of the advanced methods described in the literature (see, e.g., [25–27] and references therein). When applying the model of Delagebeaudeuf and Linh, Eq. (5-4) is used, together with setting $\phi_b/q = V_{surf}$ and $V_G = 0$ to calculate $n_{s,2DEGsg}$, and setting ϕ_b to the Schottky barrier of the gate and $V_G = V_{GS}$ to calculate $n_{s,2DEGg}$. The meaning of V_{surf} has been explained in Section 4.2.1 [see Eq. (4-23)]. This, in combination with Eq. (5-32), yields the electron concentrations in the 2DEG channel region, n_{ch}. It is n_{chsg} between source and gate, and n_{chg} in the region underneath the gate for zero I_D, i.e., for zero V_{DS}.

Next, the drain current I_D fed into the transistor is defined. Let us first determine the conditions at the source end of the channel, i.e., $x = x_1$. The potential V_1 is zero or, if the contact resistance R_{co} is taken into account, is $V_1 = I_D \times R_{co}$. Using Eq. (5-38), the velocity at x_0 is obtained knowing $n_1 = n_{chsg}$ and setting $b_1 = d_{2DEG}$ and $d_{sc1} = a_1$. From the stationary velocity- and energy-field (v–$\mathscr{E}$ and E–$\mathscr{E}$) characteristics of the channel material (GaAs in our case), the corresponding stationary values $\mathscr{E}_{1,st} = \mathscr{E}_1$ and $E_{1,st} = E_1$ are deduced. Then, starting at x_2 and moving toward the drain end, we go through the following procedure analogous to that used in Section 4.2.2:

1. Calculate the energy relaxation time at position x_i, τ_{Ei}, using Eq. (4-33) and the electron energy E_i using Eq. (4-34).
2. Set $E_{st,i} = E_i$ and deduce the stationary values of the field, velocity, and effective mass at x_i from E_i and the stationary v–$\mathscr{E}$, E–$\mathscr{E}$, and m^*–$\mathscr{E}$ characteristics of the channel material.

3. Calculate the momentum relaxation time τ_{pi} using Eq. (4-35), rearrange the momentum balance equation, and calculate the electron velocity v_i using Eq. (4-39) with

$$D_1 = 1 + \frac{v_{i-1}\tau_{P,i}}{\Delta x}$$

$$D_2 = -\frac{q\tau_{P,i}}{m^*_{st,i}}\left(-\mathcal{E}_{i-1} - \frac{\Delta x q}{\varepsilon}n_{ch} + \frac{m^*_{st,i}}{q}\frac{v^2_{i-1}}{\Delta x}\right) \quad (5\text{-}39)$$

$$D_3 = -\frac{q\tau_{p,i}\Delta x I_D}{\varepsilon m^*_{st,i}(a_1 + d_{2DEG} - d_{sc,i})W}$$

4. Calculate the electron density in the channel, n_i, from Eq. (5-38), and the electric field at x_i, $\mathcal{E}_i$, from the discretized Poisson equation (5-37).
5. Move to position x_{i+1} and proceed with step 1.

The thickness of the space–charge region at x_i, $d_{sc,i}$, is imbedded in Eq. (5-39) and needs to be calculated as follows. Near the source, $d_{sc,i}$ is assumed to be a_1, i.e., the AlGaAs layer is assumed to be completely depleted. There is no depletion in the channel near the source and n_i is n_{chsg}. When moving from the source toward the gate, x_i approaches the gate edge, i.e., the gate fringe region. Approximating the shape of the space–charge region in the fringing region by a quarter of a circle, we can estimate the extension x_{sc} of the fringing region in the x-direction based on d_{sc} in the following way. The potential in the channel at $x = x_i$ is given by

$$V_i = V_{i-1} + |\mathcal{E}_{i-1}|\Delta x \quad (5\text{-}40)$$

Such a channel potential would cause a space–change region with a thickness $d'_{sc,i}$ calculated by

$$d'_{sc,i} = \sqrt{\frac{V_i 2\varepsilon}{qn_{ch,g}} + a_1^2} = a_1 + x_{sc,i} \quad (5\text{-}41)$$

The meaning of $x_{sc,i}$ is shown in Fig. 5.12. Note that V_i includes only the voltage drop between x_i and $x = 0$ and does not include V_{GS} and ϕ_b/q. The latter two quantities have already been taken into account in the calculation of $n_{ch,g}$ and do no longer appear in any equation of our HEMT treatment. Likewise, the spacer thickness has been considered in the calculation of $n_{ch,g}$ and does not appear in Eq. (5-41) and in the following expressions.

Once x_i becomes equal or larger than $L_{SG} - x_{sc,i-1}$ (L_{SG} is the source–gate separation), see Fig. 5.12, we enter the gate region and the algorithm for the calculation of $d_{sc,i}$ changes. For this case, the electron concentration n_{ch} is changed from n_{chsg} to n_{chg}, and the electron velocity is obtained by rearranging the drift current equation (5-38)

$$v_i = \frac{I_D}{qn_{chg}(a_1 + d_{2DEG})} \quad (5\text{-}42)$$

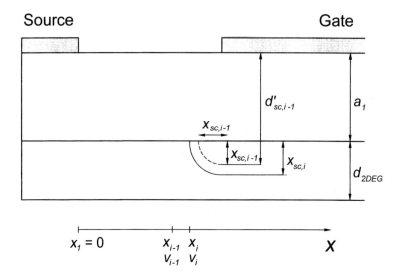

Figure 5.12. Schematic of the fringing portion of the space–charge region at the source end of the gate.

The relationship between the channel potential V_i and the space–charge region thickness $d_{sc,i}$ at a certain position x_i in the gate region (i.e., in the fringing region or underneath the gate) can be obtained as

$$V_i = \frac{qn_{i-1}(d_{sc,i}^2 - a_1^2)}{2\varepsilon} \tag{5-43}$$

One step toward the drain end, i.e., from x to $x + \Delta x$, the channel potential increases by ΔV and the thickness of the space–charge region increases by Δd_{sc}. Thus, Eq. (5-43) becomes

$$V_i + \Delta V = \frac{qn_{i-1}[(d_{sc,i} + \Delta d_{sc})^2 - a_1^2]}{2\varepsilon} \tag{5-44}$$

Subtracting Eq. (5-43) from Eq. (5-44) and rearranging the resulting equation, we obtain the quadratic equation

$$\Delta d_{sc}^2 + 2d_{sc,i}\Delta d_{sc} - \frac{2\varepsilon \Delta V}{qn_{i-1}} = 0 \tag{5-45}$$

The solution of Eq. (5-45) finally leads us to the space–charge region thickness in the gate region at position x_i

$$d_{\text{sc},i} = \sqrt{d_{\text{sc},i-1}^2 + \frac{|\mathscr{E}_{i-1}|\Delta x 2\varepsilon}{qn_{i-1}}} \tag{5-46}$$

For most semiconductors, stationary v–$\mathscr{E}$ characteristics can easily be found in the literature, but occasionally the stationary E–$\mathscr{E}$ and m^*–$\mathscr{E}$ characteristics are not available. In such cases, or when nonstationary transport effects are not of interest, a simplified HEMT model can be used. Here, the electric field at x_i is calculated by the discretized Poisson equation:

$$\mathscr{E}_i = \mathscr{E}_{i-1} + \frac{q\Delta x}{\varepsilon}(n_{\text{ch}} - n_{i-1}) \tag{5-47}$$

Using Eq. (5-47), together with the velocity deduced from the stationary v–$\mathscr{E}$ characteristics of the channel material, the electron density n_i at x_i can be obtained from Eq. (5-38). The simplified Cappy-like HEMT model does not take into account nonstationary electrom transport (velocity overshoot) and corresponds to the simplified Cappy model for the MESFET discussed at the end of Section 4.2.2.

Figure 5.13 shows the current–voltage characteristics of an AlGaAs/GaAs HEMT calculated from the complete Cappy-like HEMT model.

5.3 SMALL-SIGNAL ANALYSIS

5.3.1 Introduction

The treatment of small-signal behavior for HEMTs is basically the same as that for the MESFET. The small-signal equivalent circuit given in Fig. 4.12 and the defini-

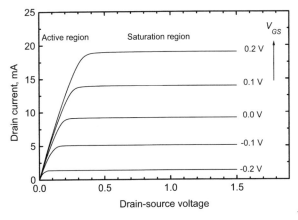

Figure 5.13. Current–voltage characteristics of an AlGaAs/GaAs HEMT calculated from the complete Cappy-like HEMT model. The transistor parameters are: $L = 0.3$ μm, $W = 200$ μm, $a_1 = 30$ mm, $d_{\text{sp}} = 3$ mm, $N_D = 2 \times 10^{18}$ cm^{-3}, Al content of the barrier layer $x = 0.24$.

tions of the circuit elements in Eqs. (4-42)–(4-48) can be applied unchanged to the HEMT. The formulas for the Y parameters, gains, and the characteristic frequencies f_T and f_{max}, i.e., Eqs. (4-74)–(4-92) are applicable to the HEMT as well. In the following, the determination of the equivalent circuit elements based on the two dc models discussed in the preceding section is presented. The modulation efficiency and the delay time of HEMTs are also discussed.

5.3.2 Modeling the Circuit Elements Based on the PHS-Like Model

The transconductance g_m of a FET has been defined in Eq. (4-42). Applying the chain rule, Eq. (4-42) can be rewritten in the form of Eq. (4-49), which is repeated here for convenience:

$$g_m = \frac{dI_D}{dp} \times \frac{dp}{ds} \times \frac{ds}{dV_{GS}}\bigg|_{V_{DS}=\text{const.}} \tag{5-48}$$

Note that this is the intrinsic transconductance. The first term on the right-hand side of Eq. (5-48) can be obtained from Eq. (5-15) as

$$\frac{dI_D}{dp} = \frac{\varepsilon_1 v_s W}{a_1 + \Delta a_1} \tag{5-49}$$

To find the second term, we differentiate Eq. (5-19) and take $dV_{DS} = 0$ according to the definition of g_m [see Eq. (4-42)], which leads to

$$0 = ds - dp + \mathscr{E}_S \cosh\left[\frac{\pi L_2}{2(a_1 + \Delta a_1)}\right] dL_2 \tag{5-50}$$

Then, setting $L_2 = L - L_1$ in Eq. (5-16) and carrying out the differentiation, we obtain

$$dL_2 = \frac{1}{\mathscr{E}_S}\left(\frac{p^2 + s^2}{2s^2} dp - \frac{s}{p} ds\right) \tag{5-51}$$

Combining Eqs. (5-50) and (5-51) yields

$$\frac{dp}{ds} = \frac{1 - \dfrac{s}{p}\cosh\left[\dfrac{\pi L_2}{2(a_1 + \Delta a)}\right]}{1 - \dfrac{s^2 + p^2}{2p^2 \cosh\left(\dfrac{s^2 + p^2}{2p^2}\right)}} \tag{5-52}$$

The third factor on the right-hand side of Eq. (5-48) is obtained from Eq. (5-11b) and given by

$$\frac{ds}{dV_{GS}} = 1 \tag{5-53}$$

Thus, the transconductance is obtained by multiplying the right-hand sides of Eqs. (5-49), (5-52), and (5-53). This results in

$$g_m = \frac{\varepsilon_1 v_S W}{a_1 + \Delta a_1} f_g \tag{5-54}$$

where f_g represents the right-hand side of Eq. (5-52).

In a similar manner, the drain resistance r_{ds} is obtained according to Eq. (4-44), together with dI_D/dp and dp/dV_{DS} for the case of constant V_{GS}:

$$r_{ds} = \frac{1}{g_{ds}} = \frac{a_1 + \Delta a_1}{\varepsilon_1 v_S W} \left\{ \frac{s^2 + p^2}{2p^2} \cosh\left[\frac{\pi L_2}{2(a_1 + \Delta a_1)}\right] - 1 \right\} \tag{5-55}$$

The gate–source and gate–drain capacitances C_{gs} and C_{gd} were defined by Eqs. (4-45) and (4-46). For the PHS MESFET model, we calculated the charge Q_{sc} stored in the space–charge region underneath the gate and the change of this charge, ΔQ_{sc}, caused by a variation of the applied voltage V_{GS} or V_{DS}. The fact that Q_{sc} and ΔQ_{sc} have the same quantity but the opposite sign as the gate charges Q_{gate} and ΔQ_{gate} easily led us to the expressions for C_{gs} and C_{gd}. The approach to calculate C_{gs} and C_{gd} in the framework of the PHS-like HEMT model is a little different. Here we calculate the charge in the 2DEG channel, Q_{ch}, and its variation ΔQ_{ch} caused by a change of V_{GS} or V_{DS}. Consider the HEMT under certain bias conditions having a number of channel electrons of N_{ch1} and a channel charge of $-qN_{ch1}$. If V_{GS} becomes more negative by $-\Delta V_{GS}$, the number of channel electrons decreases to N_{ch2}. Thus $\Delta N_{ch} = N_{ch2} - N_{ch1}$ is negative but ΔQ_{ch} is positive because every electron carries the charge $-q$. Simultaneously, the number of negative charge on the gate, N_{gate}, becomes larger and the gate charge changes by $\Delta Q_{gate} = -\Delta Q_{ch}$. Thus, the variation of the gate charge caused by change of the applied voltage can be calculated from the variation of the channel charge.

The channel charge in region 1 of the HEMT can be expressed as

$$Q_{ch1} = -qW \int_0^{L_1} n_s(x) dx \tag{5-56}$$

Expressing qn_s by Eq. (5-8), and using Eqs. (5-13), (5-14) and (5-16) to convert dx to dw, we come to

$$Q_{ch1} = \frac{\varepsilon}{a_1 + \Delta a_1} W \frac{2L_1}{s^2 - p^2} \int_s^p w^2 dw \tag{5-57}$$

Carrying out integration, Eq. (5-57) becomes

$$Q_{ch1} = -\frac{\varepsilon}{a_1 + \Delta a_1} \frac{W}{3\mathscr{E}_S} \frac{s^3 - p^3}{p} \tag{5-58}$$

The channel charge in region 2 is

$$Q_{ch2} = -\frac{\varepsilon}{a_1 + \Delta a_1} W p L_2 \qquad (5\text{-}59)$$

As discussed in the PHS MESFET model, additional positive drain charges are responsible for the voltage drop across region 2 of the transistor. These charges give rise to a gate charge component $Q_{gate,3}$ that has to be taken into account in the capacitance calculations [see Eq. (4-61)]. In the HEMT, this charge component is given by

$$Q_{gate,3} = -\varepsilon W \mathscr{E}_S \frac{2(a_1 + \Delta a_1)}{\pi} \left(\cosh \frac{\pi L_2}{2(a_1 + \Delta a_1)} - 1 \right) \qquad (5\text{-}60)$$

The different components of the gate–source capacitance are obtained by differentiating the three charge components from Eqs. (5-58)–(5-60) with respect to V_{GS} according to

$$C_{gs1,2} = \left. \frac{-dQ_{ch1,2}}{dV_{GS}} \right|_{V_{DS}=\text{const}} \qquad (5\text{-}61)$$

and

$$C_{gs3} = \left. \frac{dQ_{gate,3}}{dV_{GS}} \right|_{V_{DS}=\text{const}} \qquad (5\text{-}62)$$

The total gate–source capacitance is the sum of these three capacitances and the fringing capacitance C_{fringe} given in Eq. (4-63d):

$$C_{gs} = C_{gs1} + C_{gs2} + C_{gs3} + C_{fringe} \qquad (5\text{-}63)$$

where

$$C_{gs1} = \frac{\varepsilon}{a_1 + \Delta a_1} \frac{W}{3\mathscr{E}_S} \left(\frac{3s^2}{p} - \frac{2p^3 + s^3}{p^2} \frac{dp}{ds} \right) \qquad (5\text{-}63a)$$

$$C_{gs2} = \frac{\varepsilon}{a_1 + \Delta a_1} W \left[\left(L + \frac{p}{\mathscr{E}_S} \right) \frac{dp}{ds} - \frac{s}{\mathscr{E}_S} \right] \qquad (5\text{-}63b)$$

$$C_{gs3} = \varepsilon W \sinh\left(\frac{\pi L_2}{2(a_1 + \Delta a_1)} \right) \left(\frac{s}{p} - \frac{p^2 + s^2}{2p^2} \frac{dp}{ds} \right) \qquad (5\text{-}63c)$$

and dp/ds can be found in Eq. (5-52).

Similarly, the gate–drain capacitance C_{gd} is calculated. Without the fringing term, it is given by

$$C_{gd} = C_{gd1} + C_{gd2} + C_{gd3} \qquad (5\text{-}64)$$

with

$$C_{gd1} + C_{gd2} = \frac{\varepsilon}{a_1 + \Delta a_1} \frac{W}{1 - \frac{p^2+s^2}{p^2}\cosh\frac{\pi L_2}{2(a_1+\Delta a_1)}} \left(\frac{2p^3+s^3}{3\mathscr{E}_s p^2} - L_2 - \frac{p^2+s^2}{\mathscr{E}_s 2p} \right) \qquad (5\text{-}64a)$$

and

$$C_{gd3} = \varepsilon W \sinh\left(\frac{\pi L_2}{2(a_1+\Delta a_1)}\right) \frac{p^2+s^2}{2p^2} \frac{1}{\frac{p^2+s^2}{2p^2}\cosh\frac{\pi L_2}{2(a_1+\Delta a_1)} - 1} \qquad (5\text{-}64b)$$

The transit time τ_T (i.e., the time the electrons need to cross the channel under the gate) can be expressed as

$$\tau_T = \frac{s^3 - p^3}{3\mu_0 \mathscr{E}_s^2 p^2} + \frac{L_2}{v_S} \qquad (5\text{-}65)$$

The gate resistance R_G is another important component in the small-signal equivalent circuit. An empirical expression for the calculation of the resistance of a rectangular shape for MESFETs has been given in Section 4.3.2. For very short gate FETs, rectangular gates lead to unacceptably high values of R_G and mushroom gates are frequently used [see Section 4.6 and Fig. 4.24(b)]. Because the minimization of R_G is even more important for HEMTs (because of the higher operating frequencies), we focus on the design of mushroom gates. Figure 5.14 shows mushroom gate structures commonly used for HEMTs. Typically, the head length L_H of the mushroom gate is several times the foot print length (i.e., gate length) L, but extremely large L_H/L ratios of 10 to 20 have been reported as well [30, 31]. A quantity frequently used to describe the ohmic resistance of the gate is the so-called dc end-to-end resistance $R_{G,dc}$. For a rectangular gate,

$$R_{G,dc} = \rho_G \frac{W}{Lh} \qquad (5\text{-}66)$$

where ρ_G is the specific resistance of the gate material, W is the width of the gate, and h is the height of the gate. To estimate the resistance of more complicated gate cross sections, one can estimate the gate cross section to be composed of several rectangular gates connected in parallel. It should be noted, however, that ρ_G in Eq. (5-66) is 1.5 to 2 times larger than the bulk value of the gate material [32].

Typically, the gates of HEMTs, as well as of GaAs MESFETs, consist of a sequence of metallic materials, such as Ti/Pt/Au. Table 5.1 lists the bulk specific resistances of these individual metallic materials.

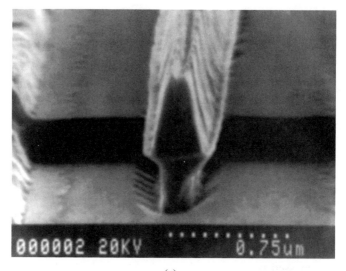

(a)

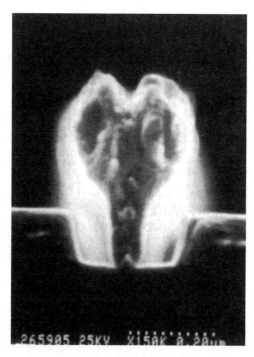

(b)

Figure 5.14. Photomicrographs of different mushroom-gate structures for HEMTs. (a) GaAs HEMT with a gate length (foodprint) of 0.2 μm. Taken from [28] © 1988 IEEE. (b) InP HEMT with a gate length (foodprint) of 50 nm. Taken from [29] © 1988 IEEE.

Table 5.1. Specific resistance (bulk values) of commonly used gate materials

Metal	Al	Ti	Pt	Au
Resistivity, Ωcm	2.7×10^{-6}	6×10^{-5}	1×10^{-5}	2.3×10^{-6}

The dc end-to-end resistance depends strongly on the gate cross section. A typical rectangular gate with a width of 1 mm and a length of 0.1 µm possesses an $R_{G,dc}$ around 1500–2000 Ω, whereas the resistance is about 150–300 Ω for a mushroom gate with a similar dimension.

The gate resistance R_G, which has to be used for small-signal considerations, is different from $R_{G,dc}$, and is given by

$$R_G = \rho_G \frac{W_f^2}{3WLh} \qquad (5\text{-}67)$$

where W_f is the gate finger width. The meaning of W_f has been explained in Section 4.6; see Figs. 4.24 and 4.30. The factor 3 in the denominator of Eq. (5-67) stems from the fact that when the gate voltage varies, charges have to be brought onto or taken out of the gate. These charges distribute along the gate, and not all charges have to move from the feed point to the very end of the gate.

5.3.3 Modeling the Circuit Elements Based on the Cappy-Like Model

The elements of the small-signal equivalent circuit of the HEMT based on the Cappy-like HEMT model from Section 5.2.2 can be calculated in exactly the same manner as described in Section 4.3.3 for the MESFET.

5.3.4 The Concept of Modulation Efficiency

At the very beginning of HEMT research, discussions were focused mainly on exploiting the high mobility of 2DEG electrons to increase transistor speed and operating frequency. It turned out, however, that mobility as well as carrier velocity are not the only factors influencing the frequency performance of HEMTs. A very useful tool to investigate the different mechanisms contributing to the cutoff frequency f_T as a figure of merit of HEMT speed is the concept of modulation efficiency introduced by Foisy et al. [33].

As a first-order approximation, consider only the intrinsic transistor and neglect all elements of the small-signal equivalent circuit except the transconductance g_m and the gate–source capacitance C_{gs}. Then [see Eq. (4-78)], the intrinsic cutoff frequency $f_{T,int}$ can be expressed by

$$f_{T,int} = \frac{g_m}{2\pi C_{gs}} \qquad (5\text{-}68)$$

According to Eqs. (4-42) and (4-45), and bearing in mind that any variation of the gate charge Q_{gate} is accompanied by a variation of the charge Q_{ch} underneath the gate of equal amount and opposite sign, we come to

$$f_{T,\text{int}} = \frac{\dfrac{dI_D}{dV_{GS}}}{2\pi \dfrac{dQ_{\text{gate}}}{dV_{GS}}} = d\frac{I_D}{2\pi dQ_{\text{gate}}} = \frac{dI_D}{2\pi |dQ_{\text{ch}}|} \tag{5-69}$$

Equation (5-69) suggests that a high cutoff frequency is achieved when the charge to be modulated to obtain a certain change of the current gain is minimized. In the following, we will elaborate the charges involved and develop a more explicit expression for $f_{T,\text{int}}$.

The maximum drain current would flow if all carriers in the 2DEG channel traveled with maximum velocity, which is, when the piecewise linear velocity-field characteristics of the PHS-like HEMT model is applied, the saturation velocity v_S. In this case, the drain current would be

$$I_D = \frac{v_S}{L} Q_{\text{VSM}} \tag{5-70}$$

where Q_{VSM} is the channel charge moving with v_S (VSM means velocity saturation model). In a transistor, however, the charges entering the channel at the source end are much slower and are accelerated only on their way toward the drain. Thus, an additional charge in the portion of the channel where $v < v_S$ is necessary to maintain the current specified in Eq. (5-70). We call this additional charge $Q_{\Delta GC}$. It is the difference between Q_{VSM} and the channel charge that is necessary to carry the current in region 1 of the transistor according to the PHS-like HEMT model. In region 1, the gradual channel approximation holds, which is the reason the subscript GC is used in the additional charge. $Q_{\Delta GC}$ is small if the low-field mobility is high.

A variation of the gate voltage not only changes the charge in the 2DEG channel, but the charge in the barrier layer as well. From Fig. 5.3, we see that there can be a considerable number of free electrons in the barrier layer. These electrons possess a very low mobility and hardly contribute to I_D. Furthermore, when V_{GS} becomes more positive, the conduction band edge in the barrier layer is shifted downward, and the energy separation between the donor level and the Fermi level becomes smaller. This leads to an increasing probability that donor states are filled with electrons (incomplete ionization caused by electrons bound to donors) and thus an additional charge component that is modulated without any contribution to the drain current has to be considered. Figure 5.15 shows the calculated sheet densities of electrons in the 2DEG, free electrons in the barrier layer, and localized electrons (bound to donors) as a function of V_{GS} and at zero V_{DS} for an AlGaAs/GaAs HEMT [33]. We denote Q_B as the charge associated with the free and localized electrons $n_{B(\text{free})}$ and $n_{B(\text{bound})}$ in the barrier layer. Q_B becomes small when the conduction band offset at the barrier/channel heterojunction is large.

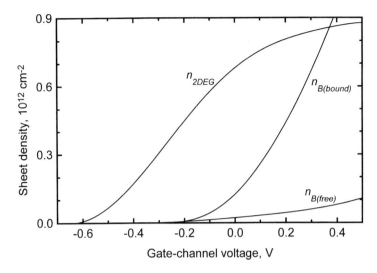

Figure 5.15. Components of the electron sheet charge in an AlGaAs/GaAs HEMT calculated by Foisy [33]. n_{2DEG} is the electron sheet density in the 2DEG, $n_{B(free)}$ is the sheet density of free electrons in the barrier layer, and $n_{B(bound)}$ is the sheet density of localized electrons in the donor states.

Thus, the charge corresponding to the maximum drain current is Q_{VSM}, and the total charge that is modulated in the HEMT, Q_{tot}, is equal to $Q_{VSM} + Q_{\Delta GC} + Q_B$. Setting $Q_{ch} = Q_{tot}$ and putting Eq. (5-70) in Eq. (5-69), we come to

$$f_{T,int} = \frac{v_S}{2\pi L} \frac{dQ_{VSM}}{d(Q_{VSM} + Q_{\Delta GC} + Q_B)} = \frac{v_S}{2\pi L} ME \quad (5-71)$$

where ME is the modulation efficiency and has a value between unity and zero. It can be seen that the charges $Q_{\Delta GC}$ and Q_B degrade the cutoff frequency from the ideal value $v_S/(2\pi L)$. Fig. 5.16 shows schematically the division of Q_{tot} into the three components.

The modulation efficiencies for a conventional AlGaAs/GaAs HEMT and a GaAs pHEMT with an $In_{0.15}Ga_{0.85}As$ channel layer calculated by Foisy et al. [33] are shown in Fig. 5.17(a). Both transistors have identical layer designs (except the channel layer) and the same gate length (1 μm). Moreover, identical transport properties were assumed for the electrons in GaAs and InGaAs ($v_S = 1.7 \times 10^7$ cm/s, μ_0 = 5000 cm^2/Vs). Clearly, the GaAs pHEMT possesses a higher modulation efficiency than the AlGaAs/GaAs HEMT. The increased modulation efficiency in the GaAs pHEMT is due mainly to the much lower Q_B in this transistor compared to that in the AlGaAs/GaAs counterpart, whereas $Q_{\Delta GC}$ is about the same in both transistors. Fig. 5.17(b) shows the calculated cutoff frequency of the two HEMTs. As expected, the GaAs pHEMT shows a considerably higher f_T than the AlGaAs/GaAs HEMT. Note that the trends of the f_T and ME curves are almost identical. Figure

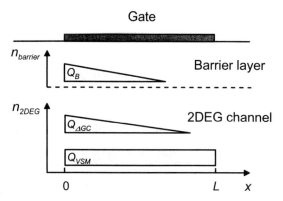

Figure 5.16. Graphic representation of the charge components Q_B, $Q_{\Delta GC}$, and Q_{VSM} in a HEMT.

5.18 shows a plot of the modulation efficiency of HEMTs versus 2DEG sheet density for four different HEMTs estimated by Nguyen et al. [34]. We see that the modulation efficiency increases with increasing sheet density, which in turn can be related directly to the conduction band offset.

The concept of modulation efficiency clearly indicates that both high-mobility electrons and a large conduction band offset are necessary for realizing high-performance HEMTs.

5.3.5 The Concept of Delay Times

The concept of delay times is frequently used to investigate the influence of the intrinsic transistor structure and several parasitic elements on the cutoff frequency of HEMTs. In general, it is applicable not only to HEMTs but to MESFETs and MOSFETs as well.

The cutoff frequency of a HEMT can be related to a characteristic time constant, known as the total delay time τ_{tot}, by

$$f_T = \frac{1}{2\pi\tau_{tot}} \qquad (5\text{-}72)$$

The total delay time consists of several components describing the delays associated with different parts of the transistor [34–36]. They are the delay time of the intrinsic transistor, τ_{int}, the charging time associated with the gate bonding pads, τ_{pad}, the charging time of the fringing capacitance, τ_{fringe}, and the delay caused by the portion of the space–charge region extending beyond the gate toward the drain, τ_{drain}:

$$\tau_{tot} = \tau_{int} + \tau_{pad} + \tau_{fringe} + \tau_{drain} \qquad (5\text{-}73)$$

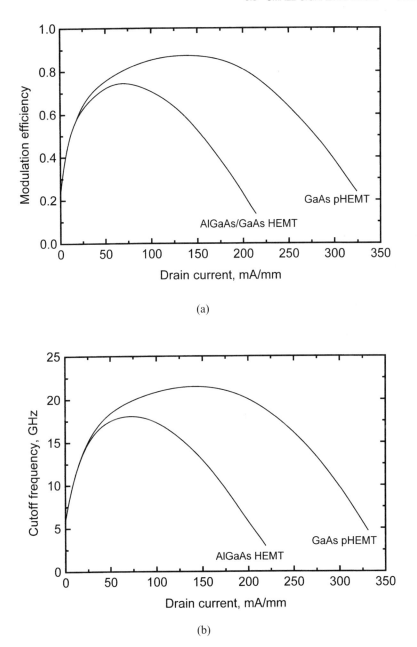

Figure 5.17. Calculated (a) modulation efficiency and (b) cutoff frequency for an AlGaAs/GaAs HEMT and a GaAs pHEMT, after Foisy et al. [33].

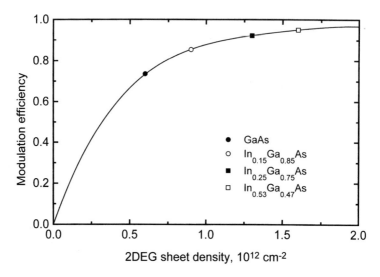

Figure 5.18. Estimated modulation efficiency as a function of 2DEG sheet density for four different material systems, after Nguyen et al. [34].

When neglecting the modulation of charges in the barrier layer, τ_{int} is the same as Eq. (5-65). If the modulation of these charges is considered, then

$$\tau_{int} = \frac{L}{v_s ME} \qquad (5\text{-}74)$$

The pad charging delay can be approximated by

$$\tau_{pad} = \frac{C_{pad}}{g_{m,ext}} \qquad (5\text{-}75)$$

where C_{pad} is the capacitance of the bonding pads and $g_{m,ext}$ is the extrinsic (i.e., terminal) transconductance. C_{pad} is typically 10 fF for a 50 × 50 µm² bonding pad [34]. The delay caused by the fringing capacitance is given by

$$\tau_{fringe} = \frac{C_{fringe}}{g_{m,int}} \qquad (5\text{-}76)$$

where $g_{m,int}$ is the intrinsic transconductance. From Eq. (4-63d) and the material data given in chapter 2, we obtain a fringing capacitance of about 0.18 pF/mm for lattice matched InP HEMTs ($Al_{0.48}In_{0.52}As$ barrier layer) and 0.17 pF/mm for GaAs pHEMTs ($Al_{0.3}Ga_{0.7}As$ barrier layer). The relationship between extrinsic and intrinsic transconductances is

$$g_{m,ext} = \frac{g_{m,int}}{1 + g_{m,int} R_S} \qquad (5\text{-}77)$$

Table 5.2. Estimated delay times and cutoff frequencies for typical GaAs pHEMTs and lattice-matched InP HEMTs with 0.15 μm gate length, 50 μm gate width, and one gate bonding pad.

	GaAs pHEMT	InP HEMT
v_S, 10^7 cm/s	2.0	2.6
R_S, Ωmm	0.45	0.25
$g_{m,ext}$, mS/mm	730	1250
τ_{int}, ps	0.76	0.58
τ_{pad}, ps	0.27	0.16
τ_{fringe}, ps	0.16	0.10
τ_{drain}, ps	0.10	0.10
τ_{tot}, ps	1.29	0.94
f_T, GHz	120	170

Data taken from [34]

where R_S is the parasitic source resistance. When the gate width of a given transistor is decreased and all other transistor dimensions are kept the same, τ_{pad} increases because the intrinsic transconductance scales with W while C_{pad} remains constant. The delay time τ_{fringe}, on the other hand, does not change because C_{fringe} scales with W. The estimation of the drain delay time is difficult. Moll et al. proposed a method to extract such a delay time from measured f_T data [35]. In Table 5.2, the delay times and cutoff frequencies for typical GaAs pHEMTs and lattice matched InP HEMTs are given.

5.4 NOISE AND POWER ANALYSIS

The framework of the noise modeling for HEMTs is very similar to that for MESFETs described in Section 4.4. Brookes developed a noise model based on the dc HEMT model presented in Section 5.2.1 [18]. The transistor equations derived in [18] are almost identical to the PHS MESFET noise model, aside from the deviation that the contribution of region 2 of the HEMT to the total noise has been neglected. The Brookes model was unable to explain the U-shape in the NF_{min} versus I_D/I_{DSS} characteristics. This was attributed to the fact that the transconductance reduction at low drain currents was not modeled properly [18]. Experience showed, however, that using the expression for the transconductance derived in Section 5.3.2, the rapid falloff of g_m at low drain currents near pinch-off can be modeled properly. Thus, when the noise contribution of region 2 is included, one can expect an U-shape of the modeled NF_{min} for HEMTs in the frame of the Brookes HEMT model similar to that obtained with the PHS noise model for MESFETs. It should be noted, however, that in the case of short-gate MESFETs with a small L/a ratio, the PHS model also fails to predict the U-shape. This is a general limitation of the two-region MESFET and HEMT models.

Aside from the Brookes model, several other noise models for the HEMT have also been reported in the literature [24, 37–41].

The power modeling of HEMTs is identical to that of MESFETs, which has been covered in detail in Section 4.6.3.

5.5 ISSUES OF AlGaAs/GaAs HEMTs

5.5.1 Transistor Structures

The cross section of a typical AlGaAs/GaAs HEMT has been shown in Fig. 5.2. The layer sequence consists of an undoped GaAs channel layer (frequently called the buffer layer) grown epitaxially on a semi-insulating GaAs substrate. This is followed by an undoped AlGaAs spacer layer and an n-type AlGaAs layer, which constitute the barrier layer. Finally, on top of the doped AlGaAs layer, a highly doped n-type GaAs cap layer is grown. The growth of the various layers can be done by either molecular beam epitaxy (MBE) or metal organic chemical vapor deposition (MOCVD).

The cap layer has a typical thickness of 30 to 50 nm and is doped with Si at $1-2 \times 10^{18}$ cm^{-3}. The purpose of the cap layer is to provide a good ohmic contact for the source and drain regions and to protect the AlGaAs surface from oxidation and depletion. The cap layer is totally removed underneath the gate.

The AlGaAs barrier layer is about 30 to 50 nm thick and doped with Si at $1-2 \times 10^{18}$ cm^{-3}. One of the most important requirements in HEMT design is that the distance between the gate and the 2DEG channel, which is a_1 in Fig. 5.2, should be small. A small a_1 ensures effective control of the channel charge by the gate voltage and the suppression of undesired short channel effects in HEMTs with very short gates. Similar to the MESFET design, the ratio of L/a_1 (which corresponds to L/a for the MESFET) should be large. Because the AlGaAs layer in HEMTs can be more heavily doped than the active layer in GaAs MESFETs without unacceptably undermining the breakdown voltage, a large L/a_1 can be much more easily obtained in HEMTs compared to MESFETs. Although an $L/a \geq 3$ is difficult to obtain in short-channel MESFET, an L/a_1 of 10 is not a problem in quarter-micron-gate HEMTs.

In most cases, the doped AlGaAs layer has a uniform doping profile. To increase the breakdown voltage, however, it is possible to use δ-doped layers (frequently also called pulse-doped or planar doped layers). A δ-doped layer consists of a Si monolayer embedded in the undoped or low-doped AlGaAs above the spacer. The Al content x in the barrier layer is typically less than 0.3. A higher Al content would result in a larger conduction band offset and thus a higher 2DEG sheet density, which is of course beneficial. Unfortunately, increasing x leads to the undesired effects discussed in Section 5.1.

The undoped AlGaAs spacer reduces the Coulomb scattering between the ionized donors in the doped AlGaAs layer and the 2DEG electrons, thus leading to an increase in the 2DEG mobility. One could therefore expect that thick spacers are beneficial for HEMT operation. On the other hand, a thicker spacer leads to a smaller 2DEG sheet density (see Section 3.4.2, Fig. 3.9), and such a trade-off needs to be carefully optimized in the HEMT design. Figure 5.19 shows the cutoff frequencies

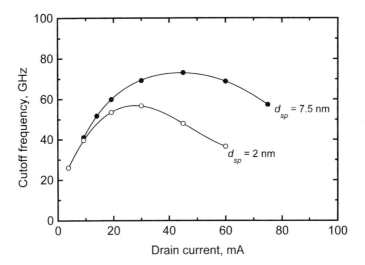

Figure 5.19. Cutoff frequency of AlGaAs/GaAs HEMTs with two different spacer thicknesses but otherwise identical structures, after [34] and Ref. 6 therein.

of 0.2 μm gate AlGaAs/GaAs HEMTs with two different spacer thicknesses d_{sp} [34]. Typically, the spacer thickness is around 2 to 5 nm.

The GaAs channel (buffer) layer is typically 0.5 to 1 μm thick and is of good crystallographic quality. The source and drain ohmic contacts must provide a low-resistivity connection between the metal contacts and the 2DEG channel. In the MESFET, the channel, i.e., the active layer, is located directly under the ohmic contacts, whereas in the HEMT the 2DEG is 30 nm or more away from the metal due to the presence of the barrier layer. Alloyed contacts of AuGe/Ni/Au are commonly used. After the metal deposition, it is thermally alloyed at about 450°C. Sometimes, nonalloyed contacts are used.

In most AlGaAs/GaAs HEMTs, the gate consists of a layer sequence of Ti/Pt/Au, but there are also a few reports on HEMTs with Al gates. Rectangular gates are fabricated by the lift-off process. The realization of mushroom gates is more complicated, and a multilayer resist technique is normally required. More details on the gate formation process in HEMTs can be found in [42].

5.5.2 Low-Noise AlGaAs/GaAs HEMTs

During the 1980 and early 1990s AlGaAs/GaAs HEMTs were widely used in low-noise applications around and above 12 GHz. Typical experimental and commercial AlGaAs/GaAs HEMTs have gate lengths ranging from 0.25 to 0.5 μm, although experimental transistors with 0.1 μm gates have been reported as well. Extrinsic transconductances in the range of 250–550 mS/mm are typical for AlGaAs/GaAs HEMTs.

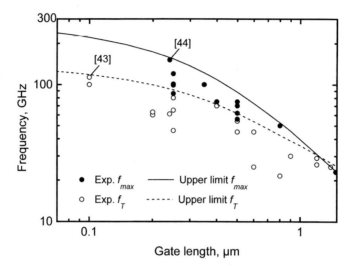

Figure 5.20. Cutoff frequency and maximum frequency of oscillation of experimental AlGaAs/GaAs HEMTs. Also shown are the upper limit fits to the state of the art in the year 2001.

Figure 5.20 shows reported cutoff frequencies f_T and maximum frequencies of oscillation f_{max} of experimental AlGaAs/GaAs HEMTs as a function of gate length. Also shown are empirical upper limits for f_T and f_{max}. The highest f_T and f_{max} values so far are 113 and 151 GHz obtained from a 0.1 µm [43] and 0.24 µm transistor [44], respectively. Note that these data were reported already in 1988.

Table 5.3 lists the 1990 state of the art minimum noise figures and associated gains of AlGaAs/GaAs HEMTs. It should be noted that no substantial improvement in the frequency and noise performance of AlGaAs/GaAs HEMTs has been made since 1990.

Fig. 5.21 compiles a large number of data on the minimum noise figure of experimental and commercial AlGaAs/GaAs HEMTs. At 60 Hz, the best minimum noise figure is below 2 dB [45]. Compared to the GaAs MESFET, at frequencies between 10 and 20 GHz, the minimum noise figure of AlGaAs/GaAs HEMTs is typically 0.2–0.3 dB lower and the associated gain is about 1–2 dB higher.

Table 5.3. Minimum noise figure and associated gain of AlGaAs/GaAs HEMTs (state of the art 1990)

Frequency, GHz	8	12	18	60
NF_{min}, dB (experimental)	0.4–1	0.5–1.2	0.5–1.5	1.8
NF_{min}, dB (commercial)		0.75–1	1–1.3	
G_a, dB (experimental)	13–16	10–15	8–13	6.4
G_a, dB (commercial)		10–12	9–10	

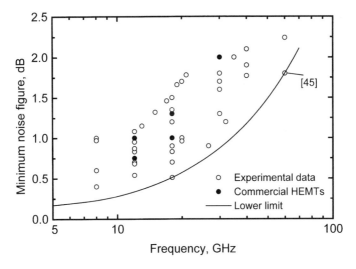

Figure 5.21. Minimum noise figure and its lower limit fit of experimental and commercially available AlGaAs/GaAs HEMTs.

During the 1980s, a lot of effort was spent on research and development to optimize AlGaAs/GaAs HEMT design and to improve transistor performance, i.e., to increase f_T, f_{max}, and gain, and to decrease the noise figure. In the 1990s, however, most R&D activities on AlGaAs/GaAs HEMTs subsided and the interest shifted to the more promising GaAs pHEMTs.

5.5.3 Power AlGaAs/GaAs HEMTs

Maximum drain currents (with slightly positive biased gate) of 200–650 mA/mm and gate–drain breakdown voltages of around 10–20 V are typical for AlGaAs/GaAs HEMTs. Recent work, however, has led to significantly improved breakdown voltages. Both gate–drain and drain–source breakdown voltages above 30 V have been obtained [46, 47], and a very high gate–drain breakdown voltage of 47 V has been realized [48]. To improve the breakdown behavior, double gate–recess structures are frequently used.

The typical output power density of AlGaAs/GaAs HEMTs is in the range of 0.5 to 1.5 W/mm at frequencies up to 40 GHz. At 1.5 GHz, a maximum output power density of 1.7 W/mm has been reported [48]. AlGaAs/GaAs HEMTs deliver higher output power density compared to GaAs MESFETs at all frequencies. Although the operating frequency of power GaAs MESFET is limited to about 30 GHz, AlGaAs/GaAs HEMTs can still deliver 0.4 W/mm output power at 60 GHz [49].

Much work in developing power AlGaAs/GaAs HEMTs was done in the five-year span from 1983 to 1988. Later, the R&D activities on this transistor type declined and the attention shifted to power GaAs pHEMTs and power InP HEMT.

The growing wireless communication market and the demand for high-power transistors for base stations, however, revived the interest in AlGaAs HEMTs in the second half of the 1990s (see, e.g., [46, 47, 50]). For example, a module consisting of four large-periphery power AlGaAs/GaAs HEMTs delivering 240 W output power at 2.14 GHz has been reported [46]. Each HEMT in the module has a gate width of 195 mm (gate finger width 870 μm) and an output power density of 0.3 W/mm. Another four-transistor module with a total gate width of 346 mm delivered 230 W output power at 2.1 GHz [50].

5.6 ISSUES OF GaAs pHEMTs

5.6.1 Transistor Structures

The basic layer structure of a GaAs pHEMT consists, from the top, of a cap layer, a barrier layer, a spacer, a channel layer, a buffer, and a substrate. The thin channel layer is composed of a pseudomorphic InGaAs alloy, which is sandwiched between the spacer and the underlying buffer. There are, however, many variations from this basic structure. In Figs. 5.22(a)–(c), three frequently used layer designs for GaAs pHEMTs are shown. Figure 5.22(a) illustrates the basic layer sequence that is often used for both low-noise and low-power pHEMTs, whereas Fig. 5.22(b) shows a layer design with two electron supply layers above and below the InGaAs channel. This design is favorable for power HEMTs because both supply layers deliver elec-

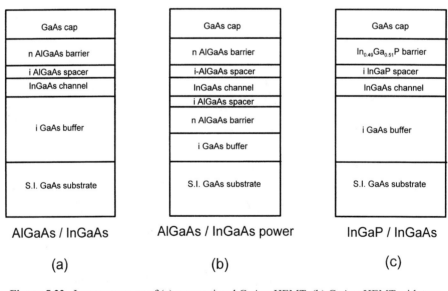

Figure 5.22. Layer sequence of (a) conventional GaAs pHEMT, (b) GaAs pHEMT with two electron supply layers, and (c) GaAs pHEMT with InGaP barrier layer.

trons to the channel, thus increasing the channel sheet density. In Fig. 5.22(c), a GaAs pHEMT with InGaP barrier layer is shown.

The GaAs cap layer is 25–50 nm thick and doped with Si at 10^{18}–10^{19} cm^{-3} to provide good source and drain ohmic contacts to the barrier and the channel. The cap is totally removed underneath the gate. The barrier layer consists of either $Al_xGa_{1-x}As$ with x between 0.2 and 0.3 or $In_{0.49}Ga_{0.51}P$. The use of an AlGaAs barrier is the rather conventional approach adopted from the AlGaAs/GaAs HEMT discussed in Section 5.5.1. The InGaP barrier layer became popular during the 1990s (see, e.g., [51–54]). $In_{0.49}Ga_{0.51}P$ is lattice-matched to GaAs, and offers a Schottky barrier of around 0.9 eV and a conduction band offset to $In_xGa_{1-x}As$ (x = 0.15–0.25) of around 0.4 eV. The bandgap of $In_{0.49}Ga_{0.51}P$ is 1.92 eV, which gives rise to a high breakdown voltage. Another advantage is that the InGaP barrier layer does not suffer from the presence of DX centers because of the absence of Al. The barrier layer is typically 25–50 nm thick. It can be homogeneously doped at 1 to 3 × 10^{18} cm^{-3} or δ-doped, where the δ-layer is located a few nm above the channel and the rest of the barrier layer adjacent to the gate is undoped. The latter option is frequently chosen in power HEMTs to increase the breakdown voltage. Between the barrier and the channel, an undoped spacer of the same material as the barrier layer is grown to reduce Coulomb scattering. The spacer thickness is about 3–4 nm.

The channel consists of pseudomorphic $In_xGa_{1-x}As$ with x typically between 0.15 and 0.25. The thickness is 10–15 nm, which is well below the critical thickness [see Fig. 3.2(b)]. In low-noise transistors, the channel is undoped, whereas for power applications, where sheet density is more important than mobility, a doped channel layer is frequently used. The HEMT structure shown in Fig. 5.22(b) has a second spacer and barrier layer underneath the channel, which can further increase the sheet density in the channel. Such a structure is sometimes used in power HEMTs. On the bottom of the layer structure, there are the undoped GaAs buffer layer up to 1 μm thick and the semi-insulating substrate. The source and drain ohmic contacts and the Schottky gate are made and designed in the same manner as GaAs MESFETs and AlGaAs/GaAs HEMTs.

Figure 5.23 shows the layout of a low-noise GaAs pHEMT. The gate length of this transistor is 0.15 μm and the gate width is 180 μm. It has three large area source bonding areas, two drain and two gate bonding pads, and four gate fingers each with a finger width of 45 μm.

5.6.2 Low-Noise GaAs pHEMTs

In general, GaAs pHEMTs show better frequency and noise performances than AlGaAs/GaAs HEMTs. Because of these advantages, combined with the fact that the two devices can be fabricated with essentially the same technology, GaAs pHEMTs have replaced AlGaAs/GaAs HEMTs in many commercial applications.

Extrinsic transconductances in the range of 300–800 mS/mm are typical for GaAs pHEMTs, and a record value of 1070 mS/mm has been reported for a 0.1 μm gate transistor [56]. Figure 5.24 shows reported cutoff frequencies f_T and maximum frequencies of oscillation f_{max} of GaAs pHEMTs. The record values for f_T and f_{max}

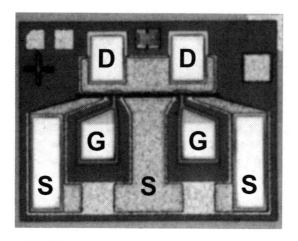

Figure 5.23. Photomicrograph of a 0.15 μm gate length, 180 μm gate width, low-noise GaAs pHEMT. Taken from [55] © 1996 IEEE.

are 152 GHz [36], and 290 GHz [57], respectively, obtained from 0.1 μm gate transistors.

The minimum noise figures versus frequency for experimental and commercial GaAs pHEMTs are shown in Fig. 5.25. The noise figures are slightly lower compared to those of AlGaAs/GaAs HEMTs at frequencies up to 60 GHz (less than 0.25 dB at 60 GHz). It should be noted that no noise data beyond 60 GHz were reported for AlGaAs/GaAs HEMTs (see Fig. 5.21).

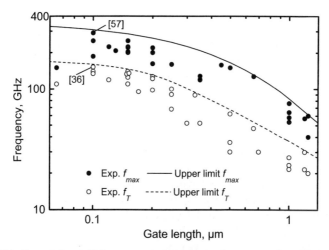

Figure 5.24. Reported cutoff frequency and maximum frequency of oscillation for experimental GaAs pHEMTs.

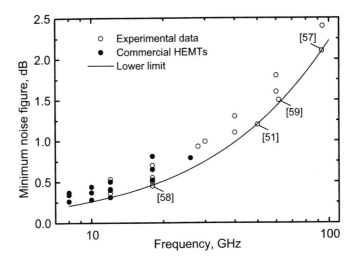

Figure 5.25. Minimum noise figure and its lower limit fit of experimental and commercially available GaAs pHEMTs.

5.6.3 Power GaAs pHEMTs

Power GaAs pHEMTs are widely used in the frequency range from 800 MHz up to several tens of GHz, and meaningful power amplification at frequencies as high as 94 GHz has been achieved. The maximum drain current I_{max} (with slightly positive biased gate) is typically 500–800 mA/mm, which is considerably higher than that of AlGaAs/GaAs HEMTs. Record I_{max} of 1000mA/mm for a 0.15 μm gate transistor has been demonstrated [60].

Figure 5.26 shows the reported output power densities versus drain–source voltage for GaAs pHEMTs. The highest output power density obtained with GaAs pHEMTs is 1.6 W/mm (V_{DS} = 12 V, f = 2 GHz) [63]. Clearly, the power density increases with increasing drain–source voltage. Note that the low-voltage range is important for power amplification in portable and battery powered devices, such as cell phones, whereas the high-voltage range is critical for base stations for mobile communications systems. In Fig. 5.27, the reported output power densities versus frequency are shown. Table 5.4 lists the reported output powers of GaAs pHEMTs at different frequencies.

5.7 ISSUES OF GaAs mHEMTs

5.7.1 Transistor Structures

It is well known that InP HEMTs outperform GaAs pHEMTs in terms of frequency limits and noise figure. The reason for the superior performance of InP HEMTs is the enhanced properties of the InGaAs channel layer with an In content of or above

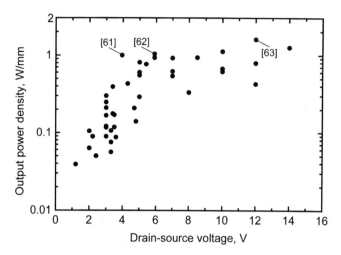

Figure 5.26. Reported output power density versus the drain–source voltage of experimental GaAs pHEMTs.

0.53. From the performance point of view, InP HEMTs could be the devices of choice for applications at microwave frequencies. Unfortunately, InP substrates are much more expensive compared to GaAs substrates. Furthermore, InP substrates are extremely brittle. Thus, although the substrate cost is only a small fraction of the total cost of integrated microwave circuits, the difficult handling of InP wafers can lead to poor process yield and increased cost. Finally, InP substrates are available only in small diameters (in the year 2000 three to four inches), which makes it hard to compete with the cost per chip of GaAs transistors fabricated on 6-inch wafers.

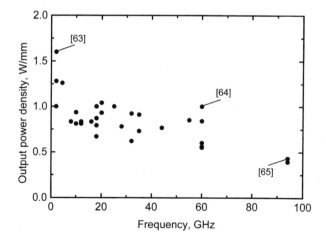

Figure 5.27. Reported output power density versus the frequency of experimental GaAs pHEMTs.

Table 5.4. Reported output power of GaAs pHEMTs

Frequency, GHz	Output power, W	Comment	Reference
2.2	140	4-transistor module	[66]
2.2	35	single transistor	[66]
12	11.2		[67]
35	0.658		[68]
60	0.225		[69]
94	0.057		[70]

Therefore it would be desirable to find a way to fabricate high-performance transistors with high In-content channels on the less brittle and larger diameter GaAs substrates.

The answer to this possibility is the concept of fabricating metamorphic GaAs HEMTs (GaAs mHEMT). Many efforts to do so were made during the late 1990s. The basic idea is to grow a thick relaxed buffer layer (much thicker than the critical thickness) on the GaAs substrate. The purpose of the buffer is twofold: first, to transform the lattice constant from that of GaAs to that of the InGaAs channel layer with high In content, and second, to protect the channel from dislocations that are unavoidable when materials with such different lattice constants are grown on each other.

The layer sequence of GaAs mHEMTs is very similar to that of GaAs pHEMTs, but different buffer layer designs have been suggested and employed. One possibility, the so-called inverse-step linearly-graded InAlAs buffer [71, 72] is shown in Fig. 5.28. The In content of the buffer is varied from 0 at the GaAs substrate interface to a certain value. Then the Al content is abruptly decreased to a value that ensures lattice matching with the channel layer. Matching is achieved if the In content y of the $In_yAl_{1-y}As$ buffer is 0.01 lower than the In content x of the $In_xGa_{1-x}As$ channel layer, i.e., $x = y + 0.01$. Other frequently used buffer layer designs are the linearly graded buffer without the inverse step [73], the linearly graded $In_xAl_yGa_{1-x-y}As$ buffer [11], and the linearly graded $Al_xGa_{1-x}As_ySb_{1-y}$ buffer [74]. The buffer is typically around 1.5 μm thick.

Using appropriately designed buffers, any desired In content in the channel can be realized. So it is possible to tailor the bandgap, and thus the conduction band offset, the channel sheet density, the breakdown behavior, as well as the transport properties in the channel. GaAs mHEMTs with In channel contents between 0.2 and 0.8 have been successfully fabricated.

The cap of GaAs mHEMTs is around 10 nm thick and consists of InGaAs with a doping concentration of 10^{18}–10^{19} cm^{-3}.

5.7.2 Performance of GaAs mHEMTs

Since the mid 1990s, numerous low-noise and power GaAs mHEMTs have been reported. Commercial devices, however, were not yet available in 2001. Extrinsic transconductances from 600 up to above 1000 mS/mm are typical for GaAs

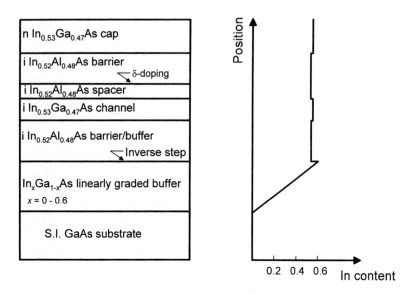

Figure 5.28. Layer sequence and position-dependent In content of an inverse-step, linearly-graded metamorphic heterostructure used in GaAs mHEMTs.

mHEMTs, and a record value of 1700 mS/mm has been reported for a 0.1 μm gate transistor [75]. The maximum drain currents are in the range of 600–900 mA/mm. In 2001, the reported record f_T and f_{max} values were 204 GHz for a 0.18 μm gate transistor [76] and 400 GHz for a 0.1 μm gate transistor [71], respectively. In regard to the power performance, an output power density of 0.92 W/mm at 35 GHz and 0.24 W/mm at 60 GHz marked the state of the art during that time [77, 78]. It should be noted, however, that these data are for small-periphery transistors delivering relatively low total output power. Gate–drain breakdown voltages of 6.3 V and 14 V have been reported for GaAs mHEMTs with $In_{0.6}Ga_{0.4}As$ and $In_{0.5}Ga_{0.5}As$ channels, respectively [79, 80].

In the year 2001, the frequency and noise performances of GaAs mHEMTs were between those of GaAs pHEMTs and InP HEMTs. Thus the concept of metamorphic layers has proven its suitability for high-performance microwave transistors. They could play an important role in future microwave electronics provided the reliability of GaAs mHEMTs can reach a level similar to that of GaAs pHEMTs.

5.8 ISSUES OF InP HEMTs

5.8.1 Transistor Structures

InP HEMTs have been widely investigated for applications at very high frequencies. In general, they outperform any other FET types in terms of operating frequency and frequency limits (f_T and f_{max}), as well as minimum noise figure. The InP HEMT typically consists of an unstrained $In_{0.53}Ga_{0.47}As$ channel lattice-matched to the InP substrate, or a pseudomorphic $In_xGa_{1-x}As$ channel ($x > 0.53$). The former will be called the lm InP HEMT and the latter InP pHEMT in the following. Figure 5.29(a) shows the layer sequence for both lm InP HEMTs and InP pHEMTs. The only difference between the lm InP HEMT and the InP pHEMT is the In content of the $In_xGa_{1-x}As$ channel layer. The layer sequence of an InP pHEMT having a composite InGaAs/InAs/InGaAs channel is given in Fig. 5.29(b). Another possible composite channel design consists of InGaAs/InP.

On top of the InP HEMT layer stack is the cap. It consists of either a single 20–40 nm thick $In_{0.53}Ga_{0.47}As$ layer doped at 10^{18}–2×10^{19} cm^{-3} or a combination of an $In_{0.52}Al_{0.48}$ layer (about 15 nm thick doped at 5×10^{18} cm^{-3}) and an $In_{0.53}Ga_{0.47}As$ layer (about 10 nm thick doped at 1 to 2×10^{19} cm^{-3}). The barrier layer is commonly $In_{0.52}Al_{0.48}As$ with a 10–20 nm thickness and is in most cases δ-doped, but there are also reports on InP HEMTs with $Ga_{0.2}In_{0.8}P$ barrier layers. Underneath the barrier layer, an undoped InAlAs spacer of about 3 nm is located.

In lm InP HEMTs, the channel is 10–20 nm thick. It is often undoped, but may be doped in power transistors to increase the electron sheet density in the channel. Another option leading to reduced impact ionization and thus higher breakdown voltage is the use of a composite channel consisting of InGaAs and InP layers. The basic idea of this composite channel is as follows. In the low-field channel region near the source, the electrons are mostly located in the high-mobility, small-bandgap InGaAs layer. As the field increases toward the drain, the energy of the electrons increases, and more and more electrons populate the InP layer. Because of the larger bandgap, the rate of impact ionization in InP is smaller compared to that

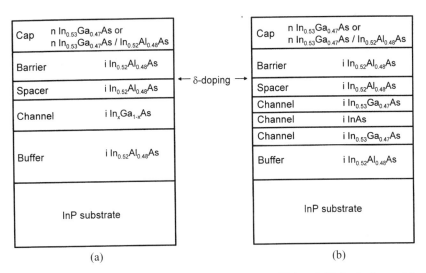

Figure 5.29. Layer sequence of (a) conventional InP HEMT ($x = 0.53$ for lattice-matched and $x > 0.53$ for pseudomorphic HEMT), and (b) InP pHEMT with composite channel.

in the InGaAs channel. In addition, while the low-field mobility of InP is smaller than that of InGaAs, the high-field transport properties, especially the saturation velocity, are better in InP.

The undoped InAlAs buffer (100–300 nm) is located below the channel. In power transistors, the buffer may contain another δ-doped layer to deliver additional electrons to the channel and thus to increase the channel sheet density. All the layers mentioned above are grown on an InP substrate by MBE or MOCVD.

The channel of InP pHEMTs has an In content typically in the range of 0.6 to 0.8. This leads to higher sheet concentrations, higher mobilities, and thus higher $\mu_0 \times n_s$ products. This, however, comes at the expense of higher impact ionization rates and lower breakdown voltages. Composite channels can be used in such devices as well. One option is a three-layer channel consisting of an n-type InP, an undoped InP, and an $In_xGa_{1-x}As$ layer. Another possible composite channel to exploit the extremely high mobilities of InAs is that shown in Fig. 5.29(b). Here, a three-layer channel consisting of $In_{0.53}Ga_{0.47}As/InAs/In_{0.53}Ga_{0.47}As$ (e.g., 2 nm/4 nm/2 nm) is used [81].

The materials for the source, drain, and gate electrodes, the formation of the electrodes, and the design rules for the electrode patterns are the same as those for GaAs MESFETs and GaAs HEMTs.

5.8.2 Low-Noise InP HEMTs

Extrinsic transconductances in the range of 700–1000 mS/mm and 1000–2000 mS/mm are typical for lm InP HEMTs and InP pHEMTs, respectively. Table 5.5 shows the highest extrinsic transconductances reported for both types of InP HEMTs.

Figures 5.30 and 5.31 show reported cutoff frequencies f_T and maximum frequencies of oscillation f_{max}, respectively, for InP HEMTs. The performance of f_T = 396 GHz for a 0.025 μm gate InP HEMT [87] and f_{max} = 600 GHz for a 0.1 μm gate InP HEMT [91] are the highest ever reported for any field-effect transistor.

InP HEMTs also possess the best high-frequency noise performance among all microwave transistors. Fig. 5.32 shows the minimum noise figure of experimental InP HEMTs together with an empirical lower limit fit. Worth mentioning are noise figures as low as 0.7 dB with 8.6 dB associated gain at 62 GHz [96], and 1.2 dB with 7.2 dB associated gain at 94 GHz [85].

Table 5.5. Record transconductances reported from experimental InP HEMTs

lm InP HEMT			InP pHEMT		
L, μm	g_m, mS/mm	Ref.	L, μm	g_m, mS/mm	Ref.
0.05	1460	[82]	0.07	2500	[81]
0.15	1300	[83]	0.03	2050	[84]
0.1	1200	[85]	0.08	2000	[86]

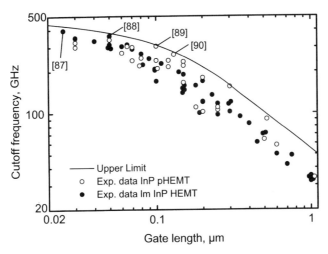

Figure 5.30. Reported cutoff frequency and its upper limit fit of experimental InP HEMTs.

InP HEMTs have some shortcomings, nevertheless. The low bandgap of the high In content channel and the resulting high-impact ionization rates lead to the onset of impact ionization even at common operating voltages. At low voltages this does not destroy the transistor, but increases the output conductance leading to low power gain. Another problem is, mentioned in Section 5.7.1, is the small-diameter, brittle, and expensive InP substrate. Finally, InP technology is less matured compared to the well-established GaAs pHEMT technology.

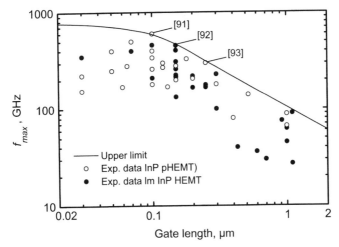

Figure 5.31. Reported maximum frequency of oscillation and its upper limit fit of experimental InP HEMTs.

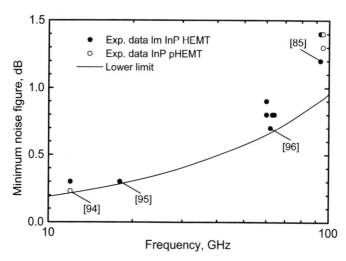

Figure 5.32. Reported minimum noise figures and its lower limit fit of experimental InP HEMTs.

5.8.3 Power InP HEMTs

The maximum drain currents I_{max} (with slightly positive biased gate) are typically 600–1000 mA/mm, and record I_{max} as high as 1250 mA/mm for lm InP HEMTs [97] and 1480 mA/mm for InP pHEMTs [98] have been reported.

Power InP HEMTs can deliver useful output powers up to 100 GHz. The high maximum currents and the low knee voltage give rise to high output power densities at low drain–source voltages and high power-added efficiencies. This is important for battery-powered mobile systems. As already mentioned, the main drawback of power InP FETs is their low breakdown voltage. This sets limits to the maximum drain–source voltage and the maximum power density.

5.9 ISSUES OF AlGaN/GaN HEMTs

5.9.1 Transistor Structures

AlGaN/GaN HEMTs attracted considerable attention soon after they were first introduced in the early 1990s due to the unique and superior properties of GaN (high bandgap, fairly high electron mobility, and high saturation velocity) and the possibility of realizing AlGaN/GaN HEMTs for very high power amplifiers operating at GHz frequencies.

Because until 2001 no useful GaN substrates existed, other substrate materials had to be used to grow GaN heterostructures. The first AlGaN/GaN HEMT was grown on sapphire substrates. Because the thermal conductivity of sapphire is quite low, SiC substrates with a higher thermal conductivity have also been considered and are becoming more popular. Nevertheless, sapphire is still being used extensively because it is less expensive compared to SiC. Since 1999, there were reports on the successful growth of AlGaN/GaN HEMT structures on Si substrates as well [99, 100]. This option is of interest not only from the perspective of substrate cost, but also because the thermal conductivity of Si is about 3 times higher than that of sapphire (at 300, K 1.3 W/cm-K compared to 0.4 W/cm-K).

In general, the layer structure of AlGaN/GaN HEMTs is similar to that of the III–V HEMTs discussed in the previous sections. It consists (from the top) of a cap, barrier, channel, and buffer layer, grown on one of the substrates mentioned above. Figures 5.33(a) and (b) show the layer sequence of two different AlGaN/GaN HEMT designs. The structure in Fig. 5.33(a) employs a doped AlGaN barrier layer above the channel, whereas an undoped barrier layer is used in the structure in Fig. 5.33(b).

The cap consists of an undoped AlGaN or GaN layer 2 to 10 nm thick, followed by the $Al_xGa_{1-x}N$ barrier layer. The Al content x is between 0.15 and 0.5, with a typical value around 0.3. A large x leads to a large conduction band offset, but for $x > 0.4$ it becomes more difficult to realize ohmic contacts and the doping efficiency decreases. The barrier layer may be either homogeneously doped (typical doping levels of 10^{18} to 2×10^{19} cm^{-3}) or δ-doped for the structure shown in Fig. 5.33(a), whereas the barrier layer is undoped for the structure in Fig. 5.33(b). In HEMTs with an undoped barrier, the 2DEG channel is induced by the polarization effects discussed in Section 3.4.4. If a doped barrier layer is used, then an undoped AlGaN spacer with a thickness around 3 nm is inserted between the barrier and channel layers.

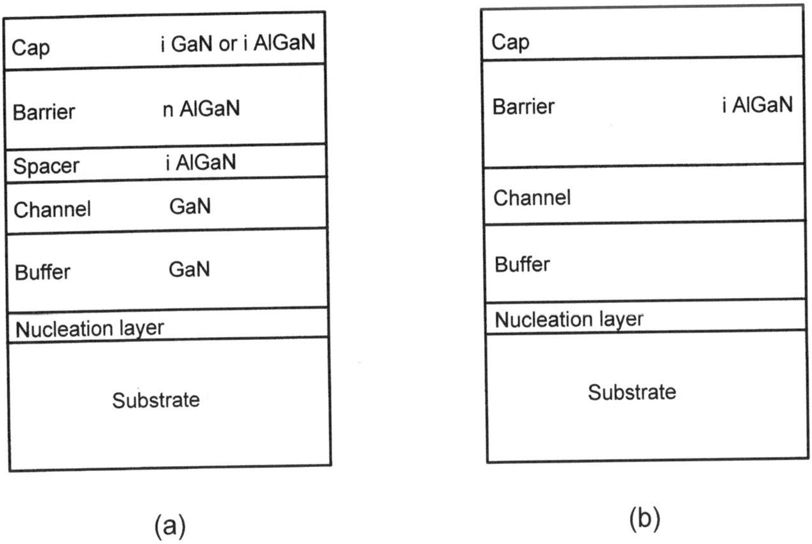

Cap	i GaN or i AlGaN
Barrier	n AlGaN
Spacer	i AlGaN
Channel	GaN
Buffer	GaN
Nucleation layer	
Substrate	

(a)

Cap	
Barrier	i AlGaN
Channel	
Buffer	
Nucleation layer	
Substrate	

(b)

Figure 5.33. Layer sequence of (a) AlGaN/GaN HEMTs with doped barrier layer, and (b) AlGaN/GaN with undoped barrier layer.

The GaN channel layer may be either intentionally doped or undoped. Doped channels are about 50 nm thick and doped at 2 to 5×10^{17} cm^{-3}, followed by a 0.5 to 2 μm thick buffer. The buffer layer and the substrate are separated by a GaN, AlGaN, or AlN nucleation layer. Channel electron mobilities of 1000–1500 cm^2/Vs and 2DEG sheet densities of 1 to 1.6×10^{13} cm^{-3} are typical for AlGaN/GaN HEMTs.

The source and drain ohmic contacts and the Schottky gate are located on top of the layer stack. The ohmic contacts typically consist of (from top to bottom) Ti/Al/Ni/Au thermally alloyed at about 800°C. The Schottky gate is made of (from the bottom) Ni/Au and can be either rectangular or mushroom shaped.

5.9.2 AlGaN/GaN HEMT Performance

Typically, AlGaN/GaN HEMTs show maximum drain currents I_{max} well above 1000 mA/mm and extrinsic transconductances g_m between 200 and 300 mS/mm, and the best reported values are 1430 mA/mm and 300 mS/mm for I_{max} and g_m, respectively [101, 102]. AlGaN/GaN HEMTs exhibit extremely high breakdown voltages. Typical values of the gate–drain breakdown voltage BV_{GD} are between 60 and 200V, and a record BV_{GD} of 284 V for a transistor with Al$_{0.5}$Ga$_{0.5}$N barrier and 3 μm gate–drain spacing has been reported [103]. Figure 5.34 shows reported cutoff frequencies f_T and maximum frequencies of oscillation f_{max} of AlGaN/GaN HEMTs. Values as high as 107 GHz for f_T and 155 GHz for f_{max} obtained from 0.15 μm and 0.12 μm AlGaN/GaN HEMTs, respectively, have been demonstrated [12, 107].

Most AlGaN/GaN HEMT developments have been focused on high-power transistors. Figure 5.35 shows the reported output power density as a function of fre-

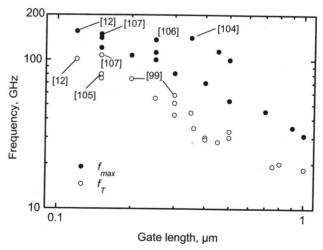

Figure 5.34. Reported cutoff frequency and maximum frequency of oscillation of AlGaN/GaN HEMTs.

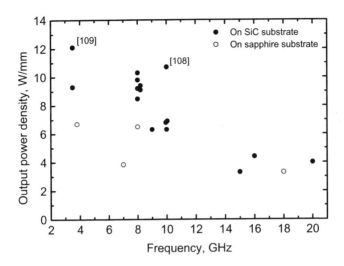

Figure 5.35. Reported output power density of AlGaN/GaN HEMTs.

quency. The power densities of 10.7 W/mm at 10 GHz and 12.1 W/mm at 3.5 GHz reported in 2001 were the highest values obtained from any type of field-effect transistor up to that time [108, 109]. Clearly AlGaN/GaN HEMTs on SiC substrates show higher output power densities than their counterparts on sapphire substrates. The reason is the higher thermal conductivity of SiC leading to less self-heating and consequently a lower channel temperature at a given power level. A critical issue in assessing reported output power data is whether the measurement has been carried out under continuous wave (CW) or pulsed conditions. The output power under pulsed conditions can be considerably higher compared to the CW case, because the average dissipated power and thus the self-heating are lower when pulsing technique is used. It has been demonstrated that a transistor with 8 mm gate width can deliver an output power of 51 W at 6 GHz under pulsed operation [110].

Although the excellent power performance is the most watched feature of AlGaN/GaN HEMTs, there were a few reports on the high-frequency noise performance of these transistors as well [12, 102, 111]. Minimum noise figures as low as 0.53, 0.72, and 1 dB have been observed at 8, 12, and 18 GHz, respectively, for a 0.12 μm-gate AlGaN/GaN HEMT [112].

A problem of AlGaN/GaN HEMTS is the frequency-dependent slump of the drain current [112]. It consists of a considerable decrease of the drain current under high-frequency operating conditions compared to dc operation. The current slump leads to a high-frequency output power lower than that expected from the dc I_D–V_{DS} characteristics. The effect is still under investigation and is currently attributed to slow-acting traps. Experiments have shown that a Si_3N_4 surface passivation considerably reduces the current slump.

REFERENCES

1. R. Dingle, H. L. Störmer, A. C. Gossard, and W. Wiegmann, Electron Mobilities in Modulation-Doped Semiconductor Heterojunction Superlattices, *Appl. Phys. Lett., 33,* pp. 665–667, 1978.
2. T. Mimura, S. Hiyamizu, T. Fujii, and K. Nanbu, A New Field-Effect Transistor with Selectively Doped GaAs/n-$Al_xGa_{1-x}As$ Heterojunctions, *Jpn. J. Appl. Phys., 19,* pp. L225–L227, 1980.
3. S. Moskowitz, High-Mobility Transistors: Breakthrough or Ballyhoo? *Microwaves, 19,* pp. 20–21, Oct. 1980.
4. P. C. Chao, S. C. Palmateer, P. M. Smith, U. K. Mishra, K. H. G. Duh, and J. C. M. Hwang, Millimeter-Wave Low-Noise High Electron Mobility Transistors, *IEEE Electron Device Lett., 6,* pp. 531–533, 1985.
5. A. W. Swanson, J. Herb, and M. Young, First Commercial HEMT Challenges GaAs FETs, *Microwaves & RF, 24,* pp. 107–110, Nov. 1985. See also: A. W. Swanson, J. Herb, and M. Young, Toshiba's HEMT Device Will Follow Goulds's to Market, *Electronics, 58,* p. 13, Issue 47 1985.
6. A. Ketterson, W. T. Masselink, J. S. Gedymin, J. Klem, W. Kopp, H. Morkoc, and K. R. Gleason, Characterization of InGaAs/AlGaAs Modulation-Doped Field Effect Transistors, *IEEE Trans. Electron Devices, 33,* pp. 564–571, 1986.
7. T. Henderson, M. Aksun, C. Peng, H. Morkoc, P. C. Chao, P. M. Smith, K. H. G. Duh, and L. F. Lester, Microwave Performance of a Quarter-Micrometer Gate Low-Noise Pseudomorphic InGaAs/AlGaAs MODFET, *IEEE Electron Device Lett., 7,* pp. 645–647, 1986.
8. K. Hirose, K. Ohata, T. Mizutani, T. Itoh, and M. Ogawa, 700 mS/mm 2DEGFETs Fabricated from High Electron Mobility MBE-Grown n-AlInAs/GaInAs Heterostructures, *Proc. GaAs and Related Compounds,* pp. 529–532, 1985.
9. J. M. Kuo, B. Calvic, and T. Y. Chang, New Pseudomorphic MODFET Utilizing $Ga_{0.47-u}In_{0.53+u}/Al_{0.48+u}In_{0.52-u}$ Heterostructures, *IEDM Tech. Dig.,* pp. 460–463, 1986.
10. K. Inoue, J. C. Harmand, and T. Matsumo, High-Quality $In_xGa_{1-x}As$/InAlAs Modulation-Doped Heterostructures Grown Lattice-Mismatched on GaAs Substrates, *J. Crystal Growth, 111,* pp. 313–317, 1992.
11. W. E. Hoke, P. J. Lemonias, J. J. Mosca, P. S. Lyman, A. Torabi, P. F. Marsh, R. A. McTaggart, S. M. Lardizabad, and K. Helzar, Molecular Beam Epitaxial Growth and Device Performance of Metamorphic High Electron Mobility Transistor Structures Fabricated on GaAs Substrates, *J. Vac. Sci. Technology, B17,* pp. 1131–1135, 1999.
12. W. Lu, J. Yang, M. A. Khan, and I. Adesida, AlGaN/GaN HEMTs on SiC with over 100 GHz f_T and Low Microwave Noise, *IEEE Trans. Electron Dev., 48,* pp. 581–585, 2001.
13. Y.-F. Wu, D. Kapolnek, J. P. Ibbertson, P. Parikh, B. P. Keller, and U. K. Mishra, Very-High Power Density AlGaN/GaN HEMTs, *IEEE Trans. Electron Dev., 48,* pp. 586–590, 2001.
14. M. Zeuner, T. Hackbarth, G. Höck, D. Behammer, and U. König, High-Frequency SiGe-n-MODFET for Microwave Applications, *IEEE Microwave and Guided Wave Letters, 9,* pp. 410–412, 1999.
15. F. Aniel, N. Zerounian, R. Adde, M. Zeuner, T. Hackbarth, and U. König, Low Tem-

perature Analysis of 0.25 μm T-Gate Strained Si/Si$_{0.55}$Ge$_{0.45}$ n-MODFET's, *IEEE Trans. Electron Devices, 47,* pp. 1477–1483, 2000.

16. B. R. Bennet, A. S. Bracker, R. Magno, J. B. Boos, R. Bass, and D. Park, Monolithic Integration of Resonant Interband Tunneling Diodes and High Electron Mobility Transistors in the InAs/GaSb/AlSb Material System, *J. Vac. Sci. Technol., B18,* pp. 1650–1652, 2000.

17. J. B. Boos, B. R. Bennett, W. Kruppa, D. Park, M. J. Yang, and B. V. Shanabrook, AlSb/InAs HEMTs Using Modulation InAs(Si)-Doping, *Electron. Lett., 34,* pp. 403–404, 1998.

18. T. M. Brookes, The Noise Properties of High Electron Mobility Transistors, *IEEE Trans. Electron Devices, 33,* pp. 52–57, 1986.

19. D. Delagebeaudeuf and N. T. Linh, Metal-(n) AlGaAs-GaAs Two-Dimensional Electron Gas FET, *IEEE Trans. Electron Devices, 29,* pp. 955–960, 1982.

20. M. D. Feuer, Two-Layer Model for Source Resistance in Selectively Doped Heterojunction Transistors, *IEEE Trans. Electron Devices, 32,* pp. 7–11, 1985.

21. H. H. Berger, Contact Resistance on Diffused Resistors, *ISSCC Dig.,* pp. 160–161, 1969. More details can be found in: H. Berger, *Modelbeschreibung planarer Ohmscher Metall-Halbleiterkontakte,* Ph.D. Thesis, Techn. Hochschule Aachen, 1970.

22. G. K. Reeves and H. B. Harrison, Obtaining the Specific Contact Resistance from Transmission Line Model Measurements, *IEEE Electron Device Lett., 3,* pp. 111–113, 1982.

23. Y. Ando and T. Itoh, Accurate Modeling for Parasitic Source Resistance in Two-Dimensional Electron Gas Field-Effect Transistors, *IEEE Trans. Electron Devices, 36,* pp. 1036–1044, 1989.

24. A. Cappy, A. Vanoverschelde, M. Schortgen, C. Versnaeyen, and G. Salmer, Noise Modeling in Submicrometer-Gate Two-Dimensional Electron-Gas Field-Effect Transistors, *IEEE Trans. Electron Devices, 32,* pp. 2787–2796, 1985.

25. Y.-H. Byun, K. Lee, and M. Shur, Unified Charge Control Model and Subthreshold Current in Heterostructure Field-Effect Transistors, *IEEE Electron Device Lett., 11,* pp. 50–53, 1990.

26. G. George and J. R. Hauser, Improved Analytic MODFET Charge-Control Model, *Solid-State Electron., 36,* pp. 481–482, 1993.

27. A. Majumdar, A Complete Charge Control Model for HEMTs, *Solid-State Electron., 41,* pp. 1825–1826, 1997.

28. C. Yuen, C. K. Nishimoto, M. W. Glenn, Y.-C. Pao, R. A. LaRue, R. Norton, M. Day, I. Zubeck, S. G. Bandy, and G. A. Zdasiuk, A Monolithic Ka-Band HEMT Low-Noise Amplifier, *IEEE Trans. Microwave Theory Tech., 36,* pp. 1930–1937, 1988.

29. L. D. Nguyen, A. S. Brown, M. A. Thompson, L. M. Jelloian, L. E. Larson, and M. Matloubian, 650-Å Self-Aligned-Gate Pseudomorphic Al$_{0.48}$In$_{0.52}$As/Ga$_{0.2}$In$_{0.8}$As High Electron Mobility Transistor, *IEEE Electron Device Lett., 13,* pp. 143–145, 1992.

30. J. H. Lee, H.-S. Yoon, B.-S. Park, S.-J. Meang, C.-W. Lee, H.-T. Choi, C.-E. Yun, and C.-S. Park, Noise Performance of Pseudomorphic AlGaAs/InGaAs/GaAs High Electron Mobility Transistors with Wide Head T-Shaped Gate Recessed by Electron Cyclotron Resonance Plasma Etching, *Jpn. J. Appl. Phys., 38,* Pt.1, pp. 654–657, 1999.

31. Y. Yamashita, A. Endoh, K. Shinohara, M. Higashiwaki, K. Hikosaka, T. Mimura, S. Hiyamizu, and T. Matsui, Ultra-Short 25-nm-Gate Lattice Matched InAlAs/InGaAs

HEMTs within the Range of 400 GHz Cutoff Frequency, *IEEE Electron Device Lett., 22,* pp. 367–369, 2001.

32. H. Fukui, Determination of the Basic Device Parameters of a GaAs MESFET, *Bell Syst. Techn. J., 58,* pp. 771–797, 1979.

33. M. Foisy, P. J. Tasker, B. Hughes, and L. F. Eastman, The Role of Inefficient Charge Modulation in Limiting the Current-Gain Cutoff Frequency of the MODFET, *IEEE Trans. Electron Devices, 35,* pp. 871–878, 1988.

34. L. D. Nguyen, L. E. Larson, and U. K. Mishra, Ultra-High-Speed Modulation-Doped Field-Effect Transistors: A Tutorial Review. *Proc. IEEE, 80,* pp. 494–518, 1992.

35. N. Moll, M. R. Hueschen, and A. Fischer-Colbrie, Pulse-Doped AlGaAs/InGaAs Pseudomorphic MODFET's, *IEEE Trans. Electron Devices, 35,* pp. 879–886, 1988.

36. L. D. Nguyen, P. J. Tasker, D. C. Radulescu, and L. F. Eastman, Characterization of Ultra-High-Speed Pseudomorphic AlGaAs/InGaAs (on GaAs) MODFET's, *IEEE Trans. Electron Devices, 36,* pp. 2243–2248, 1989.

37. A. van der Ziel and E. N. Wu, Thermal Noise in High Electron Mobility Transistors, *Solid-State Electron., 26,* pp. 383–384, 1983.

38. E. N. Wu and A. van der Ziel, Induced-Gate Thermal Noise in High Electron Mobility Transistors, *Solid-State Electron.,* 26, pp. 639–642, 1983.

39. G. W. Wang, Y. K. Chen, J. B. Kuang, and L. F. Eastman, MODFET Noise Model and Properties with Hot-Electron Effects, *IEEE Trans. Electron Devices, 36,* pp. 1847–1850, 1989.

40. Y. Ando and T. Itoh, DC, Small-Signal, and Noise Modeling of Two-Dimensional Electron Gas Field-Effect Transistors Based on Accurate Charge-Control Characteristics, *IEEE Trans. Electron Devices, 37,* pp. 67–78, 1990.

41. A. F. M. Anwar and K.-W. Liu, A Noise Model for High Electron Mobility Transistors, *IEEE Trans. Electron Devices, 41,* pp. 2087–2092, 1994.

42. P. C. Chao, Gate Formation Technologies, in R. L. Ross, S. P. Svensson, and P. Lugli (eds.), *Pseudomorphic HEMT Technology and Applications,* Kluwer, Dordrecht, 1996, pp. 93–107.

43. A. N. Lepore, H. M. Levy, R. C. Tiberio, P. J. Tasker, H. Lee, E. D. Wolf, L. F. Eastman, and E. Kohn, 0.1 μm Gate Length MODFETs with Unity Current Gain Cutoff Frequency Above 110 GHz, *Electron. Lett., 24,* pp. 364–366, 1988.

44. I. Hanyu, S. Asai, M. Nunokawa, K. Joshin, Y. Hirachi, S. Ohmura, Y. Aoki, and T. Aigo, Super Low-Noise HEMTs with a T-Shaped WSi_x Gate, *Electron. Lett., 24,* pp. 1327–1328, 1988.

45. K.-H. G. Duh, S.-M. J. Liu, L. F. Lester, P. C. Chao, P. M. Smith, M. B. Das, B. R. Lee, and J. Ballingall, Ultra-Low-Noise Characteristics of Millimeter-Wave High Electron Mobility Transistors, *IEEE Electron Device Lett., 9,* pp. 521–523, 1988.

46. K. Inoue, K. Ebihara, H. Haematsu, T. Igarashi, H. Takahashi, and J. Fukaya, A 240 W Push-Pull GaAs Power FET for W-CDMA Base Stations, *MTT-S Dig.,* pp. 1719–1722, 2000.

47. N. Sakura, K. Matsunaga, K. Ishikura, I. Takenaka, K. Asano, N. Iwata, M. Kanamori, and M. Kuzuhara, 100W L-Band GaAs Power FP-HFET Operated at 30V, *MTT-S Dig.,* pp. 1715–1718, 2000.

48. K. Asano, Y. Miyoshi, K. Ishikura, Y. Nashimoto, M. Kuzuhara, and M. Mizuta, Nov-

el High Power AlGaAs/GaAs HFET with a Field-Modulating Plate Operated at 35V Drain Voltage, *IEDM Tech. Dig.,* pp. 59–62, 1998.

49. P. Saunier and J. W. Lee, High-Efficiency Millimeter-Wave GaAs/GaAsAs Power HEMT's, *IEEE Electron Device Lett., 7,* pp. 503–505, 1986.

50. K. Matsunaga, K. Ishikura, I. Takenaka, W. Contrara, A. Wakejima, K. Ota, M. Kanamoi, and M. Kuzuhara, A Low-Distortion 230 W GaAs Power FP-HFET Opearted at 22 V for Cellular Base Stations, *IEDM Tech. Dig.,* pp. 393–396, 2000.

51. M. Takikawa and K. Joshin, Pseudomorphic n-InGaP/InGaAs/GaAs High Electron Mobility Transistors for Low-Noise Amplifiers, *IEEE Electron Device Lett., 14,* pp. 406–408, 1993.

52. M. Chertouk, S. Bürkner, K. Bachem, W. Pletschen, S. Kraus, J. Braunstein, and G. Tränkle, Advantages of Al-Free GaInP/InGaAs pHEMTs for Power Applications, *Electron. Lett., 34,* pp. 590–592, 1998.

53. M. Zaknoune, O. Schuler, S. Piotrowicz, F. Mollot, D. Theron, and Y. Crosnier, High-Power V-Band $Ga_{0.51}In_{0.49}P/In_{0.2}Ga_{0.8}As$ Pseudomorphic HEMT Grown by Gas Source Molecular Beam Epitaxy, *IEEE Microwave and Guided Wave Lett., 9,* pp. 28–30, 1999.

54. M. Zaknoune, O. Schuler, F. Mollot, D. Theron, and Y. Crosnier, 0.1µm $Ga_{0.51}In_{0.49}P/In_{0.2}Ga_{0.8}As$ pHEMT Grown by GSMBE with High DC and RF Performances, *Electron. Lett., 35,* pp. 501–502, 1999.

55. H. Takenaka and D. Ueda, 0.15 µm T-Shaped Gate Fabrication for GaAs MODFET Using Phase Shift Lithography, *IEEE Trans. Electron Devices, 43,* pp. 238–244, 1996.

56. F. Diette, D. Langrez, J. L. Codron, E. Delos, D. Theron, and G. Salmer, 1510mS/mm 0.1µm Gate Length Pseudomorphic HEMTs with Intrinsic Current Gain Cutoff Frequency of 220 GHz, *Electron. Lett., 32,* pp. 848–850, 1996.

57. K. L. Tan, R. M. Dia, D. C. Streit, T. Lin, T. Q. Trinh, A. C. Han, P. H. Liu, P.-M. D. Chow, and H. C. Yen, 94-GHz 0.1-µm T-Gate Low-Noise Pseudomorphic InGaAs HEMT's, *IEEE Electron Device Lett., 11,* pp. 585–587, 1990.

58. J.-H. Lee, H.-S. Yoon, C.-S. Park, and H.-M. Park, Ultra Low Noise Characteristics of AlGaAs/InGaAs Pseudomorphic HEMT's with Wide Head T-Shaped Gate, *IEEE Electron Device Lett., 16,* pp. 271–273, 1995.

59. K. L. Tan, R. M. Dia, D. C. Streit, L. K. Shaw, A. C. Han, M. D. Sholley, P. H. Liu, T. Q. Trinh, T. Lin, and H. C. Yen, 60 GHz Pseudomorphic $Al_{0.25}Ga_{0.75}As/In_{0.28}Ga_{0.72}As$ Low-Noise HEMTs, *IEEE Electron Device Lett., 12,* pp. 23–25, 1991.

60. A. Tessmann, O. Wohlgemut, R. Reuter, W. H. Haydl, H. Massler, and A. Hülsmann, A Coplanar 148 GHz Cascode Amplifier MMIC Using 0.15 µm GaAs pHEMTs, *MTT-S Dig.,* pp. 991–994, 2000.

61. P. Saunier and H. Q. Tserng, AlGaAs/InGaAs Heterostructure with Doped Channels for Discrete Devices and Monolithic Amplifiers, *IEEE Trans. Electron Devices, 36,* pp. 2231–2235, 1989.

62. R. Actis, K. B. Nichols, W. F. Kopp, T. J. Rogers, and F. W. Smith, High-Performance 0.15-µm-Gate-Length pHEMTs Enhanced with a Low-Temperature-Grown GaAs Buffer, *IEEE MTT-S Dig.,* pp. 445–448, 1995.

63. W. Marsetz, A. Hülsmann, K. Köhler, M. Demmler, and M. Schlechtweg, GaAs pHEMT with 1.6W/mm Output Power Density, *Electron. Lett., 35,* pp. 748–749, 1999.

64. P. M. Smith, P. C. Chao, J. M. Ballingall, and A. W. Swanson, Microwave and mm-

Wave Power Amplification Using Pseudomorphic HEMTs, *Microwave Journal, 33,* pp. 71–86, May 1990.

65. P. M. Smith, L. F. Lester, P.-C. Chao, P. Ho, R. P. Smith, J. M. Ballingall, and M.-Y. Kao, A 0.25-μm Gate-Length Pseudomorphic HFET with 32-mW Output Power at 94 GHz, *IEEE Electron Device Lett., 10,* pp. 437–439, 1989.

66. I. Takaneka, K. Ishikura, H. Takahashi, K. Asano, and M. Kanamori, Low Distortion High Power GaAs Pseudomorphic Heterojunction FETs for L/S-Band Digital Cellular Base Stations, *MTT-S Dig.,* pp. 1711–1714, 2000.

67. J. Udomoto, S. Chaki, M. Komaru, T. Kunii, Y. Kohno, S. Goto, K. Gotoh, A. Inoue, N. Tanino, T. Takagi, and O. Ishihara, An 11 W Ku-Band Heterostructure FET with WSi/Au T-Shaped Gate, *MTT-S Dig.,* pp. 339–342, 1995.

68. P. M. Smith, L. F. Lester, D. W. Ferguson, P. C. Chao, P. Ho, M. Kao, J. M. Ballingall, and R. P. Smith, Ka-Band High Power Pseudomorphic Heterostructure FET, *Electron. Lett., 25,* pp. 639–640, 1989.

69. R. Lai, M. Wojtowicz, C. H. Chen, M. Biedenbender, H. C. Yen, D. C. Streit, K. L. Tan, and P. H. Liu, High-Power 0.15-μm V-Band Pseudomorphic InGaAs-AlGaAs-GaAs HEMT, *IEEE Microwave and Guided Wave Lett., 3,* pp. 363–365, 1993.

70. M.-Y. Kao, P. M. Smith, P. Ho, P.-C. Chao, K. H. G. Duh, A. A. Jabra, and J. M. Ballingall, Very High Power-Added Efficiency and Low-Noise 0.15-μm Gate-Length Pseudomorphic HEMT's, *IEEE Electron Device Lett., 10,* pp. 580–582, 1989.

71. M. Zaknoune, Y. Cordier, S. Bollaert, D. Ferre, D. Theron, and Y. Crosnier, 0.1 μm High Performance Metamorphic $In_{0.32}Al_{0.68}As/In_{0.33}Ga_{0.67}As$ HEMT on GaAs Using Inverse Step InAlAs Buffer, *Electron. Lett., 35,* pp. 1670–1671, 1999.

72. Y. Cordier, M. Zaknoune, S. Bollaert, and A. Cappy, Charge Control and Electron Transport Properties in $In_xAl_{1-y}As/In_xGa_{1-x}As$ Metamorphic HEMTs: Effect of Indium Content, *Proc. InP and Related Materials,* pp. 102–105, 2000.

73. M. Zaknoune, B. Bonte, C. Gaquiere, Y. Cordier, Y. Druelle, D. Theron, and Y. Crosnier, InAlAs/InGaAs Metamorphic HEMT with High Current Density and High Breakdown Voltage, *IEEE Electron Device Lett., 19,* pp. 345–347, 1998.

74. D.-W. Tu, S. Wang, J. S. M. Liu, K. C. Hwang, W. Kong, P. C. Chao, and K. Nicols, High-Performance Double-Recessed InAlAs/InGaAs Power Metamorphic HEMT on GaAs Substrate, *IEEE Microwave and Guided Wave Lett., 9,* pp. 458–460, 1999.

75. D. M. Gill, B. C. Kane, S. P. Svensson, D.-W. Tu, P. N. Uppal, and N. E. Byer, High-Performance, 0.1 μm InAlAs/InGaAs High Electron Mobility Transistors on GaAs, *IEEE Electron Device Lett., 17,* pp. 328–330, 1996.

76. D. C. Dumka, W. E. Hoke, P. J. Lemonias, G. Cueva, and I. Adesida, Metamorphic $In_{0.52}Al_{0.48}As/In_{0.53}Ga_{0.47}As$ HEMTs on GaAs Substrate with f_T Over 200 GHz, *IEDM Tech. Dig.,* pp. 783–786, 1999.

77. C. S. Whelan, P. F. Marsh, W. E. Hoke, R. A. McTaggart, C. P. McCarroll, and T. E. Kazior, GaAs Metamorphic (MHEMT): An Attractive Alternative to InP HEMTs for High Performance Low Noise and Power Applications, *Proc. InP and Related Materials,* pp. 337–340, 2000.

78. C. Gaquire, S. Bollaert, M. Zaknoune, Y. Cordier, D. Theron, and Y. Crosnier, Influence on Power Performance at 60 GHz of Indium Composition in Metamorphic HEMTs, *Electron. Lett., 35,* pp. 1489–1451, 1999.

79. P. F. Marsh, S. Kang, R. Wohlert, P. M. McIntosh, W. E. Hoke, R. A. McTaggart, S.

M. Lardizabal, R. E. Leoni III, C. S. Whelan, P. J. Lemonias, and T. E. Kazior, Millimeter-Wave Low Noise Metamorphic HEMT Amplifiers and Devices on GaAs Substrates, *Proc. GaAs IC Symp.*, pp. 221–223, 1999.

80. K. Higuchi, H. Matsumoto, T. Mishima, and T. Nakamura, Optimum Design and Fabrication of InAlAs/InGaAs HEMT's on GaAs with Both High Breakdown Voltage and High Maximum Frequency, *IEEE Trans. Electron Devices, 46*, 99. 1312–1318, 1999.

81. D. Xu, T. Suemitsu, J. Osaka, Y. Umeda, Y. Yamane, Y. Ishii, T. Ishii, and T. Tamamura, Depletion- and Enhancement-Mode Modulation-Doped Field-Effect Transistors for Ultrahigh-Speed Applications: An Electrochemical Fabrication Technology, *IEEE Trans. Electron Devices, 47*, pp. 33–43, 2000.

82. L. D. Nguyen, A. S. Brown, M. A. Thompson, and L. M. Jelloian, 50-nm Self-Aligned-Gate Pseudomorphic AlInAs/GaInAs High Electron Mobility Transistors, *IEEE Trans. Electron Devices, 39*, pp. 2007–2014, 1992.

83. P. C. Chao, A. J. Tessmer, K.-H. G. Duh, P. Ho, M.-Y. Kao, P. M. Smith, J. M. Ballingall, S.-M. J. Liu, and A. A. Jabra, W-Band Low-Noise InAlAs/InGaAs Lattice-Matched HEMT's, *IEEE Electron Device Lett., 11*, pp. 59–61, 1990.

84. D. Xu, T. Suemitsu, J. Osaka, Y. Umeda, Y. Yamane, Y. Ishii, T. Ishii, and T. Tamamura, An 0.03-μm Gate-Length Enhancement-Mode InAlAs/InGaAs/InP MODFET with 300 GHz f_T and 2 S/mm Extrinsic Transconductance, *IEEE Electron Device Lett., 20*, pp. 206–208, 1999.

85. K. H. G. Duh, P. C. Chao, S. M. J. Liu, P. Ho, M. Y. Kao, and J. M. Ballingall, A Super Low-Noise 0.1 μm T-Gate InAlAs-InGaAs-InP HEMT, *IEEE Microwave and Guided Wave Lett., 1*, pp. 114–116, 1991.

86. T. P. Chin, Y. C. Chen, M. Barsky, M. Wojtowicz, R. Grundbacher, R. Lai, D. C. Streit, and T. R. Block, High Performance InP High Electron Mobility Transistors by Valved Phosphorus Cracker, *J. Vac. Sci. Technol., B18*, pp. 1642–1644, 2000.

87. Y. Yamashita, E. Endoh, K. Shinohara, M. Higashiwaki, K. Hikosaka, T. Mimura, S. Hiyamizu, and T. Matsui, Ultra-Short 25-nm-Gate Lattice-Matched InAlAs/InGaAs HEMTs within the Range of 400 GHz Cutoff Frequency, *IEEE Electron Device Lett., 22*, pp. 367–369, 2001.

88. A. Endoh, Y. Yamashita, M. Higashiwaki, K. Hikosaka, T. Mimura, S. Hiyamizu, and T. Matsui, High f_T 50-nm-Gate Lattice-Matched InAlAs/InGaAs HEMTs, *Proc. InP and Related Mat.*, pp. 87–90, 2000.

89. D. C. Streit, T. R. Block, M. Wojtowicz, D. Pascua, R. Lai, G. I. Ng, P.-H. Liu, and K. L. Tan, Graded-Channel InGaAs-InAlAs-InP High Electron Mobility Transistors, *J. Vac. Sci. Technol., B13*, pp. 774–776, 1995.

90. D. Xu, H. Heiß, S. Kraus, M. Sexl, G. Böhm, G. Tränkle, G. Weimann, and G. Abstreiter, High-Performance Double-Modulation-Doped InAlAs/InGaAs/InAs HFET's, *IEEE Electron Device Lett., 18*, pp. 323–326, 1997.

91. P. M. Smith, S.-M. J. Liu, M.-Y. Kao, P. Ho, S. C. Wang, K. H. G. Duh, S. T. Fu, and P. C. Chao, W-Band High Efficiency InP-Based Power HEMT with 600 GHz f_{max}, *IEEE Microwave and Guided Wave Lett., 5*, pp. 230–232, 1995.

92. P. Ho, M. Y. Kao, P. C. Chao, K. H. G. Duh, J. M. Ballingall, S. T. Allen, A. J. Tessmer, and P. M. Smith, Extremely High Gain 0.15μm Gate-Length InAlAs/InGaAs/InP HEMTs, *Electron. Lett., 27*, pp. 325–327, 1991.

93. K. Schimpf, M. Sommer, M. Horstmann, M. Hollfelder, A. van der Hart, M. Marso, P.

Kordos, and H. Lüth, 0.1-μm T-Gate Al-Free InP/InGaAs/InP pHEMT's for W-Band Applications Using a Nitrogen Carrier for LP-MOCVD Growth, *IEEE Electron Device Lett., 18,* pp. 144–146, 1997.

94. T. Hwang, P. Chye, and P. Gregory, Super Low Noise Pseudomorphic InGaAs Channel InP HEMTs, *Electron. Lett., 29,* pp. 10–11, 1993.

95. P. C. Chao, A. J. Tessmer, K. H. G. Duh, P-Ho, M.-Y. Kao, P. M. Smith, J. M. Ballingall, S.-M. J. Liu, and A. A. Jabra, W-Band Low-Noise InAlAs/InGaAs Lattice-Matched HEMT's, *IEEE Electron Device Lett., 11,* pp. 59–62, 1990.

96. M.-Y. Kao, K. H. G. Duh, P. Ho, and P.-C. Chao, An Extremely Low-Noise InP-Based HEMT with Silicon Nitride Passivation, *IEDM Tech. Dig.,* pp. 907–910, 1994.

97. P. Chevalier, X. Wallart, F. Mollot, B. Bonte, and R. Fauquermbergue, Composite Channel HEMTs for Millimeter-Wave Powerr Applications, *Proc. InP and Related Materials,* pp. 207–210, 1998.

98. S. Piotrowicz, C. Gaquiere, B. Bonte, E. Bourcier, D. Theron, X. Wallart, and Y. Crosnier, Best Combination Between Power Density, Efficiency, Gain, at V-Band with an InP-Based PHEMT Structure, *IEEE Microwave and Guided Wave Lett., 8,* pp. 10–12, 1998.

99. E. M. Chumbes, A. T. Schremer, J. A. Smart, D. Hogue, J. J. Komiak, and J. R. Shealy, Microwave Performance of AlGaN/GaN High Electron Mobility Transistors on Si(111) Substrates, *IEDM Tech Dig.,* pp. 397–400, 1999.

100. E. M. Chumbes, A. T. Schremer, J. A. Smart, Y. Wang, N. C. Mac Donald, D. Hogue, J. J. Komiak, S. J. Lichwalla, R. E. Leoni III, and J. R. Shealy, AlGaN/GaN High Electron Mobility Transistors on Si(111) Substrates, *IEEE Trans. Electron Devices, 48,* pp. 420–426, 2001.

101. A. T. Ping, Q. Chen, J. W. Yang, M. A. Khan, and I. Adesida, DC and Microwave Performance of High-Current AlGaN/GaN Heterostructure Field Effect Transistors Grown on p-Type SiC Substrtaes, *IEEE Electron Device Lett., 19,* pp. 54–56, 1998.

102. N. X. Nguyen, M. Micovic, W.-S. Wong, P. Hashimioto, P. Janke, D. Harvey, and C. Nguyen, Robust Low Microwave Noise GaN MODFETs with 0.60dB Noise Figure at 10 GHz, *Electron. Lett., 36,* pp. 469–471, 2000.

103. U. K. Mishra, Y.-F. Wu, B. P. Keller, S. Keller, and S. P. Denbaars, GaN Microwave Electronics, *IEEE Trans. Microwave Theory Techn., 46,* pp. 756–761, 1998.

104. R. Welch, T. Jenkins, B. Neidhard, L. Kehias, T. Quach, P. Watson, R. Worley, M. Barsky, R. Sandhu, and M. Wojtowicz, Low Noise Hybrid Amplifier Using AlGaN/GaN Power HEMT Devices, *Dig. GaAs IC Symp.,* pp. 153–155, 2001.

105. M. Micovic, A. Kurdoghlian, P. Janke, P. Hashimoto, D. W. S. Wong, J. S. Moon, L. McCray, and C. Nguyen, AlGaN/GaN Heterojunction Field Effect Transistors Grown by Nitrogen Plasma Assisted Molecular Beam Epitaxy, *IEEE Trans. Electron Devices, 48,* pp. 591–596, 2001.

106. W. Kumar, W. Lu, R. Schwindt, J. Van Hove, P. Chow, and I. Adesida, 0.25μm Gate-Length, MBE-Grown AlGaN/GaN HEMTs with High Current and High f_T, *Electron. Lett., 37,* pp. 858–859, 2001.

107. V. Kumar, W. Lu, F. A. Khan, R. Schwindt, A. Kuliev, J. Yang, M. Asif Khan, and A. Adesida, High Performance 0.15 μm Recessed Gate AlGaN/GaN HEMTs on Sapphire, *IEDM Tech. Dig.,* pp. 573–576, 2001.

108. V. Tilak, B. Green, V. Kaper, H. Kim, T. Prunty, J. Smart, J. Shealy, and L. F. East-

man, Influence of Barrier Thickness on the High-Power Performance of AlGaN/GaN HEMTs, *IEEE Electron Device Lett., 22,* pp. 504–506, 2001.
109. J. W. Palmour, S. T. Sheppard, R. P. Smith, S. T. Allen, W. L. Pribble, T. J. Smith, Z. Ring, J. J. Sumakeris, A. W. Saxler, and J. W. Milligan, Wide Bandgap Semiconductor Devices and MMICs for RF Power Applications, *IEDM Tech. Dig.,* pp. 385–388, 2001.
110. Y.-F. Wu, P. M. Chavarkar, M. Moore, P. Parikh, B. P. Keller, and U. K. Mishra, A 50-W AlGaN/GaN HEMT Amplifier, *IEDM Tech. Dig.,* pp. 375–376, 2000.
111. A. T. Ping, E. Piner, J. Redwing, M. Asif Khan, and I. Adesida, Microwave Noise Performance of AlGaN/GaN HEMTs, *Electron. Lett., 36,* pp. 175–176, 2000.
112. B. M. Green, K. K. Chu, E. M. Chumbes, J. A. Smart, J. R. Shealy, and L. F. Eastman, The Effect of Surface Passivation on the Microwave Characteristics of Undoped AlGaN/GaN HEMT's, *IEEE Electron Device Lett., 21,* pp. 268–270, 2000.

CHAPTER 6

MOSFETs

6.1 INTRODUCTION

The term MOSFET stands for *M*etal-*O*xide-*S*emiconductor *F*ield-*E*ffect *T*ransistor. In principle, MOSFETs can be fabricated on any semiconductor material, yet the silicon MOSFET is actually the only MOSFET type currently used. The reason for this is the unrivaled properties of the Si/SiO_2 interface, which is of major importance for MOSFET operation. Until now, no other interface between a semiconductor and an insulator comparable to the Si/SiO_2 interface in terms of quality, stability, and reliability could be realized. Therefore, this chapter concentrates on the silicon MOSFET, called MOSFET hereafter, and its use in microwave electronics.

To date, the MOSFET enjoys the largest market share in VLSI electronics, and most microprocessors and memory circuits are based on this transistor type. Despite its growing popularity in mainstream electronics since 1980, the MOSFET has until recently been considered as a slow device not suitable for high-frequency operation. Indeed, due to the relatively low electron mobility in the channel, the MOSFET is much slower than a III–V FET with comparable dimensions. Continuous scaling of VLSI circuits over the years, however, has led MOSFETs with deep-submicron gates to provide a sufficiently high-speed performance for many microwave applications in the lower GHz range. Furthermore, submicron MOSFET technology is more matured and less expensive than any other semiconductor technology. Thus, during the 1990s, the interest in microwave MOSFETs grew considerably and the device became an interesting and attractive option for large-volume, civil communications markets where cost is of major concern [1–6]. In the late 1990s and early 2000s, a variety of different MOS integrated circuits operating around 0.9 GHz had been realized [7, 8], but there were also reports of MOS ICs successfully designed for 2.4 and 5 GHz [9, 10]. In 2001, MOS circuits with frequency capability ranging from 900 MHz to 2.4 GHz were commercially available [11].

Experimental n-channel MOSFETs with a gate length of 60 nm showing a cutoff frequency f_T of 245 GHz [12] have been reported and low minimum noise figures of 0.5 dB at 2 GHz and 1.2 dB at 12 GHz [13, 14] have been demonstrated. A major problem of microwave MOSFETs is the large resistivity of the polysilicon gate, which gives rise to a considerable noise level and limits the gain at high frequencies and thus the maximum frequency of oscillation f_{max}. Up to the year 2000, MOSFETs with high cutoff frequencies suffered from relatively low maximum frequencies of oscillation. The record f_{max} at that time of 66 GHz [15] was much lower than the record f_T. This is opposite to the trend found in other types of microwave FETs where f_{max} is typically much larger than f_T. Since 2001, considerable progress on f_{max} has been made, and MOSFETs with maximum frequencies of oscillation exceeding 100 GHz have emerged [12, 16–18].

For microwave power applications, a special version of the MOSFET, the so-called LDMOSFET (Laterally Diffused MOSFET), is widely used in commercial applications. Around the year 2000, the LDMOSFET became the dominant power device in the base stations of 2.5 GHz mobile communication systems [19, 20].

Both n-channel MOSFETs (nMOSFET) and p-channel MOSFETs (pMOSFET) are used. The former has the advantage of higher speed, and the latter is needed to construct complementary MOS (CMOS) circuits needed in digital logic circuits. Figures 6.1(a) and (b) show the basic structures of the conventional bulk nMOSFET and pMOSFET, respectively. The nMOSFET consists of a p-type Si substrate frequently called the bulk. The substrate has a backside contact at which the bulk potential can be applied. On the top, the surface between the drain and source regions is covered by a thin SiO_2 layer called the gate oxide (hatched area). The contact on top of the gate oxide is commonly formed by n^+-type polysilicon (for short, poly Si or poly). Other gate materials, such as metals, can be used as well. As the name implies, polysilicon is not a single-crystal. Rather, it consists of many regions, each with a regular atomic structure inside. At the boundaries between these regions the regularity is broken. Beside the gate and the gate oxide, the heavily doped n^+-type source and drain regions are implanted into the substrate. The poly gate and the source and drain regions are often doped during the same implantation step. Thus, the gate acts as an implantation mask and the source and drain regions are self-aligned to the gate. Due to the lateral diffusion of the dopants during the subsequent annealing and other processing steps, the source and drain regions extend laterally underneath the gate. On top of the source and drain regions, the ohmic source and drain contacts are deposited.

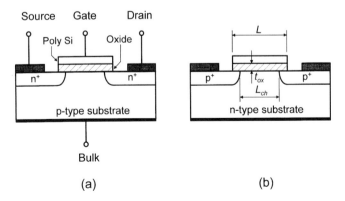

Figure 6.1. Cross sections of conventional bulk MOSFETs. (a) nMOSFET. (b) pMOSFET.

The structure of the pMOSFET shown in Fig. 6.1(b) is very similar to that of the nMOSFET, only the conductivity type of the regions is opposite, i.e., the substrate is of the n-type, and the source, drain, and the gate are of the p-type. The gate length of a MOSFET is L, the gate width (not shown in the figure) is W, and the thickness of the oxide is t_{ox}. The distance between the source–bulk and bulk–drain metallurgical junctions is called the channel length L_{ch}. Thus L is larger than L_{ch}, and the difference of the two depends on the degree of the lateral diffusion of the source and drain dopants.

Another increasingly popular MOSFET version is the SOI (*S*ilicon *o*n *I*nsulator) MOSFET. The SOI nMOSFET in Fig. 6.2 consists of an insulating SiO_2 layer deposited on the surface of a Si substrate. On top of the SiO_2 layer, a monocrystalline Si layer is situated and it acts in a similar way to the bulk of the conventional bulk MOSFETs shown in Fig. 6.1. The main difference is that it is much thinner and has no backside contact. Thus, its potential is floating, whereas the bulk potential in a bulk MOSFET is defined by the potential applied to the bulk contact. Currently, SOI wafers are more expensive compared to Si wafers. This drawback is compensated, however, by the fact that undesired short channel effects can be more effectively suppressed in very short gate-length SOI MOSFETs. Consequently there is a widespread consensus that SOI MOSFETs possess a higher scaling flexibility and potential than conventional bulk MOSFETs. For microwave applications, SOI MOSFET technology offers the additional advantage of having an insulating substrate, which is important for microwave integrated circuits containing passive elements such as inductors and capacitors.

The operational mechanism of a MOSFET is as follows. Consider the nMOSFET shown in Fig. 6.1(a) and assume that the source and the bulk are connected to the ground. The applied drain–source voltage V_{DS} is positive and the gate–source voltage V_{GS} can be either positive or negative. First, let us neglect the effect of the gate. A drain current in this structure has to pass through two pn junctions, namely the source–bulk and the bulk–drain junctions connected in series. Because the

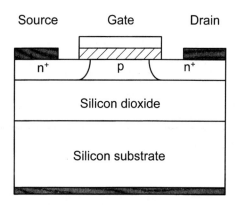

Figure 6.2. Cross section of an SOI nMOSFET.

source and bulk are grounded and V_{DS} is positive, the source–bulk junction is in equilibrium and the bulk–drain junction is reverse-biased. As a result, only the negligibly small reverse current of the bulk–drain junction flows between the drain and source and the drain current I_D is practically zero.

Next we include the effect of the gate. In the literature, the interface between the gate oxide and the semiconductor is commonly designated as the surface, and we will adopt this designation. A positive gate–source (and thus gate–bulk) voltage results in an electric field perpendicular to the surface and in the direction from the gate into the substrate. It repels holes away from the surface and attracts electrons toward it. If V_{GS} is large enough, the electron density near the surface can become as large as or even larger than the acceptor concentration in the substrate, while the hole density near the surface is practically decreased to zero. Thus the conductivity type of the region near the surface between the source and drain is transformed from the originally p-type to n-type. This results in a conductive path between the source and drain, and a substantial drain current can flow. The amount of the drain current depends on both V_{DS} and V_{GS}. The quantitative description of MOSFET characteristics has been discussed in detail in many excellent textbooks [21–25].

In the following, we will develop models to describe the dc and small-signal behavior of MOSFETs, with special emphasis being placed on microwave nMOSFETs. The treatments apply generally to pMOSFETs as well, provided we exchange the doping type in the device (p-type instead of n-type and vice versa) and change the polarities of the applied voltages (minus instead of plus and vice versa). The two-terminal MOS structure is first considered. This is followed by the derivation of a dc MOSFET model and the discussion of the elements of the small-signal equivalent circuit. We will not follow the typical approaches found in existing MOS textbooks, but rather present an approach that is especially suited and useful for microwave MOSFETs.

6.2 TWO-TERMINAL MOS STRUCTURE

6.2.1 Qualitative Description

The two-terminal MOS structure is the heart of MOSFETs. For nMOSFETs, such a structure is given in Fig. 6.3. It consists of a p-type Si substrate (doping concentration N_A) with the bulk contact, the gate oxide, and the poly gate. Early MOSFETs actually had metal gates (e.g., aluminum gates), but almost all modern MOSFETs employ poly gates. Consider the case of a grounded bulk contact. When a gate potential V_G is applied, the voltage drop across the structure is V_{GB}.

Figure 6.4(a) shows the energy band diagrams of the poly gate, oxide, and substrate when they are still separated from each other. The vacuum energy level E_{vac} is chosen as the reference energy. The n$^+$ poly gate is heavily doped so that its Fermi level E_F is located slightly above the conduction band edge E_C. The electron affinity χ_{Si} (the separation between E_{vac} and E_C) is about 4.05 eV for the poly and for the monocrystalline Si substrate. Thus, the work function of the poly gate, ϕ_G, is

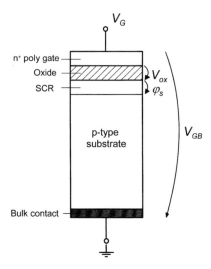

Figure 6.3. Two-terminal MOS structure with a p-type substrate.

approximately equal to χ_{Si}, whereas the work function of the substrate, ϕ_b, depends on the doping concentration N_A. The properties of the SiO$_2$ layer are described by the bandgap $E_{G,SiO2}$ of around 9 eV and the electron affinity χ_{SiO2} of 0.95 eV.

When the three layers of the MOS structure are brought together and no external voltage is applied, the Fermi levels in the poly gate and Si substrate are aligned at the same position. Thus E_F becomes a straight line across the entire structure and the bands have to be adjusted to E_F in the same manner as in the pn junction (see Section 2.4). The resulting energy band diagram is shown in Fig. 6.4(b). We see that the conduction band of the substrate is bent downward near the surface, where the separation between E_C and E_F is decreased and the separation between E_V and E_F is increased. The band bending at the surface is designated by $q\varphi_s$. According to Eqs. (2-18) and (2-19), the electron and hole densities n_s and p_s at the surface become larger and smaller, respectively, than their values n_b and p_b in the bulk. Thus, in the bulk near the surface a space–charge region (designated SCR in Fig. 6.3) is created. This condition is called depletion because the majority carriers are depleted from the surface region. Because of the extremely high doping concentration of the n$^+$ poly, the conduction band bending in the gate is negligible.

A voltage applied to the gate will alter the band bendings, thus affecting the carrier concentrations in the surface region. The band bendings, carrier densities, and charge densities in the substrate for different gate–bulk voltages V_{GB} are shown in Fig. 6.5. From the top to bottom, the figure shows the band diagram, carrier concentration, and net charge distribution as a function of depth y. The origin of the y-axis is located at the surface. A sufficiently large negative V_{GB} leads to upward band bendings at the surface, an increase in the surface majority carrier density p_s, and a decrease in the surface minority carrier density n_s. This condition is called accumu-

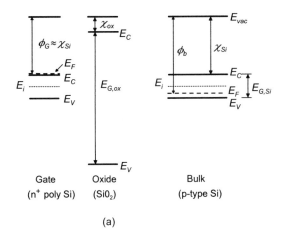

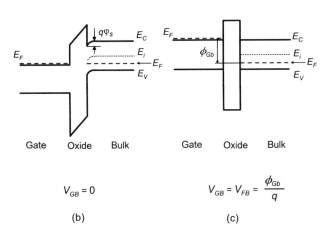

Figure 6.4. Energy band diagrams of a two-terminal MOS structure for the cases of poly gate, oxide, and substrate (a) separated from each other, (b) brought in contact. (no applied voltage), and (c) flat band case. χ (SiO$_2$) = 0.95 eV, χ (Si) = 4.05 eV, E_G (SiO$_2$) ≈ 9 eV, E_G (Si) = 1.12 eV.

lation and shown in Fig. 6.5(a). A slightly negative V_{GB} results in the disappearance of the band bendings, resulting in flat bands. Under this condition, the electron and hole concentrations in the entire substrate are constant and the conditions $n_s = n_b$ and $p_s = p_b$ hold. This is the so-called flat band case as shown in Figs. 6.4(c) and 6.5(b), and V_{GB} for this case is the flat band voltage V_{FB}. For a V_{GB} larger (more positive) than the flat band voltage, the bands at the surface are bent downward. Holes are repelled from the surface, leaving behind the uncompensated acceptor ions. Thus, a space–charge region with a thickness d_{sc} arises. At the same time, the electron concentration near the surface is increased above n_b. The fraction of the ex-

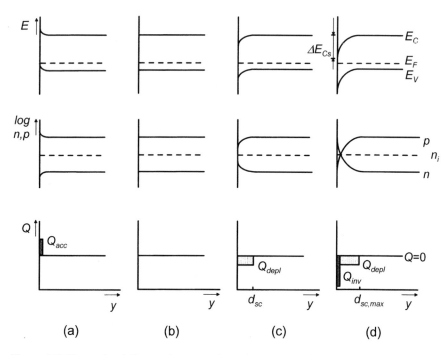

Figure 6.5. Energy band diagram (upper part), carrier concentration (middle part), and charge distribution (lower part) in the substrate of a two-terminal MOS structure. The origin of the y-axis is the surface, i.e., the Si/SiO$_2$ interface. (a) Accumulation, $V_{GB} < 0$, (b) flat band, $V_{GB} = V_{FB} < 0$, (c) depletion, $V_{GB} > V_{FB}$, and (d) onset of strong inversion, $V_{GB} = V_{th} > 0$.

cess electrons to the total net charge, however, is still negligibly small. This condition is called depletion and depicted in Fig. 6.5(c).

An even more positive gate–bulk voltage reinforces the effects described above, i.e., n_s and d_{sc} increase while p_s decreases. When n_s becomes equal to p_s, the conductivity type of the surface is changing from the original p-type (the Si substrate is doped with acceptors) to the n-type. This condition is called the onset of weak inversion. Finally, at a sufficiently large V_{GB}, the surface electron concentration becomes equal to the hole concentration in the bulk ($n_s = p_b$). This important condition defines the onset of *strong inversion* and is shown in Fig. 6.5(d). The gate–bulk voltage causing the onset of strong inversion is called the threshold voltage V_{th}. When V_{GB} is increased beyond V_{th}, n_s is increased further while the thickness of the space–charge region remains almost the same as when $V_{GB} = V_{th}$. We call the extension of the space–charge region at the onset of strong inversion the maximum thickness $d_{sc,max}$. The inversion region where $n > p$ is restricted to a thin layer underneath the surface called the inversion channel, or, for short, the channel. The channel extends only a few nanometers from the surface into the substrate.

Because no current flows through the insulating oxide, the mass action law is valid for all conditions discussed above. For the operation of microwave MOSFETs, only the depletion and inversion conditions are of interest. In the following, an expression for the threshold voltage of the two-terminal MOS structure will be derived.

6.2.2 Derivation of the Threshold Voltage

The derivation of the threshold voltage V_{th} in a two-terminal MOS structure consists of two main steps with several substeps. To clarify the procedure, we first specify these steps.

Step (1) Determination of the threshold voltage V'_{th} of the ideal MOS structure. The different work functions of the gate and the substrate, and the influence of oxide charges are neglected. This is done by finding:
(1a) the band bending and the corresponding surface potential at the onset of strong inversion,
(1b) the electric field at the surface, and
(1c) the electric field and the voltage drop in the oxide.
Step (2) Derivation of the threshold voltage V_{th} taking into account the work function difference and oxide charges. This is done by
(2a) including the effect of the work function difference, and
(2b) including the effect of oxide charges.

For simplicity, we neglect the difference between the effective densities of states in the conductive and valance bands, and set $N_C = N_V$. We start with step (1a) utilizing the condition of $n_s = p_b$ for the onset of strong inversion. In the band diagram this means that, at the surface, the separation between the conduction band edge E_{Cs} and the Fermi level E_F must be equal to the separation between the Fermi level and the valence band edge E_{Vb} in the bulk. This condition is fulfilled when the band bending at the surface, ΔE_{Cs} [see Fig. 6.5(d)], is

$$\Delta E_{Cs} = -E_G + 2(E_F - E_{Vb}) = -2(E_{ib} - E_F) \tag{6-1}$$

where E_{ib} is the intrinsic Fermi energy in the bulk, and the subscripts s and b designate the surface and the bulk, respectively. Assuming complete ionization and rearranging Eq. (2-23), one obtains

$$\Delta E_{Cs} = -2k_B T \ln \frac{N_A}{n_i} \tag{6-2}$$

where k_B is the Boltzmann constant, T is the temperature, and n_i is the intrinsic free-carrier concentration. The corresponding surface potential φ_s can be calculated using the relationship between energy and potential given in Eq. (2-50). This yields

$$\varphi_s = -\frac{\Delta E_{Cs}}{q} = 2\frac{k_B T}{q} \ln \frac{N_A}{n_i} \tag{6-3}$$

where q is the elementary charge. Now we proceed to step (1b) and calculate the field at the surface. In general, the electric field at the surface is caused by the net charge inside the substrate consisting of free carriers (electrons and holes), and ionized acceptors in the space–charge region. Extensive analytical derivations, e.g., [24, 25], and numerical calculations, e.g., [21], indicated, however, that the contribution of the free carriers to the field is much smaller than that of the ionized acceptors. Thus, we can approximate the net charge in the substrate by the charge of the ionized acceptors in the space–charge region. Because of the requirement of charge neutrality, the same amount of charge, but with opposite sign, must reside at the gate.

The thickness of the space–charge region at the onset of strong inversion can be calculated in analogy to Eq. (2-122) by

$$d_{sc,max} = \sqrt{\frac{\varepsilon_{Si}\varphi_s}{qN_A}} \tag{6-4}$$

where ε_{Si} is the dielectric constant of Si. Thus the net charge stored in the space–charge region per unit area, Q_{depl}/A, is given by

$$\frac{Q_{depl}}{A} = -qN_A d_{sc,max} = -\sqrt{2qN_A \varepsilon_{Si}\varphi_s} \tag{6-5}$$

This, together with the dielectric displacement $\mathscr{D}_{s,Si}$ on Si side of the surface, leads to the electric field $\mathscr{E}_{s,Si}$ at the surface:

$$\mathscr{E}_{s,Si} = \frac{\mathscr{D}_{s,Si}}{\varepsilon_{Si}} = \frac{1}{\varepsilon_{Si}}\frac{Q_{depl}}{A} = -\sqrt{\frac{2qN_A\varphi_s}{\varepsilon_{Si}}} \tag{6-6}$$

We continue with step (1c) and develop an expression for the electric field in the oxide. At the surface, the continuity of the dielectric displacement is required, i.e., $\mathscr{D}_{s,ox} = \mathscr{D}_{s,Si}$ at $y = 0$:

$$\varepsilon_{ox}\mathscr{E}_{ox} = \varepsilon_{Si}\mathscr{E}_{s,Si} \tag{6-7}$$

where ε_{ox} is the dielectric constant of the oxide and $\mathscr{E}_{ox}$ is the electric field in the oxide. The field in the oxide is constant and can be calculated from Eqs. (6-7) and (6-6):

$$\mathscr{E}_{ox} = -\frac{1}{\varepsilon_{ox}}\sqrt{2qN_A\varepsilon_{Si}\varphi_s} \tag{6-8}$$

The definition of the electric field, $\mathscr{E} = -dV/dy$, together with Eq. (6-8), yields the voltage drop V_{ox} across the oxide:

$$V_{ox} = -\mathscr{E}_{ox}t_{ox} = \frac{t_{ox}}{\varepsilon_{ox}}\sqrt{2qN_A\varepsilon_{Si}\varphi_s} \tag{6-9}$$

From Fig. 6.3 we see that V_{GB} is the sum of the voltage drop across the space–charge region, i.e., the band bending at the surface ΔE_{Cs}, and the voltage drop across the oxide V_{ox}. Thus the threshold voltage V'_{th} for the ideal two-terminal MOS structure is

$$V'_{th} = \varphi_s + V_{ox} = \varphi_s + \frac{t_{ox}}{\varepsilon_{ox}}\sqrt{2qN_A\varepsilon_{Si}\varphi_s} \qquad (6\text{-}10)$$

Step 1 of the derivation of the threshold voltage is now completed, and we can proceed to step 2 to consider the effects of the different work functions and oxide charges on V_{th}. An inspection of Fig. 6.4(b) reveals that for zero applied gate–bulk voltage, the bands in the substrate near the surface and in the oxide are already bent downward due to the different work functions of the gate and bulk materials. Considering the band diagrams in Fig. 6.4, it becomes clear that the total band bending (band bending in the oxide plus band bending in the bulk) is equal to the work function difference ϕ_{Gb}:

$$\phi_{Gb} = \phi_G - \phi_b = -\frac{E_G}{2} - (E_{ib} - E_F) = -\frac{E_G}{2} - k_B T \ln\frac{N_A}{n_i} \qquad (6\text{-}11)$$

Thus, the conditions in a MOS structure with $\phi_G \neq \phi_b$ are identical to those in an ideal structure (with $\phi_G = \phi_b$) to which a positive voltage equal to $-V_{FB} = -\phi_{Gb}/q$ is applied. Note that V_{FB} and ϕ_{Gb} are negative in our n$^+$-poly/oxide/p-bulk structure. This means that $-\phi_{Gb}$ given in (6-11) has to be subtracted from V'_{th} in (6-10) to get the threshold voltage of the real MOS structure. We obtain

$$V_{th} = \frac{k_B T}{q} \ln\frac{N_A}{n_i} + \frac{t_{ox}}{\varepsilon_{ox}}\sqrt{4k_B T N_A \varepsilon_{Si} \ln\frac{N_A}{n_i}} - \frac{E_G}{2q} \qquad (6\text{-}12)$$

So far, we have assumed that there are no charges in the oxide and at the Si/SiO$_2$ interface. In real MOS structures, such charges, called oxide charges, are present. Detailed discussions on the origin and nature of oxide charges, and on the approach of how to model the influence of these charges on V_{th}, can be found in [23–25]. The final formula for the threshold voltage including these charges is

$$V_{th} = \frac{k_B T}{q} \ln\frac{N_A}{n_i} + \frac{t_{ox}}{\varepsilon_{ox}}\sqrt{4k_B T N_A \varepsilon_{Si} \ln\frac{N_A}{n_i}} - \frac{E_G}{2q} - Q_{ox}\frac{t_{ox}}{\varepsilon_{ox}} \qquad (6\text{-}13)$$

where Q_{ox} is the equivalent oxide charge per unit area. This expression looks slightly different from V_{th} given in [21–25], but it can be reduced to the familiar expression in [21–25] by incorporating the oxide capacitance per unit area, $C_{ox} = \varepsilon_{ox}/t_{ox}$ and carrying out a few algebraic manipulations. Finally, it is worth reiterating that V_{th} given by Eq. (6-13) is the voltage V_{GB} causing the onset of strong inversion in two-terminal MOS structures.

6.3 DC ANALYSIS

6.3.1 Introduction

In this section, a dc MOSFET model closely related to the PHS-like HEMT model will be derived. It is called the PHS-like MOSFET model and is valid for the conventional bulk MOSFET. The major advantage of this model compared to the common textbook MOSFET models is that it properly describes the important effect of velocity saturation. The physics of SOI MOSFETs is more complicated because of the floating body. More information on SOI MOSFETs and, more specifically, on microwave SOI MOSFETs can be found in [26–29].

First, the operation of the bulk nMOSFET is discussed qualitatively. Figure 6.6(a) shows the transistor having a zero drain–source voltage V_{DS} and a positive gate–source voltage V_{GS} larger than the threshold voltage ($V_{GS} > V_{th}$). The source and bulk are both grounded. As we know from the preceding discussions, a $V_{GS} > V_{th}$ induces an inversion channel at the surface between source and drain. This channel creates an n-type conductive path between source and drain. Since V_{DS} is zero, no driving force exists in the channel to move electrons from source to drain and consequently the drain current I_D is zero. Furthermore, underneath the channel, the space–charge region extends into the substrate. The n^+ source and drain regions are surrounded by space–charge regions as well. Because the potential difference among the source, drain, and bulk terminals is zero, the thickness of the space–charge regions associated with the source and drain regions depends only on the built-in voltages of the source–bulk and the bulk–drain pn junctions [see Eq. (2-93)]. The electron concentration in the channel, on the other hand, depends on the gate–channel voltage, which is constant across the entire channel and equal to V_{GS} due to the absence of V_{DS}.

Now we apply a positive V_{DS} to the transistor. This voltage leads to several effects shown in Fig. 6.6(b). First, it induces a channel field in the $-x$ direction, driving electrons from the source toward the drain thus giving rise to a drain current. Second, the gate–channel voltage, which is V_{GS} at the source end of the channel and is $V_{GS} - V_{DS}$ at the drain end, is decreasing from the source to the drain. This leads to an x-dependent inversion charge in the channel decreasing toward the drain as well. Applying a

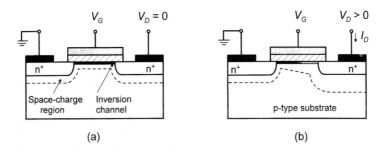

Figure 6.6. Cross section of an nMOSFET with $V_{GS} > V_{th}$ and (a) $V_{DS} = 0$ and (b) $V_{DS} > 0$.

positive V_{GS}, on the other hand, can be viewed as bringing positive charges onto the gate. Because the poly gate is heavily doped, we can consider it as an equipotential plane. To ensure charge neutrality in the transistor, charges of the same amount but opposite sign must exist in the semiconductor underneath the gate oxide. This opposite charge is the sum of the negative inversion free-carrier (electron) charge and the negative charge of the acceptor ions in the space–charge region below the channel. The decrease of the inversion charge toward the drain must be compensated by a larger (more negative) charge in the space–charge region, thereby causing the space–charge region to become thicker when moving from the source toward the drain. Finally, the bulk–drain pn junction is reverse-biased, and the thickness of the space–charge region of this junction depends on the applied V_{DS}.

To describe the dc MOSFET behavior accurately, all the physics mentioned above should be taken into account. This has been done in [23–25] by a model frequently called the bulk charge theory. Unfortunately, the effect of velocity saturation has not been satisfactorily included in this theory. Velocity saturation plays an important role and needs to be considered in microwave short-channel MOSFET modeling. A simpler model is the square law theory, in which the x-dependent inversion channel charge is taken into account, whereas the widening of the space–charge region in x-direction is neglected. In the following, we will develop a MOSFET model based on the square law theory and include the effect of velocity saturation. For simplicity, we will only consider the case of both the source and bulk grounded, but the same approach can be extended to the case of nonzero bulk potential [24].

6.3.2 PHS-like MOSFET Model

The basic idea of the PHS-like MOSFET model is to divide the channel into two regions: a low-field region (region 1) where the carrier velocity increases with increasing field and a high-field region (region 2) where the magnitude of the field is larger than the critical field $\mathscr{E}_S$ and the carriers travel with the saturation velocity v_S. We apply piecewise velocity–field (v–$\mathscr{E}$) characteristics with $v = -\mu_0 \mathscr{E}$ in region 1 ($|\mathscr{E}| < \mathscr{E}_S$) and $v = v_S$ in region 2 ($|\mathscr{E}| > \mathscr{E}_S$). This is exactly the same approach used in the PHS-like MESFET and HEMT models.

Hofstein and Warfield [30] first developed such a two-region MOSFET model. Although their work was rarely cited, many recent MOSFET models [31–36] were derived more or less based on the ideas presented in it. We follow the Hofstein and Warfield model to describe the conditions in region 1, but apply a different approach to describe the potential distribution in region 2. Our approach is closely related to the one for analyzing region 2 in the PHS-like MESFET and HEMT models. Figure 6.7 shows the transistor model with applied voltages and the two regions in the channel.

The creation of an inversion layer in a two-terminal MOS structure was qualitatively discussed in Section 6.2.1. For the calculation of the drain current, however, the amount of the inversion charge is necessary. The actual distribution of the inversion charge, including the quantization effect prominent in MOS devices, can be ac-

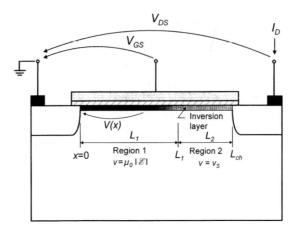

Figure 6.7. Two-region model MOSFET with the applied voltages V_{GS} and V_{DS}.

curately described by self-consistent solutions of the Poisson and Schrödinger equations. Figure 6.8 shows the electron concentration versus depth of a two-terminal MOS structure calculated in this way. When comparing this with the electron concentration in an AlGaAs/GaAs heterostructure (see Fig. 5.3), several similarities become obvious. In both cases, a two-dimensional electron gas is created: at the AlGaAs/GaAs interface in Fig. 5.3 and at the SiO$_2$/Si interface in Fig. 6.8. Furthermore, the electron concentration is small near the interface and peaks a few nm

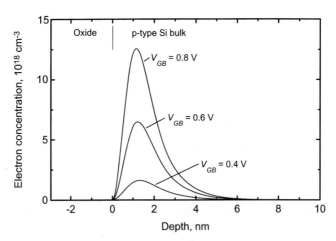

Figure 6.8. Electron concentration near the surface as a function of distance into the substrate calculated using a self-consistent solution of the Schrödinger and Poisson equations. Depth = 0 corresponds to the Si/SiO$_2$ interface. Features of the MOS structure: n$^+$ poly gate, t_{ox} = 3 nm, and N_A = 5 × 10^{17} cm^{-3}.

away from the interface, around 3 nm in the AlGaAs/GaAs structure and around 1.5 nm in the MOS structure. Therefore, we can expect a similar mathematical description of the electron density in the two structures. This is indeed the case, and the electron sheet density n_s in MOS structures with a gate voltage above the threshold voltage (which is the bias range important for microwave MOSFET operation) can be approximated by

$$qn_s = \frac{\varepsilon_{ox}}{t_{ox} + \Delta t_{ox}}(V_{GB} - V_{th}) \tag{6-14}$$

where Δt_{ox} is a correction term accounting for the quantization. Comparing Eq. (6-14) with Eq. (5-7) it becomes clear that Δt_{ox} corresponds to Δa_1 for the HEMT. In the following, Eq. (6-14) will be used as the starting point for the derivation of the drain current of the MOSFET. We will not describe every single step in detail because the derivation is basically the same as that for the drain current in HEMTs.

We first consider the intrinsic transistor and assume that the external applied voltages are equal to the intrinsic voltages. This means that the potential at $x = 0$ (intrinsic source, see Fig. 6.7) is zero and the potential at $x = L_{ch}$ (intrinsic drain) is V_{DS}. Equations (5-1) to (5-22) for the HEMT model are still valid, and the steps leading to Eq. (5-22) are to be repeated here. The only differences are that the gate length L of the HEMT is replaced by the channel length L_{ch} for the MOSFET and that the low-field mobility μ_0 in Eqs. (5-1) and (5-2) is replaced by the effective mobility μ_{eff} in the channel. The physics of the effective mobility will be covered later in Section 6.3.3.

The drain current in region 1 has the form of

$$I_D = \frac{\varepsilon_{ox}}{t_{ox} + \Delta t_{ox}} \frac{\mu_{eff} W}{2L_1}(s^2 - p^2) \tag{6-15}$$

and the drain current in region 2 is

$$I_D = \frac{\varepsilon_{ox}}{t_{ox} + \Delta t_{ox}} W p v_s \tag{6-16}$$

where s and p are the potentials defined by Eqs. (5-11b) and (5-11c). The length L_1 of region 1 can be calculated by

$$L_1 = \frac{1}{\mathscr{E}_S}\frac{s^2 - p^2}{2p} \tag{6-17}$$

The drain–source voltage is the sum of the voltage drops across regions 1 and 2:

$$V_{DS} = V_1 + V_2 = s - p + \frac{2(t_{ox} + \Delta t_{ox})\mathscr{E}_S}{\pi}\sinh\frac{\pi L_2}{2(t_{ox} + \Delta t_{ox})} \tag{6-18}$$

where L_2 is the length of region 2 given by $L_{ch} - L_1$. Equation Eq. (6-18) was derived in the same manner as (4-18) for the MESFET and Eq. (5-19) for the HEMT based on the method described by Grebene and Ghandi [37], but it differs from the expression of Hofstein and Warfield developed for the MOSFET. Equation (6-18) is better suited for MOSFET modeling, as it is derived based on the assumption of a two-dimensional potential distribution in region 2 underneath the gate, instead of a one-dimensional distribution assumed by Hofstein and Warfield. The drain–source saturation voltage V_{DSS} can be modeled in the same way as Eq. (5-22) for the HEMT. Because of the similarities discussed above it is not surprising to see that the MOSFET and HEMT possess very similar current–voltage characteristics (for the HEMT characteristics, see Fig. 5.13).

To take into account the effect of the extrinsic MOSFET, the voltage drops across the parasitic source and drain resistances R_S and R_D have to be included in the dc analysis. The relationship between the terminal (extrinsic) and the intrinsic voltages is given by Eqs. (4-21) and (4-22). A lot of work has been done to model R_S and R_D in MOSFETs [38–42]. However, an accurate determination of R_S and R_D for the MOSFET by analytical means is almost impossible. Figure 6.9 shows the flow of electrons from the source terminal to the inversion channel through the different components of the parasitic source resistance R_S. First, the electrons leave the source contact and spread out across the n$^+$ source region. The resistive component for this flow is the contact resistance R_{co}. Then the electrons move more or less homogeneously distributed throughout the n$^+$ source region toward the gate. This is modeled by the source sheet resistance R_{sh}. Near the end of the source region, the current paths are squeezed and enter the thin accumulation region underneath the surface. The corresponding resistive component is the spreading resistance R_{sp}. Finally, the resistance associated with the accumulation layer underneath the gate edge is R_{acc}. The sum of the four resistive components mentioned above is the para-

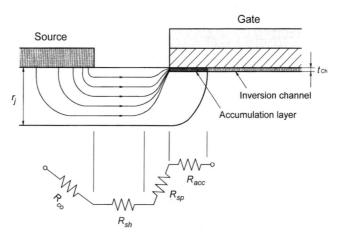

Figure 6.9. Pattern of carrier flow in the source region of a MOSFET, and the resistive components in such a region.

sitic source resistance R_S. Similar conditions exist at the drain side of the MOSFET. Apparently, the description of the two parasitic resistances is a complex two-dimensional problem.

The method proposed in [38, 39] avoids a sophisticated two-dimensional analysis, yet gives a good estimate of R_S and R_D. Let us focus on R_S. For the case of a large source contact and homogeneously doped source, the contact resistance is given by

$$R_{co} = \frac{1}{W}\sqrt{\frac{\rho_{co}}{q\mu_0 N_D r_j}} \tag{6-19}$$

where ρ_{co} is the specific contact resistance (typically 10^{-6}–10^{-7} Ωcm^2 for metal–Si and silicide–Si junctions), r_j and N_D are the depth and the doping concentration of the source region, and μ_0 is the low-field mobility in this region. The sheet resistance R_{sh} is

$$R_{sh} = \frac{1}{q\mu_0 N_D}\frac{L_{SG}}{Wr_j} \tag{6-20}$$

where L_{SG} is the separation between the source contact and the gate. The spreading resistance can be approximated by [38]

$$R_{sp} = \frac{1}{q\mu_0 N_D}\frac{2}{\pi W}\ln(0.75\frac{r_j}{t_{ch}}) \tag{6-21}$$

where t_{ch} is the channel thickness on the order of a few nm. The channel thickness can be estimated from Fig. 6.8. The current flows automatically in a pattern that minimizes the resistance to be surmounted, which results in about equal values for R_{sp} and R_{acc} [39–41]. Thus, $R_{sp} \approx R_{acc}$. The value of the parasitic source resistance R_S is the sum of R_{co}, R_{sh}, R_{sp}, and R_{acc}.

6.3.3 Effective Mobility

The free-carrier mobility in the inversion channel of a MOSFET is described by the so-called effective mobility μ_{eff}. It can be considerably lower than its bulk counterpart μ_0. The drain current in a MOSFET flows in the close vicinity of the surface, where the crystal structure is significantly distorted. Figure 6.8 indicates that most of the inversion electrons reside within a distance of only about 3 nm away from the surface. Thus it is not surprising that the channel mobility is influenced by the surface properties. Actually, the surface induces additional scattering mechanisms that result in the lower effective channel mobility. The main additional scattering mechanisms are surface acoustic phonon scattering and surface roughness scattering. Moreover, the effective mobility depends on the electric field perpendicular to the surface as well. The more positive the gate–source voltage is, the larger is the perpendicular field, and the more the electrons of the inversion layer are attracted to-

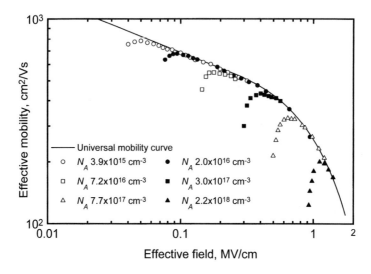

Figure 6.10. Measured effective electron mobilities in the channel of an nMOSFET for different substrate doping levels. Data taken from [50].

ward the surface. The flow of electrons in the channel is therefore subjected to more surface scattering when V_{GS} is increased.

A lot of work has been done to measure the effective mobility and to develop reliable models for μ_{eff} [43–51]. Experiments showed that the mobility of electrons measured as a function of the effective field (the field perpendicular to the semiconductor surface) always falls below a characteristic curve independent of the substrate doping and gate oxide thickness. The same has been observed for holes. This is called the universal mobility behavior, and it is clearly illustrated in Figure 6.10, which shows measured effective electron mobilities in the channel versus the effective field for different substrate doping concentrations [50]. The effective field $\mathscr{E}_{eff}$ is the average perpendicular field experienced by the electrons in the channel and can be expressed as

$$\mathscr{E}_{eff} = \frac{1}{\varepsilon_{Si}}\left(|Q_{depl}| + \frac{1}{\eta}|Q_{inv}|\right) \qquad (6\text{-}22)$$

where η is equal to 2, Q_{depl} is the depletion charge per square unit, i.e., the charge of the acceptor ions in the space–charge region per square unit, and Q_{inv} is the inversion charge per square unit, i.e., the charge of the electrons in the inversion channel per square unit. The depletion charge is

$$Q_{depl} = -qN_A d_{sc,max} \qquad (6\text{-}23)$$

with $d_{sc,max}$ has been given in (6-4), and the inversion charge per unit area is

$$Q_{inv} = -qn_s \tag{6-24}$$

where n_s is the electron sheet density in the inversion channel from Eq. (6-14).

Baccarani and Wordeman [44] proposed the empirical expression

$$\mu_{eff} = 3.25 \times 10^4 \mathcal{E}_{eff}^{-1/3} \times \frac{cm^2}{Vs} \tag{6-25}$$

for the effective electron mobility. It describes the effective mobility fairly well up to an effective field of about 0.3 MV/cm but is not able to reproduce the rapid fall-off characteristics at higher fields. Furthermore, Eq. (6-25) cannot explain the peak effective mobility as a function of the effective field. A better agreement with the measured mobility is obtained using the following two-region mobility model. Below an effective field of about 0.3 MV/cm, μ_{eff} can be expressed as

$$\mu_{eff} = a \mathcal{E}_{eff}^b \tag{6-26}$$

where a and b are fitting parameters. Above 0.3 MV/cm, it is approximated by

$$\mu_{eff} = c + d \exp\frac{-\mathcal{E}_{eff} + e}{f} + g \exp\frac{-\mathcal{E}_{eff} + e}{h} \tag{6-27}$$

where c, d, e, f, g, and h are fitting parameters as well. Equations (6-26) and (6-27), in which the effective field is in MV/cm, hold for both electrons and holes. The fitting parameters extracted from the measured mobility data reported in [50] are listed in Table 6.1. The universal mobility curve calculated in this way is included in Fig. 6.10.

The experimental data from [50] show that for each substrate doping N there is a specific effective field $\mathcal{E}_m$, at which the effective mobility reaches its maximum value. The relationship between N and $\mathcal{E}_m$ can be described by the simple fitting function

$$\mathcal{E}_m = rN^s \tag{6-28}$$

Table 6.1. Fitting parameters needed in Eqs. (6-26) and (6-27)

Parameter	Electrons	Holes	Units
a	380	72	cm²/Vs
b	−0.258	−0.4	—
c	−56.3	33	cm²/Vs
d	540	370	cm²/Vs
e	0.28	−0.025	MV/cm
f	1.16	0.078	MV/cm
g	41.9	135	cm²/Vs
h	1.51	0.58	MV/cm

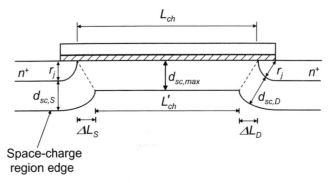

Figure 6.11. Schematic showing two portions of the space–charge region underneath the gate: space–charge region caused by the gate field (between the two dashed lines), and space–charge region caused by the source–bulk and the bulk–drain pn junctions (outside the

where r and s are fitting parameters. For p-type substrates $r = 7.25 \times 10^{-10}$ MV/cm and $s = 0.5$ to get $\mathcal{E}_m$ in MV/cm, whereas for n-type substrates $r = 5.54 \times 10^{-10}$ MV/cm and $s = 0.503$. It has been shown that the experimental hole inversion layer mobility can be reproduced properly with η set to 3 in the expression for the effective field given in Eq. (6-22) [50, 51].

6.3.4 Modifications of the Threshold Voltage

Equation (6-13) is a frequently used expression for V_{th}, but actually it should be used only as a first-order approximation. In short-channel MOSFETs having a gate length in the deep submicron range and a high substrate doping density, two physical effects neglected in Eq. (6-13) are critical to the threshold voltage. First, in short-channel MOSFETs, V_{th} is not independent of the channel length as Eq. (6-13) suggested. Second, in MOSFETs with a heavily doped substrate, the energy quantization in the channel influences V_{th}.

Let us consider the MOSFET in Fig. 6.11, in which the depletion charge under the gate is generated partially by the gate field (between the two dashed lines), and partially by the space–charge regions of the source–bulk and the bulk–drain pn junctions (outside the dashed lines). Therefore, a less gate charge, or in other words a smaller gate–source voltage, is necessary to create the space–charge region underneath the gate oxide compared to the two-terminal MOS structure considered earlier. As a consequence, a smaller gate–source voltage is necessary to induce the inversion charge corresponding to the onset of strong inversion. This leads to a reduction in the threshold voltage as the channel length is decreased, rather than a constant as described by Eqs. (6-12) and (6-13). An improved V_{th} model including this effect was developed in [52] and is given below.

The magnitude of the total depletion charge stored in the space–charge region underneath the gate oxide, neglecting the effect of the two pn junctions, is given by

6.3 DC ANALYSIS

$$|Q_{depl}| = qN_A d_{sc,max} L_{ch} W \qquad (6\text{-}29)$$

This is the case of the two-dimensional MOS structure. Taking into account the depletion charge associated with the two pn junctions and assuming that only the charge inside the two dashed lines in Fig. 6.11 is caused by the gate, the magnitude of this charge, $|Q'_{depl}|$ is

$$|Q'_{depl}| = qN_A d_{sc,max} \frac{L_{ch} + L'_{ch}}{2} W \qquad (6\text{-}30)$$

where $L'_{ch} = L_{ch} - \Delta L_S - \Delta L_D$, and ΔL_S and ΔL_D are defined in Fig. 6.11. Dividing Eq. (6-30) by Eq. (6-29) yields

$$\frac{Q'_{depl}}{Q_{depl}} = 1 - \frac{\Delta L_S + \Delta L_D}{2 L_{ch}} \qquad (6\text{-}31)$$

Using this expression and Eq. (6-13), an improved threshold voltage model accounting for the channel length dependence can be derived:

$$V_{th} = \frac{k_B T}{q} \ln \frac{N_A}{n_i} + \frac{t_{ox}}{\varepsilon_{ox}} \sqrt{4 k_B T N_A \varepsilon_{Si} \ln \frac{N_A}{n_i} \left(1 - \frac{\Delta L_S + \Delta L_D}{2 L_{ch}}\right)} - \frac{E_G}{2} - Q_{ox} \frac{t_{ox}}{C_{ox}} \qquad (6\text{-}32)$$

Expressions for ΔL_S and ΔL_D are needed, which can be obtained by geometrical considerations. Figure 6.12 shows the details of the MOSFET region near source. The following relation holds for the triangular area in the figure:

$$(r_j + \Delta L_S)^2 + d_{sc,max}^2 = (r_j + d_{sc,S})^2 \qquad (6\text{-}33)$$

where $d_{sc,S}$ is the extension of the space–charge region of the source–bulk pn junction into the p-type substrate, which can be calculated from Eq. (2-93). Solving for

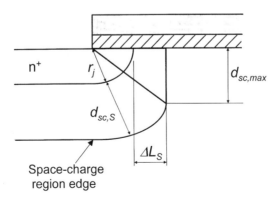

Figure 6.12. Detailed depiction of the source region for modeling the threshold voltage.

ΔL_S in Eq. (6-33) results in

$$\Delta L_S = -r_j + \sqrt{(r_j + d_{sc,S})^2 d_{sc,max}^2} \qquad (6\text{-}34)$$

ΔL_D associated with the drain junction can be calculated in the same way, but the wider space–charge region caused by the positive drain potential needs to be accounted for. This is done by replacing $(V_{bi} - V_{pn})$ in Eq. (2-96) with $(V_{bi} + V_{DS})$. Equation (6-32) shows that the threshold voltage depends on the channel length and that V_{th} decreases for smaller channel length provided the substrate doping and the oxide thickness are not varied.

The second effect influencing the threshold voltage stems from the quantization of the energy levels in the inversion channel. The quantization, which has been discussed in Section 3.4.2 for the 2DEG in AlGaAs/GaAs HEMTs, causes the lowest allowed energy states for electrons in the conduction band of the inversion layer to be located at higher energy levels than those in the bulk material. This means that more energy, or in other words a higher voltage, is necessary to induce a certain electron density near the surface. Thus, the threshold voltage is increased when the quantization effect is significant and is taken into account. The threshold voltage shift by quantization is negligible in MOS structures with low substrate doping but increases considerably for high substrate dopings. The effect of quantization on the threshold voltage has been investigated [53], and the following empirical formula for the threshold voltage shift ΔV_{th} has been suggested:

$$\Delta V_{th} = V_{th}(Q) - V_{th}(nQ) = f + t_{ox}g \qquad (6\text{-}35)$$

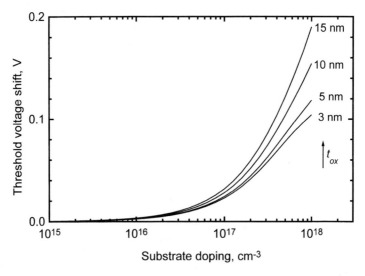

Figure 6.13. Calculated threshold voltage shift caused by the quantization effect in nMOSFETs versus substrate doping.

where $V_{th}(Q)$ and $V_{th}(nQ)$ are the threshold voltages taking into account and neglecting the quantization effect, respectively, t_{ox} is the oxide thickness in nm, and f and g are functions describing the dependence of ΔV_{th} on the substrate doping density. These functions are

$$f_p = 0.006 + 2.276\,N_A - 23.803\,N_A^2 + 110.198\,N_A^3 - 170.128\,N_A^4 \quad (6\text{-}36)$$

$$g_p = 0.075\,N_A - 0.035\,N_A^2 + 0.008\,N_A^3 - 0.063\,N_A^4 \quad (6\text{-}37)$$

for p-type substrates, and

$$f_n = -1.664\,N_D + 13.272\,N_D^2 - 53.092\,N_D^3 + 74.902\,N_D^4 \quad (6\text{-}38)$$

$$g_n = -0.063\,N_D + 0.098\,N_D^2 - 0.757\,N_D^3 + 1.64\,N_D^4 \quad (6\text{-}39)$$

for n-type substrates. Here, N_A and N_D are the bulk doping concentrations in 10^{19} cm^{-3}. Figure 6.13 shows the deviation of the threshold voltage from its classical value for nMOSFETs as a function of substrate doping density for different gate oxide thicknesses calculated using Eqs. (6-35)–(6-39).

6.4 SMALL-SIGNAL ANALYSIS

Several different approaches are used for the small-signal analysis of MOSFETs, and the three most well-known methods will be discussed below.

6.4.1 MESFET/HEMT-Like Equivalent Circuit

The small-signal equivalent circuit for the MESFET and HEMT given in Fig. 4.12 is applicable to the small-signal modeling of the MOSFET as well. This is not surprising, as the expressions derived in Section 6.3.2 for the calculations of the dc behavior and the drain current of an nMOSFET are very similar to those of a HEMT. To illustrate this point, we list in Table 6.2 the corresponding equations in the PHS-like HEMT and PHS-like MOSFET models. These equations are in fact identical except for the following two areas:

Table 6.2. Corresponding dc equations in the PHS-like HEMT model and PHS-like MOSFET model

Quantity	Equation for PHS-like HEMT model	Equation for PHS-like MOSFET model
I_D	(5-14)	(6-15)
I_D	(5-15)	(6-16)
L_1	(5-16)	(6-17)
V_{DS}	(5-19)	(6-18)

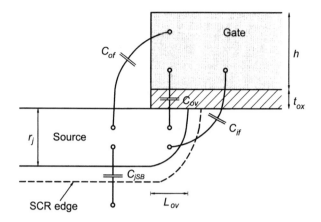

Figure 6.14. Schematic showing the source side of a MOSFET with the parasitic capacitances C_{ov}, C_{if}, C_{of}, and C_{jSB}.

1. The low-field mobility μ_0 in Eq. (5-14) for HEMTs and the effective mobility μ_{eff} in Eq. (6-15) for MOSFETs
2. The term $(a_1+\Delta a_1)$ in Eqs. (5-14), (5-15), and (5-19) for HEMTs and $(t_{ox}+\Delta t_{ox})$ in Eq. (6-15), (6-16), and (6-18) for MOSFETs

This similarity gives rise to a nice scenario wherein there is no need to derive expressions for the transconductance g_m, drain resistance r_{ds}, gate–source capacitance C_{gs}, gate–drain capacitance C_{gd}, and transit time τ_T. Instead, we simply take Eqs. (5-48)–(5-65) and replace, as mentioned above, μ_0 with μ_{eff}, and $(a_1 + \Delta a_1)$ with $(t_{ox} + \Delta t_{ox})$. As in the MESFET and HEMT small-signal modeling, no reliable expression for the charging resistance R_i does exist.

The expressions for the gate–source capacitance of MESFETs and HEMTs have a fringing term C_{fringe} [see Eqs. (4-63) and (5-63)], that accounts for the capacitance associated with charges in the fringing space–charge region near the gate edge. This component does not exist in a MOSFET. There are, however, other parasitic capacitances that have to be considered. Figure 6.14 shows the parasitic capacitances C_{ov}, C_{if}, C_{of}, and C_{jSB} near the source region of the MOSFET. The overlap of the source and gate regions gives rise to an overlap capacitance C_{ov}. Furthermore, there are the stray capacitances between the poly gate and the n^+ source region, which we call the inner and the outer fringing capacitances C_{if} and C_{of}. These parasitic capacitances C_{ov}, C_{if}, and C_{of} appear on both sides of the gate, i.e., between gate and source and between gate and drain.

As a first-order approximation, C_{ov} can be modeled as a parallel plate capacitance and given by

$$C_{ov} = \frac{\varepsilon_{ox} L_{ov} W}{t_{ox}} \tag{6-40}$$

where L_{ov} is the length of the overlap. The inner and outer fringing capacitances have been studied by Shrivastava and Fitzpatrick [54] based on electrostatic considerations. The resulting expressions are

$$C_{if} = \frac{2\varepsilon_{Si}W}{\pi}\ln\left(1 + \frac{r_j}{2t_{ox}}\right) \tag{6-41}$$

and

$$C_{of} = \frac{2\varepsilon_{ox}W}{\pi}\ln\left(1 + \frac{h}{t_{ox}}\right) \tag{6-42}$$

where h is the height of the poly gate. For a typical MOSFET design with $h/t_{ox} \approx 40$ and $r_j/t_{ox} \approx 20$, the following useful estimate of the combined effect of the three parasitic capacitances has been suggested [24]:

$$C_{par} = C_{ov} + C_{if} + C_{of} \approx \varepsilon_{ox}W\left(\frac{L_{ov}}{t_{ox}} + 7\right) \tag{6-43}$$

Here, a relative dielectric constant of 11.9 for Si and of 3.9 for SiO_2 are used. Although the overlap adds a parasitic capacitance that deteriorates the small-signal performance, a certain minimum overlap on the order of $L_{ov} \approx (2-3) \times t_{ox}$ is needed to avoid reliability problems [55]. This requirement, together with Eq. (6-43), results in a typical C_{par} of about 0.3 fF per μm gate width, independent of the other transistor make-ups, such as the gate length, oxide thickness, etc. [24].

The parasitic source–bulk and drain–bulk junction capacities C_{jSB} (shown in Fig. 6.14) and C_{jDB} may also be considered in the small signal equivalent circuit. They can be calculated based on the concept of parallel plate capacitors. The space–charge regions of the source–bulk and bulk–drain pn junctions act the dielectric and the area of the junctions is the area of the parallel plates. However, the inclusion of C_{jSB} and C_{jDB} makes the small-signal equivalent circuit more complicated, and the two capacitances can be neglected in a first-order model.

The gate resistance R_G of MOSFETs can be modeled in the same way as that of HEMTs [see Eq. (5-67)]. For MOSFETs, however, the resistivity of the gate material, ρ_G, is not typically used, but rather the sheet resistance R_{sh} of the gate material given in $\Omega/\square$ is considered. The two can be related as

$$R_{sh} = \frac{\rho_G}{h} \tag{6-44}$$

For a multifinger MOSFET in which the gate fingers are connected on one end only, the gate resistance has the form of

$$R_G = R_{sh}\frac{W_f^2}{3WL} \tag{6-45}$$

where W_f is the gate finger width. When the gate fingers are connected on both ends, the factor of 3 in the denominator has to be replaced with 12. The sheet resistance of typical poly gates is 10 $\Omega/\square$ or more [56], which is too high for microwave MOSFETs (and in some cases for MOSFETs in VLSI circuits as well). Furthermore, the poly gates are of rectangular shape, making it difficult to obtain a low gate resistance for very short gates. The reduction of R_G is of major importance to fully exploit the high-frequency potential of microwave MOSFETs.

To reduce the gate resistance, silicides (e.g., Ni, Co, or Pt silicide) are frequently deposited on top of the poly gate, and gate finger widths smaller than those in the III–V FETs are commonly used in microwave MOSFETs. Gate sheet resistances on the order of 2–5 $\Omega/\square$ have been achieved with silicided gates. Another option to further reduce R_G is to use either metal overgates on top of the poly gate [57], pure metal gates [13], or mushroom-shaped gates [13, 58]. Microwave MOSFETs with mushroom metal gates have been reported in [13], showing sheet resistances of 1.5 $\Omega/\square$ and 0.35 $\Omega/\square$ for W/TiN and Al/TiN gates, respectively.

The MESFET/HEMT-like equivalent circuit has been successfully used to describe the small-signal behavior of microwave MOSFETs with gate lengths ranging from 0.15 to 1 μm [59–63]. An extension of the MESFET/HEMT-like equivalent circuit shown in Fig. 6.15 has also been used to model both SOI and bulk MOSFETs with a 70 nm gate and f_T and f_{max} values up to 114 GHz and 135 GHz, respectively [16]. The results demonstrate the suitability of the MESFET/HEMT-like equivalent circuit for the small signal analysis of microwave MOSFETs.

Table 6.3 shows the values of the equivalent circuit components extracted from the measured S parameters for two microwave MOSFETs.

The small-signal figures of merit, such as the current and power gains, as well as the characteristic frequencies f_T and f_{max}, can be calculated using the same approaches described in Section 4.3.4.

6.4.2 Transmission Line Model

In the small-signal equivalent circuit given in Fig. 6.15, the physical effects taking place in MOSFETs are described exclusively by lumped elements. Actually, the gate and the channel of a MOSFET act like an RC transmission line. This is demonstrated in Fig. 6.16. Several studies have analyzed the small signal behavior of MOSFETs, taking into account the distributed nature of the gate and the channel [64–67]. For example, expressions for the Y parameters of the intrinsic MOSFET having one gate finger have been derived in [66, 67]. The equivalent circuit used for such an analysis is shown in Fig. 6.17. This circuit differs from the equivalent circuit in Fig. 4.12 in the following two respects:

1. The gate resistance is an integral part of the intrinsic transistor in Fig. 6.17.
2. The intrinsic transistor is divided into several small subregions, each of which is described by the distributed gate resistance $R'_g = R_g/W$, gate–source capacitance $C'_{gs} = C_{gs}/W$, gate–drain capacitance $C'_{gd} = C_{gd}/W$, and charging resistance $R'_i = R_i/W$.

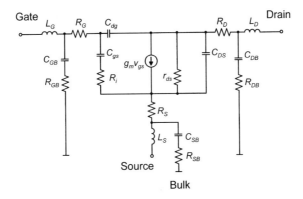

Figure 6.15. Extended small-signal equivalent circuit for the MOSFET.

It is important to point out that R'_g is not the lumped gate resistance R_G given in Eq. (6-45). Instead, R'_g is calculated by

$$R'_g = R_{sh}\frac{W}{L} \tag{6-46}$$

It should be noted that the gate resistance in the MESFET/HEMT-like equivalent circuit already includes the influence of distributed effects using the factor of 3 in the denominator of Eq. (6-45).

Table 6.3. Components of the small-signal equivalent circuit of two microwave MOSFETs

	MOSFET 1 [60]	MOSFET 2 [61]
Gate (finger number × length × width), μm × μm	1 × 0.35 × 50	4 × 0.25 × 50
f_T, GHz	23	23.6
f_{max}, GHz	56	32
C_{gs}, pF	0.055	0.16
C_{gd}, pF	0.015	0.057
C_{ds}, pF	—	0.045
g_m, mS	9.3	29.5
R_i, Ω	29.6	1
r_{ds}, Ω	1000	305
R_G, Ω	8.4	4.1
R_S, Ω	3.1	1.1
R_D, Ω	3.1	10.6
L_G, pH	—	0.1
L_S, pH	—	17.4
L_D, pH	—	0.1

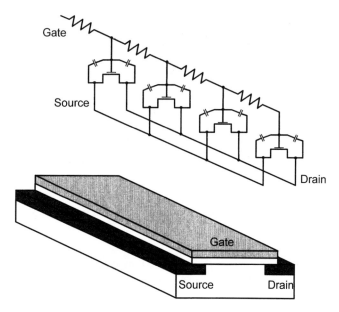

Figure 6.16. Distributed elements of the gate of a MOSFET.

Based on the transmission line concept, a rather lengthy derivation leads to the following expressions for the Y parameters of the intrinsic MOSFET [67]:

$$y_{11} = \frac{s(C_{gs} + C_{gd}) + s^2 C_{gs} C_{gd} R_i}{1 + sR_i C_{gs}} \frac{\tanh \Theta}{\Theta} \quad (6\text{-}47)$$

$$y_{12} = -sC_{gd} \frac{\tanh \Theta}{\Theta} \quad (6\text{-}48)$$

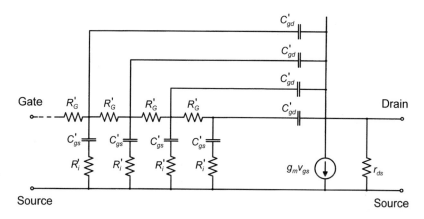

Figure 6.17. Distributed small-signal equivalent circuit of the intrinsic MOSFET.

$$y_{21} = \left(\frac{g_m}{1 + sR_i C_{gs}} - sC_{gd}\right)\frac{\tanh \Theta}{\Theta} \qquad (6\text{-}49)$$

$$y_{22} = \frac{1}{r_{ds}} + sC_{gd} + \frac{\left(\dfrac{g_m}{1 + sR_i C_{gs}} - sC_{gd}\right)C_{gd}}{C_{gd} + \dfrac{C_{gs}}{1 + sR_i C_{gs}}}\left(1 - \frac{\tanh \Theta}{\Theta}\right) \qquad (6\text{-}50)$$

where

$$\Theta = \sqrt{sR_g C_{gd}\frac{sR_g C_{gs}}{1 + sR_i C_{gs}}} \qquad (6\text{-}51)$$

and

$$s = j\omega = j2\pi f \qquad (6\text{-}52)$$

where j is the imaginary unit, ω is the angular frequency, and f is the frequency. It has been shown that such a model can successfully describe the small signal behavior of MOSFETs, and close agreement between the measured and modeled Y parameters has been obtained in the frequency range up to 15 GHz [67]. A drawback of the distributed MOSFET model is that the expressions for the Y parameters and the resulting expressions for the current and power gains are rather cumbersome. Moreover, the distributed model in the form of Eqs. (6-46)–(6-52) is applicable only for single-finger MOSFETs. In the case of multifinger transistors, the two-port networks of the individual fingers have to be connected in parallel, and the overall Y parameters have to be calculated. It has been shown, however, that the distributed effect can be included in a simplified MESFET/HEMT-like model, as shown in Fig. 6.18 for the intrinsic MOSFET, without a significant loss of accuracy [68]. The only difference of this and the intrinsic equivalent circuit in Fig. 4.12 is the addition of the gate capacitance C_g in Fig. 6.18 given by

$$C_g = \frac{C_{gs} + C_{gd}}{5} \qquad (6\text{-}53)$$

6.4.3 Compact Models

The physics-based models discussed so far give insights into the MOSFET operation and structure optimization. They, however, require iterative computations, which is not desirable for circuit simulations. Circuit designers prefer compact models, such as the BSIM, MOS Model 9, and EKV, in which simple and empirical equations are incorporated to calculate the dc currents and the elements of the equivalent circuit. The commonly used compact MOSFET models were not de-

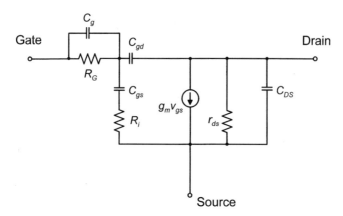

Figure 6.18. Simplified MESFET/HEMT-like small-signal equivalent circuit of the intrinsic MOSFET accounting for the distributed effect.

veloped specifically for microwave circuits but rather for digital and low-frequency analog circuits. Their application in the GHz range therefore may lead to incorrect results, and many recent studies have appeared in literature to address this issue and to develop improved compact MOS models for microwave applications [69–74].

6.5 NOISE AND POWER ANALYSIS

Modeling the high-frequency noise behavior of MOSFETs has been conducted since 1960s [75–77]. Because of the growing interests in microwave MOSFETs, research activities in MOSFET noise modeling have increased since the mid-1990s [36, 78–80]. The majority of the MOSFET noise models were developed based on the approach similar to that for the noise modeling of GaAs MESFET discussed in Section 4.4. Like the treatments for MESFETs and HEMTs, the PHS-like MOSFET dc model can be used as the basis for the derivation of a high-frequency noise model for MOSFETs. Specifically, such a model can be obtained by combining the equations of the PHS-like MOSFET dc model with the descriptions of the MESFET noise model given in [81] and [82]. Another option is to include noise considerations into the framework of the MOSFET distributed model, taking into account the distributed nature of the gate as reported in [65–67]. Although expressions for the drain and gate noise voltages and currents have been reported, an analytical formula for the minimum noise figure based on the distributed MOSFET model was not yet available in 2001.

In regard to the power performance, the MOSFET can be analyzed in the same way as the MESFET, and the treatment has been described in detail in Section 4.5.

6.6 ISSUES OF SMALL-SIGNAL LOW-NOISE MOSFETs

6.6.1 Transistor Structures

The driving force in MOSFET development has been and still is Si VLSI industry. Major semiconductor companies and research labs around the world expended a lot of effort to make MOSFETs smaller, i.e., to scale the transistors, to cram more transistors onto a single Si chip, and to keep pace with Moore's law, which governed the evolution of semiconductor electronics during the past several decades. MOSFET scaling not only led to an increased circuit complexity, but to faster transistors and circuits as well. An indication of increased circuit speed is the continuous rise in the clock frequency of microprocessors. The increasing speed of MOSFETs had stimulated the interest of the microwave community for considering these devices as a cost-effective alternative to the more expensive III–V microwave transistors. A 1 μm gate length Si nMOSFET with a cutoff frequency f_T of 17 GHz and a maximum frequency of oscillation f_{max} of 20 GHz was reported in 1984 [83]. The magic 100 GHz f_T mark had been passed in the early 1990s with 90 nm gate length nMOSFETs showing an f_T of 118 GHz [84], and MOSFETs with cutoff frequencies exceeding 200 GHz emerged in 2001 [12]. Though most of the progress on MOSFET high-speed performance still comes from the work on Si VLSI, the activities on microwave MOSFETs have evolved dynamically recently.

There are two major research directions on the small-signal low-noise microwave MOSFETs. The first is the development of Si-based integrated solutions for the mass markets of mobile communication systems and wireless networks, like mobile phones and Bluetooth. Here, the integration levels striven for are rather low and the operating frequencies range from 900 MHz to about 5 GHz. The second direction is the development of mixed-signal circuits and systems-on-chip. For these applications, the expected levels of integration are very high and the envisaged operating frequencies are much higher (see Table 1.11).

Both the conventional bulk and SOI MOSFETs are under investigation for small-signal low-noise applications. The basic structures of these transistors have been shown in Figs. 6.1 and 6.2. In short-channel MOSFETs, however, the source and drain regions are modified to suppress the short-channel effects. As shown in Fig. 6.19, the modified source and drain frequently consist of two regions [85–87]. Directly adjacent to the channel and underneath the gate, additional shallow regions are added to the heavily doped and deep source and drain regions. Furthermore, the channel underneath the gate can be inhomogeneously doped with retrograde or so-called halo profiles. The gate oxide is very thin and is only a few nm for MOSFETs with subquarter micron gates. For MOSFETs in VLSI circuits, a lower limit for the oxide thickness around 1.5 nm to avoid unacceptable tunneling currents from the gate through the thin oxide to the channel and reliability degradation is currently well accepted. There are, however, reports on MOSFET with oxides thinner than 1.5 nm as well [88].

As mentioned above, the poly gate of MOSFETs can be covered with a highly conductive silicide film to reduce the gate resistance. The same material can be used to reduce the parasitic source and drain resistances as well (see Fig. 6.19). Silicidation for the gate and for the source and drain contacts can be realized during the same fabrication step by incorporating the spacer next to the poly gate, as shown in Fig. 6.19.

6.6.2 Si MOSFET Performance

When considering the evolution of the characteristic frequencies f_T and f_{max} of MOSFETs, one issue is worth mentioning. For many years, the maximum frequency of oscillation f_{max} of short-gate MOSFETs has not been able to keep up with the improvement of f_T. In the year 2000, for nMOSFETs the record f_T was 245 GHz, compared to the record f_{max} of only 66 GHz for $L = 0.5$ μm [15] and 60 GHz for $L = 80$ nm [57]. The main reason for this is the high gate resistance R_G which degrades the power gain and limits f_{max} [see Eqs. (4-88)–(4-92)]. Reducing R_G is imperative in microwave MOSFETs, because R_G not only limits the gain attainable at a certain frequency, but also sets a lower limit for the minimum noise figure NF_{min}. Only in the year 2001 were nMOSFETs with f_{max} clearly exceeding 100 GHz presented. Table 6.4 lists the state of the art of high-speed MOSFETs in terms of f_T and f_{max}.

Figure 6.20 shows reported cutoff frequencies of Si nMOSFETs and pMOSFETs as a function of gate length. The plot indicates a steady increase of f_T with decreasing gate length, a trend also seen from microwave MESFETs and HEMTs. Also included in the figure is the empirical upper limit indicating the year 2001 state of the art. The maximum frequency of oscillation depends on L as well, as illustrated in Fig. 6.21. Prior to the year 2000, the reported record f_{max} for both nMOSFETs and pMOSFETs were obtained from devices with a rather long gate of 0.5 μm. Since 2001, the increased activities to optimize f_{max} have led to remarkable progress.

It should be noted that in the literature both the gate length L and the channel length L_{ch} are used to describe the critical transistor dimension. Thus, when comparing MOSFET performances, one has to examine carefully the length used to characterize the transistor.

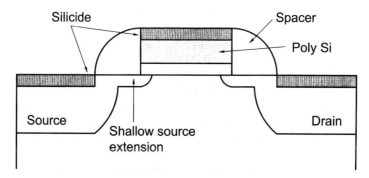

Figure 6.19. Cross section of a microwave MOSFET with shallow source/drain regions and silicided source, drain, and gate.

Table 6.4. State of the art of high-speed MOSFETs in terms of f_T and f_{max}

MOSFET type	Gate length, nm	f_T, GHz	f_{max}, GHz	Ref.
nMOSFET bulk	60	245		[12]
nMOSFET SOI	70	114	135	[16]
nMOSFET bulk	180	70	150	[17]
nMOSFET SOI	80	120	185	[14]
nMOSFET SOI	50	178	193	[18]
nMOSFET SOI	70	141	98	[89]
nMOSFET SOI	80	140	60	[57]
nMOSFET SOI	500	25	66	[15]
pMOSFET bulk	65 (L_{ch})	67		[84]
pMOSFET bulk	150	65		[12]

The best reported minimum noise figures NF_{min} of experimental Si microwave MOSFETs are given in Fig. 6.22. Up to 10 GHz, MOSFET noise figures below 1 dB have been obtained. In regard to the dc performance of MOSFETs, maximum on-currents and transconductances above 2 A/mm and 1.9 S/mm, respectively, were demonstrated [88]. It should be noted, however, that these transistors were not designed for microwave applications.

Microwave MOSFETs are well suited for applications at lower GHz frequencies. Although III–V FETs possess lower noise figures and higher power gains, the performance of well-designed MOSFETs is adequate for many applications in wireless communications systems up to 2.5 GHz. For such a mass market, the low cost and high level of maturity of MOS technology are clearly advantageous.

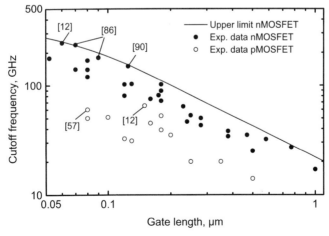

Figure 6.20. Reported cutoff frequencies of microwave MOSFETs as a function of the gate length.

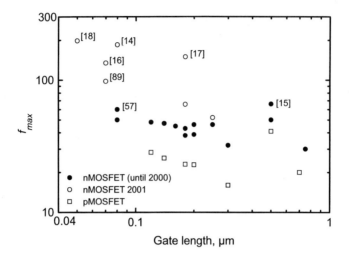

Figure 6.21. Reported maximum frequencies of oscillation f_{max} of microwave MOSFETs as a function of the gate length.

Recent advances in the high-speed performance have paved the way for the successful applications of MOSFETs at even higher frequencies, and their use in communication systems operating beyond 5 GHz seems possible in the not too distant future.

6.7 ISSUES OF POWER MOSFETs

In principle, power MOSFETs can be realized with the standard MOSFET structure shown in Fig. 6.1. Experimental power amplifiers based on this concept have been reported (see, e.g., [8, 91, 92]). A main motivation of this work was to realize an entire radio system, including power amplifier for cellular phones, using the cost-effective CMOS technology, perhaps even on a single chip. To achieve the required performance at 900 MHz, 1.8 GHz, or 2.4 GHz, deep submicron CMOS is necessary. But several shortcomings are currently in place, especially that the breakdown voltage of short-gate MOSFETs is low.

As shown in Table 6.5, there is a clear tendency toward decreasing supply volt-

Table 6.5. Examples of CMOS microwave power amplifiers

Technology	Supply voltage	Operating frequency	Ref.
0.25 μm	1.8 V	900 MHz	[8]
0.35 μm	2.0 V	1.9 GHz	[91]
0.8 μm/1 μm	2.5 V	850 MHz	[92]

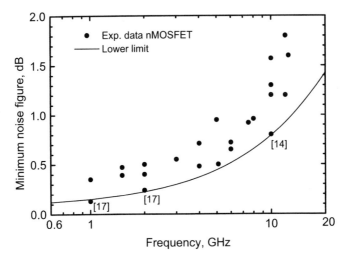

Figure 6.22. Reported minimum noise figures of microwave MOSFETs as a function of the frequency. Also included is the lower limit fit indicating the year 2001 state of the art of the noise performance of microwave nMOSFETs.

age with decreasing gate length in CMOS power amplifiers. Because mobile electronics are designed for low power consumption and low voltage operation, the problem of low breakdown voltage should not prevent the use of MOSFETs in low-voltage power amplifiers for handheld systems.

High-power microwave amplifiers as used in the base stations for mobile communication systems, on the other hand, are designed for maximum operating (and thus breakdown) voltage to deliver maximum output power. For such applications, conventional MOSFETs with a device structure shown in Figs. 6.1 and 6.2 are not suitable, and a different MOSFET structure called the LDMOSFET (Laterally Diffused MOSFET) is used [93–98]. The cross section of a typical high-power LDMOSFET is shown in Fig. 6.23. It consists of a p^+-Si substrate on top of which a lightly doped p^--layer is grown epitaxially. Like a conventional MOSFET, the drain and source regions are heavily doped n-type regions. Although the n^+-source region extends to the gate, the n^+-drain region is spatially separated from the gate. The conductive connection between the n^+-drain and the channel region underneath the gate is realized by an n^--LDD (Lightly Doped Drain) region, which is frequently called the drift region. This is the region mainly contributing to a high breakdown voltage. The p-type body region overlaps a portion of the channel near the source and the n^+-source region. It is realized by ion implantation followed by a diffusion step. The designation laterally diffused MOSFET stems from the fact that the doping of the p-type body region is partially done by the lateral diffusion of the implanted acceptors after the source implantation during a diffusion step. The portion of the p-type body region directly underneath the gate serves as the channel. Thus the channel length is considerably smaller than the gate length. Finally, there is the highly doped p-type sinker extending through the entire epitaxial p-layer down to the p$^+$-substrate. As can be seen in Fig. 6.23, the source contact is extended beyond the n$^+$-source region to the sinker. By this, the source terminal is connected to the grounded substrate, and no bond wires are needed to for the source region. Thus, the lead inductance is drastically minimized and the frequency performance of the LDMOSFET is improved. Typically, the gate length of LDMOSFETs is in the range of 0.3–1 μm and the gate oxide is several tens of nanometers thick. As such, the technology, especially the lithography and the oxidation for the gate oxide, of the LDMOSFET is much more relaxed than that of the small-signal microwave MOSFET.

Depending on the specific designs, typical f_T and f_{max} of LDMOSFETs are around 5 to 15 GHz, and the drain–source breakdown voltage BV_{DS} is around 20 to 40 V. Maximum BV_{DS} in excess of 70 V has been reported [97]. As a rule of thumb, the breakdown voltage of a power FET should be 2–3 times the operation voltage. This means that the current LDMOSFETs are capable of operations ranging from 10 to 30 V. Figure 6.24 shows reported cutoff frequencies f_T of Si LDMOSFET as a function of the source-drain breakdown voltage BV_{DS}. The reported f_{max} is in the same range as f_T. Although LDMOSFETs possess much lower f_T and f_{max} than small-signal microwave MOSFETs, the frequency performance of LDMOSFETs is absolutely sufficient for high-power amplification in the frequency bands of up to 2.5 GHz used for mobile communication systems. High-power LDMOSFETs with output powers exceeding 100 W have emerged, e.g., 120 W output power at 2 GHz with a power gain of 10.6 dB have been reported in the late 1990s [98]. Transistors with a maximum output power up to 220 W at 2 GHz became commercially available in 2001 [20]. Currently, Si LDMOSFETs have superseded all other transistor

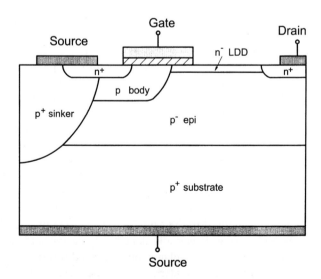

Figure 6.23. Cross section of a high-power Si LDMOSFET.

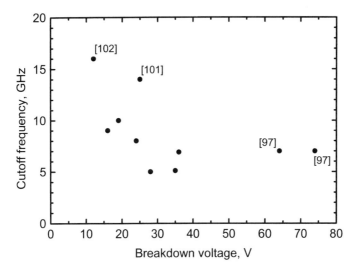

Figure 6.24. Reported cutoff frequencies f_T of LDMOSFET as a function of the drain–source breakdown voltage BV_{DS}.

types and clearly dominate the market for base station applications [19, 20].

Besides the base stations, there is a second envisaged application for LDMOSFETs. As mentioned above, mobile radios require both low-noise and low-power amplifiers, and the LDMOSFET is an alternative to the conventional power MOSFET for mobile transceivers in 900 MHz to 2.5 GHz communication systems [96, 99–101]. Because the operating voltage is relatively low in these systems, the requirements on the breakdown voltage of the LDMOSFET are less stringent. In this application, the LDMOSFET does not use the p^+-type sinker mentioned above, and the source contact is located on top of the wafer. The compatibility of the LDMOS technology with standard CMOS technology is inevitable, however. Both bulk and SOI LDMOSFETs for transceiver applications have been reported.

REFERENCES

1. N. Camilleri, J. Costa, D. Lovelace, and D. Ngo, Silicon MOSFETs, the Microwave Technology for the 90s, *IEEE MTT-S Dig.,* pp. 545–548, 1993.
2. T. H. Lee, CMOS RF: No Longer an Oxymoron, *Proc. GaAs IC Symp.,* pp. 244–247, 1997.
3. T.E. Kolding, J. H. Mikkelsen, and T. Larsen, CMOS Technology Adjusts to RF Applications, *Microwaves & RF, 37,* pp. 79–88, June 1998.
4. T. Manku, Microwave CMOS—Device Physics and Design, *IEEE J. Solid-State Circuits, 34,* pp. 277–285, 1999.
5. L. E. Frenzel, RF, Wireless, and Optical Technologies Become the Hot Topics, *Electronic Design, 49,* pp. 95–100, Febr. 21, 2001.

6. P. H. Woerlee, M. J. Knitel, R. v. Langevelde, D. B. M. Klaassen, L. F. Tiemeijer, A. J. Scholten, and A. T. A. Zegers-van Duijnhoven, RF-CMOS Performance Trends, *IEEE Trans. Electron Devices, 48,* pp. 1776–1782, 2001.
7. H. Darabi and A. A. Abidi, A 4.5-mW 900-MHz CMOS Receiver for Wireless Paging, *IEEE J. Solid-State Circuits, 35,* pp. 1085–1096, 2000.
8. C. Yoo and Q. Huang, A Common-Gate Switched 0.9-W Class-E Power Amplifier with 41% PAE in 0.25-μm CMOS, *IEEE J. Solid-State Circuits, 36,* pp. 823–830, 2001.
9. K. Yamamoto, T. Heima, A. Furukawa, M. Ono, Y. Hashizume, H. Komurasaki, S. Maeda, H. Sato, and N. Kato, A 2.4-GHz-Band 1.8-V Operation Single-Chip Si-CMOS T/R-MMIC Front-End with a Low Insertion Loss Switch, *IEEE J. Solid-State Circuits, 36,* pp. 1186–1197, 2001.
10. H. Samavati, H. R. Rategh, and T. H. Lee, A 5-GHz CMOS Wireless LAN Receiver Front End, *IEEE J. Solid-State Circuits, 35,* pp. 765–772, 2000.
11. J. Browne, Chip Contains Full Bluetooth Solution, *Microwaves & RF, 40,* p. 139, April 2001.
12. H. S. Momose, E. Morifujii, T. Yoshitomi, T. Ohguro, M. Saito, and H. Iwai, Cutoff Frequency and Propagation Delay Time of 1.5-nm Gate Oxide CMOS, *IEEE Trans. Electron Devices, 48,* pp. 1165–1174, 2001.
13. A. Chatterjee, R. A. Chapman, G. Dixit, J. Kuehne, S. Hattangady, H. Yang, G. A. Brown, R. Aggarwal, U. Erdogan, Q. He, M. Hanratty, D. Rogers, S. Murtaza, S. J. Fang, R. Kraft, A. L. P. Rotondaro, J. C. Hu, M. Terry, W. Lee, C. Fernando, A. Konecni, G. Wells, D. Frystak, C. Bowen, M. Rodder, and I.-C. Chen, Sub–100nm Gate Length Metal Gate NMOS Transistors Fabricated by a Replacement Gate Process, *IEDM Tech. Dig.,* pp. 821–824, 1997.
14. T. Hirose, Y. Momiyama, M. Kosugi, H. Kano, Y. Watanabe, and T. Sugii, A 185 GHz f_{max} SOI DTMOS with a New Metallic Overlay-Gate for Low-Power RF Applications, *IEDM Tech. Dig.,* pp. 943–945, 2001.
15. R. A. Johnson, P. R. dela Houssaye, C. E. Chang, P.-F. Chen, M. E. Wood, G. A. Garcia, I. Lagnado, and P. M. Asbeck, Advanced Thin-Film Silicon-on-Sapphire Technology: Microwave Circuit Applications, *IEEE Trans. Electron Devices, 45,* pp. 1047–1054, 1998.
16. T. Matsumoto, S. Maeda, K. Ota, Y. Hirano, K. Eikyu, H. Sayama, T. Iwamatsu, K. Yamamoto, T. Katoh, Y. Yamaguchi, T. Ipposhi, H. Oda, S. Maegawa, Y. Inoue, and M. Inuishi, 70 nm SOI-CMOS of 135 GHz f_{max} with Dual Offset-Implanted Source-Drain Extension Structure for RF/Analog and Logic Applications, *IEDM Tech. Dig.,* pp. 219–222, 2001.
17. L. F. Tiemeijer, H. M. J. Boots, R. J. Havens, A. J. Scholten, P. H. W. de Vreede, P. H. Woerlee, A. Heringa, and D. B. M. Klassen, A Record High 150 GHz f_{max} Realized at 0.18 μm Gate Length in an Industrial RF-CMOS Technology, *IEDM Tech. Dig.,* pp. 223–226, 2001.
18. S. Narashima, A. Ajmera, H. Park, D. Schepis, N. Zamdmer, K. A. Jenkins, J.-O. Plouchart, W. H. Lee, J. Mezzapelle, J. Bruley, B. Doris, J. W. Sleight, S. K. Fung, S. H. Ku, A. C. Mocuta, I. Yang, P. V. Gilbert, K. P. Muller, P. Agnello, and J. Welser, High Performance Sub-40nm CMOS Devices on SOI for the 70nm Technology Node, *IEDM Tech. Dig.,* pp. 625–628, 2001.
19. G. Heftman, Wireless Semi Technology Heads into New Territory, *Microwaves & RF, 39,* pp. 31–40, Feb. 2000.

20. J. Browne, More Power Per Transistor Translates into Smaller Amplifiers, *Microwaves & RF, 40,* pp. 132–136, Jan. 2001.
21. A. S. Grove, *Physics and Technology of Semiconductor Devices,* Wiley, New York, 1967.
22. S. M. Sze, *Physics of Semiconductor Devices,* 2nd edition, Wiley, New York, 1981.
23. R. F. Pierret, *Field Effect Devices,* Addison-Wesley, Reading, MA, 1990.
24. Y. Taur and T. H. Ning, *Fundamentals of Modern VLSI Devices,* Cambridge University Press, Cambridge, 1998.
25. Y. Tsividis, *Operation and Modeling of the MOS Transistor,* 2nd edition, WCB McGraw-Hill, Boston, 1999.
26. J.-P. Colinge, *Silicon-on-Insulator Technology: Materials to VLSI,* Kluwer, Boston, 1997.
27. J. P. Colinge, Silicon-on-Insulator Technology: Past Achievements and Future Prospects, *MRS Bull., 23,* pp. 16–19, 1998.
28. Special Issue on SOI Integrated Circuits and Devices, *IEEE Trans. Electron Devices, 45,* pp. 997–1161, 1998.
29. J. B. Kuo and K.-W. Su, *CMOS VLSI Engineering: Silicon-on-Insulator (SOI),* Kluwer, Boston, 1998.
30. S. R. Hofstein and G. Warfield, Carrier Mobility and Current Saturation in the MOS Transistor, *IEEE Trans. Electron Devices, 12,* pp. 129–138, 1965.
31. Y. A. El-Mansy and A. R. Boothroyd, A Simple Two-Dimensional Model for IGFET Operation in the Saturation Region, *IEEE Trans. Electron Devices, 24,* pp. 254–262, 1977.
32. K.-Y. Toh, P.-K. Ko, and R. G. Meyer, An Engineering Model for Short-Channel MOS Devices, *IEEE J. Solid-State Circuits, 23,* pp. 950–958, 1988.
33. K. Sonoda, K. Taniguchi, and C. Hamaguchi, Analytical Device Model for Submicrometer MOSFET's, *IEEE Trans. Electron Devices, 38,* pp. 2662–2668, 1991.
34. H.-C. Chow and W.-S. Feng, An Improved Analytical Model for Short-Channel MOSFET's, *IEEE Trans. Electron Devices, 39,* pp. 2626–2629, 1992.
35. K. Takeuchi and M. Fukuma, Effects of the Velocity Saturated Region on MOSFET Characteristics, *IEEE Trans. Electron Devices, 41,* pp. 1623–1627, 1994.
36. D. P. Triantis, A. N. Birbas, and D. Kondis, Thermal Noise Modeling for Short-Channel MOSFETs, *IEEE Trans. Electron Devices, 43,* pp. 1950–1955, 1996.
37. A. Grebene and S. K. Ghandi, General Theory for Pinched Operation of the Junction-Gate FET, *Solid-State Electron., 12,* pp. 573–589, 1969.
38. G. Baccarani and G. A. Sai-Halasz, Spreading Resistance in Submicron MOSFETs, *IEEE Electron Device Lett., 4,* pp. 27–29, 1983.
39. K. K. Ng and W. T. Lynch, Analysis of the Gate-Voltage Dependent Series Resistance of MOSFETs, *IEEE Trans. Electron Devices, 33,* pp. 965–972, 1986.
40. J. M. Pimbley, Two-Dimensional Current Flow in the MOSFET Source-Drain, *IEEE Trans. Electron Devices, 33,* pp. 986–996, 1986.
41. K. K. Ng and W. T. Lynch, The Impact of Intrinsic Series Resistance on MOSFET Scaling, *IEEE Trans. Electron Devices, 34,* pp. 503–511, 1987.
42. K. Y. Lim and X. Zhou, A Physically-Based Semi-Empirical Series Resistance Model

for Deep-Submicron I-V Modeling, *IEEE Trans. Electron Devices, 47,* pp. 1300–1302, 2000.
43. K. Yamaguchi, Field-Dependent Mobility Model for Two-Dimensional Numerical Analysis of MOSFET's, *IEEE Trans. Electron Devices, 26,* pp. 1074, 1979.
44. G. Baccarani and M. R. Wordeman, Transconductance Degradation in Thin-Oxide MOSFETs, *IEEE Trans. Electron Devices, 30,* pp. 1295–1304, 1983.
45. N. D. Arora and G. S. Gildenblat, A Semi-Empirical Model of the MOSFET Inversion Layer Mobility for Low-Temperature Operation, *IEEE Trans. Electron Devices, 34,* pp. 89–93, 1987.
46. C. Lombardi, S. Manzini, A. Saporito, and M. Vanzi, A Physically Based Mobility Model for Numerical Simulation of Nonplanar Devices, *IEEE Trans. Computer-Aided Design, 7,* pp. 1164–1171, 1988.
47. D. Vasileska and D. K. Ferry, Scaled Silicon MOSFET's: Universal Mobility Behavior, *IEEE Trans. Electron Devices, 44,* pp. 577–583, 1997.
48. M. Darwish, J. L. Lentz, M. R. Pinto, P. M. Zeitzoff, T. J. Krutsick, and H. H. Vuong, An Improved Electron and Hole Mobility Model for General Purpose Device Simulation, *IEEE Trans. Electron Devices, 44,* pp. 1529–1538, 1997.
49. S. Villa, A. L. Lacaita, L. M. Perron, and R. Bez, A Physically-Based Model of the Effective Mobility in Heavily-Doped n-MOSFET's, *IEEE Trans. Electron Devices, 45,* pp. 110–115, 1998.
50. S. Takagi, M. Iwase, and A. Toriumi, On Universality of Inversion-Layer Mobility in n- and p-Channel MOSFETs, *IEDM Tech. Dig.,* pp. 398–401, 1988.
51. S. Takagi, A. Toriumi, M. Iwase, and T. Tango, On the Universality of Inversion Layer Mobility in Si MOSFET's: Part I – Effects of Substrate Impurity Concentration, *IEEE Transactions Electron Devices, 41,* pp. 2357–2362, 1994.
52. L. D. Lau, A Simple Theory to Predict the Threshold Voltage of Short-Channel IGFET's, *Solid-State Electron., 17,* pp. 1059–1063, 1974.
53. S. Jallepalli, J. Bude, W. K. Shih, M. R. Pinto, C. M. Maziar, and A. F. Tasch, Effects of Quantization on the Electrical Characteristics of Deep Submicron p- and n-MOSFETs, Symp. *VLSI Technol. Dig.,* pp. 138–139, 1996.
54. R. Shrivastava and K. Fitzpatrick, A Simple Model for the Overlap Capacitance of a VLSI MOS Device, *IEEE Trans. Electron Devices, 29,* pp. 1870–1875, 1982.
55. T. Y. Chan, A. T. Wu, P. K. Ko, and C. Hu, Effects of the Gate-to-Drain/Source Overlap on MOSFET Characteristics, *IEEE Electron Device Lett., 8,* pp. 326–328, 1987.
56. S. J. Hillenius, VLSI Process Integration, in S. M. Sze (ed.), *VLSI Technology,* McGraw-Hill, New York, 1988.
57. Y. Momiyama, T. Hirose, H. Kurata, K. Goto, Y. Watanabe, and T. Sugii, A 140 GHz f_t and 60 GHz f_{max} DTMOS Integrated with High-Performance SOI Logic Technology, *IEDM Tech. Dig.,* pp. 451–454, 2000.
58. H.-C. Lin, R. Lin, W.-F. Wu, R.-P. Yang, M.-S. Tsai, T.-S. Chao, and T.-Y. Huang, A Novel Self-Aligned T-Shaped Gate Process for Deep Submicron Si MOSFET's Fabrication, *IEEE Electron Device Lett., 19,* pp. 26–28, 1998.
59. A. Caviglia, R. C. Potter, and L. J. West, Microwave Performance of SOI n-MOSFET's and Coplanar Waveguides, *IEEE Electron Device Lett., 12,* pp. 26–27, 1991.
60. A. E. Schmitz, R. H. Walden, L. E. Larson, S. E. Rosenbaum, R. A. Metzger, J. R.

Behnke, and A. Macdonald, A Deep-Submicrometer Microwave/Digital CMOS/SOS Technology, *IEEE Electron Device Lett., 12,* pp. 16–17, 1991.

61. M. H. Hanes, A. K. Agarwal, T. W. O'Keeffe, H. M. Hobgood, J. R. Szedon, T. J. Smith, R. R. Siergiej, P. G. McMullin, H. C. Nathanson, M. C. Driver, and R. N. Thomas, MICROX™—An All-Silicon Technology for Monolithic Microwave Integrated Circuits, *IEEE Electron Device Lett., 14,* pp. 219–221, 1993.

62. S. P. Voingescu, S. W. Tarasewicz, T. MacElwee, and J. Ilowski, An Assessment of the State-of-the Art 0.5 μm Bulk CMOS Technology for RF Applications, *IEDM Tech. Dig.,* pp. 721–724, 1995.

63. T. Tanaka, Y. Momiyama, and T. Sugii, F_{max} Enhancement of Dynamic Threshold-Voltage MOSFET (DTMOS) Under Ultra-Low Supply Voltage, *IEDM Tech. Dig.,* pp. 423–426, 1997.

64. L.-S. Kim and R. W. Dutton, Modeling of the Distributed Gate RC Effect in MOSFET's, *IEEE Trans. Computer Aided Design, 8,* pp. 1365–1367, 1989.

65. B. Razavi, R.-H. Yan, and K. F. Lee, Impact of Distributed Gate Resistance on the Performance of MOS Devices, *IEEE Trans. Circuits Syst.—I: Fundamental Theory and Applications, 41,* pp. 750–754, 1994.

66. E. Abou-Allam and T. Manku, A Small-Signal MOSFET Model for Radio Frequency Applications, *IEEE Trans. Computer Aided Design of Integrated Circuits and Systems, 16,* pp. 437–447, 1997.

67. E. Abou-Allam and T. Manku, An Improved Transmission-Line Model for MOS Transistors, *IEEE Trans. Circuits Syst.—II: Analog and Digital Signal Processing, 46,* pp. 1380–1387, 1999.

68. S. F. Tin, A. A. Osman, and K. Mayaram, Comments on "A Small-Signal MOSFET Model for Radio Frequency IC Applications," *IEEE Trans. Computer Aided Design of Integrated Circuits and Systems, 17,* pp. 372–374, 1998.

69. W. Liu, R. Gharpurey, M. C. Chang, U. Erdogan, R. Aggarwal, and J. P. Mattia, RF MOSFET Modeling Accounting for Distributed Substrate and Channel Resistances with Emphasis on the BSIM3v3 SPICE Model, *IEDM Tech. Dig.,* pp. 309–312, 1997.

70. A. J. Scholten, L. F. Tiemeijer, P. W. H. de Vreede, and D. B. M. Klaassen, A Large Signal Non-Quasi-Static MOS Model for RF Circuit Simulation, *IEDM Tech. Dig.,* pp. 163–166, 1999.

71. S. H.-M. Jen, C. C. Enz, D. R. Pehlke, M. Schröter, and B. J. Sheu, Accurate Modeling and Parameter Extraction for MOS Transistors Valid up to 10 GHz, *IEEE Trans. Electron Devices, 46,* pp. 2217–2227, 1999.

72. C. C. Enz and Y. Cheng, MOS Transistor Modeling for RF IC Design, *IEEE Trans. Solid-State Circuits, 35,* pp. 186–201, 2000.

73. R. van Langefelde, L. F. Tiemeijer, R. J. Havens, M. J. Knitel, R. F. M. Roes, P. H. Woerlee, and D. B. M. Klaassen, RF Distortion in Deep-Submicron CMOS Technologies, *IEDM Tech. Dig.,* pp. 807–810, 2000.

74. J.-S. Goo, W. Liu, H.-C. Choi, K. R. Green, Z. Yu, T. H. Lee, and R. W. Dutton, The Equivalence of van der Ziel and BSIM 4 Models in Modeling the Induced Gate Noise of MOSFETs, *IEDM Tech. Dig.,* pp. 811–814, 2000.

75. A. G. Jordan and N. A. Jordan, Theory of Noise in Metal Oxide Semiconductor Devices, *IEEE Trans. Electron Devices, 12,* pp. 148–156, 1965.

76. E. Halladay and A. van der Ziel, On the High Frequency Excess Noise and Equivalent Circuit Representation of the MOS-FET with n-Type Channel, *Solid-State Electronics, 12*, pp. 161–176, 1969.
77. A. van der Ziel, *Noise Sources, Characterization, and Measurement,* Prentice-Hall, Englewood Cliffs, NJ, 1970.
78. D. P. Triantis, A. N. Birbas, and S. E. Plevridis, Induced Gate Noise in MOSFETs Revisited: The Submicron Case, *Solid-State Electronics, 41,* pp. 1937–1942, 1997.
79. C. H. Chen and M. J. Deen, High Frequency Noise of MOSFETs. I. Modeling, *Solid-State Electronics, 42,* pp. 2069–2081, 1998.
80. C. H. Park and Y. J. Park, Modeling of Thermal Noise in Short-Channel MOSFETs at Saturation, *Solid-State Electronics, 44,* pp. 2053–2057, 2000.
81. H. Statz, H. A. Haus, and R. A. Pucel, Noise Characteristics of Gallium Arsenide Field-Effect Transistors, *IEEE Trans. Electron Devices, 21,* pp. 549–562, 1974.
82. R. A. Pucel, H. A. Haus, and H. Statz, Signal and Noise Properties of Gallium Arsenide Microwave Field-Effect Transistors, *Advan. Electron. Electron Phys., 38,* pp. 195–265, 1975.
83. D. C. Shaver, Microwave Operation of Submicrometer Channel-Length Silicon MOSFET's, *IEEE Electron Device Lett., 6,* pp. 36–39, 1984.
84. Y. Taur, S. Wind, Y. J. Mii, Y. Lii, D. Moy, K. A. Jenkins, C. L. Chen, P. J. Coane, D. Klaus, J. Bucchignano, M. Rosenfield, M. G. R. Thomson, and M. Polcari, High Performance 0.1 μm CMOS Devices with 1.5 V Power Supply, *IEDM Tech. Dig.,* pp. 127–130, 1993.
85. Y. Mii, S. Rishton, Y. Taur, D. Kern, T. Lii, K. Lee, K. A. Jenkins, D. Quinlan, T. Brown Jr., D. Danner, F. Sewell, and M. Polcari, Experimental High Performance Sub-0.1μm Channel MOSFET's, *IEEE Electron Device Lett., 15,* pp. 28–30, 1994.
86. H. S. Momose, E. Morifuji, T. Yoshitomi, T. Ohguro, M. Saito, T. Morimoto, Y. Katsumata, and H. Iwai, High-Frequency AC Characteristics of 1.5 nm Gate Oxide MOSFETs, *IEDM Tech. Dig.,* pp. 105–108, 1996.
87. T. Ohguro, H. Naruse, H. Sugaya, H. Kimijima, E. Morifuji, T. Yoshitomi, T. Morimoto, H. S. Momose, Y. Katsumata, and H. Iwai, 0.12 μm Raised Gate/Source/Drain Epitaxial Channel NMOS Technology, *IEDM Tech. Dig.,* pp. 927–930, 1998.
88. G. Timp, J. Bude, K. K. Bourdelle, J. Garno, A. Ghetti, H. Gossmann, M. Green, G. Forsyth, Y. Kim, R. Kleiman, F. Klemens, A. Kornblit, C. Lochstampfor, W. Mansfield, S. Moccio, T. Sorsch, D. M. Tennant, W. Timp, and R. Tung, The Ballistic Nano-Transistor, *IEDM Tech. Dig.,* pp. 55–58, 1999.
89. N. Zamdmer, A. Ray, J.-O. Plouchart, L. Wagner, N. Fong, K. A. Jenkins, W. Jin, P. Smeys, I. Yang, G. Shahidi, and F. Assaderaghi, A 0.13-μm SOI CMOS Technology for Low-Power Digital and RF Applications, *Tech. Dig. Symp. VLSI Technology,* pp. 85–86, 2001.
90. C. Wann, F. Assaderaghi, L. Shi, K. Chan, S. Cohen, H. Hovel, K. Jenkins, Y. Lee, D. Sadana, R. Viswanathan, S. Wind, and Y. Taur, High-Performance 0.07-μm CMOS with 9.5-ps Gate Delay and 150 GHz f_T, *IEEE Electron Device Lett., 18,* pp. 625–627, 1997.
91. K.-C. Tsai and P. R. Gray, A 1.9-GHz 1-W CMOS Class-E Power Amplifier for Wireless Communications, *IEEE J. Solid-State Circuits, 34,* pp. 962–970, 1999.
92. D. Su and W. McFarland, A 2.5-V 1-W Monolithic CMOS RF Power Amplifier, *Proc. Custom Integrated Circuits Conf.,* pp. 189–192, 1997.

93. I. Yoshida, 2-GHz Si Power MOSFET Technology, *IEDM Tech. Dig.*, pp. 51–54, 1997.
94. G. Ma, W. Burger, and X. Ren, High Efficiency Submicron Gate LDMOS Power FET for Low Voltage Wireless Communications, *IEEE MTT-S Dig.*, pp. 1303–1306, 1997.
95. I. Yoshida, M. Katsueda, Y. Maruyama, and I. Kohjiro, A Highly Efficient 1.9-GHz Si High-Power MOS Amplifier, *IEEE Trans. Electron Devices, 45,* pp. 953–956, 1998.
96. Y. Tan, M. Kumar, J. K. O. Sin, J. Cai, and J. Lau, A LDMOS Technology Compatible with CMOS and Passive Components for Integrated RF Power Amplifiers, *IEEE Electron Device Lett., 21,* pp. 82–84, 2000.
97. J. Cai, C. Ren, N. Balasubramanian, and J. K. O. Sin, A Novel High Performance Stacked LDD RF LDMOSFET, *IEEE Electron Device Letters, 22,* pp. 236–236, 2001.
98. A. Wood, W. Brakensiek, C. Dragon, and W. Burger, 120 Watt, 2GHz, Si LDMOS RF Power Transistor for PCS Base Station Applications, *IEEE MTT-S Dig.,* pp. 707–710, 1998.
99. K. Shenai, E. McShane, and S. K. Leong, Lateral RF SOI Power MOSFETs with f_T of 6.9 GHz, *IEEE Electron Device Lett., 21,* pp. 500–502, 2000.
100. An SOI LDMOS/CMOS/BJT Technology for Integrated Power Amplifiers Used in Wireless Transceiver Applications, *IEEE Electron Device Lett., 22,* pp. 136–138, 2001.
101. J. G. Fiorenza, D. A. Antoniadis, and J. A. del Alamo, RF Power LDMOSFET on SOI, *IEEE Electron Device Lett., 22,* pp. 139–141, 2001.
102. Y. Hoshino, M. Morikawa, S. Kamohara, M. Kawakami, T. Fujioka, Y. Matsunaga, Y. Kusakari, S. Ikeda, I. Yoshida, and S. Shimizu, High Performance Scaled Down Si LDMOSFET with Thin Gate Bird's Beak Technology for RF Power Amplifiers, *IEDM Tech. Dig.,* pp. 205–208, 1999.

CHAPTER 7

SILICON BIPOLAR JUNCTION TRANSISTORS

7.1 INTRODUCTION

The **B**ipolar **J**unction **T**ransistor (BJT) was the first three-terminal semiconductor device capable of delivering signal amplification. From this point of view, the importance of such a device can hardly be overemphasized. After World War II, researchers at Bell Labs were trying to make field-effect devices. Although this attempt was unsuccessful, it later led to the birth of the BJT in late 1947 [1]. The first laboratory BJTs, and the first commercial BJTs available later, were made of germanium (Ge). This changed in the 1960s, when most work was focused on silicon BJTs.

Since the early days of BJT development, device engineers have expended a lot of effort to improve transistor speed and operating frequency. The first transistors capable of amplifying signals at frequencies around 1 GHz were Ge BJTs developed in the late 1950s. Soon after that, the investigation of Si and GaAs BJTs started. The work on GaAs BJTs had only a limited success, and by 1968 the interest in these transistors had faded and the research activities stopped [2]. Research and development of Si BJTs for microwave applications, on the other hand, led to much more fruitful results. By 1963, Si BJTs became competitive with Ge BJTs, and several years later the Si BJT was the dominant device for microwave applications [3]. Up to now, the Si BJT has been the only "pure" BJT type playing a significant role in microwave electronics. Therefore, the discussions in the remainder of this chapter shall be restricted to Si BJTs, and the term BJT is used exclusively for the Si BJT.

In 1970, state of the art BJTs showed maximum frequencies of oscillation f_{max} in excess of 15 GHz and minimum noise figures NF_{min} of 1.3, 2.6, and 4 dB at frequencies of 1, 2, and 4 GHz, respectively [3]. Power BJTs delivered output powers of 100, 20, and 5 W at 1.2, 2, and 4 GHz, respectively [3]. More recently, other microwave transistors such as MESFETs, HEMTs, and HBTs came into use, but the BJT remained an important microwave device for the frequency range up to 4 GHz and for other high-speed applications, such as optical fiber transmission systems with high data rates and computer networks.

The structure of a typical high-speed BJT of the mid-1970s is shown in Fig. 7.1. It consists, from the bottom to the top, of a p-type substrate, a highly doped n^+-type

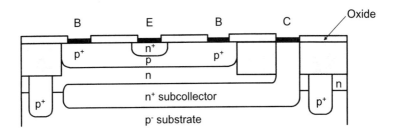

Figure 7.1. Cross section of a conventional double-diffused BJT used during the mid 1970s. After [4].

subcollector, a lightly doped n-type collector, a p-type base, and a highly doped n^+-type emitter. The emitter and base regions were formed by a double-diffusion process. The minimum lateral dimension, which is the emitter width W_E, was around 2.5 to 3 µm, and the thickness of the base was about 250 nm. Although the technology was mature and widely used, several shortcomings associated with such a BJT design and fabrication process became apparent in the late 1970s. Important innovations leading to a breakthrough in BJT technology and to the continuing improvement of the high-speed performance of BJTs were developed during the 1980s. They include [4]

- Introduction of ion implantation to form the base region (and in some transistors to form the emitter as well)
- Development of double polysilicon BJTs using polycrystalline Si for the emitter and base contacts
- Introduction of self-aligned transistor structures
- Use of selectively implanted collectors

Figure 7.2 shows the cross section of a modern microwave BJT fabricated employing these features with a base layer thickness in the range of 30 to 80 nm. More information on the evolution of BJTs can be found in the excellent reviews by Ning [4], Nakamura and Nishizawa [5], and Warnock [6].

Like field-effect transistors, BJT high-frequency performance has been steadily improved in the past several decades. Although the gate delay τ_d is not a typical figure of merit of microwave transistors, its trend serves as an excellent indicator for the evolution of high-speed BJT development. Figure 7.3 shows a compilation of the emitter-coupled logic (ECL) gate delay times reported over the past twenty years. The rapid decrease in gate delay during the 1980s can be attributed to the introduction of the double polysilicon transistor technology mentioned above and to the successful scaling of BJTs. During the 1990s, however, the improvement slowed down for two reasons. First, both the vertical and lateral BJT scaling approached practical limits, and second, since the early 1990s, research and develop-

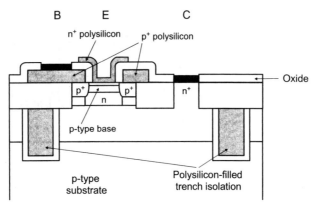

Figure 7.2. Cross section of a modern double-polysilicon, self-aligned BJT with implanted base. After [4].

ment of Si-based high-speed bipolar devices has been shifted more and more from Si BJTs to SiGe HBTs [7]. This is true not only for digital circuits but for microwave transistors as well.

In the year 2000, microwave BJTs fabricated in the research lab possessed a record cutoff frequency f_T of 100 GHz and record maximum frequency of oscillation f_{max} of 101 GHz [8]. At 3, 6, and 10 GHz, minimum noise figures as low as 0.7, 1.3, and 1.7 dB, respectively, together with associated power gains of 17, 14, and 11 dB, respectively, have been obtained [9]. Commercial low-noise BJTs with f_T in excess of 40 GHz and with noise figures below 1 dB at 1 GHz and below 2 dB at 6

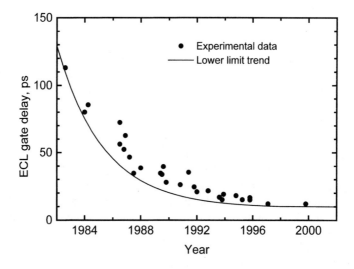

Figure 7.3. Trend of ECL gate delay times from 1982 to 2000.

GHz were available. For high-power applications, multicell power BJTs with output powers of 900 W at 1 GHz, 280 W at 1.4 GHz, and 110 W at 3 GHz (under pulsed operation) could be found on the market.

The operation and the controlling mechanism of a BJT differ significantly from those of the field-effect transistor. Furthermore, terminologies used to designate the BJT terminals, important dimensions, and operation modes are different as well. In general, bipolar transistors can be classified into npn and pnp transistors. The structure and the basic operation of npn and pnp BJTs are very similar. Actually, only the conductivity types and the polarity of the applied voltages of the two transistors are interchanged, a situation similar to the case of n- and p-channel MOSFETs. We will concentrate our discussions on npn transistors, which are more important for microwave applications due to their higher operating frequencies, but the same concepts and physics apply generally to pnp transistors.

Figure 7.4 shows a simplified structure and circuit symbol of the npn BJT. The transistor consists of an n-type emitter, a p-type base and an n-type collector, and thus contains two pn junctions, namely the emitter–base and the collector–base junctions. Typically, the doping concentration in the emitter, N_{DE}, is much higher than the base doping concentration N_{AB}, and the base is more heavily doped than the collector whose doping concentration is N_{DC}. We designate the locations of the metallurgical emitter–base and collector–base junctions as x_{jEB} and x_{jCB}, respectively, and the total thickness of the simplified transistor from the emitter contact to the collector contact as x_{tot}. The origin of the x-axis is located at the emitter contact. Compared to the BJT structures in Figs. 7.1 and 7.2, it is clear that the simplified BJT structure is only one-dimensional, and the heavily doped subcollector has been omitted.

There are two space–charge regions (shaded areas in Fig. 7.4) associated with the emitter–base and the collector–base junctions in the BJT. The space–charge region thickness depends on the doping concentrations of the two regions adjacent to the junction and on the applied voltages. The edges of the emitter–base space–charge region are x_{EB} and x_{BE}, whereas the edges of the collector–base space–charge region are x_{BC} and x_{CB}.

The bias conditions are defined by the applied base–emitter voltage V_{BE}, collector–emitter voltage V_{CE}, and collector–base voltage V_{CB}. According to the Kirchhoff voltage law, the relation among these voltages is

$$V_{CB} + V_{BE} - V_{CE} = 0 \tag{7-1}$$

The terminal currents are the emitter current I_E, base current I_B, and collector current I_C. The currents are considered to be positive when positive carriers flow into the transistor. This means that an electron current flowing out of the transistor is positive, whereas an electron current flowing into the transistor is negative. The transistor can be seen as a node for which, according to the Kirchhoff current law, the relation

$$I_E + I_C + I_B = 0 \tag{7-2}$$

holds.

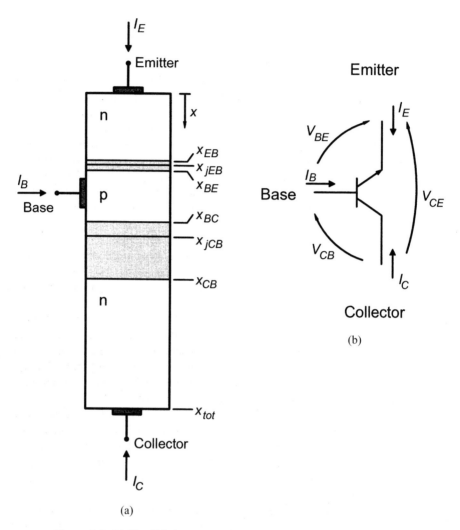

Figure 7.4. (a) Simplified structure of a BJT. (b) Circuit symbol of a BJT.

Depending on the choice of the polarity of the applied voltages, four different BJT operation modes exist. In Table 7.1, the operation modes and the corresponding polarities of the applied voltages are listed for an npn transistor. For microwave applications where the BJT acts as an amplifying device, mainly the forward active and, to some extent, the saturation modes are of interest.

We now discuss the basic concept and operation of the simplified BJT structure from Fig. 7.4. Figure 7.5 shows the electron and hole concentrations n and p in the transistor under the thermal equilibrium condition, i.e., no voltages applied. If complete ionization of impurity dopants is assumed, then the majority carrier concentra-

7.1 INTRODUCTION

Table 7.1. Different operation modes for the BJT

Operation mode	Emitter–base junction bias	V_{BE}	Collector–base junction bias	V_{CB}
Forward active (normal)	forward	> 0	reverse	> 0
Reverse active (inverse)	reverse	< 0	forward	< 0
Cutoff	reverse	< 0	reverse	> 0
Saturation	forward	> 0	forward	< 0

tion is equal to the doping concentration (e.g., $n_{0E} = N_{DE}$ in the emitter) in a region of interest. Consequently, the minority carrier concentration can be calculated using the mass action law [see Eq. (2-24)]. The distance between x_{BE} and x_{BC} is the thickness of the quasineutral base, w_B. The distance between the emitter contact ($x = 0$) and x_{EB} is the thickness of the quasineutral emitter, w_E, and the distance between x_{CB} and the collector contact at $x = x_{tot}$ is the thickness of the quasineutral collector, w_C. We call these regions quasineutral regions because, under the nonequilibrium condition (i.e., with applied voltages), these regions maintain their charge neutrality condition somewhat (quasineutrality) even with the presence of injected excess minority carriers. The quasineutrality becomes questionable, however, when the level of minority carrier injection is high.

From this coordinate system, the thickness of the emitter layer (including quasineutral emitter and the emitter–base space–charge region on the emitter side) is x_{jEB}, the thickness of the base layer (including quasineutral base and emitter–base and collector–base space–charge regions on the base side) is $x_{jCB} - x_{jEB}$, and the thickness of the collector layer (including quasineutral collector and collector–base space–charge region on the collector side) is $x_{tot} - x_{jCB}$.

Consider now the BJT biased under the forward active mode, in which the emitter–base junction is forward biased and the collector–base junction is reverse-biased (both V_{BE} and V_{CB} are positive). Under this condition, the barrier potential at the forward-biased emitter–base junction becomes smaller compared to its equilibrium value, the emitter–base space charge region becomes narrower, and electrons are injected from the emitter into the base. On the other hand, the space–charge region thickness of the reverse-biased collector–base junction is increased, and the minority concentration at x_{BC} becomes very small. The operation of the BJT is governed to a large extent by the minority carrier concentrations. Figure 7.6 shows the minority carrier concentrations in the quasineutral emitter, base, and collector regions. In Fig. 7.6(a) a BJT with a small w_B is shown. In the quasineutral base there is a large gradient of the electron concentration. The electrons injected into the base at the emitter–base junction will move toward the collector–base junction by diffusion. A few electrons may recombine with holes and thus will disappear in the quasineutral base, but the great majority of electrons will reach the collector–base junction by diffusion. Experiencing the large electric field at the reverse-biased collector–base junction, these electrons will then be swept across the collector–base space–charge region and enter the collector region. Once there, they become the

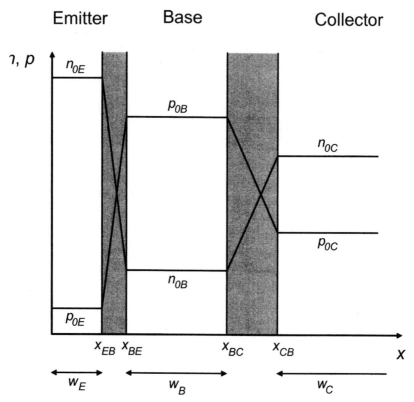

Figure 7.5. Equilibrium carrier concentrations in a BJT having a step doping profile. The shaded areas are the space–charge regions of the emitter–base and collector–base junctions.

majority carriers, travel to the collector contact, and constitute the collector current I_C. The quasineutral base of the transistor from Fig. 7.6(a) is so thin that the diffusion tail of the emitter–base junctions extends to the collector–base space–charge region. This condition is imperative for proper transistor operation.

Figure 7.6(b) shows the minority carrier concentrations in a transistor with a relatively large w_B. Here, all injected electrons recombine in the quasineutral base before reaching the collector–base junction, and the equilibrium condition [$n_B(x) = n_{0B}$] holds in a portion of the quasineutral base. In this case, the collector current is only the reverse-biased current of the collector–base junction, which is much smaller compared to that in Fig. 7.6(a). This is not desirable, and such a device structure will not operate like a bipolar transistor. Thus, a small w_B is necessary for proper BJT operation. If, however, the base is too thin and/or the base doping is too low, the emitter–base and the collector–base space–charge regions can touch each other, thus resulting in a zero thickness of the quasineutral base. This is known as base punch-through and must be avoided, as it would result in a large uncontrollable electron flow from the emitter to the collector. Thus, having a minimal thickness of

7.1 INTRODUCTION   341

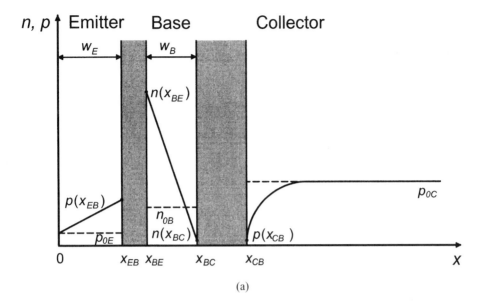

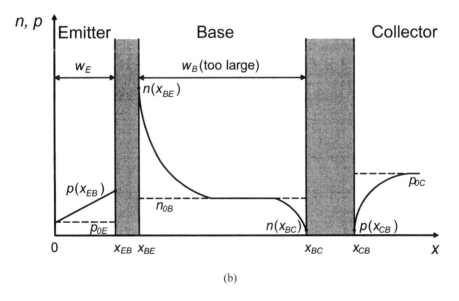

Figure 7.6. Minority carrier concentration in (a) a BJT with a thin base and (b) a BJT with a thick base. Both transistors are biased in the forward active mode.

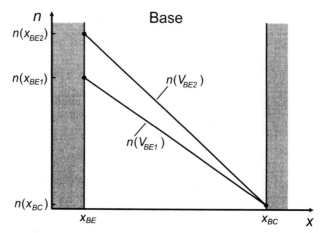

Figure 7.7. Electron distribution in the quasineutral base of a BJT for two different base–emitter voltages ($V_{BE1} < V_{BE2}$).

the quasineutral base but without inducing base punch-through is a key consideration in the design of high-performance BJTs.

Besides the electron injection from the emitter into the base, holes are also injected from the base into the emitter due to the lowering of the barrier potential in the forward-biased emitter–base space–charge region. This is frequently called back-injection. For proper BJT operation, the electron injection into the base should be much larger than the back-injection of holes into the emitter. This condition is achieved by using a large ratio of emitter to base doping density.

Next we take a look at how the collector current can be controlled. Consider the minority carrier distribution in the base shown in Fig. 7.7. The lower curve is for an applied base–emitter voltage V_{BE1}, which results in an electron concentration $n(x_{BE1})$ at x_{BE}. When V_{BE} is increased from V_{BE1} to V_{BE2}, the emitter–base space–charge region becomes slightly narrower (not shown) and, according to Eq. (2-105),* the electron concentration at x_{BE} increases to $n(x_{BE2})$. Because V_{BE} is in the exponent of Eq. (2-105), a small increase of V_{BE} results in a large increase of $n(x_{BE})$. Thus, if V_{CB} is held constant, increasing V_{BE} leads to a larger electron density gradient in the quasineutral base. The electron current passing through the quasineutral base is mainly a diffusion current, and its amount is proportional to the gradient of the carrier concentration [see Eq. (2-68)]. Consequently, the change of V_{BE} from V_{BE1} to V_{BE2} leads to a considerable increase of I_C. This, in essence, is the main mechanism controlling the collector current in the BJT.

The base current I_B is constituted by holes supplied from the base terminal and essentially comprises four components. The first one is the aforementioned injec-

---

*The voltage V_{pn} in Eq. (2-105) has to be replaced by V_{BE}.

tion of holes from the base to emitter across the forward-biased emitter–base junction. The second component is the holes compensating for the loss of holes due to electron–hole recombination in the quasineutral base. For a BJT with a thin base, this component is very small. The third component is the holes that compensate the recombination in the space–charge region of the forward-biased emitter–base junction. Finally, there is a small base current component resulting from the holes generated in the reverse-biased collector–base junction. It is normally much smaller than the other base current components mentioned above.

Figure 7.8 shows the band diagram of a BJT. Under the equilibrium condition (no voltages applied) the Fermi level E_F is flat throughout the device, and its position with respect to the conduction band in the emitter and in the collector, and with respect to the valence band in the base, indicates that $N_{DE} \gg N_{AB} \gg N_{DC}$ [see Fig. 7.8(a)]. In the forward active mode shown in Fig. 7.8(b), the potential barrier at the forward-biased emitter–base junction is lowered. If the emitter is grounded, then the bands in the base and in the collector are to be pulled downward.

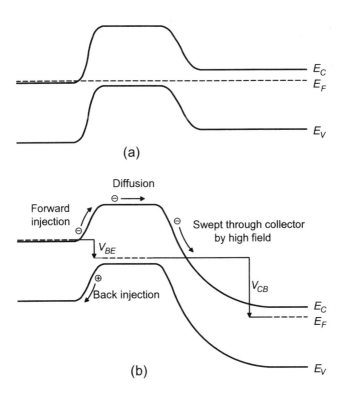

Figure 7.8. Band diagram of a BJT (a) under equilibrium condition and (b) biased in the forward active mode.

7.2 DC ANALYSIS

A dc model describing the operation principle of BJTs with special emphasis on microwave applications will be developed in this section. A variety of different physical effects may occur in BJTs, depending on the specific device design and operating conditions. This makes the development of a comprehensive analytical BJT model very difficult. Therefore, we will start with the development of a crude but simple first-order model and then include the important second-order effects in the model individually.

7.2.1 First-Order Model Development

To develop a first-order BJT model, we consider the npn BJT shown in Fig. 7.4, having abrupt emitter–base and base-collector junctions and uniformly doped emitter, base, and collector regions. This simplified structure permits a clear focus on many phenomena that are important in transistor operation without mathematical complications. The transistor is operated in the forward active mode.

We first derive expressions for the electron diffusion current density J_{nB} in the quasineutral base assuming low-level injection. Under this condition, the level of the injected minority carrier density is much smaller than the doping density. The junction areas of the emitter–base and the collector–base junctions are assumed to be equal. Note that in practical BJTs, the active area of the collector–base junction is larger than that of the emitter–base junction. The first step is the determination of the thicknesses of the quasineutral emitter and base regions, w_E and w_B, respectively. Because $N_{DE} \gg N_{AB}$ and $N_{AB} \gg N_{DC}$, we can assume that the space–charge region of the emitter–base junction resides entirely in the base, and that the space–charge region of the collector–base junction resides entirely in the collector. The thickness of the emitter–base space–charge region, d_{scEB}, is given in Eq. (2-98):

$$d_{scEB} = \sqrt{\frac{2\varepsilon(V_{biEB} - V_{BE})}{q} \frac{N_{DE} + N_{AB}}{N_{DE}N_{AB}}} \quad (7\text{-}3)$$

where V_{biEB} is the built-in voltage of the emitter–base junction obtained from Eq. (2-86). The thicknesses w_B and w_E are then given by

$$w_B = x_{jCB} - x_{jEB} - d_{scEB} \quad (7\text{-}4)$$

and

$$w_E = x_{jEB} \quad (7\text{-}5)$$

To analyze the electron diffusion current density at $x = x_{BE}$, $J_{nB}(x_{BE})$, the knowledge of the electron distribution versus x in the quasineutral base, $n_B(x)$, is needed. This is conventionally obtained by solving the steady-state continuity equation in the quasineutral base [10–12], which for an npn BJT is given by

$$D_{nB}\frac{d^2 n_B(x)}{dx^2} - \frac{n_B(x) - n_{0B}}{\tau_{nB}} = 0 \tag{7-6}$$

where n_{0B} is the equilibrium electron concentration in the quasineutral base under the equilibrium condition, D_{nB} is the minority electron diffusion coefficient, and τ_{nB} is the minority electron lifetime in the base. The boundary conditions for solving Eq. (7-6) are the minority carrier densities at the edges of the quasineutral base, i.e., $n_B(x_{BE})$ and $n_B(x_{BC})$. For the low-level injection condition, we obtain from Eq. (2-105)

$$n_B(x_{BE}) = n_{0B} \exp\frac{V_{BE}}{V_T} = \frac{n_i^2}{N_{AB}} \exp\frac{V_{BE}}{V_T} \tag{7-7}$$

and analogously

$$n_B(x_{BC}) = n_{0B} \exp-\frac{V_{CB}}{V_T} \approx 0 \tag{7-8}$$

Solving Eq. (7-6) based on these boundary conditions leads to a formidable expression for $n_B(x)$, that contains several hyperbolic terms and results in an unwieldy formula for $J_{nB}(x_{BE})$. Fortunately, this can be simplified considering a characteristic feature of a very thin base layer and thus a very small w_B. If $w_B \ll L_{nB}$, where L_{nB} is the minority carrier diffusion length in the quasineutral base defined in Section 2.4, $n_B(x)$ is a linear function with respect to x. Thus, combining Eqs. (2-68), (7-7) and (7-8), and exploiting the physics that the minority current in the quasineutral base is driven predominantly by diffusion (the electric field is negligible in the quasineutral base), $J_{nB}(x_{BE})$ can be derived as

$$J_{nB}(x_{BE}) = qD_{nB}\frac{dn_B}{dx} = qD_{nB}\frac{n(x_{BC}) - n(x_{BE})}{w_B} = -\frac{qD_{nB}}{w_B}\frac{n_i^2}{N_{AB}}\exp\frac{V_{BE}}{V_T} \tag{7-9}$$

We already mentioned that not all electrons entering the quasineutral base at x_{BE} arrive at x_{BC} because a few electrons may disappear due to recombination. This is modeled by the base transport factor α_B given by

$$\alpha_B = \frac{1}{\cosh\left(\frac{w_B}{L_{nB}}\right)} \approx 1 - \frac{1}{2}\left(\frac{w_B}{L_{nB}}\right)^2 \tag{7-10}$$

Note that α_B becomes unity for the case of zero recombination in the quasineutral base. In properly designed BJTs, α_B is very close to unity. The collector current density J_C can be approximated as the electron current density J_n that reaches the collector–base space–charge region. Thus,

$$J_C = J_{nB}(x_{BC}) = J_{nB}(x_{BE})\alpha_B \qquad (7\text{-}11)$$

Equation (7-11) results in a negative collector current density. This indicates that the negatively charged electrons move in the positive x-direction, which is equivalent to a motion of positive carriers in the negative x-direction and thus to a negative current density as discussed above.

The base current density J_B is obtained by summing the following components: 1) the hole injection current density $J_{pE}(x_{EB})$ from the base to emitter; 2) the recombination current density J_{scr} in the emitter–base space–charge region; and 3) the recombination current density J_{rB} in the quasineutral base. The hole current density $J_{pE}(x_{EB})$ is

$$J_{pE}(x_{EB}) = -qD_{pE}\frac{dp_E}{dx} = -qD_{pE}\frac{p(x_{EB}) - p(0)}{w_E} \qquad (7\text{-}12)$$

where D_{pE} is the hole diffusion coefficient in the emitter. The hole density at $x = x_{EB}$ is obtained from Eq. (2-106):

$$p_E(x_{EB}) = p_{0E}\exp\frac{V_{BE}}{V_T} = \frac{n_i^2}{N_{DE}}\exp\frac{V_{BE}}{V_T} \qquad (7\text{-}13)$$

In the frame of the first-order model, the hole density at the emitter contact at $x = 0$ is assumed to be equal to its equilibrium density, i.e., $p(0) = p_{0E}$. In a thin quasineutral emitter, where electron–hole recombination is negligibly small, the hole density is distributed linearly. This, together with the physics that the minority current in the quasineutral emitter is driven only by diffusion, yields

$$J_B = J_{pE}(x_{EB}) + J_{scr} + J_{rB}$$

$$= -qD_{pE}\frac{p(x_{EB}) - p_{0E}}{w_E} - \frac{q(x_{BE} - x_{EB})n_i \exp\left(\frac{V_{BE}}{2V_T}\right)}{2\tau_{scr}} + J_{nB}(x_{BE})(1 - \alpha_B) \qquad (7\text{-}14)$$

Here τ_{scr} is the carrier lifetime in the emitter–base space–charge region. In deriving J_{scr}, the Shockley–Read–Hall recombination process has been used and assumed to be the dominant recombination mechanism in the space–charge region. Moreover, a single-level trap located in the middle of the energy bandgap was considered [10].

The collector and base currents I_C and I_B are obtained by multiplying J_C and J_B to the negative junction area A:

$$I_C = -J_C A \qquad (7\text{-}15)$$

and

$$I_B = -J_B A \qquad (7\text{-}16)$$

The negative signs in Eqs. (7-15) and (7-16) give positive collector and base currents. This corresponds to the definition of the currents in Eq. (7-2). The base current consists of holes (positive charges) flowing into the base, i.e., into the transistor, and should therefore be positive. The collector current is constituted by electrons (negative charges) leaving the transistor, and is therefore positive as well. Finally, the emitter current I_E is obtained from Eq. (7-2).

The dc current gain is the ratio of the output current to the input current of a transistor biased under a certain circuit configuration. At this point, the common–base and the common–emitter configurations shown in Fig. 7.9 are of interest. The dc common–base current gain α_0 is defined as

$$\alpha_0 = \left| \frac{I_C}{I_E} \right| \tag{7-17}$$

and the dc common–emitter current gain β_0 is

$$\beta_0 = \frac{I_C}{I_B} \tag{7-18}$$

On the other hand, the corresponding small-signal current gains α and β are

$$\alpha = \left| \frac{dI_C}{dI_E} \right| \tag{7-19}$$

and

$$\beta = \frac{dI_C}{dI_B} \tag{7-20}$$

The current gain α is always smaller than but close to unity, whereas β is typically much larger than unity. In a properly designed BJT, β can amount to several hundreds. In general, the current gains α and β should be large.

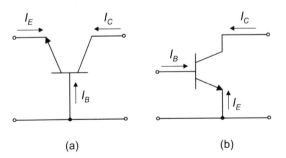

Figure 7.9. (a) BJT in the common–base configuration and (b) BJT in the common–emitter configuration.

Equations (7-3)–(7-20) constitute the first-order dc BJT model. Several important effects taking place in BJTs, however, have been neglected so far. These second-order effects will be incorporated in the first-order model in the following sections.

7.2.2 Extensions of the First-Order Model

In this section, the simple first-order model will be extended to include several second-order effects important for modern microwave BJTs.

A. Minority Carrier Mobility and Lifetime. The performance of bipolar transistors is strongly determined by the properties of the minority carriers in the emitter and base. The free-carrier mobility models given in Section 2.3 are valid only for majority carriers and have been used in the first-order model developed in Section 7.2.1 to calculate the diffusion coefficients. The mobility of minority carriers in Si can be considerably higher than that of majority carriers, especially at high doping levels exceeding 10^{17} cm^{-3}. Because the diffusion coefficient and the mobility are related to each other by the Einstein relation, the minority carrier diffusion coefficient is different from the majority carrier diffusion coefficient as well. The expression given in Eq. (2-51), along with the parameters μ_{min}, μ_{max}, N_{ref}, and α given in Table 7.2, can be used to calculate the mobilities of minority electrons and holes in Si.

The minority carrier lifetimes, τ_n and τ_p, are needed to calculate the minority carrier diffusion lengths L_n and L_p. They can be approximated using the following empirical formula

$$\tau = \frac{1}{aN + bN^2} \qquad (7\text{-}21)$$

proposed in [14, 15]. Here, a and b are fitting parameters given in Table 7.3, and N is the doping concentration in the region of interest.

B. High-Level Injection. In the first-order model, the minority carrier concentrations at the edges of the emitter–base space–charge region were derived assuming low-level injection. To account for the effect of high-level injection, the approach of arbitrary injection levels given in Section 2.4 can be utilized. Using Eq. (2-103), the electron concentration at $x = x_{BE}$ for all levels of injection is

Table 7.2. Parameters for the calculation of the minority carrier mobilities

Carrier type	μ_{min}	μ_{max}	N_{ref}	α
	cm^2/Vs	cm^2/Vs	cm^{-3}	—
Minority electrons	200	1430	5.3×10^{16}	0.68
Minority holes	122	480	1.4×10^{17}	0.7

Taken from [13].

Table 7.3. Parameters for the calculation of the minority carrier lifetimes using Eq. (7-21)

Carrier type	a, cm^3/s	b, cm^6/s
Minority electrons	3.45×10^{-12}	0.95×10^{-31}
Minority holes	7.80×10^{-13}	1.80×10^{-31}

$$n(x_{BE}) = -\frac{N_{AB}}{2} + \sqrt{\left(\frac{N_{AB}}{2}\right)^2 + n_i^2 \exp\left(\frac{V_{BE}}{V_T}\right)} \quad (7\text{-}22)$$

and, similarly, the hole concentration at $x = x_{EB}$ can be found in Eq. (2-104) as

$$p(x_{EB}) = -\frac{N_{DE}}{2} + \sqrt{\left(\frac{N_{DE}}{2}\right)^2 + n_i^2 \exp\left(\frac{V_{BE}}{V_T}\right)} \quad (7\text{-}23)$$

C. Accurate Calculation of the Thicknesses of the Quasineutral Emitter and Base. In an asymmetrically doped pn junction, the space–charge region extends mainly into the lightly doped region, but a small portion of the space–charge region does exist in the heavily doped region. To account for this effect, Eqs. (7-4) and (7-5) of the simple first-order model have to be replaced by

$$w_B = x_{BC} - x_{BE} = x_{jCB} - x_{jEB} - x_{pBE} - x_{pCB} \quad (7\text{-}24)$$

and

$$w_E = x_{EB} = x_{jEB} - x_{nEB} \quad (7\text{-}25)$$

The thicknesses x_{nEB} and x_{pBE} are the extensions of the emitter–base space–charge region into the emitter and the base, respectively, and x_{pCB} is the extension of the collector–base space–charge region into the base. These thicknesses can be calculated using Eqs. (2-96) and (2-97). For example, x_{pCB} is given by

$$x_{pCB} = \sqrt{\frac{2\varepsilon(V_{biCB} + V_{CB})}{q} \frac{N_{DC}}{N_{AB}(N_{DC} + N_{AB})}} \quad (7\text{-}26)$$

where V_{biCB} is the built-in voltage of the collector–base junction. Note that the positive collector–base voltage V_{CB} is added to V_{biCB} in Eq. (7-26) because the collector–base junction is reverse-biased and the space–charge region becomes thicker than under the thermal equilibrium condition.

D. Early Effect. The collector current in the simple first-order model is not a function of V_{CE}, a characteristic inconsistent with the experimental results showing a slight increase of I_C with increasing V_{CE} in the forward active operation. Consider the electron concentration in the quasineutral base of a BJT under a fixed V_{BE} and two different collector–base voltages V_{CB1} and V_{CB2}, as shown in Fig. 7.10. According to Eq. (7-1), V_{BE} and V_{CB1} correspond to a certain V_{CE1}. If V_{CB1} is decreased by ΔV_{CB} to V_{CB2}, then V_{CE1} is reduced by the same ΔV_{CB} to V_{CE2} because V_{BE} is fixed. From Eq. (7-26), we see that a smaller V_{CB} results in a smaller x_{pCB}, a thicker quasineutral base, and a smaller slope of the electron concentration (see Fig. 7.10). At the same time, the base transport factor in Eq. (7-20) becomes slightly lower. Thus, the electron current density J_{nB} and the collector current I_C at $V_{CB} = V_{CB2}$ ($V_{CE} = V_{CE2}$) differ from those at V_{CB1} ($V_{CE} = V_{CE1}$). This effect was first analyzed by Early [16] and is called the base width modulation or the Early effect. A measure of the influence of the Early effect on BJT operation is the Early voltage V_A, which is a positive voltage commonly defined as

$$V_A = \left| \frac{J_C}{dJ_C/dV_{CB}} \right|_{V_{BE}} = \left| \frac{J_C}{(dJ_C/dw_B)(dw_B/dV_{CB})} \right|_{V_{BE}} \quad (7\text{-}27)$$

One can determine V_A either by deriving expressions for dJ_C/dw_B and dw_B/dV_{CB} from the model equations given above, which is rather lengthy, or simply by calculating the collector current for two different values of V_{CB} while fixing V_{BE} and applying

$$V_A \approx \left| \frac{J_C}{\Delta J_C/\Delta V_{CB}} \right|_{V_{BE}} \quad (7\text{-}28)$$

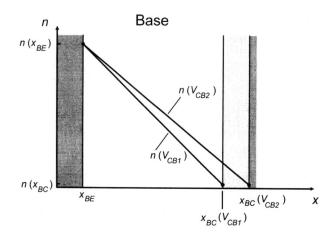

Figure 7.10. Electron distribution in the quasineutral base of a BJT for two different collector–base voltages ($V_{CB1} > V_{CB2}$).

The Early voltage can also be extrapolated from the I_C versus V_{CE} curve (for fixed V_{BE}). The intersection of the extrapolated line with the voltage axis (i.e., zero I_C) is $-V_A$, as shown in Fig. 7.11. It should be noted that typically the output characteristics of a BJT are collector currents plotted as a function of the collector–emitter voltage for a fixed base current, not fixed base–emitter voltage as defined by Eqs. (7-27) and (7-28). Thus, V_A determined from the I_C versus V_{CE} characteristics with fixed I_B is not exactly the same as but closely related to V_A modeled in Eq. (7-27). In general, a high Early voltage is desired for BJTs.

E. Saturation of Carrier Velocity. Another improvement of the simple first-order model is the inclusion of carrier velocity saturation. In the first-order model, the electron concentration at the collector side of the quasineutral base (at $x = x_{BC}$) was calculated using Eq. (7-8). This approach results in a zero $n(x_{BC})$ even for small values of V_{CB}, which would require an infinite electron velocity at x_{BC} to maintain the current continuity. The velocity of a carrier traveling through a semiconductor, however, cannot be higher than the saturation velocity (if nonstationary carrier transport is neglected). Using the more generalized expression for the electron concentration of the form of Eq. (7-22) does not solve the problem either because it would result in $n(x_{BC}) = 0$ as well. The solution to this dilemma is as follows. Neglecting the influence of electron–hole recombination in the base, a reasonable assumption for a narrow base, the electron current density must be constant in the entire base. Furthermore, because the maximum allowed electron velocity is the saturation velocity v_S, the relation

$$|J_{nB}(x_{BE})| \approx |J_{nB}(x_{BC})| \leq qn(x_{BC})v_S \tag{7-29}$$

holds. Thus, the minimum electron density at x_{BC} necessary to maintain the collector current is

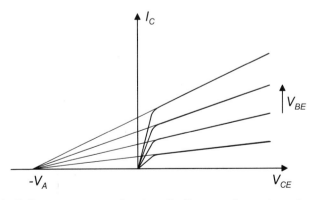

Figure 7.11. Collector current as a function of collector–emitter voltage showing the Early voltage. The parameter is the emitter–base voltage.

$$n(x_{BC}) = \frac{|J_{nB}(x_{BE})|}{qv_S} \tag{7-30}$$

Inserting Eq. (7-30) into Eq. (7-9) yields a more accurate description for the collector current density:

$$J_{nB}(x_{BE}) = J_C = -\frac{qD_{nB}n(x_{BE})}{w_B + \dfrac{D_{nB}}{v_S}} \tag{7-31}$$

F. Parasitic Resistances. The effect of voltage drops on the parasitic emitter, base, and collector resistances R_E, R_B, and R_C, respectively, can be critical to BJT modeling. We have assumed so far that the terminal voltage between the base and emitter contacts is equal to the junction voltage across the emitter–base junction, V_{BE}, and likewise the terminal voltage between the collector and base contacts is equal to the junction voltage across the collector–base junction, V_{CB}. However, the terminal voltages of the BJT differ from their corresponding intrinsic junction voltages because of the voltage drops across the parasitic resistances.

Figure 7.12 shows the cross section of a typical BJT with distributed parasitic resistances. In this realistic two-dimensional transistor structure, the n$^+$-type subcollector is included. The y-direction perpendicular to the surface from the emitter to

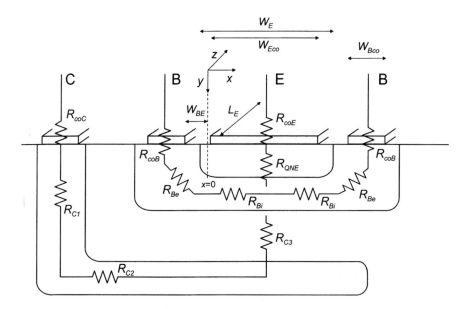

Figure 7.12. Cross section of a BJT showing the different components of the parasitic emitter, base, and collector resistances.

the collector corresponds to the x-direction in the one-dimensional structure used in our previous discussions (see, e.g., Fig. 7.4). The extension of the emitter in the x-direction is the emitter lateral width W_E, and the extension of the emitter contact in the z-direction is called the emitter length L_E. Typically $W_E < L_E$. Note that this coordinate system is different from that in FETs, where the extension in the x-direction is the gate length L and the extension of the gate in the z-direction is the gate width W.

The resistance between the contact and the underlying semiconductor layer is the contact resistance. We have to distinguish between the so-called vertical-flow and horizontal-flow contacts. The emitter and collector contacts of the BJT in Fig. 7.12 are vertical-flow contacts, as the currents flow vertically out of these contacts. The contact resistance for vertical-flow contacts, $R_{co,v}$, is calculated by

$$R_{co,v} = \frac{\rho_{co}}{A_{co}} \qquad (7\text{-}32)$$

where ρ_{co} is the specific contact resistance and A_{co} is the area of the contact. For the emitter contact, for example, A_{co} is the emitter area $A_{Eco} = W_{Eco} \times L_E$. The specific contact resistance between metal and highly doped Si is typically on the order of 1–2×10^{-7} Ω-cm^2. In the case of a polysilicon emitter, to be dealt with later in this section, the same value can be assumed for the specific contact resistance between the metal and polysilicon. The base contact in Fig. 7.12, on the other hand, is a horizontal-flow contact. Here, the current leaves the contact vertically, but it is bent into the horizontal direction in the semiconductor underneath the contact. The contact resistance R_{coB} of the horizontal-flow base contact is given by

$$R_{coB} = \frac{\sqrt{R_{sh}\rho_{co}}}{L_E} \coth\left(W_{Bco}\sqrt{\frac{R_{sh}}{\rho_{co}}}\right) \qquad (7\text{-}33)$$

where R_{sh} is the sheet resistance of the base layer underneath the contact and W_{Bco} is the lateral width of the base contact. The sheet resistance is defined as the resistance of a semiconductor layer having a length equal to the width. Thus, the sheet resistance of the base is

$$R_{sh} = \frac{\rho_B}{w_B} \qquad (7\text{-}34)$$

where ρ_B is the resistivity of the base.

The total parasitic emitter resistance R_E is the sum of the emitter contact resistance, R_{coE}, and the resistance of the quasineutral emitter, R_{QNE}:

$$R_E = R_{coE} + R_{QNE} = \frac{\rho_{co}}{A_{Eco}} + \frac{1}{q\mu_{nE}N_{DE}}\frac{w_E}{A_E} \qquad (7\text{-}35)$$

where μ_{nE} is the electron mobility in the emitter and w_E is the thickness of the quasineutral emitter.

The base resistance consists of the base contact resistance R_{coB}, the extrinsic base resistance R_{Be}, and the intrinsic base resistance R_{Bi}. Note that there are two base contacts in Fig. 7.12. We first develop a method to calculate the intrinsic base resistance based on the approach given in [10, 17]. The base current on the left-hand side of the BJT in Fig. 7.12 flows from the left base contact through the extrinsic portion of the base and enters the thin intrinsic base layer. At $x = 0$, all holes delivered by the base contact are present, but the number of holes decreases with increasing x because at any x position holes can be injected into the emitter or be annihilated by recombination. Thus the base current is x-dependent and becomes zero at the center of the base at $x = W_E/2$. Because the transistor shown in Fig. 7.12 is symmetrical in the x-direction, we first consider only the one-half structure between $x = 0$ and $W_E/2$. The position-dependent base current $I_B(x)$ can be assumed linearly dependent of x:

$$I_B(x) = \frac{I_B}{2}\left(1 - \frac{2x}{W_E}\right) \qquad (7\text{-}36)$$

where I_B is the total base current of the transistor and $I_B/2$ flows into the left base contact. The resistance of an incremental piece of the base, dR, with incremental length dx is

$$dR = \rho_{Bi}\frac{dx}{L_E w_B} \qquad (7\text{-}37)$$

where ρ_{Bi} is the resistivity of the intrinsic base given by

$$\rho_{Bi} = \frac{1}{q\mu_{pB}N_{AB}} \qquad (7\text{-}38)$$

and μ_{pB} is the hole mobility in the intrinsic base. Applying the Ohm's law, the voltage drop across the incremental base piece is

$$dV_B(x) = \frac{I_B}{2}\left(1 - \frac{2x}{W_E}\right)\frac{\rho_{Bi}}{L_E w_B}dx \qquad (7\text{-}39)$$

The voltage drop between 0 and any position x is obtained by indefinite integration of Eq. (7-39) using the boundary condition $V_B(0) = 0$. The result of the integration is

$$V_B(x) = \frac{I_B}{2}\frac{\rho_{Bi}}{L_E w_B}\left(x - \frac{x^2}{W_E}\right) \qquad (7\text{-}40)$$

The averaged voltage drop V_{Bav} between 0 and $W_E/2$ can be obtained by integrating Eq. (7-40) from 0 to $W_E/2$ and dividing the result by $W_E/2$. Applying this definition we obtain

$$V_{\text{Bav}} = \frac{1}{W_E/2} \int_0^{W_E/2} V_B(x) dx = \frac{1}{12} \frac{\rho_{\text{Bi}} W_E}{L_E w_B} I_B \qquad (7\text{-}41)$$

Applying the Ohm's law again, we get the intrinsic resistance of one half of the base extending from $x = 0$ to $x = W_E/2$ as

$$R_{\text{Bi}} = \frac{V_{\text{Bav}}}{I_B/2} = \frac{1}{6} \frac{\rho_{\text{Bi}} W_E}{L_E w_B} \qquad (7\text{-}42)$$

Because the transistor is symmetrical in the x-direction, the right-hand side base has the same resistance as the left-hand side base, and the two intrinsic base resistances are in parallel. Thus, the total intrinsic base resistance is

$$R_{\text{Bi}} = \frac{1}{12} \frac{\rho_{\text{Bi}} W_E}{L_E w_B} \qquad (7\text{-}43)$$

The same approach can be used for transistors with only one base terminal, and R_{Bi} for this case is

$$R_{\text{Bi}} = \frac{1}{3} \frac{\rho_{\text{Bi}} W_E}{L_E w_B} \qquad (7\text{-}44)$$

The extrinsic base resistance R_{Be}, on the other hand, can be modeled by considering only the geometry and doping density of the extrinsic base region. Thus

$$R_{\text{Be}} = \rho_{\text{Be}} \frac{W_{\text{BE}}}{L_E w_{\text{Be}}} \qquad (7\text{-}45)$$

where ρ_{Be} is the resistivity of the extrinsic base, W_{BE} is the horizontal separation between the base contact and the n-type emitter region as shown in Fig. 7.12, and w_{Be} is the thickness of the extrinsic base. Finally, the total base resistance is given by

$$R_B = \frac{R_{\text{coB}} + R_{\text{Be}}}{n} + R_{\text{Bi}} \qquad (7\text{-}46)$$

where n is the number of base contacts ($n = 2$ in Fig. 7.12).

The collector resistance R_C consists of the collector contact resistance R_{coC} modeled by Eq. (7-32), and the three resistance components R_{C1}, R_{C2}, and R_{C3} (see Fig. 7.12). Component R_{C1} describes the resistance of the vertical part of the collector between the contact and the subcollector, and R_{C2} is the resistance of the subcollector. These two components can be calculated based on the dimensions and the doping concentrations of the corresponding regions. Component R_{C3} is the resistance of the undepleted part of the collector (i.e., quasineutral collector). The thickness of the undepleted collector, x_{coll}, is given by $x_{\text{coll}} = x_{\text{tot}} - x_{\text{CB}}$ (see Fig. 7.4), and

$$R_{C3} = \frac{1}{q\mu_{nC}N_{DC}} \frac{x_{coll}}{A_E} \quad (7\text{-}47)$$

where μ_{nC} is the electron mobility in the collector. We again assumed the collector area is equal to the emitter area A_E.

Considering the voltage drops across the parasitic resistances, the relations between the external voltages (subscript ext) and the intrinsic junction voltages (subscript int) are

$$V_{BE,ext} = V_{BE,int} + I_B R_B + (I_B + I_C)R_E \quad (7\text{-}48)$$

$$V_{CE,ext} = V_{CE,int} + I_C R_C + (I_B + I_C)R_E \quad (7\text{-}49)$$

and

$$V_{CB,ext} = V_{CB,int} + I_C R_C + I_B R_B \quad (7\text{-}50)$$

We have now arrived at a point where, when given the device make-up (e.g., doping concentrations, thicknesses and lateral widths of the different layers, etc.), I_E, I_C, and I_B can be calculated as a function of the applied voltages V_{BE} and V_{CE}. Two different plots are frequently used to characterize the dc behavior of BJTs. The first is the logarithmic plot of I_C and I_B versus V_{BE} for fixed V_{CB}, conventionally referred to as the Gummel plot. Fig. 7.13(a) illustrates the Gummel plot obtained from the model for a typical microwave BJT. The smaller slope in the base current at low voltages is due to the recombination current in the emitter–base space–charge region, which is proportional to $\exp(V_{BE}/2V_T)$ as can be seen from Eq. (7-14). The second plot to characterize the dc behavior is the linear plot of the collector current versus V_{CE} for fixed I_B, conventionally called the current–voltage output characteristics of BJT biased in the common–emitter configuration. Figure 7.13(b) shows such characteristics obtained from the model developed above.

G. Bandgap Narrowing. In heavily doped semiconductors, the donors and acceptors do not create discrete energy levels E_D and E_A in the forbidden gap as in lightly doped semiconductors described in Section 2.2.1, but rather there are narrow impurity bands with a continuous distribution of energy levels. This is due to the fact that at a relatively high doping concentration, the distance between neighboring dopant atoms is very small. These closely spaced atoms interact with each other, and the originally discrete energy level is transformed into a band. At an even higher doping concentration, the impurity band can overlap the conduction band edge in an n-type material, or the valence band in a p-type material. Thus, for an n-type material, not only the energy states of the conduction band but also the unfilled states of the impurity band can be occupied by free electrons traveling through the semiconductor. The same holds for the valence band and the holes in a p-type material. As a result, the effective bandgap is decreased, or it is apparently narrower than that

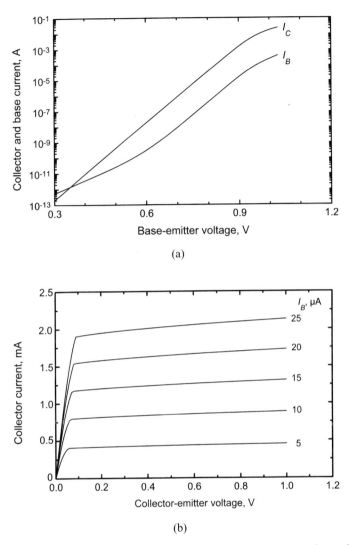

Figure 7.13. (a) Gummel plot and (b) common emitter output current–voltage characteristics of a BJT calculated using the model described below. The transistor parameters are $w_B = 50$ nm; $N_{AB} = 2 \times 10^{18}$ cm^{-3}, $N_{DE} = 10^{20}$ cm^{-3}.

of the lightly doped material. This effect is described by the bandgap narrowing ΔE_{BGN}.

A lot of work has been done in the past to quantify ΔE_{BGN} as a function of the doping concentration, and the following empirical expressions have been developed as a useful guide to estimate the bandgap narrowing. For n-type Si, the bandgap narrowing can be calculated by [15]

$$\Delta E_{BGN} = 18.7 \ln\left(\frac{N_D}{7 \times 10^{17} \text{ cm}^{-3}}\right) \times \text{meV} \quad (7\text{-}51)$$

for doping concentrations larger than 7×10^{17} cm^{-3}, and $\Delta E_{BGN} = 0$ otherwise. For p-type Si, the expression [14, 18]

$$\Delta E_{BGN} = 9\left\{\ln\left(\frac{N_A}{10^{17} \text{ cm}^{-3}}\right) + \sqrt{\left[\ln\left(\frac{N_A}{10^{17} \text{ cm}^{-3}}\right)\right]^2 + 0.5}\right\} \times \text{meV} \quad (7\text{-}52)$$

holds for $N_A > 10^{17}$ cm^{-3}, and $\Delta E_{BGN} = 0$ otherwise. Figure 7.14 shows the bandgap narrowing calculated from Eqs. (7-51) and (7-52) for n- and p-type Si as a function of the doping concentration.

An important implication of the bandgap narrowing is the increase of the intrinsic carrier concentration n_i, because a smaller E_G leads to a larger n_i [see Eq. (2-24)]. This in turn affects the minority carrier concentrations and all other parameters related to n_i. Taking into account the bandgap narrowing, the mass action law Eq. (2-24) is modified to

$$n_0 p_0 = n_{ie}^2 = n_i^2 \exp\left(\frac{\Delta E_{BGN}}{k_B T}\right) \quad (7\text{-}53)$$

where n_0 and p_0 are the equilibrium carrier concentrations, n_{ie} is the effective intrinsic carrier density taking the bandgap narrowing into account, and n_i is the intrinsic carrier concentration without bandgap narrowing. In modern microwave BJTs, the

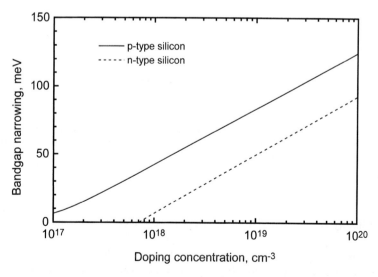

Figure 7.14. Bandgap narrowing in Si as a function of doping concentration calculated from Eqs. (7-51) and (7-52).

emitter doping can be extremely high (above 10^{20} cm^{-3}) and the base doping can exceed 10^{18} cm^{-3}. Thus, the effect of bandgap narrowing is critical to the accurate modeling of such devices.

H. Quasisaturation Operation and Base Pushout. So far, we have considered a simplified n$^+$-p-n structure (except in Fig. 7.12). Furthermore, the model developed employed the depletion approximation, which assumes that the free carrier densities n and p are zero in the emitter–base and collector–base space–charge regions. In practical BJTs, however, a heavily doped n$^+$ subcollector is located underneath the lightly doped collector, and free carriers exist in the space–charge regions.

To explain the quasisaturation operation and base pushout, the effect of the n$^+$ subcollector needs to be included and the depletion approximation needs to be relaxed for the collector–base space–charge region. Consider the BJT structure shown in Fig. 7.15. It consists of an n$^+$ emitter, a p-type base, an n-type collector with a thickness w_{coll}, and a heavily doped n$^+$ subcollector. The applied external V_{CB} is chosen to reverse bias the collector–base junction for $I_C = 0$. Now a positive V_{BE} is applied, the emitter–base junction is forward-biased, and a collector current I_C flows. If V_{CB} is held constant and I_C is increased (by increasing V_{BE}), two tendencies can take place in the collector–base space–charge region. One tendency causes the space–charge region to shrink (contraction tendency), whereas the other causes the space–charge region to expand (expansion tendency).

The physical mechanism underlying the contraction tendency is as follows. The collector resistance R_{C3} and the collector current I_C give rise to a voltage drop across the undepleted collector region (quasineutral collector) outside the space–charge region. If the external V_{CB} is held constant and I_C is increased, this voltage drop will be increased and the reverse junction voltage across the collector–base space–charge region will be reduced, thereby resulting in a smaller collector–base space–charge region thickness. This is called the contraction tendency. For a sufficiently large collector current, the collector–base junction can become forward biased even though a reverse voltage is applied between the base and collector terminals. Under such a condition, the BJT is said to operate in the quasisaturation mode. This mode is insignificant in low-power microwave BJTs, because the collector region in these devices is typically thin and is relatively highly doped, and thus the voltage drop in the

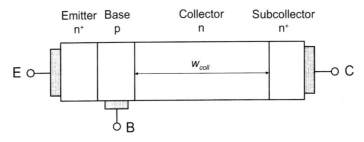

Figure 7.15. Simplified BJT structure including a heavily doped subcollector.

quasineutral collector is relatively small. It can be important, however, in power bipolar transistors in which the collector layer is thicker and lightly doped.

The expansion tendency, on the other hand, results from the injection of electrons in the collector–base space–charge region. Consider again the BJT in Fig. 7.15 operated in the forward active mode with a constant V_{CB}. Because the doping ratio N_{AB}/N_{DC} is much larger than unity, it can be assumed that the collector–base space–charge region resides entirely in the collector and that its extension into the base is negligible. In Section 7.2.2.E we stressed that in the forward active mode the electron density at the base side of the collector–base junction, $n(x_{BC})$, can be considerable. These electrons travel with the saturation velocity through the collector–base space–charge region and add to the positively charged donor ions in the region. As the collector current is increased, more electrons associated with the collector current pass through the collector–base space–charge region as well. This results in a decrease in the net space–charge density in the region. Since V_{CB} is constant, the space–charge region will need to expand to maintain the same applied voltage. This is called the expansion tendency. A sufficiently large I_C will ultimately cause the collector–base space–charge region to expand to the boundary of the highly doped subcollector.

Clearly, the first-order BJT model model becomes invalid when the contraction and expansion tendencies are considered. Kirk [19] and Bowler and Lindholm [20] have developed a model to estimate the thickness of the collector–base space–charge region, d_{scCB}, accounting for both the expansion and contraction tendencies:

$$d_{scCB} = d_{scCB0}\sqrt{\frac{1 - J_C/J_2}{1 - J_C/J_1}} \quad (7\text{-}54)$$

where d_{scCB0} is d_{scCB} at $J_C = 0$ and can be modeled using the conventional depletion model, and J_1 and J_2 are given by

$$J_1 = qN_{DC}v_S \quad (7\text{-}55)$$

and

$$J_2 = \frac{qN_{DC}\mu_{nC}(-V_{CB} + V_{biCB})}{w_{coll}} \quad (7\text{-}56)$$

Here, v_S is the electron saturation velocity and μ_{nC} is the electron mobility in the collector. J_1 is the collector current density that would flow if all free carriers in the collector outside the space–charge region moved with the saturation velocity. J_2 is the current density below which the transistor is operated in the forward active region, i.e., below J_2 the collector–base junction is reverse-biased, whereas above J_2 it becomes forward biased. For $J_1 > J_2$ the contraction tendency dominates, and if $J_1 < J_2$ the expansion tendency prevails [20].

Let us focus on the expansion tendency because it is the prevailing tendency in typical microwave BJTs. As mentioned earlier, a sufficiently large collector current density can cause the collector–base space–charge region to expand all the way to the boundary of the n⁺ subcollector ($d_{scCB} = w_{coll}$). In this case, the entire collector becomes the space–charge region, and the electric field extends from $x = 0$ to $x = w_{coll}$ (here, the origin of the x-axis is located at the collector–base pn junction). This is shown qualitatively in Fig. 7.16, curve (a). The magnitude of the field is maximum at the metallurgical collector–base junction and decreases to zero at the n/n⁺ interface. As the collector current density is increased further, the field distribution changes, as shown by the curves (b) to (d) in Fig. 7.16. Because V_{CB} is fixed, the areas under the curves are required to be constant. Since the space–charge region cannot extend further because of the heavily doped subcollector [this can be checked by Eq. (5-35)], the areas for the different current levels can only be constant when the field at the collector–base junction decreases and the field at the n/n⁺ interface increases.

At a certain J_C, denoted as J_K, the field at the collector–base metallurgical junction becomes zero [see curve (e)]. At this point, there is no longer a potential barrier for the holes at the collector–base junction. Thus holes can be injected from the base into the collector, and the conductivity of the collector region adjacent to the base changes from n-type to p-type. This suggests nothing but an apparent extension of the base into the collector, and the effective base thickness w_{Beff} is increased by Δw_B. This phenomenon is called the current–induced base pushout or the Kirk effect, and w_{Beff} can be estimated by [11]

$$w_{Beff} = w_B + \Delta w_B = w_B + w_{coll}\left(1 - \frac{J_K - J_1}{J_c - J_1}\right) \quad \text{for } J_C \geq J_K \quad (7\text{-}57)$$

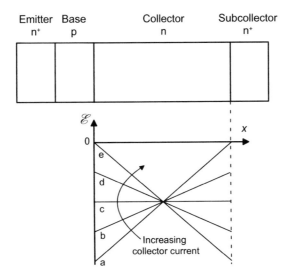

Figure 7.16. Qualitative illustration of the field distribution in the collector of a BJT.

where

$$J_K = qv_s\left[N_{DC} + \frac{2\varepsilon(V_{CB} + V_{biCB})}{qw_{coll}^2}\right] \qquad (7\text{-}58)$$

It is important to point out that the base pushout decreases the collector current (compared to the current obtained from the first-order model) and increases the base transit time τ_B. The meaning and calculation of the base transist time will be discussed in Section 7.3.2. The base pushout is a main factor contributing to the experimentally observed rapid falloff of the current gain and cutoff frequency at high current densities.

The estimation given above can serve as an approximation for the base pushout behavior. A more accurate treatment, however, requires a self-consistent numerical solution of the basic semiconductor equations, i.e., continuity, current, and Poisson

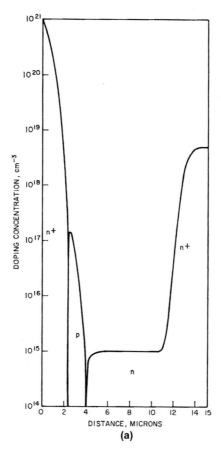

Figure 7.17. (a) Doping profile of the BJT considered. Taken from [21] © 1969 IEEE.

equations. This has been done in [21]. Figure 7.17(a) shows the doping profile and the vertical dimensions of the investigated BJT structure, and in Fig. 7.17(b) the simulated field distributions are illustrated for different collector current levels. At small J_C, the maximum field is located precisely at the collector–base metallurgical junction. At high J_C, however, the maximum field moves toward the n/n$^+$ interface, and the field in a large part of the collector is practically zero. Thus, a large portion of the collector becomes part of the base due to the base pushout. At very high current levels, the effective base can extend all the way to the subcollector and the effective base thickness approaches $w_B + w_{coll}$.

I. Effect of Nonuniform Doping Concentration.

Our previous treatment has been based on the assumption that the doping concentrations N_{DE} and N_{AB} in the emitter and base regions are uniform. In modern microwave BJTs, the base is formed by ion implantion, and the emitter doping is achieved by outdiffusion from

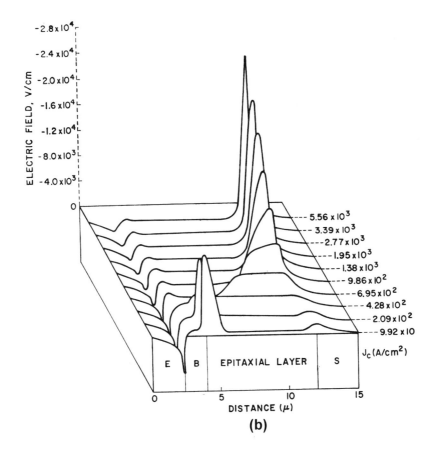

Figure 7.17. (b) Field distribution in the BJT simulated for different collector densities illustrating the base pushout. Taken from [21] © 1969 IEEE.

a heavily n-doped polysilicon emitter during a drive-in annealing. This automatically leads to nonuniform emitter and base doping profiles. A doping profile typical for microwave BJTs is shown in Fig. 7.18, illustrating the nonuniform doping densities in the emitter and base.

The influence of the nonuniform base doping profile on BJT dc behavior is the main focus of this section. The net base doping profile can be approximated by a Gaussian distribution as

$$N_{AB}(x) = N_{ABmax} \exp\left(-\frac{x^2 \ln(N_{ABmax}/N_{DC})}{w_B^2}\right) \quad (7\text{-}59)$$

Here, the x-axis is perpendicular to the semiconductor surface and into the bulk. The origin of the x-axis is located at the emitter–base metallurgical junction, which is the point where $N_{AB}(x) = N_{DE}(x)$. The base doping profile described by Eq. (7-59) shows a peak acceptor concentration N_{ABmax} at the emitter–base junction and decreases to the collector doping density N_{DC} at the collector–base junction.

The doping gradient gives rise to a hole diffusion in the base in the $+x$ direction. This leads to a negative space charge (uncompensated acceptor ions) near the emitter and a positive space charge (extra holes) near the collector. Thus a built-in field $\mathscr{E}$ directed in the $-x$ direction is established in the quasineutral base, thereby creating a driving force for a hole drift in the same direction. Under the thermal equilibrium condition (zero applied voltage), diffusion and drift compensate each other, and we can write

$$J_{pB} = q\mu_{pB}N_{AB}(x)\mathscr{E} - qD_{pB}\frac{dN_{AB}(x)}{dx} = 0 \quad (7\text{-}60)$$

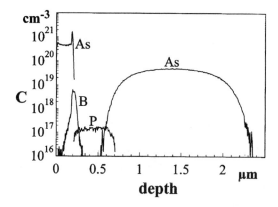

Figure 7.18. Doping profile of a state of the art microwave BJT with an f_T of 52 GHz. Taken from [9] © 2001 IEEE.

where J_{pB} is the hole current density in the base, μ_{pB} is the hole mobility in the base, and D_{pB} is the hole diffusion coefficient in the base. Rearranging this expression, we get

$$\mathcal{E}(x) = \frac{k_B T}{q} \frac{1}{N_{AB}(x)} \frac{dN_{AB}(x)}{dx} \qquad (7\text{-}61)$$

Due to the presence of the built-in field $\mathcal{E}(x)$, the equation for the electron current density in the base [see Eq. (7-9)] has to be modified to include both the drift and the diffusion terms:

$$J_{nB} = q\mu_{nB}\mathcal{E}(x)n(x) + qD_{nB}\frac{dn(x)}{dx} \qquad (7\text{-}62)$$

Inserting the built-in field from Eq. (7-61), applying the Einstein relation, and carrying out some algebraic manipulations, Eq. (7-62) becomes

$$J_{nB}\frac{N_{AB}(x)}{qD_{nB}}dx = d[N_{AB}(x)n(x)] \qquad (7\text{-}63)$$

Integrating Eq. (7-63) over the quasineutral base, i.e., from $x = x_{BE}$ to $x = x_{BC}$, yields

$$J_{nB} = -\frac{qD_{nB}n_i^2 \exp\dfrac{V_{BE}}{V_T}}{\displaystyle\int_{x_{BE}}^{x_{BC}} N_{AB}(x)dx} \qquad (7\text{-}64)$$

Here we used Eqs. (7-7) and (7-8) to express the electron densities at x_{BE} and x_{BC}. The term in the denominator is conventionally referred to as the Gummel number for the base, and it gives the total integrated base dose in cm^{-2}. The same approach can be used to model the base current, including the effect of nonuniform emitter doping density [22].

An important effect relevant to the nonuniform base doping profile has been neglected. If the maximum base doping density N_{ABmax} is higher than 10^{17} cm^{-3}, which is typical for microwave BJTs, bandgap narrowing occurs. As described in Eq. (7-52) and Fig. 7.14, the bandgap narrowing increases with increasing doping density. Thus, the bandgap is smallest at the emitter–base junction where the doping is highest, and increases toward the collector–base junction. This creates a second built-in field in the base in the direction opposite to the first built-in field caused by the nonuniform doping profile. Both fields tend to compensate each other, and Eqs. (7-60)–(7-64) actually overestimate the effect of a nonuniform base doping on BJT performance.

J. Avalanche Breakdown. When the reverse voltage across the collector–base junction is increased, the field inside the collector–base space–charge region increases as well. If the field is increased beyond a critical value, considerable impact ionization takes place, as discussed in Section 2.6. Then, avalanche multiplication occurs and the transistor operates in the breakdown region. Breakdown leads to a dramatic increase in the collector current and can destroy the transistor. The avalanche multiplication factor M is commonly used to describe the breakdown behavior. It is defined as the ratio of the current with impact ionization to the current without impact ionization. For a pn junction, the multiplication factor is frequently expressed by the empirical formula

$$M = \frac{1}{1 - (V/BV)^m} \qquad (7\text{-}65)$$

where BV is the breakdown voltage of the junction, V is the applied voltage, and m is an empirical constant between 3 and 6 depending on the material of the junction. Similarly, considering the reverse-biased collector–base junction of a BJT, M is given by

$$M = \frac{1}{1 - (V_{CB}/BV_{CB})^m} \qquad (7\text{-}66)$$

The collector–base breakdown voltage BV_{CB} can be estimated from Eq. (2-126) and Figs. 2.26 and 2.27.

The breakdown behavior of bipolar transistors is more complicated than that of pn junctions and is commonly characterized by the following two breakdown voltages:

1. Collector–base breakdown voltage BV_{CB0} (V_{CB} causing the breakdown when the emitter is open, i.e., $I_E = 0$)
2. Collector–emitter breakdown voltage BV_{CE0} (V_{CE} causing the breakdown when the base is open, i.e., $I_B = 0$)

Intuitively, one would expect BV_{CE0} to be larger than BV_{CB0} because, according to Eq. (7-1), V_{CE} is the sum of V_{BE} and V_{CB}. This is not the case, however.

Consider the BJT with an open emitter as shown in Fig. 7.19(a). The collector current prior to the onset of breakdown, i.e., at low V_{CB}, is denoted I_{CB0} and the collector current at an arbitrary collector–base voltage is

$$I_C = MI_{CB0} \qquad (7\text{-}67)$$

Combining Eqs. (7-67) and (7-66), it becomes clear that I_C increases sharply when V_{CB} approaches BV_{CB0}.

Now consider the BJT with an open base shown in Fig. 7.19(b). Because the base is neither connected to the ground nor to a defined potential, the base potential is floating somewhere between the emitter and the base potentials. As a result, the applied voltage V_{CE} causes the collector–base junction to be reverse-biased and the

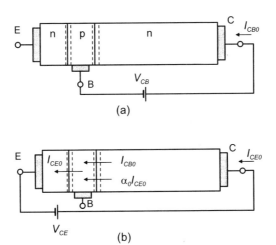

Figure 7.19. BJT under (a) open emitter configuration and (b) open base configuration.

emitter–base junction to be slightly forward-biased, with the largest portion of V_{CE} dropping across the collector–base junction. The reverse bias causes the current I_{CB0} to flow across the collector–base junction, and the low forward voltage across the emitter–base junction causes a small current I_{CE0} to flow across the emitter–base junction. Part of I_{CE0} arrives at the collector, namely $\alpha_0 I_{CE0}$, which contributes to the total collector current I_{CE0} prior to the onset of breakdown. Thus

$$I_{CE0} = \alpha_0 I_{CE0} + I_{CB0} \qquad (7\text{-}68)$$

In analogy to Eq. (7-67), the collector current at an arbitrary collector–emitter voltage is

$$I_C = M I_{CE0} = M(\alpha_0 I_C + I_{CB0}) \qquad (7\text{-}69)$$

Solving this equation for I_C yields

$$I_C = \frac{M I_{CB0}}{1 - \alpha_0 M} \qquad (7\text{-}70)$$

Equation (7-70) suggests that the collector current of a BJT with an open base will increase sharply when $\alpha_0 M$ approaches unity:

$$\alpha_0 M = 1 \qquad (7\text{-}71)$$

Because, as mentioned above, the largest portion of the applied V_{CE} drops across the collector–base junction, we can approximate $V_{CB} \approx V_{CE}$. Using this approximation, putting M from Eq. (7-66) into Eq. (7-71), and solving for BV_{CE0}, which is V_{CE} at the onset of breakdown, we get

$$BV_{CE0} = BV_{CB0}(1 - \alpha_0)^{1/m} \quad (7\text{-}72)$$

Applying Eqs. (7-2), (7-17), and (7-18) to express α_0 in terms of β_0 and assuming α_0 to be unity, a reasonable approximation for properly designed BJTs, we obtain

$$BV_{CE0} = \frac{BV_{CB0}}{\sqrt[m]{\beta_0}} \quad (7\text{-}73)$$

From Eqs. (7-72) and (7-73), it is apparent that BV_{CE0} is smaller than BV_{CB0} and that there is a trade-off between the current gain β_0 and the breakdown voltage BV_{CE0}. Figure 7.20 shows the I–V characteristics of a BJT including the breakdown voltages BV_{CB0} und. BV_{CE0}.

K. Current Crowding. The intrinsic base resistance R_{Bi} and x-dependent potential $V_B(x)$ in the base have been discussed in Section 7.2.2.F. The x-dependence (the direction of the x-axis is defined in Fig. 7.12) of the base potential V_B also leads to a variation in the emitter–base junction voltage V_{BE} in the x-direction. $V_{BE}(x)$ is largest at $x = 0$ (at the emitter edge) and decreases toward the center of the base ($x = W_E/2$). Because the carrier injection depends exponentially on V_{BE} [see Eq. (7-7)], the electron current density J_n injected from the emitter into the base becomes x-dependent as well. It is largest near $x = 0$ and decreases toward the center. This effect, which is illustrated qualitatively in Fig. 7.21, is called current crowding and results in a nonuniform current distribution across the emitter–base junction. In transistors with an extremely large base resistance, the current crowding can lead to a situation

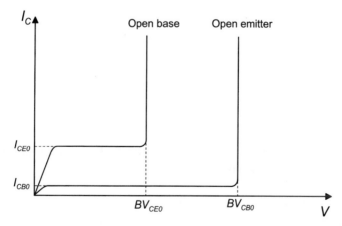

Figure 7.20. Current–voltage characteristics of a BJT. Open base: I_C versus V_{CE}. Open emitter: I_C versus V_{CB}.

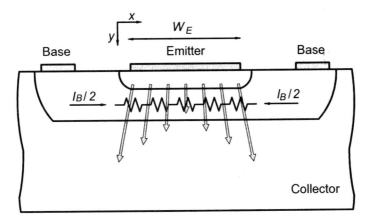

Figure 7.21. Cross section of a BJT illustrating the occurrence of current crowding.

where the center region does not contribute to the collector current at all. Liu developed a method to calculate the x-dependence of V_{BE}, J_B, and J_E for GaAs HBTs [17] that can be applied to Si BJTs as well. The derivation is lengthy and will not be repeated here. Finally it leads to the so-called effective emitter width W_{Eeff} which is a useful figure of merit to quantify the effect of current crowding. Consider a BJT carrying a certain emitter current I_E. This transistor has an emitter width W_E and is subject to the current crowding. A comparative transistor in which current crowding is assumed not to occur would need a smaller emitter width W_{Eeff} to carry the same emitter current. The ratio between W_{Eeff} and W_E is called the dc emitter utilization EU_{dc} and can be expressed as [17]

$$EU_{dc} = \frac{W_{Eeff}}{W_E} = \frac{2\int_0^{W_E/2} J_E(x)dx}{W_E J_E(0)} = \frac{\sin c \times \cos c}{c} \qquad (7\text{-}74)$$

where c is a constant to be determined from the following transcendental equation:

$$c \times \tan c = \frac{q}{k_B T} \frac{I_E}{w_B} \frac{\rho_B}{4(1+\beta_0)} \frac{W_E}{L_E} \qquad (7\text{-}75)$$

ρ_B is the base resistivity in Ω-cm. Equation (7-75) has to be solved numerically within the limits of $\pi/2$ and 0 for c.

L. Finite Surface Recombination Velocity and Polysilicon Emitter.

The hole current density in the emitter has been previously calculated using Eq. (7-12) assuming that equilibrium conditions exist at the emitter contact, i.e., the hole concentration at the emitter contact, $p_E(0)$, is equal to the equilibrium hole concentration p_{0E}. This would require an infinite surface recombination velocity S_M at the emitter contact. In reality, S_M at a metal/n$^+$–Si interface has a finite value on the order of 1 to 3×10^5 cm/s. The assumption of an infinite surface recombination velocity at the contact would overestimate the gradient of the hole density in the quasi-neutral emitter and the hole current density, and thus underestimate the current gain β_0. The hole current density at the emitter contact can be expressed in terms of the surface recombination velocity as

$$J_p(0) = -q[p_E(0) - p_{0E}]S_M \tag{7-76}$$

Combining Eq. (7-76) with Eq. (7-12) and solving for $p_E(0)$, we get

$$p_E(0) = \frac{\dfrac{D_{pE}}{w_E} p_E(x_{EB}) + p_{0E} S_M}{S_M + \dfrac{D_{pE}}{w_E}} \tag{7-77}$$

This boundary condition can be incorporated into the first-order BJT model for more accurate calculation of the base current.

Around 1980, many experiments on the use of heavily doped polysilicon for the emitter contact were carried out. During that time it was common belief that the n$^+$ polysilicon behaves like a metal and that the polysilicon/silicon interface acts like an ohmic contact. This would mean that the holes injected from the base recombine at the polysilicon emitter with a rate determined by the recombination velocity of a metal/Si interface. The experiments revealed, however, that only a portion of the holes recombine at the interface, while a considerable portion of the holes recombine in the polysilicon [23]. This effect can be interpreted by the two-region model shown in Fig. 7.22, together with the assumption that the surface recombination velocity S_P at the polysilicon/silicon interface is much smaller than

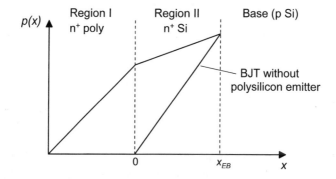

Figure 7.22. Two-region model of an n$^+$-polysilicon/n$^+$-silicon emitter. Shown also is the hole concentration in the BJT without polysilicon emitter.

Table 7.4. Measured current gains of bipolar transistors having the same lateral dimensions and base design (base sheet resistance about 7 kΩ/□) but different emitter configurations

Emitter stack	Current gain
Al/silicide/n$^+$–Si	40–60
Al/n$^+$–Si	65–70
Al/n$^+$–polysilicon/n$^+$–Si	145–170
Al/n$^+$–polysilicon	≈ 400

Data taken from [4].

S_M. The limited hole recombination at the polysilicon/silicon interface, i.e., at $x = 0$, thus leads to an increased hole concentration at $x = 0$. This, in turn, gives rise to a smaller gradient of the hole concentration in the emitter, thereby resulting in a smaller hole injection from the base to the emitter, a smaller base current, and a higher current gain β_0.

The current gains of BJTs having the same dimension and base design but different emitter configurations have been studied in [4, 23]. The experimental results, listed in Table 7.4, indicate that BJTs with the polysilicon emitter possess considerably higher current gains compared to BJTs with the conventional metal–Si system.

It should be noted that besides the two-region model, there are other models reported to explain the behavior of polysilicon emitters. For example, several models consider the effects of a thin interoxide layer located between the polysilicon and silicon on hole transport and recombination. The issue is still quite controversial, and not a single model has succeeded in becoming universally accepted. Key papers on this topic up to 1989 were collected and reviewed in [24]. Later works can be found in [25, 26] and references therein. For a first-order treatment, we suggest using Eq. (7-77) and replacing S_M with S_P to calculate the hole concentration at the polysilicon/Si interface. A key in this approach is the proper choice of S_P. An impression on the order of S_P can be obtained from Figs. 7.23 and 7.24. Figure 7.23 shows S_P as a function of the polysilicon thickness in the absence of an interoxide layer. The influence of the interoxide thickness on S_P is shown in Fig. 7.24 for a polysilicon thickness of 300 nm. Frequently, S_P is used as a fitting parameter in BJT modeling to adjust the simulated BJT currents to measured data.

7.3 SMALL-SIGNAL ANALYSIS

7.3.1 Small-Signal Equivalent Circuit

The small-signal behavior of BJTs can be modeled, analogously to FETs, using a small-signal equivalent circuit. Figure 7.25 shows the widely used π-model for a BJT biased in the common–emitter configuration. Each element of the equivalent circuit describes a physical effect in the transistor. The transistor is biased with the dc voltages V_{BE}, V_{CB}, and V_{CE}, resulting in the dc collector, emitter, and base cur-

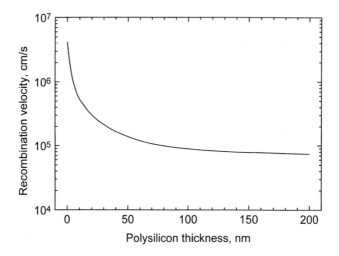

Figure 7.23. Calculated recombination velocity at the n^+-polysilicon/silicon interface as a function of polysilicon thickness. Data taken from [26].

rents I_C, I_E, and I_B. When ac voltages v_{BE}, v_{CB}, or v_{CE} are superimposed onto the dc bias, ac currents are induced. In the following, we will give physical insight into the different elements of the π-model equivalent circuit and provide approximated expressions to estimate the values of these elements. These expressions will be derived based on the simple first-order model presented in Section 7.2.1, but taking into account the Early effect and the fact that the emitter–base and the

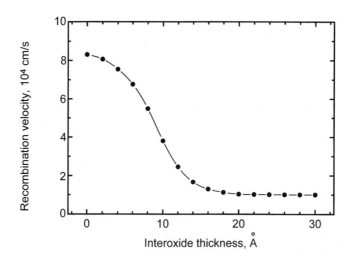

Figure 7.24. Recombination velocity at a polysilicon/n^+-silicon interface as a function of interoxide thickness (for a polysilicon thickness of 300 nm). Data taken from [26].

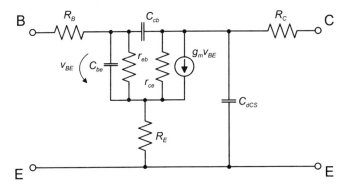

Figure 7.25. Small-signal equivalent circuit for the π-model of a BJT biased in the common–emitter configuration.

collector–base space–charge regions extend into both the highly and low-doped regions adjacent to the junctions.

The transconductance g_m describes the variation of the collector current with respect to a change in the base–emitter voltage:

$$g_m = \left.\frac{dI_C}{dV_{BE}}\right|_{V_{CE}=\text{const.}} = -A \left.\frac{dJ_C}{dV_{BE}}\right|_{V_{CE}=\text{const.}} \quad (7\text{-}78)$$

In properly designed BJTs, the base transport factor α_B is near unity, and we can approximate J_C by the electron diffusion current density J_{nB} at the base side of the emitter–base space–charge region, which has been given in Eq. (7-9). Taking the derivative of J_{nB} with respect to V_{BE} and inserting the result in Eq. (7-78), we arrive at

$$g_m = \frac{I_C}{V_T} \quad (7\text{-}79)$$

Due to the Early effect, the collector current depends also on the collector–emitter voltage. This is modeled by the resistance r_{ce}, defined as

$$r_{ce} = \left.\frac{dV_{CE}}{dI_C}\right|_{V_{BE}=\text{const.}} \quad (7\text{-}80)$$

When the base–emitter voltage is fixed, a variation in the emitter–collector voltage, ΔV_{CE}, gives rise to the same change in the base-collector voltage ($dV_{CE} = dV_{CB}$). This, together with (7-28), leads to

$$r_{ce} = \frac{V_A}{I_C} \quad (7\text{-}81)$$

The emitter–base resistance r_{eb} takes care of the variation in the base current due to a change in the emitter–base voltage:

$$r_{eb} = \left.\frac{dV_{BE}}{dI_B}\right|_{V_{CE}=\text{const.}} \qquad (7\text{-}82)$$

The resistance r_{eb} is frequently denoted r_π. For medium and large base–emitter voltages, the hole injection from the base to the emitter constitutes the dominating component for the base current. The hole injection density J_{pE} at $x = x_{EB}$ has been given in Eq. (7-12). Putting this equation into Eq. (7-82) and bearing in mind that $\exp(V_{BE}/V_T) \gg 1$ in most cases, we have

$$r_{eb} = \frac{V_T}{I_B} = \frac{\beta_0}{g_m} \qquad (7\text{-}83)$$

The two capacitive elements C_{cb} and C_{eb} in Fig. 7.25 are the total capacitances for the collector–base and emitter–base junctions, respectively. In general, the total capacitance of a pn junction consists of two parallel capacitance components, namely the junction depletion capacitance C_d and the diffusion capacitance C_D. For BJTs in the forward active mode, the collector–base junction is reverse-biased, and the diffusion capacitance is negligible. Thus C_{cb} consists only of the junction depletion capacitance and is given by

$$C_{cb} = C_{dBC} = \frac{\varepsilon A}{d_{scCB}} \qquad (7\text{-}84)$$

where A is the junction area and d_{scCB} is the thickness of the collector–base space–charge region [see (2-98)]. Because the emitter–base junction is forward-biased, both the junction depletion capacitance and the diffusion capacitance have to be considered:

$$C_{eb} = C_{dEB} + C_{DEB} \qquad (7\text{-}85)$$

The emitter–base junction depletion capacitance C_{dEB} is calculated in analogy to Eq. (7-84) as

$$C_{dEB} = \frac{\varepsilon A}{d_{scEB}} \qquad (7\text{-}86)$$

The diffusion capacitance C_{DEB} is the variation in the amount of the minority carrier charge in the quasineutral emitter and in the quasineutral base with respect to the change in the emitter–base voltage:

$$C_{DEB} = \left.\frac{dQ_{pE}}{dV_{BE}}\right|_{V_{CE}=\text{const.}} + \left.\frac{d|Q_{nB}|}{dV_{BE}}\right|_{V_{CE}=\text{const.}} \qquad (7\text{-}87)$$

Here, Q_{pE} is the excess hole charge in the n-type quasineutral emitter and is related to the area under the curve for the hole concentration between $x = 0$ and $x = x_{EB}$ in Fig. 7.6(a), and $|Q_{nB}|$ is the magnitude of the excess electron charge in the p-type quasineutral base and is related to the area under the curve for the electron concentration between $x = x_{BE}$ and $x = x_{BC}$. Thus,

$$Q_{pE} = qA \frac{w_E[p_E(x_{EB}) - p_E(0)]}{2} \tag{7-88}$$

and

$$|Q_{nB}| = qA \frac{w_B[n_B(x_{BE}) - n_B(x_{BC})]}{2} \tag{7-89}$$

In carrying out the derivation of Q_{pE} and Q_{nB} with respect to V_{BE} in the frame of the first-order model, one has to bear in mind that

- $p_E(0)$ is equal to the equilibrium hole concentration in the emitter, i.e., not dependent on V_{BE}, and $n_B(x_{BC})$ is zero.
- The electron and hole concentrations $n_B(x_{BE})$ and $p_E(x_{EB})$ are calculated from Eqs. (2-105) and (2-106), i.e., for the low-level injection case.
- The thicknesses of the quasineutral base and of the quasineutral emitter, w_B and w_E, depend on V_{BE}. They are calculated from Eqs. (7-24) and (7-25), respectively.

From this, we obtain

$$\left.\frac{dQ_{pE}}{dV_{BE}}\right|_{V_{CE}=\text{const.}} = \frac{qA}{2} \left\{ \frac{x_{nEB}}{2(V_{biEB} - V_{EB})} [p_E(x_{EB}) - p_{0E}] + w_E \frac{p_E(x_{EB})}{V_T} \right\} \tag{7-90}$$

and

$$\left.\frac{d|Q_{nB}|}{dV_{BE}}\right|_{V_{CE}=\text{const.}} = \frac{qA}{2} \left[\frac{n_B(x_{EB})}{2} \left(\frac{x_{pCB}}{V_{biCB} + V_{CB}} + \frac{x_{pBE}}{V_{biEB} - V_{BE}} \right) + w_B \frac{n_B(x_{BE})}{V_T} \right] \tag{7-91}$$

The capacitances C_{cb} and C_{be} are frequently denoted C_μ and C_π, respectively. Note that the bias conditions for the circuit elements given in Eqs. (7-78)–(7-91) are the intrinsic junction voltages.

The third capacitive element in Fig. 7.25 is the depletion capacitance C_{dCS} associated with the space–charge region between the n$^+$ subcollector and the p-type substrate. It can be modeled as a parallel plate capacitor in analogy to C_{cb}. The voltage across the junction, which is necessary to determine the extension of the space–charge region, is the potential difference between the subcollector and ground, i.e., between the subcollector and the emitter for a transistor biased in the common–emitter configuration. The parasitic emitter, base, and collector resistances, R_E, R_B, and R_C, in Fig. 7.25 have already been modeled previously in Section 7.2.2.

Besides the π-model, sometimes a different transistor model called the T-model is used for the small-signal analysis. Figure 7.26 shows the T-model of a BJT biased in the common–emitter configuration. While the amplification of the transistor is modeled by the current source $g_m \times v_{BE}$ in the π-model, it is described in the T-model by the current source $\beta \times i_B$. Other elements in the T-model have physical meanings similar to those in the π-model.

Applying the two-port theory to the small-signal equivalent model, the frequency-dependent small signal parameters of the BJT (Y, H, or S parameters; see Section 1.2) can be determined. These parameters can then be used to calculate the small signal current gains and the power gains in a manner similar to that discussed in Section 4.3.4 for the MESFET.

7.3.2 Delay Time Analysis

Here, an alternative option to describe the dynamic behavior of BJTs operated in the forward active mode will be discussed. It is the delay time analysis and is related to the charging and discharging of the emitter–base diffusion capacitance C_{DEB} and the emitter–base and collector–base junction depletion capacitances C_{dEB} and C_{dCB}. The basic idea of such an analysis is to determine the different delay times needed to rearrange the mobile charges in the transistor when the applied voltage varies by a small amount. The total delay time is the sum of all these delay times and is called the emitter–collector delay time τ_{EC}.

We start with the delay times related to the C_{DEB}. The total amount of charge associated with C_{DEB}, Q_{DEB}, is given by

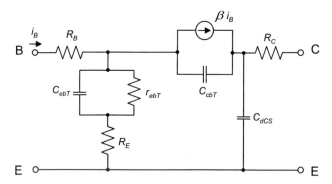

Figure 7.26. Small-signal equivalent circuit for the T-model of a BJT biased in the common–emitter configuration.

$$Q_{DEB} = |Q_{pE}| + |Q_{nB}| + |Q_{BC}| \tag{7-92}$$

where Q_{pE} is the hole charge in the n-type quasineutral emitter and Q_{nB} is the electron charge in the p-type base. The third component, Q_{BC}, takes into account the fact that the collector–base space–charge region is not fully depleted, but rather there is a collector current constituted by electrons traveling with a certain velocity passing through the collector–base space–charge region. This effect has been discussed in Section 7.2.2.E. Note that we use the absolute values for the charges in Eq. (7-92), otherwise the charge of the positively charged minority holes and the negatively charged minority electrons would partially compensate each other. The charge components in Eq. (7-92) can be written as the product of the charging current, which is the collector current, and three different delay times:

$$Q_{DEB} = (\tau_E + \tau_B + \tau_{BC})I_C \tag{7-93}$$

where τ_E is the emitter delay time, τ_B is the base transit time, and τ_{BC} is the transit time related to the collector–base space–charge region. The charge of the minority holes in the quasineutral emitter, Q_{pE}, has been given in Eq. (7-88). Assuming the dominant component of the base current is the hole injection from the base to the emitter and combining Eqs. (7-12), (7-16), (7-18) and (7-88), we get

$$Q_{pE} = I_C \frac{w_E^2}{2D_{pE}\beta_0} \tag{7-94}$$

This, together with Eq. (7-93) and the assumption of a thin emitter, leads to

$$\tau_E = \frac{w_E^2}{2D_{pE}\beta_0} \tag{7-95}$$

A different expression $\tau_E = \tau_{pE}/(2\beta_0)$, where τ_{pE} is the hole lifetime in the emitter, can also be seen in the literature. It is valid for BJTs with a thick emitter. Equation (7-95), on the other hand, is derived based on the assumption of a shallow emitter with $w_E \ll L_{pE}$, which is fulfilled in modern microwave BJTs.

If electron–hole recombination in the quasineutral base is neglected, Eqs. (7-9) and (7-15) are used to express the collector current, and Eq. (7-89) is used for the minority electron charge in the base, then the base transit time τ_B is given by

$$\tau_B = \frac{w_B^2}{2D_{nB}} \tag{7-96}$$

The collector–base space–charge region transit time τ_{BC} can be approximated by [27]

$$\tau_{BC} = \frac{d_{scCB}}{2v_S} \tag{7-97}$$

where d_{scCB} is the thickness of the collector–base space–charge region given in Eq. (2-98).

Two further delay times contribute to the emitter–collector delay time. They have to do with the time necessary to charge or discharge the emitter–base and collector–base junction depletion capacitances C_{dEB} and C_{dCB}. Consider a BJT with a certain dc bias and a variation dV_{BE} in the base–emitter voltage. In the above analysis, the delay time has been defined as the charge component of interest divided by the collector current [see Eq. (7-93)]. Obviously, we can express the delay time as the variation of the charge stored in the depletion capacitance of interest divided by the variation of the collector current caused by dV_{BE} as well. Thus, for the delay time τ_{EB} related to the emitter–base depletion capacitance, we obtain

$$\tau_{EB} = \frac{dQ_{dEB}}{dI_C} = \frac{dQ_{dEB}}{dV_{BE}} \frac{dV_{BE}}{dI_C} = \frac{C_{dEB}}{g_m} \qquad (7\text{-}98)$$

If the emitter and collector are short-circuited, a variation in V_{BE} leads to the same change in V_{CB} but with the opposite sign. Thus the delay time τ_{CB} associated with the collector–base depletion capacitance is

$$\tau_{CB} = \frac{dQ_{dCB}}{dI_C} = \frac{dQ_{dCB}}{dV_{BE}} \frac{dV_{BE}}{dI_C} = \frac{C_{dCB}}{g_m} \qquad (7\text{-}99)$$

Equation (7-99) is valid for negligible parasitic resistances R_E and R_C. If the effect of these resistances is included, a variation in V_{BE} does not change V_{CB} by the same amount, and τ_{CB} becomes

$$\tau_{CB} = \frac{dQ_{dCB}}{dI_C} = C_{dCB}\left(R_E + R_C + \frac{1}{g_m}\right) \qquad (7\text{-}100)$$

The total emitter–collector transit time is the sum of the five delay times discussed above:

$$\tau_{EC} = \tau_E + \tau_B + \tau_{BC} + \tau_{EB} + \tau_{CB} \qquad (7\text{-}101)$$

7.3.3 Cutoff Frequency and Maximum Frequency of Oscillation

Once the elements of the small-signal equivalent circuit are determined, the frequency-dependent small-signal current gain h_{21} and the frequency-dependent unilateral power gain U can be calculated from the small-signal equivalent circuit. Then, the cutoff frequency f_T and the maximum frequency of oscillation f_{max} can easily be determined. This is done by extrapolating h_{21} and U (calculated at not too low frequencies) assuming the –20 dB/dec slope versus frequency as discussed in Section 4.3.4.

A second and frequently used approach to find f_T is based on the delay times described in the previous section. The cutoff frequency is related to the emitter–collector transit time as

$$f_T = \frac{1}{2\pi\tau_{EC}} \tag{7-102}$$

The concept of the above equation is similar to that of Eq. (5-72). It indicates that a certain time is needed to adjust the charge distribution from the emitter to the collector after the BJT is subject to a small variation in the applied voltage. Such a delay time is τ_{EC} and the corresponding frequency is f_T.

The unilateral power gain of a BJT can also be expressed in terms of f_T by

$$U = \frac{\alpha_B^2 \omega_T}{4R_B C_{cb} \omega^2} \tag{7-103}$$

where ω is the angular frequency equal to $2 \times \pi \times f$, ω_T is equal to $2 \times \pi \times f_T$, and α_B is the base transport factor given in Eq. (7-10). A detailed derivation of Eq. (7-103) can be found in [17]. If the base transport factor is approximately unity, which is the case of properly designed BJTs, and bearing in mind that f_{max} is the frequency at which U becomes unity, then Eq. (7-103) leads to

$$f_{max} = \sqrt{\frac{f_T}{8\pi R_B C_{cb}}} \tag{7-104}$$

Since the current gain h_{21} and unilateral power gain U depend on the dc operating conditions, f_T and f_{max} are bias dependent as well. This bias-dependence is typically represented in the plots of f_T and f_{max} versus I_C for a constant V_{CB}, as shown in Fig. 7.27 for a state of the art Si BJT. Because the collector current depends on the base–emitter voltage, and thus on the base current, it is also feasible to plot f_T and f_{max} as functions of V_{BE} or I_B.

7.4 NOISE ANALYSIS

Two noise components are important for the high-frequency noise in microwave BJTs: the thermal noise and shot noise. Thermal noise occurs because of the random motion of electrons in a resistive material such as a piece of semiconductor. This type of noise has already been dealt with in the discussions of the noise generated in the channel of MESFETs in Section 4.4. For a semiconductor with a resistance R, the mean square thermal noise voltage $\langle v_{th}^2 \rangle$ generated in the material is given by

$$\langle v_{th}^2 \rangle = 4k_B TR\Delta f \tag{7-105}$$

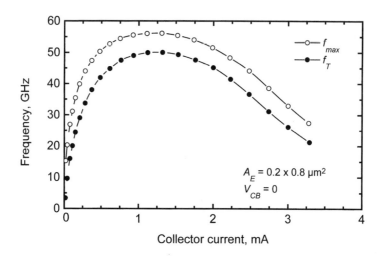

Figure 7.27. Characteristic frequencies f_T and f_{max} of a state of the art microwave BJT as a function of collector current. Data taken from [9].

where Δf is the bandwidth. Applying Ohm's law, the corresponding mean square thermal noise current $\langle i_{th}^2 \rangle$ is

$$\langle i_{th}^2 \rangle = \frac{4k_B T \Delta f}{R} \quad (7\text{-}106)$$

Shot noise, on the other hand, arises from the fluctuation of the number of charged particles. When the number of particles crossing a plane in the device fluctuates, the current fluctuates as well, and the shot noise is induced. Shot noise is a dominant component for the total noise of pn junctions. The mean square shot noise current can be expressed as [28, 29]

$$\langle i_{sh}^2 \rangle = 2qI\Delta f \quad (7\text{-}107)$$

where I is the dc current passing through the device. For the case of a forward-biased emitter–base junction, the dc current crossing the junction is given by [see Eq. (2-117)]

$$I = I_S \left(\exp \frac{V_{EB}}{V_T} - 1 \right) \quad (7\text{-}108)$$

The derivative of the current I with respect to the base–emitter voltage, which is the ac emitter conductance g_{eb} of the junction, and thus the reciprocal of the ac resistance r_{eb}, is

$$\frac{dI}{dV_{BE}} = g_{eb} = \frac{1}{r_{eb}} = \frac{I_S}{V_T}\left(\exp\frac{V_{BE}}{V_T} - 1\right) \quad (7\text{-}109)$$

For most forward voltages, the exponential term is much larger than unity, and Eq. (7-109) becomes

$$g_{eb} = \frac{1}{r_{eb}} = \frac{I_S}{V_T}\exp\frac{V_{BE}}{V_T} = \frac{I}{V_T} \quad (7\text{-}110)$$

Combining Eqs. (7-110) and (7-107), we obtain

$$\langle i_{sh}^2 \rangle = 2\frac{k_B T}{r_{eb}}\Delta f \quad (7\text{-}111)$$

and the corresponding mean square shot noise voltage is

$$\langle v_{sh}^2 \rangle = 2k_B T r_{eb}\Delta f \quad (7\text{-}112)$$

The preceding analysis has suggested that, for a fixed bandwidth, both the thermal noise current (and voltage) and the shot noise current (and voltage) are independent of frequency. This type of noise is called white noise.

Having a good understanding of the noise mechanisms in a BJT, we can now proceed to develop a noise model and an expression for the noise figure NF. The most widely used model to describe the high-frequency noise of BJTs was developed based on the small-signal T-model by Hawkins [30]. Figure 7.28 shows the

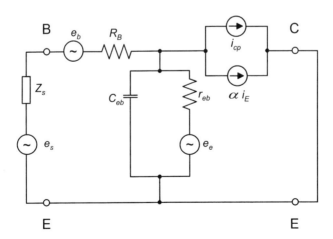

Figure 7.28. Equivalent circuit of the BJT noise model based on the T-model.

equivalent circuit for the noise model for a BJT biased in the common–emitter configuration. It consists of the BJT itself and of the signal source with the source impedance Z_S. Comparing the Hawkins model with the T-model in Fig. 7.26, it is clear that for the noise analysis only the base resistance R_B and the ac emitter resistance r_{eb} are considered, whereas all other resistances are neglected. Furthermore, all capacitances except for the emitter–base capacitance C_{eb} are omitted. The T-model in Fig. 7.28 consists of four noise sources, e_s, e_b, e_e, and i_{cp}, the first of which is related to the signal source, whereas the other three describe the noises generated in the transistor. The noise voltage sources e_s and e_b represent the thermal noise of the resistance of the signal source and of the base resistance, respectively. They are, in accordance with Eq. (7-105),

$$\langle e_s^2 \rangle = 4k_B T R_S \tag{7-113}$$

and

$$\langle e_b^2 \rangle = 4k_B T R_B \tag{7-114}$$

with R_S being the real part of the source impedance Z_S. The source impedance consists of the real part R_S and the imaginary part X_S according to $Z_S = R_S + jX_S$, where j is the imaginary unit.

The shot noises produced in the transistor are modeled by the noise voltage source e_e and the noise current source i_{cp}

$$\langle e_e^2 \rangle = 2k_B T r_{eb} \tag{7-115}$$

$$\langle i_{cp}^2 \rangle = 2k_B T (\alpha_{(0)} - |\alpha|^2) r_{eb} \tag{7-116}$$

where $\alpha_{(0)}$ is the low-frequency value of the frequency-dependent common–base current gain α. The relation between $\alpha_{(0)}$ and α is

$$\alpha = \frac{\alpha_{(0)}}{1 + j \dfrac{f}{f_B}} \tag{7-117}$$

where f is the frequency, and f_B is the cutoff frequency of the base alone given by

$$f_B = \frac{1}{2\pi \tau_B} \tag{7-118}$$

The collector noise current is strongly correlated with the emitter noise current. This is modeled by the current source αi_E. The portion of the collector current noise not correlated with the emitter noise current is i_{cp}.

A straightforward but rather lengthy circuit analysis can be found in [30], which leads to the following expression for the BJT noise figure:

$$NF = 1 + \frac{R_B}{R_S} + \frac{r_{eb}}{2R_S} + \left(\frac{\alpha_{(0)}}{|\alpha|^2} - 1\right)\frac{(R_S + R_B + r_{eb})^2 + X_S^2}{2r_{eb}R_S}$$

$$+ \frac{\alpha_{(0)}}{|\alpha|^2}\frac{r_{eb}}{2R_S}[\omega^2 C_{eb}^2 X_S^2 - 2\omega C_{eb}X_S + \omega^2 C_{eb}^2(R_S + R_B)^2] \quad (7\text{-}119)$$

The minimum noise figure NF_{min} is obtained by first taking the derivative of Eq. (7-119) with respect to X_S and setting dNF/dX_S to zero. Then the resulting equation is differentiated with respect to R_S and dNF/dR_S is set to zero. The final result is

$$NF_{min} = a\frac{R_B + R_{opt}}{r_{eb}} + \left(1 + \frac{f^2}{f_B^2}\right)\frac{1}{\alpha_{(0)}} \quad (7\text{-}120)$$

where R_{opt} is the real part of optimum source impedance given by

$$R_{opt} = \sqrt{R_B^2 - X_{opt}^2 + \left(1 + \frac{f^2}{f_B^2}\right)\frac{r_{eb}(2R_B + r_{eb})}{\alpha_{(0)}a}} \quad (7\text{-}121)$$

and a is

$$a = \left[\left(1 + \frac{f^2}{f_B^2}\right)\left(1 + \frac{f^2}{f_E^2}\right) - \alpha_{(0)}\right]\frac{1}{\alpha_{(0)}} \quad (7\text{-}122)$$

X_{opt} in Eq. (7-121) is the imaginary part of the optimum source impedance given by

$$X_{opt} = \left(1 + \frac{f^2}{f_B^2}\right)\frac{2\pi f C_{eb} r_{eb}^2}{\alpha_{(0)}a} \quad (7\text{-}123)$$

and f_E in Eq. (7-123) is the cutoff frequency of the emitter alone that is calculated by

$$f_E = \frac{1}{2\pi r_{eb}C_{eb}} \quad (7\text{-}124)$$

Figure 7.29 compares the minimum noise figures calculated from the Hawkins noise model Eqs. (7-119)–(7-124) and obtained from measurements. For the Si BJT investigated in [30], current-independent values for $\alpha_{(0)} = 0.98$, $R_B = 11\ \Omega$, and $f_B = 23$ GHz were given. The ac emitter resistance r_{eb} was calculated from Eq. (7-110) and the emitter–base capacitance C_{eb} was obtained from Eq. (7-124).* Clearly the Hawkins model compares favorably with the experimental noise figures both qualitatively and quantitatively. Note that the characteristic shape of the BJT NF_{min} ver-

*If the details of the BJT design are known, the emitter cutoff frequency f_E can be calculated using Eq. (7-124). Otherwise, f_E is obtained from $1/f_E = 1/f_\alpha - 1/f_B$, with f_α extracted from Fig. 3 in [30].

384   SILICON BIPOLAR JUNCTION TRANSISTORS

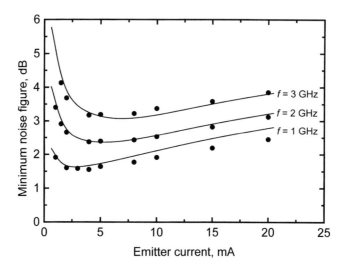

Figure 7.29. Comparison of the BJT minimum noise figures calculated using the Hawkins noise model and obtained from measurements. Experimental data taken from [30].

sus I_E curve is similar to that of the FET NF_{min} versus I_D/I_{DSS} curve. For both transistor types, NF_{min} shows a pronounced minimum at low currents.

7.5 POWER ANALYSIS

The idealized dc output characteristics of a BJT biased in the common–emitter configuration is shown in Fig. 7.30. Qualitatively, it looks very similar to the output characteristic of a MESFET shown in Fig. 4.22. For both transistor types, the output I–V characteristics show two characteristic regions. The first region below the knee voltage V_k shows a large slope, i.e a large increase of the current when the voltage is increased. In the second region above V_k, the current rises only slightly when the voltage is increased. The maximum emitter–collector voltage that can be applied to the BJT is BV_{CE0} [see Eq. (7-73)] and corresponds to $V_{m,max} = BV_{GD}$ for the MESFET. This similarity allows us to follow the power analysis of MESFETs presented in Section 4.5, provided I_{max} and $V_{m,max}$ are replaced by I_{Cmax} and BV_{CE0}, respectively. For a BJT operating as a common–emitter, class A amplifier, we obtain

$$R_L = \frac{BV_{CE0} - V_k}{I_{Cmax}} \qquad (7\text{-}125)$$

for the load in the case of power matching [see Eq. (4-109)]

$$P_{out,max} = \frac{I_{Cmax}(BV_{CE0} - V_k)}{8} = \frac{(BV_{CE0} - V_k)^2}{8R_L} \qquad (7\text{-}126)$$

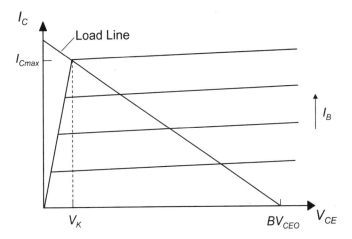

Figure 7.30. Output *I–V* characteristics of a BJT including the load line for maximum output power.

for the maximum output power [see Eq. (4-112)]

$$P_{dc} = \frac{I_{Cmax}(BV_{CE0} + V_k)}{4} \quad (7\text{-}127)$$

for the dc power dissipated in the BJT [see Eq. (4-115)]

$$PAE = \frac{1}{2}\frac{BV_{CE0} - V_k}{BV_{CE0} + V_k} \quad (7\text{-}128)$$

for the power-added efficiency [see Eq. (4-116)], and

$$P_{BJT} = \frac{1}{8}I_{Cmax}BV_{CE0} + \frac{3}{8}I_{Cmax}V_k \quad (7\text{-}129)$$

for the total power (i.e., dc power plus RF power) dissipated in the BJT. Equations (7-127)–(7-129) are valid when the transistor is biased for maxmimum output power.

In power BJTs, thermal behavior and self-heating significantly affect the transistor performance and therefore should be considered. Two thermal effects are important and will be discussed qualitatively below. The first is the temperature dependence of the common–emitter current gain β. The current gain β increases with increasing temperature and can be estimated by the empirical expression [31]

$$\beta(T) = \beta(T_A)[1 + 0.007(T - T_A)] \quad (7\text{-}130)$$

where T_A is the ambient temperature and T is the actual temperature of the transistor. Consider a single-finger BJT with a constant base current I_B fed into the transistor. The base current results in a certain collector current I_C. The collector current, together with the applied collector–emitter voltage V_{CE}, leads to a dc power, P_{dc}, dissipated in the transistor in the form of heat. Consequently, the transistor temperature increases above the ambient temperature and, because of the aforementioned temperature dependence of the current gain, the collector current increases as well. The result is an even higher transistor temperature and a further increase of I_C. This phenomenon is called thermal runaway, which gives rise to an unstable thermal behavior of the BJT and can destroy the transistor.

The second thermal effect takes place in multifinger power BJTs. The multifinger design is frequently used to minimize the self-heating effect in power transistors (see Sections 2.7 and 4.6.3). In multifinger transistors, the removal of the heat generated in the individual fingers is most effective in the outermost fingers and least effective for the center fingers due to a mechanism called thermal coupling. This results in more significant self-heating in the center fingers and therefore a nonuniform temperature distribution among the multiple fingers. Furthermore, because the center fingers become hottest, they contribute most to the collector current increase. This in turn increases the temperature of these fingers even further. Thus, the thermal runaway is in essence "accelerated" by the multifinger structure.

Nonuniform temperature distribution and thermal runaway can be limited by proper transistor design (choice of appropriate finger dimensions and finger spacings, thinning of the substrate). A more effective means to reduce the thermal problems is the use of an additional emitter resistor called the emitter ballasting resistor, as shown in Fig. 7.31. The voltage drop across the ballasting resistor increases with

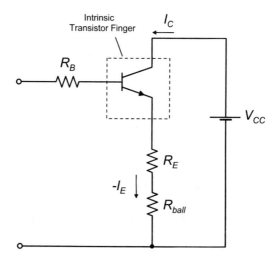

Figure 7.31. Emitter ballasting resistor of an individual emitter finger in a multifinger power BJT.

increasing collector current, which consequently decreases the base–emitter voltage and the base current leading to a reduced collector current. Thus the increase of I_C (caused by the temperature-induced increase of β) is compensated and limited to a reasonable value. It should be noted, however, that the ballasting resistor deteriorates the frequency performance of the transistor and that the use of the ballasting resistor presents a trade-off between the high-frequency performance and the thermal stability.

The effect of self-heating in BJTs with a single emitter finger or with a multifinger design can be analyzed as follows. Based on the method described in Section 2.7, the temperature of the emitter fingers is calculated, e.g., by Eqs. (2-129)–(2-132), using an initial current and bias conditions. The dissipated power P_{diss} in Eq. (2-129) is determined using either Eq. (7-128) for a BJT operated as a class A amplifier, or using the corresponding formula for a different amplifier class. From the calculated finger currents the new finger temperatures can be obtained. This iterative procedure is repeated until the solution converges. The transistor currents are calculated from an appropriate dc model described in Section 7.2. Note that the temperature-dependent parameters covered in Chapter 2 need to be used in this approach.

7.6 ISSUES OF Si BJTs

7.6.1 Transistor Structures

Today, for both digital and microwave applications, the most widely used BJT structure is the double-polysilicon, self-aligned transistor shown in Fig. 7.2. The characteristic features of this structure are a polysilicon emitter contact, a polysilicon base contact (to which the emitter contact is self-aligned), and a selectively implanted collector that is sometimes called pedestal collector.

Adjacent transistors on a chip of an integrated circuit have to be isolated from each other. This can be done either by deep trenches (shown in Fig. 7.2) or by reverse-biased pn junctions (shown in Fig. 7.1). Figure 7.32 shows the cross section of a high-speed BJT with trench isolation. This isolation scheme requires only little chip area because the lateral dimension of the trench is defined by lithography and can be made as small as the smallest transistor lateral dimension. This technology, however, is complex and expensive. The pn isolation, on the other hand, is simple and inexpensive but needs more chip space. This should not be a problem for microwave circuits where most of the chip area is consumed not by the active devices but by passive components like capacitors and inductors.

The advantages of using the polysilicon emitter have been addressed in Section 7.2.2. The benefit gained from the polysilicon base contact is of totally different nature. As can be seen from Figs. 7.1 and 7.2, the base of a BJT consists of the intrinsic and extrinsic parts. The intrinsic base is located directly underneath the emitter window and is the main region governing the transistor operation. The resistance of the intrinsic base, R_{Bi}, has been modeled in Section 7.2.2. The only purpose of the extrinsic base is to connect the intrinsic base to the external base terminals. To re-

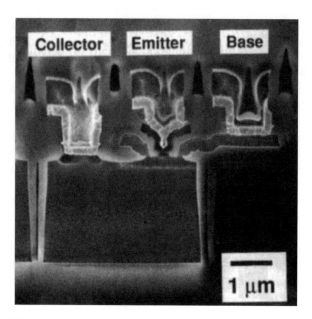

Figure 7.32. SEM cross-sectional view of a fabricated high-speed BJT. Taken from [34] © 1995 IEEE.

duce its contribution to the total base resistance, the extrinsic base is commonly heavily doped. Although the extrinsic base has no beneficial effect on the actual transistor action, it does add a considerable contribution to the collector–base capacitance C_{cb}. Therefore the area of the pn junction between the extrinsic base and the collector should be minimized. This can effectively be achieved by the polysilicon base contact, which is p^+ doped and serves as the source of the dopants for the extrinsic base. From Fig. 7.2 we see that only a portion of the polysilicon is in contact with the p^+ extrinsic base, whereas the larger portion is located on top of a thick oxide layer. It is important to prevent the doping of the extrinsic base from extending into the intrinsic base.

The self-alignment between the emitter and base is realized as follows. After the polysilicon base contact is deposited and patterned, the entire surface of the polysilicon (both the horizontal planes and the vertical edges) is covered with a thin oxide layer from an oxidation step. The oxide on the vertical edges of the polysilicon serves as a sidewall spacer. Then the polysilicon emitter is deposited into the window between the sidewall spacers on the left and the right of the polysilicon base (see Fig. 7.2). From this process, the area of the extrinsic base is further reduced, and the ratio of the total area of the collector–base pn junction (including both extrinsic and intrinsic base) to the area of emitter–base junction (intrinsic base only) can be as low as 3:1 [32], comparing to a ratio of 10:1 or more in the conventional BJT structure in Fig. 7.1 [4].

The aim of the selectively implanted collector is to further reduce the collector–base capacitance. The intrinsic collector, which is the portion of the collector directly underneath the emitter window, is designed to have a certain doping concentration to meet the targets of transistor speed and breakdown voltage. The extrinsic collector located underneath the extrinsic base, on the other hand, does not have to have the same doping because it does not contribute to the actual transistor operation. Thus it can be more highly or less doped than the intrinsic collector. A lower doping is more advantageous as it would decrease the extrinsic collector–base capacitance. This concept is realized through the selectively implanted collector. In this process, a lightly doped epitaxial collector layer is first grown. After the emitter window is opened, an ion implantation is carried out through the window to adjust the doping in the intrinsic collector to the desirable level. The choice of the appropriate doping of the intrinsic collector is not straightforward. Studies have shown that to suppress the base pushout (see Section 7.2.2), the collector current density should be smaller than one third of the current density J_1 given in Eq. (7-55) [33]. To meet this requirement and to obtain a collector current density satisfying the requirement of circuit design, the collector doping density should be high. This, however, leads to a large C_{cb} and low breakdown voltage. Such a trade-off complicates the design of the collector.

The multi-emitter finger structure is commonly used in advanced BJTs. This layout is also called the interdigitated structure. A multifinger layout of a BJT in which the collector is not located at the surface but rather at the backside of the chip is shown in Fig. 7.33, although multifinger layouts of transistors having all terminals

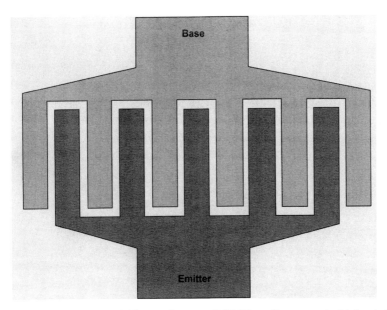

Figure 7.33. Layout of a multifinger microwave BJT. The collector terminal is located at the back side of the chip.

inculding the collector at the surface can be found (see Fig. 8.30). Both small-signal and power BJTs utilize multifinger layouts, but the number of parallel emitter fingers, and thus the total emitter area, is much larger in power BJTs than in small-signal BJTs. Moreover, power transistors frequently consist of several identically designed multifinger cells that are connected in parallel to form a large BJT. At a given frequency, the number of fingers increases with increasing targeted output power. Besides the multifinger structure, other layouts such as the overlay [3] and mesh structures [3, 35] have also been considered.

7.6.2 BJT Performance

Let us first focus on the dc performance of BJTs. Figures 7.34(a) and (b) show the Gummel plot and the common–emitter I–V output characteristics, respectively, of

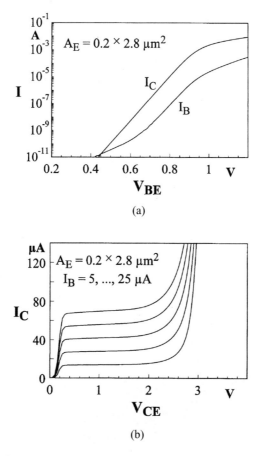

Figure 7.34. (a) Gummel plot and (b) common emitter output I–V characteristics of an advanced microwave BJT. Taken from [9] © 2001 IEEE.

an advanced double-polysilicon self-aligned microwave BJT for low-noise applications. The Gummel plot shows several characteristic features:

- The slope of the base current in the low-voltage range ($V_{BE} < 0.65$ V) is smaller than at higher base–emitter voltages because of the recombination in the emitter–base space–charge region.
- The I_C and I_B curves are almost parallel in the mid-voltage range.
- In the high-voltage range ($V_{BE} > 0.9$ V) the slopes of both I_C and I_B decrease. This effect is caused by the effects of high-level injection, base pushout, and the increasing voltage drops across the parasitic resistances. In the high-voltage range, the gap between the I_C and I_B curves also becomes narrower, thus suggesting a reduced common–emitter current gain β_0.

These characteristics are commonly seen not only in low-noise BJTs but in BJTs in general. The current gain β_0 can be easily extracted from the Gummel plot. Figure 7.35 shows the β_0 versus I_C characteristics for three different high-speed BJTs, where device 1 is a BJT designed for digital applications [34], device 2 is a power microwave BJT [35], and device 3 is the low-noise BJT considered in Fig. 7.34. In general, a high β_0 over a large I_C range, i.e., a wide plateau of the β_0 versus I_C curve, is one of the design targets for BJTs.

The output characteristics given in Fig. 7.34(b) exhibit a sharp increase in the collector current at high collector–emitter voltages ($BV_{CE0} = 2.7$ V) caused by the avalanche breakdown. Furthermore, a finite and small slope of the I_C versus V_{CE} curves between 0.5 and 2 V caused by the Early effect can be observed. The Early voltage for this transistor is around 23 V.

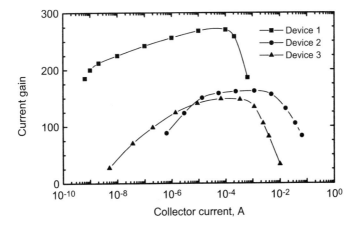

Figure 7.35. Common–emitter current gain β_0 as a function of collector current for three different BJTs. Data taken from [9, 34, 35].

Typical current-dependent cutoff frequency f_T and maximum frequency of oscillation f_{max} have been shown in Fig. 7.27. Both frequencies possess a similar current dependence and peak at about the same current level. The values of f_T and f_{max} depend on the collector–base voltage as well. This is shown in Fig. 7.36 for the case of f_{max} and for two different V_{CB}. The transistor considered here is the same as that in Fig. 7.27.

At the current level where f_T peaks (around 1.3 mA for the transistor in Fig. 7.27), the dynamic behavior of BJTs is commonly dominated by the base transit time τ_B and the collector–base transit time τ_{BC}. It has been discussed in Section 7.3.2 that the base transit time depends on the base thickness and the collector–base transit time is a function of the doping and the thickness of the collector. Depending on the collector design, f_T can be either smaller or larger than f_{max}.

Figure 7.37 shows reported cutoff frequencies of experimental microwave BJTs as a function of the base thickness. In spite of the scattering of the data, the general trend is clear and as expected: a thinner base leads to a higher cutoff frequency. For BJTs with a very thin base, however, f_T is dominated less by τ_B and more by τ_{BC}. Reported maximum frequencies of oscillation of experimental microwave BJTs as a function of base thickness are shown in Fig. 7.38. No clear dependence of f_{max} on w_B can be observed. This is not surprising, as f_{max} depends predominately on C_{cb} and thus on the design of the collector. In Table 7.5 the state of the art of BJTs in terms of f_T and f_{max} is summarized. It can be seen that a bipolar transistor can be optimized for either f_T or f_{max}, but not both due to the design trade-off concerning the collector doping. A high collector doping is beneficial to f_T because the delay time τ_{BC} becomes small. At the same time, however, a high collector doping will lead to a large collector–base capacitance C_{cb}, which degrades f_{max}.

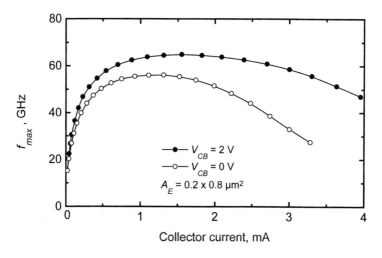

Figure 7.36. Maximum frequency of oscillation of a BJT as a function of collector current for two different collector–base voltages. Data taken from [9].

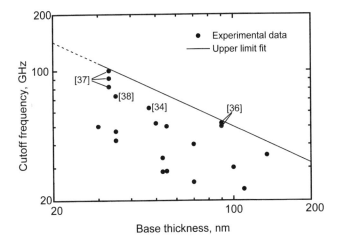

Figure 7.37. Reported cutoff frequencies of microwave BJTs as a function of base thickness.

It has been addressed earlier that the collector design has a strong influence on the breakdown behavior of BJTs. A thick and lightly doped collector results in larger breakdown voltages BV_{CE0} and BV_{CB0}. At the same time, such a collector leads to a large τ_{BC} and thus a relatively low f_T. This trade-off is illustrated in Fig. 7.39, which shows the dependence of f_T on the breakdown voltage of experimental BJTs.

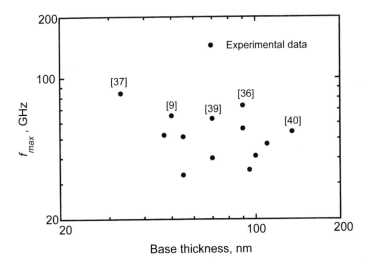

Figure 7.38. Reported maximum frequencies of oscillation of microwave BJTs as a function of base thickness.

Table 7.5. State of the art of microwave BJTs in terms of f_T and f_{max}

f_T, GHz	f_{max}, GHz	Reference
52	65	[9]
52	56	[36]
50	73	[36]
63	52	[34]
82	84	[37]
100	—	[8]
72	101	[8]
82	92	[8]

Clearly, BJTs with a high cutoff frequency suffer from a low breakdown voltage, whereas BJTs with a high breakdown voltage possess a low f_T.

7.6.3 Low-Noise BJTs

The noise performance of BJTs has been improved continuously over the past years. Although the noise of BJTs in the microwave range is generally higher compared to that of III–V FETs, the minimum noise figure NF_{min} of BJTs in the frequency range up to about 5 GHz is sufficiently low for many applications. The reported minimum noise figures of laboratory BJTs as a function of frequency are compiled in Fig. 7.40. Minimum noise figures below 1.5 dB at frequencies up to 6

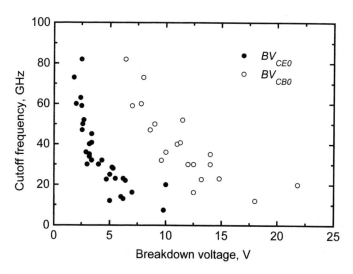

Figure 7.39. Reported cutoff frequency of BJTs as a function of collector–emitter breakdown voltage BV_{CE0} and collector–base breakdown voltage BV_{CB0}.

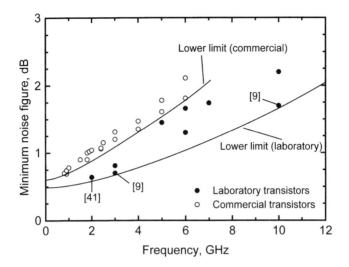

Figure 7.40. Minimum noise figures of laboratory and commercial microwave BJTs as a function of frequency.

GHz and below 2 dB at 10 GHz have been demonstrated. Also shown in the figure are the minimum noise figures of commercial BJTs. Advanced commercial BJTs with $NF_{min} < 1$ dB at 1.8 GHz and with $NF_{min} < 2$ dB at 6 GHz are available. Figure 7.41 shows the minimum noise figures of a state of the art low-noise BJT as a function of the collector current. The typical U-shape with a pronounced minimum NF_{min} is clearly indicated.

In the future, we do not expect significant improvements in the frequency and noise limits of BJTs, as most recent research efforts on bipolar devices have been spent on SiGe HBTs. Evidence for this trend is the fact that the number of journal papers dealing with theoretical and experimental work on microwave BJTs in the late 1990s and early 2000s was much smaller than those on SiGe HBTs. Nevertheless, the technology of microwave BJTs is very mature, and there are still many commercial applications for such devices, particularly in microwave systems with operating frequencies up to 5 GHz.

7.6.4 Power BJTs

Microwave power BJTs face a similar situation to that of low-noise BJTs. The research activities on power BJTs for microwave applications have faded, but commercial applications for such devices are still promising.

In principle, it is possible to design BJTs with BV_{CE0} values of several hundreds of volts, and such transistors are commercially available. The cutoff frequency of these BJTs, however, is much lower than 1 GHz. Thus, BJTs with extremely high breakdown voltages cannot be used in microwave amplifiers. Typical power BJTs

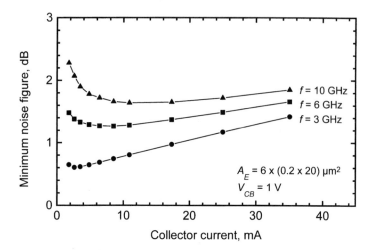

Figure 7.41. Minimum noise figures obtained from a state of the art low-noise BJT. Data taken from [9].

for operation between 1 and 3 GHz have breakdown voltages BV_{CE0} and BV_{CB0} of about 20–70 V. Commercial transistors with output powers (under pulsed operation) up to 1 KW at 1 GHz and more than 100 W around 3 GHz are available. Figure 7.42 shows the output power data of microwave power BJTs commercially available in the year 2001.

Microwave power BJTs normally utilize a multifinger, multicell structure with

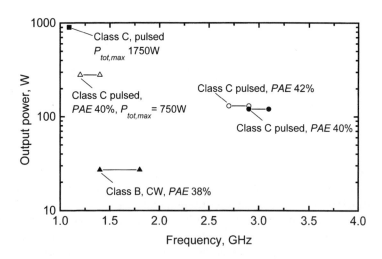

Figure 7.42. Output power of commercial microwave power BJTs as a function of frequency. Data taken from [42].

long and narrow emitter stripes. As has been shown in a recent investigation [35], a mesh structure instead of a multifinger structure can also be used to enhance the performance of microwave power BJTs. The mesh structure consists of a large number of emitter spots. Whereas in a multi-emitter finger BJT the L_E / W_E ratio can be quite large, it is much smaller in BJTs employing the emitter spot design, e.g., 2 μm/0.8 μm in [35]. The emitter spots act electrically like emitter fingers, but the lateral dimensions are different. From the device physics point of view, an emitter spot acts like an emitter finger, but with a reduced self-heating effect due to the smaller dimension than its finger counterpart.

No significant improvement over the power performance of BJTs is expected in the future due to the matured technology and the shift of interest from BJTs to heterostructure devices.

REFERENCES

1. W. B. Shockley, The Path to the Conception of the Junction Transistor, *IEEE Trans. Electron Devices, 23,* pp. 597–620, 1976. Reprinted in *IEEE Trans. Electron Devices,* 31, pp. 1523–1546, 1984.

2. H. T. Yuan, W.V. McLevige, and H. D. Shih, GaAs Bipolar Digital Integrated Circuits, in N. G. Einspruch and W. R. Wisseman (eds.), *VLSI Electronics—Microstructure Science,* vol. 11, pp. 173–213, Academic Press, Orlando, 1985.

3. H. F. Cooke, Microwave Transistors: Theory and Design, *Proc. IEEE, 59,* pp. 1163–1181, 1971.

4. T. H. Ning, History and Future Perspective of the Modern Silicon Bipolar Transistor, *IEEE Trans. Electron Devices, 48,* pp. 2485–2491, 2001.

5. T. Nakamura and H. Nishizawa, Recent Progress in Bipolar Transistor Technology, *IEEE Trans. Electron Devices, 42,* pp. 390–398, 1995.

6. J. D. Warnock, Silicon Bipolar Device Structures for Digital Applications: Technology Trends and Future Directions, *IEEE Trans. Electron Devices, 42,* pp. 377–389, 1995.

7. J. D. Cressler, SiGe HBT Technology: A New Contender for Si-Based RF and Microwave Circuit Applications, *IEEE Trans. Microwave Theory Tech., 46,* pp. 572–589, 1998.

8. Y. Kiyota, E. Ohue, T. Onai, K. Washio, M. Tanabe, and T. Inada, Lamp-Heated Rapid Vapor-Phase Doping Technology for 100-GHz Si Bipolar Transistors, *Proc. BCTM,* pp. 173–176, 1996.

9. J. Böck, H. Knapp, K. Aufinger, T. F. Meister, M. Wurzer, S. Boguth, and L. Treitinger, High-Performance Implanted Base Silicon Bipolar Technology for RF Applications, *IEEE Trans. Electron Devices, 48,* pp. 2514–2519, 2001.

10. A. S. Grove, *Physics and Technology of Semiconductor Devices,* Wiley, New York, 1967.

11. S. M. Sze, *Physics of Semiconductor Devices,* Wiley, New York, 1981.

12. D. A. Neamen, *Semiconductor Physics and Devices—Basic Principles,* Irwin, Homewood, 1992.

13. D. Nuernbergk, *Simulation des Transportverhaltens in Si/Si$_{1-x}$Ge$_x$/Si-Heterobipolartransistoren,* H. Utz Verlag, Munich, 1999.

14. S. E. Swirhun, Y.-H. Kwark, and R. M. Swanson, Measurement of Electron Lifetime, Electron Mobility, and Bandgap Narrowing in Heavily Doped p-Type Silicon, *IEDM Tech. Dig.*, pp. 24–27, 1986.
15. J. del Alamo, S. Swirhun, and R. M. Swanson, Simultaneous Measurement of Hole Lifetime, Hole Mobility, and Bandgap Narrowing in Heavily Doped n-Type Silicon, *IEDM Tech. Dig.*, pp. 290–293, 1985.
16. J. M. Early, Effects of Space-Charge Layer Widening in Junction Transistors, *Proc. IRE, 40*, pp. 1401–1406, 1952.
17. W. Liu, *Fundamentals of III–V Devices—HBTs, MESFETs, and HFETs/HEMTs*, Wiley, New York, 1999.
18. J. W. Slotboom and H. D. de Graaff, Measurement of Bandgap Narrowing in Si Bipolar Transistors, *Solid-State Electron., 19*, pp. 857–862, 1976.
19. C. T. Kirk, Jr., A Theory of Transistor Cutoff Frequency (f_T) Falloff at High Current Densities, *IRE Trans. Electron Devices 19*, pp. 164–174, 1962.
20. D. L. Bowler and F. A. Lindholm, High Current Regimes in Transistor Collector Regions, *IEEE Trans. Electron Devices, 20*, pp. 257–263, 1973.
21. H. C. Poon, H. K. Gummel, and D. L. Scharfetter, High Injection in Epitaxial Transistors, *IEEE Trans. Electron Devices, 16*, pp. 455–457, 1969.
22. J. J. Liou, *Advanced Semiconductor Device Physics and Modeling*, Artech House, Boston, 1994.
23. T. H. Ning and R. D. Issac, Effect of Emitter Contact on Current Gain of Silicon Bipolar Devices, *IEEE Trans. Electron Devices, 27*, pp. 2051–2055, 1980.
24. A. K. Kapoor and D. J. Roulston (eds.), *Polysilicon Emitter Bipolar Transistors*, IEEE Press, New York, 1989.
25. I. R. C. Post, P. Ashburn, and G. Wolstenholme, Polysilicon Emitters for Bipolar Transistors: A Review and Re-Evaluation of Theory and Experiment, *IEEE Trans. Electron Devices, 39*, pp. 1717–1731, 1992.
26. N. F. Rinaldi, On the Modeling of Polysilicon Emitter Bipolar Transistors, *IEEE Trans. Electron Devices, 44*, pp. 395–403, 1997.
27. R. G. Meyer and R. S. Muller, Charge Control Analysis of the Collector-Base Space-Charge-Region Contribution to Bipolar-Transistor Time Constant τ_T, *IEEE Trans. Electron Devices, 34*, pp. 450–452, 1987.
28. A. van der Ziel, *Noise Sources, Characterization, and Measurement*, Prentice-Hall, Englewood Cliffs, 1970.
29. A. van der Ziel, *Noise in Solid State Devices and Circuits*, Wiley, New York, 1986.
30. R. J. Hawkins, Limitations of Nielsen's and Related Noise Equations Applied to Microwave Bipolar Transistors, and a New Expression for the Frequency and Current Dependent Noise Figure, *Solid-State Electron., 20*, pp. 191–196, 1977.
31. W. Liu, *Handbook of III–V Heterojunction Bipolar Transistor*, Wiley, New York, 1998.
32. T. H. Ning, R. D. Isaac, P. M. Solomon, D. D. Tang, H. N. Yu, G. C. Feth, and S. K. Wiedmann, Self-Aligned Bipolar Transistors for High-Performance and Low-Power-Delay VLSI, *IEEE Trans. Electron Devices, 28*, pp. 1010–1013, 1981.
33. Y. Taur and T. H. Ning, *Fundamentals of Modern VLSI Devices*, Cambridge University Press, Cambridge, 1998.
34. T. Uchino, T. Shiba, T. Kikuchi, Y. Tamaki, A. Watanabe, and Y. Kiyota, Very-High-

Speed Silicon Bipolar Transistors with In-Situ Doped Polysilicon Emitter and Rapid Vapor-Phase Doping Base, *IEEE Trans. Electron Devices, 42,* pp. 406–412, 1995.
35. F. Carrara, A. Scuderi, G. Tontodonato, and G. Palmisano, A Very High Efficiency Silicon Bipolar Transistor, *IEDM Tech. Dig.,* pp. 883–886, 2001.
36. K. Washio, E. Ohue, M. Tanabe, and T. Onai, Self-Aligned Metal/IDP Si Bipolar Technology with 12-ps ECL and 45-GHz Dynamic Frequency Divider, *IEEE Trans. Electron Devices, 44,* pp. 2078–2082, 1997.
37. E. Ohue, Y. Kiyota, T. Onai, M. Tanabe, and K. Washio, 100-GHz f_T Si Homojunction Bipolar Technology, Symp. *VLSI Technol. Dig.,* pp. 106–107, 1996.
38. E. F. Crabbe, B. S. Meyerson, J. M. C. Storck, and D. L. Harame, Vertical Profile Optimization of Very High Frequency Epitaxial Si- and SiGe-Base Bipolar Transistors, *IEDM Tech. Dig.,* pp. 83–86, 1993.
39. C. Yoshino, K. Inou, S. Matsuda, H. Nakajima, Y. Tsuboi, H. Naruse, H. Sugaya, Y. Katsumata, and H. Iwai, A 62.8 GHz f_{max} LP-CVD Epitaxially Grown Silicon Base Bipolar Transistor with Extremely High Early Voltage of 85.7 V, *Symp. VLSI Technol. Dig.,* pp. 131–132, 1995.
40. S. Niel, O. Rozeau, L. Allioud, C. Hernandez, P. Llinares, M. Guillermet, J. Kirtsch, A. Monroy, J. de Pontcharra, G. Auvert, B. Blanchard, M. Mouis, G. Vincent, and A. Chantre, A 54 GHz f_{max} Implanted Base 0.35 µm Single-Polysilicon Bipolar Technology, *IEDM Tech. Dig.,* pp. 807–810, 1997.
41. K. Aufinger, J. Böck, T. F. Meister, and J. Popp, Noise Characteristics of Transistors Fabricated in an Advanced Silicon Bipolar Technology, *IEEE Trans. Electron Devices, 43,* pp. 1533–1538, 1996.
42. See website http://www.semiconductor.philips.com/catalog.

CHAPTER 8

HETEROJUNCTION BIPOLAR TRANSISTORS

8.1 INTRODUCTION

The *H*eterojunction *B*ipolar *T*ransistor (HBT) is a bipolar transistor with a basic structure similar to that of a Si BJT discussed in Section 7.1. Like the BJT, it has three terminals, namely emitter, base, and collector, and consists of either an npn or pnp layer sequence. The main difference between the two devices is that in an HBT, the emitter and base are made of different materials, with the bandgap in the emitter being larger than that in the base. Thus, the emitter–base junction of an HBT is a heterojunction.

The basic idea to exploit the properties of a heterojunction in bipolar semiconductor devices was first proposed in a patent application in 1948 by W. Shockley, which stated: "What is claimed is: ... (2) A device as set forth in claim 1 in which one of the separated zones is of a semiconductor material having a wider energy gap than that of the material in the other zones" [1]. In 1957, H. Kroemer published a pioneering paper describing the basic theory of the current gain of HBTs [2]. It took, however, many years to put the ideas of Shockley and Kroemer successfully into operational devices. The main reason for this delay was of a technological nature. Although the advantages of heterojunctions were well recognized, it was not possible at the that time to fabricate such junctions with a sufficiently high interface quality for proper device operation. The first ray of hope was AlGaAs/GaAs HBTs grown by liquid phase epitaxy (LPE) reported in the 1970s and early 1980s, e.g., [3, 4]. The major breakthrough came from the introduction of two more advanced epitaxial growth techniques: molecular beam epitaxy (MBE) and metal–organic chemical vapor deposition (MOCVD). Thanks to these two techniques, since the 1980s, AlGaAs/GaAs HBTs operating at microwave frequencies have been widely investigated and their performance continuously improved. Besides GaAs-based HBTs, InP HBTs and SiGe HBTs were also investigated during the 1980s and 1990s. InP HBTs are attractive because these transistors offer higher frequency limits than GaAs HBTs and are better suited for low-power operation. SiGe HBTs, on the other hand, paved the way for multi-GHz applications based on the well-established and cost-effective Si technology. In the late 1990s, HBTs made of the wide-bandgap material GaN started to gain the interest of several research groups. The experimental results on GaN HBTs reported in 2001, however, concerned only the dc performance, and no experimental verification of the microwave properties of GaN HBTs was available.

GaAs HBTs with AlGaAs or InGaP emitters have clearly emerged as viable devices for microwave power amplifiers since the 1990s and are commercially available. Experimental GaAs HBTs having cutoff frequencies f_T in excess 150 GHz [5] and a record maximum frequency of oscillation f_{max} of 350 GHz [6] have been reported. For InP HBTs, the reported record f_T and f_{max} values are 341 GHz and 300 GHz, respectively, [7, 8]. A special version of the InP HBT, called the transferred substrate HBT, with extremely reduced parasitics showed a unilateral power gain U of 21 dB at 100 GHz [9], and the extrapolation of U with the –20 dB/dec slope results in a f_{max} above 1 THz for such a transistor. This is the highest f_{max} ever obtained from any three-terminal semiconductor device. SiGe HBTs became commercially available in the late 1990s and were mainly used in wireless communications applications. Laboratory SiGe transistors with 210 GHz f_T [10] and 285 GHz f_{max} [11] have been realized. In the frequency range up to 12 GHz, SiGe HBTs show excellent noise figures of less than 1 dB [12, 13]. This is the best noise performance for all bipolar transistors.

As mentioned above, the basic structure of HBTs is similar to that of BJTs. While the BJT consists of only one semiconductor material (Si), the HBT is formed of layers of different semiconductor materials. HBTs can be either single heterojunction bipolar transistor (SHBTs) or double heterojunction bipolar transistors (DHBTs). The base and collector layers of the SHBT are made of the same semiconductor material, thereby leading to an emitter–base heterojunction and a collector–base homojunction. For DHBTs, on the other hand, base and collector are made of different materials, and both the emitter–base junction and the collector–base junction are heterojunctions. It is important to mention that only the junctions adjacent to the base, i.e., the emitter–base and collector–base metallurgical junctions, are of primary importance to HBT operation. Table 8.1 shows the different layer sequences commonly used in SHBTs and DHBTs.

HBTs can be further classified into the spike HBT, which has a spike conduction band at the emitter–base heterointerface, and the smooth HBT, which has a smooth conduction band. Most HBTs are formed with materials having the same lattice constant, and the heterojunctions are lattice-matched. In SiGe HBTs, however, the

Table 8.1. Layer sequences of some typical SHBTs and DHBTs

HBT type	Emitter	Base	Collector	Subcollector	Substrate	Comment
AlGaAs/GaAs	AlGaAs	GaAs	GaAs	GaAs	GaAs	SHBT
InGaP/GaAs	InGaP	GaAs	GaAs	GaAs	GaAs	SHBT
InAlAs/InGaAs	InAlAs	InGaAs	InGaAs	InGaAs	InP	SHBT
InAlAs/InGaAs	InAlAs	InGaAs	InP	InGaAs	InP	DHBT
InP/InGaAs	InP	InGaAs	InGaAs	InGaAs	InP	SHBT
InP/InGaAs	InP	InGaAs	InP	InGaAs	InP	DHBT
SiGe	Si	SiGe	Si	Si	Si	DHBT

SiGe base is a strained layer because the lattice constant of SiGe is larger than that of Si. Finally, an HBT can either be an npn or pnp transistor. Because almost all HBTs are of the npn type, the remaining discussions in this chapter will be focussed on npn HBTs.

Although there is a variety of different material combinations and HBT types, the basic operation principle of all of these HBTs is essentially the same. When the emitter–base heterojunction of an HBT is forward-biased, electrons are injected from the emitter into the base. The amount of the forward bias voltage V_{BE} applied to the emitter–base heterojunction governs the magnitude of electron injection and thus the collector current I_C. The major difference in the operations of BJTs and HBTs is the different free-carrier transport mechanisms in the emitter–base homojunction and heterojunction.

Figures 8.1(a)–(c) show the energy band diagrams of the emitter–base junctions of a BJT and two HBTs. On the left-hand side of the figures are the band diagrams of the emitter and base regions being separated from each other. The right-hand side figures show the band diagrams of the junctions when the emitter and base are in contact with each other and a forward bias is applied. The band diagram of the BJT shown in Fig. 8.1(a) reveals that, because of the same bandgap of the emitter and the base, the potential barriers seen by electrons and holes at the junction are the same. Therefore, an injection of electrons from the emitter to base will be accompanied by a considerable injection of holes from the base to emitter. As discussed in Section 7.2, a sufficiently large common–emitter current gain β_0 can be obtained from this device only by having an emitter doping density much higher than the base doping density.

In Figs. 8.1(b) and (c), the bandgaps for the emitter and base are E_{GE} and E_{GB}, respectively, and the bandgap difference ΔE_G is $E_{GE} - E_{GB}$. Such a bandgap difference gives rise to the conduction and valence band offsets ΔE_C and ΔE_V. The main difference of the two heterojunctions in Figs. 8.1(b) and 8.1(c) lies in the different positions of the conduction band edges and thus the different electron affinities. The two materials in Fig. 8.1(b) have the same electron affinity, and the junction is a smooth heterojunction with the conduction band offset equal to zero and the valence band offset equals to the bandgap difference. In other words, the potential barrier for electrons is smaller than that for holes by ΔE_G. Thus, the undesired hole injection from the base to the emitter is effectively suppressed and an acceptable current gain can be obtained, even if the base doping is equal to or higher than the emitter doping. This offers the device engineer a new degree of freedom in transistor design. Smooth emitter–base heterojunctions are typical for SiGe HBTs (for arbitrary Ge contents in the SiGe base) and $In_{0.49}Ga_{0.51}P$/GaAs HBTs. $In_{0.49}Ga_{0.51}P$ is lattice-matched to GaAs, and such a heterojunction pair is commonly used in InGaP/GaAs HBTs.

Fig. 8.1(c) shows a spike heterojunction with a conduction band spike at the heterointerface. This is typical for abrupt AlGaAs/GaAs HBTs, abrupt InP/InGaAs HBTs, and abrupt InAlAs/InGaAs HBTs. Here, the word "abrupt" has nothing to do with the doping profile, but rather it means that the material composition changes abruptly at the emitter–base heterointerface, e.g., changes from $Al_{0.3}Ga_{0.7}As$ to GaAs at the heterointerface. In the case of a spike heterojunction, the emitter and

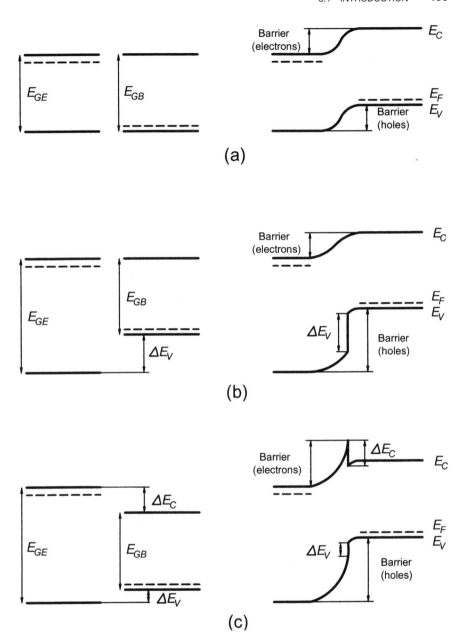

Figure 8.1. Energy band diagrams for (a) a homojunction, (b) a smooth heterojunction, and (c) a spike heterojunction.

the base materials have different electron affinities, and the bandgap difference is shared between the conduction and valence band offsets. Again, the potential barrier for electrons is smaller than for holes, but the difference between the two potential barriers is smaller than ΔE_G ($\Delta E_G - \Delta E_C = \Delta E_V$). Nevertheless, the hole injection from the base into the emitter is still strongly suppressed.

The conduction band spike in AlGaAs/GaAs and InP-based HBTs can be lowered or even totally removed by using a graded layer and/or setback layer. These features will be discussed in more detail in Section 8.2.4.

The emitter injection efficiency G_e has the general form of

$$G_e = \frac{I_n}{I_p} \tag{8-1}$$

where I_n and I_p are the electron and hole diffusion currents crossing the forward-biased emitter–base junction, respectively, and [14, 15]

$$G_e \propto \frac{N_{DE}}{N_{AB}} \exp \frac{\Delta E_G}{k_B T} \quad \text{for a smooth heterojunction} \tag{8-2}$$

and

$$G_e \propto \frac{N_{DE}}{N_{AB}} \exp \frac{\Delta E_V}{k_B T} \quad \text{for a spike heterojunction} \tag{8-3}$$

For a BJT, the proportionality becomes

$$G_e \propto \frac{N_{DE}}{N_{AB}} \tag{8-4}$$

Here, N_{DE} is the emitter doping density, N_{AB} is the base doping density, k_B is the Boltzmann constant, and T is the temperature. It is clear that, due to the valence band offset, a larger emitter injection efficiency can be obtained in HBTs than in BJTs with comparable emitter and base doping concentrations. A large emitter injection efficiency automatically leads to a large common–emitter current gain β_0, and HBTs with current gains of several thousands can easily be realized.

However, a very large current gain is typically not the primary design goal for microwave transistors. Instead, a more important issue is that an acceptable current gain is achievable with $N_{DE} \ll N_{AB}$, and the base dopant concentration can be made very high. A high base doping density is desirable for microwave transistor operation. First, it allows for a very thin base without running the risk of having base punch-through. A narrow base, in turn, leads to a small base transit time and thus to a high cutoff frequency. Second, the high base doping density leads to a small intrinsic base resistance R_{Bi}, which gives rise to a high maximum frequency of oscillation and an improved high-frequency noise performance. A low emitter doping density is also beneficial for high-speed operation. It leads to a smaller emitter–base space–charge region capacitance and therefore a smaller emitter–base delay time τ_{EB}. To take the above-mentioned advantages, HBTs frequently employ a base doping density exceeding the emitter doping density.

The material composition in the base may be either homogeneous or graded. A graded material composition results in a graded bandgap in the base. If the bandgap is larger at the emitter–base junction and decreases toward the collector–base junction, a built-in field arises in the base which enhances the electron transport across the base. Consequently, the base transit time is reduced and the cutoff frequency is further increased.

In III–V HBTs, $N_{AB} \gg N_{DE}$ and both the doping densities are typically uniform. For SiGe HBTs, both $N_{AB} \gg N_{DE}$ and $N_{AB} \ll N_{DE}$ (i.e., similar to the doping profile of Si BJTs) are used. Further, the SiGe base is frequently graded with a small Ge content near the emitter–base junction and an increasing Ge content toward the base–collector junction. Therefore, the bandgap in the base decreases toward the collector junction and a built-in field exists.

8.2 DC ANALYSIS

The development of models to describe HBT dc behavior will be presented in this section. From the previous section, we know that there are a large number of different HBTs. It will be difficult to develop a universal dc model for all these devices. Our approach will be instead to analyze the dc characteristics in a manner that provides insights into the HBT operation and to concentrate on important device physics leading to analytical solutions.

As mentioned in Chapter 7, the minority carrier mobilities, diffusion coefficients, and carrier life times are critical parameters for properly describing the dc currents in a bipolar transistor. Furthermore, bandgap narrowing taking place in heavily doped semiconductors can influence the carrier transport and carrier densities. Therefore, the minority carrier and bandgap narrowing parameters will first be reviewed. Then HBTs having smooth and spike emitter–base heterojunctions will be considered. Finally, some physical effects that are important for HBT operation are discussed. Some of these effects can hardly be modeled analytically, but they will at least be described phenomenologically and illustrated by the results of numerical device simulations whenever available.

8.2.1 Minority Carrier and Bandgap Narrowing Parameters

Transport of minority free carriers—electrons—in the base is a main mechanism governing the dc behavior of HBTs. Unfortunately, very little is known about the minority carrier transport for many semiconductors. When the values of minority carrier parameters are absent, a conventional and reasonable approach is to approximate them with the majority carrier parameters, which are more readily available in the literature.

For GaAs, the low-field minority electron mobility can be modeled by [14, 16]

$$\mu_n = \left(1 + \frac{N_A}{3.98 \times 10^{15} + N_A/641}\right)^{1/3} \times \frac{cm^2}{Vs} \quad (8\text{-}5)$$

where N_A is the acceptor concentration in cm^{-3}. In a strained SiGe layer, the carrier mobilities are anisotropic, and the electron transport perpendicular to the heterointerface, i.e., parallel to the growth direction of the SiGe base, is to be considered. Figure 8.2 shows the low-field electron mobility in this direction as a function of the base doping concentration and for three different Ge contents. Similarly, Fig. 8.3 shows the low-field minority electron mobility as a function of the Ge content for four different base doping levels, again in the direction parallel to the growth direction [17–19]. Data for the majority and minority hole mobility in strained SiGe can be found in [17–19] as well. The minority electron diffusion coefficient can be calculated from the low-field electron mobility and the Einstein relation [see Eq. (2-70)].

The minority electron lifetime in GaAs is given by [16]

$$\tau_n = \left(\frac{N_A}{10^{10}\ cm^{-3}} + \frac{N_A^2}{1.6 \times 10^{29}\ cm^{-3}}\right)^{-1} \times s \quad (8\text{-}6)$$

Since no reliable τ_n data for SiGe is available, the use of Si minority electron lifetime given in Eq. (7-21) is recommended. The minority electron lifetime in InGaAs, which is important for the modeling of InP HBTs, has been investigated in [20]. Empirical fits to the data presented in [20] lead to the lifetime characteristics given

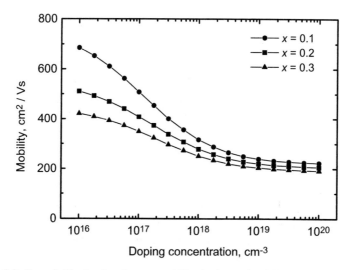

Figure 8.2. Low-field minority electron mobility in the strained Si$_{1-x}$Ge$_x$ layer as a function of the doping concentration.

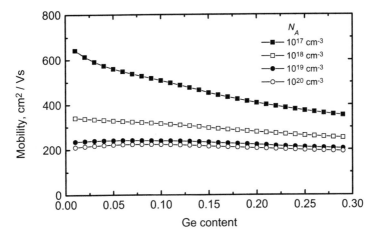

Figure 8.3. Low-field minority electron mobility in the strained $Si_{1-x}Ge_x$ layer as a function of Ge content x.

in Fig. 8.4, where both the radiative and nonradiative recombination processes were assumed active. The resulting lifetime τ_n is given by

$$\frac{1}{\tau_n} = \frac{1}{\tau_{rad}} + \frac{1}{\tau_{nonrad}} \tag{8-7}$$

where τ_{rad} and τ_{nonrad} are the radiative and nonradiative lifetimes, respectively.

Since most microwave HBTs have a heavily doped p-type base, bandgap narrowing ΔE_{BGN} in p-type GaAs, InGaAs, and SiGe should be considered. It can be expressed as [21, 22]

$$\Delta E_{BGN}[meV] = a\left(\frac{N_A}{10^{18}}\right)^{1/2} + b\left(\frac{N_A}{10^{18}}\right)^{1/3} + c\left(\frac{N_{AB}}{10^{18}}\right)^{1/4} \tag{8-8}$$

where a, b, and c are fitting parameters given in Table 8.2.

Table 8.2. Values of fitting parameters for calculating bandgap narrowing in p-type GaAs, $In_{0.53}Ga_{0.47}As$, and $Si_{1-x}Ge_x$ (x is the Ge content)

	p-GaAs [21]	p-$In_{0.53}Ga_{0.47}As$ [22]	$Si_{1-x}Ge_x$ [21]*
a	3.88	3.65	$5.07 \times (1 + 0.18x)$
b	9.71	9.2	$11.1 \times (1 - 0.35x)$
c	12.19	11.3	$15.2 \times (1 - 0.54x)$

*The fitting parameters for $Si_{1-x}Ge_x$ are valid for $x \leq 0.3$.

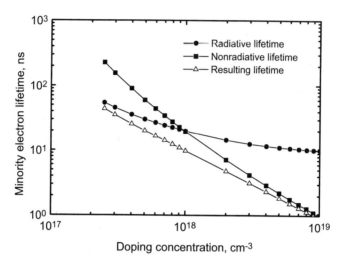

Figure 8.4. Minority electron lifetime versus the doping density in the InGaAs material.

8.2.2 HBTs with Smooth Emitter–Base Heterojunction

Models for the dc currents in smooth HBTs will be developed in this section, and the formulas will be derived based on the approximations of low-level injection, neglegible effect of electron saturation velocity in the collector–base junction, and no parasitic resistances. An extension of the model taking into account arbitrary injection levels, velocity saturation, and parasitic resistances can be obtained in a similar manner as described for the BJT in Section 7.2.2. The effects of avalanche breakdown and current crowding on HBT performance can be treated using the approaches given in Section 7.2.2 as well.

The energy band diagram of a smooth emitter–base heterojunction has been shown in Fig. 8.1(b). The characteristic feature is that the conduction band has no spike at the heterointerface because the conduction band offset ΔE_C is zero (or nearly zero). The one-dimensional structure of such an HBT is shown in Fig. 8.5. The same coordinate system used for the BJT has been adopted here (see Fig. 7.4) so that the expressions derived for the HBT are compatible with those for the BJT in Chapter 7.

A. Homogeneous Doping, Constant Bandgap in the Base. The collector current of an HBT having a smooth heterojunction, homogeneously doped emitter and base regions, and a constant bandgap in the base (i.e., a homogeneous material composition in the base) can be modeled in close analogy to the collector current of a Si BJT based on the drift–diffusion theory described in Section 7.2. We simply repeat the steps leading to Eq. (7-9) and obtain the following expression for the electron diffusion current density at x_{BE}, $J_{nB}(x_{BE})$:

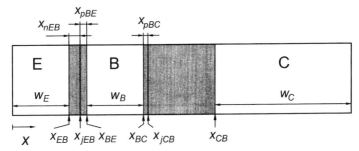

Figure 8.5. One-dimensional structure of an HBT showing the spatial coordinates important for the dc modeling.

$$J_{nB}(x_{BE}) = qD_{nB}\frac{dn}{dx} = qD_{nB}\frac{n(x_{BC}) - n(x_{BE})}{w_B} = -\frac{qD_{nB}}{w_B}\frac{n^2_{iB}}{N_{AB}}\exp\frac{V_{BE}}{V_T} \quad (8\text{-}9)$$

where D_{nB} is the electron diffusion coefficient in the base, $n(x_{BC})$ and $n(x_{BE})$ are the electron concentrations at the edges of the quasineutral base, w_B is the thickness of the quasineutral base, n_{iB} is the intrinsic free-carrier concentration in the base, V_{BE} is the base–emitter applied voltage, and V_T is the thermal voltage. Because the base and emitter in an HBT are made of different semiconductors having different bandgaps, the intrinsic carrier concentration in the base is different from that in the emitter, n_{iE}. Their values can be calculated from Eq. (2-24) using the appropriate E_G, N_C, and N_V for the emitter and base materials. The electron current density at x_{BC}, $J_{nB}(x_{BC})$, has been given in Eq. (7-11). Since the base layer in microwave HBTs is commonly very thin, the base transport factor α_B given in Eq. (7-10) can be assumed to be unity. This yields $J_{nB}(x_{BE}) = J_{nB}(x_{BC}) = J_C$, where J_C is the collector current density. Finally, the collector current I_C is obtained from Eq. (7-15).

Next we consider the base current I_B of the HBT. It essentially consists of the four components I_{Bp}, $I_{B,surf}$, $I_{B,bulk}$, and $I_{B,scr}$, which will be discussed in the following. The current I_{Bp} is constituted by the flow of holes injected from the base to emitter. Such an injection results in the hole current density $J_{pE}(x_{EB})$ at $x = x_{EB}$, which is obtained combining Eqs. (7-12) and (7-13) as

$$J_{pE}(x_{EB}) = -qD_{pE}\frac{dp}{dx}\bigg|_{x_{EB}} = -\frac{qD_{pE}}{w_E}\frac{n^2_{iE}}{N_{DE}}\left(\exp\frac{V_{BE}}{V_T} - 1\right) \quad (8\text{-}10)$$

where D_{pE} is the hole diffusion coefficient in the emitter and w_E is the thickness of the quasineutral emitter. The thicknesses w_B and w_E are calculated by Eqs. (7-24) and (7-25). The extensions of the emitter–base space–charge region into the emitter, x_{nEB}, and into the base, x_{pBE}, and the extension of the collector–base space–charge region into the base, x_{pBC}, are determined from Eqs. (3-17) and (3-18) as

$$x_{nEB} = \left[\frac{2\varepsilon_E\varepsilon_B N_{AB}(V_{bi,EB} - V_{BE})}{qN_{DE}(\varepsilon_E N_{DE} + \varepsilon_B N_{AB})}\right]^{1/2} \quad (8\text{-}11)$$

$$x_{\text{pBE}} = \left[\frac{2\varepsilon_E\varepsilon_B N_{\text{DE}}(V_{\text{bi,EB}} - V_{\text{BE}})}{qN_{\text{AB}}(\varepsilon_E N_{\text{DE}} + \varepsilon_B N_{\text{AB}})} \right]^{1/2} \tag{8-12}$$

and

$$x_{\text{pBC}} = \left[\frac{2\varepsilon_C\varepsilon_B N_{\text{DC}}(V_{\text{bi,CB}} + V_{\text{CB}})}{qN_{\text{AB}}(\varepsilon_C N_{\text{DC}} + \varepsilon_B N_{\text{AB}})} \right]^{1/2} \tag{8-13}$$

where ε_E, ε_B, and ε_C are the dielectric constants of the emitter, base, and collector, respectively, N_{DC} is the collector doping density, and $V_{\text{bi,EB}}$ and $V_{\text{bi,CB}}$ are the built-in voltages of the emitter–base and collector–base junctions, respectively, given by Eqs. (3-11) and (3-12).

For the case of $V_{\text{BE}} > 3 \times V_T$, the exponential term in Eq. (8-10) becomes much larger than unity, and Eq. (8-10) is reduced to

$$J_{\text{pE}}(x_{\text{EB}}) = -\frac{qD_{\text{pE}}}{w_E} \frac{n_{iE}^2}{N_{\text{DE}}} \exp\frac{V_{\text{BE}}}{V_T} \tag{8-14}$$

Multiplying Eq. (8-14) by the (negative) emitter–base junction area A_E yields the base current component I_{Bp}.

At this point, we go back to the emitter injection efficiency G_e defined in Eq. (8-1). Combining Eqs. (8-9), (8-14), and (2-24), we can express G_e as

$$G_e = \frac{J_{\text{nB}}}{J_{\text{pE}}} = \frac{D_{\text{nB}}}{D_{\text{pE}}} \frac{w_E N_{\text{DE}}}{w_B N_{\text{AB}}} \times \frac{N_{\text{CB}} N_{\text{VB}}}{N_{\text{CE}} N_{\text{VE}}} \times \exp\frac{\Delta E_G}{k_B T} \tag{8-15}$$

where N_{CB} and N_{CE} are the effective densities of states in the conduction bands of the base and the emitter, and N_{VB} and N_{VE} are the effective density of states in the valence bands of the base and the emitter, respectively. For a BJT, the emitter injection efficiency looks similar, but the last two factors are equal to unity because of the same emitter and base materials. Equation (8-15) shows the same basic proportionality as Eq. (8-2).

The second component of the base current, $I_{\text{B,surf}}$, is the surface recombination current taking place at the exposed extrinsic base region. This component is only important for the GaAs-based HBT because of the high GaAs surface recombination velocity on the order of 10^6 cm/s. It is important to emphasize that the current component $I_{\text{B,surf}}$ is proportional to the emitter perimeter but not the emitter area. For an GaAs HBT without surface passivation, as shown in Fig. 8.6(a), a considerable amount of electrons injected from the emitter to the base will recombine with holes at the exposed GaAs surface. This, together with a large perimeter-to-area ratio emitter, can cause $I_{\text{B,surf}}$ to become the dominating component for the base current and the main factor for the current gain degradation. To suppress such a recombination current, the extrinsic base surface of GaAs HBTs is frequently passivated using a thin, fully depleted ledge as shown in Fig. 8.6(b) for an AlGaAs/GaAs HBT. A properly designed ledge can effectively suppress surface recombination, as

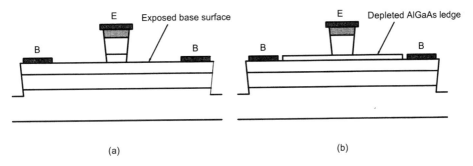

Figure 8.6. GaAs HBTs with (a) unpassivated extrinsic base and (b) passivated extrinsic base using fully depleted AlGaAs ledge.

it can prevent the electrons from reaching the extrinsic surface. The modeling of the surface recombination current is fairly complicated. More information on the estimation of $I_{B,surf}$ and on the ledge design can be found in [14]. Fortunately, the effect of surface recombination is negligible in SiGe HBTs and InP HBTs. The reason is that the surfaces of Si, SiGe, and InGaAs possess a very small surface recombination velocity of about 10^3 cm/s.

The third component of I_B is the base bulk recombination current $I_{B,bulk}$. It is related to the recombination in the quasineutral base and is constituted by the flow of holes to supplement the loss of holes in the quasineutral base due to recombination. This current can be estimated by [14]

$$I_{B,bulk} = \frac{qn(x_{BE})w_B}{2\tau_n} A_E \tag{8-16}$$

where τ_n is the minority electron lifetime in the base and A_E is the emitter area.

The last base current component, $I_{B,scr}$, is the electron–hole recombination current in the emitter–base space–charge region. Similar to the Si BJT, it is proportional to the emitter area and depends on the base–emitter voltage as

$$I_{B,scr} \propto \exp \frac{V_{BE}}{2V_T} \tag{8-17}$$

The detailed modeling of $I_{B,scr}$ is not straightforward and has been presented in [23].

As the components of I_B (except I_{Bp}) are very sensitive to the processing of HBTs, empirical base current models based on the measured data for a given technology may be developed. Note that the conventional assumption for the BJT that the hole injection from the base to emitter is the dominant component for the base current is not valid for HBTs in general.

B. Homogeneous Doping, Graded Bandgap in the Base (Graded Base). A graded bandgap in the base of an HBT is be achieved by varying the material composition of the base, e.g., by varying the Ge content in the base of a SiGe HBT. A graded base with the bandgap decreasing toward the collector, as shown in Fig. 8.7, is beneficial for improving the HBT performance. In the base, the separation between the Fermi level E_F and the valence band edge E_V is mainly determined by the base doping density. Thus, for a constant base doping, this separation is nearly constant throughout the base, and the bandgap grading results in a gradient mainly of the conduction band edge. This gives rise to a built-in field acting as an additional driving force for accelerating the electrons passing across the quasineutral base toward the collector.

Modeling the HBT having a graded base is not an easy task, as it requires the knowledge of the free-carrier transport parameters as a function of the position-dependent material composition. An approximation is feasible, however. Consider an HBT with a bandgap decreasing linearly throughout the base. We use a modified coordinate system where the origin of the X-axis is located at x_{BE}. Thus, the quasineutral base extends from $X = 0$ to $X = w_B$ (see Fig. 8.7). For simplicity, we assume that the electron diffusion coefficient in the base, D_{nB}, is constant and does not depend on the bandgap, i.e., on the material composition. The same holds for the effective densities of states in the conduction and valence bands. Furthermore, the base transport factor given in Eq. (7-10) is assumed unity. Then, using Eqs.(7-9) and (7-11) to express the collector current density J_C taking into account the fact that the intrinsic free-carrier concentration n_{iB} is not constant in the graded base, we arrive at

$$J_C = -\frac{qD_{nB}}{N_{AB}} \exp\left(\frac{V_{BE}}{V_T}\right) \times \frac{1}{\int_0^{w_B} \frac{dX}{n_{iB}^2(X)}} \tag{8-18}$$

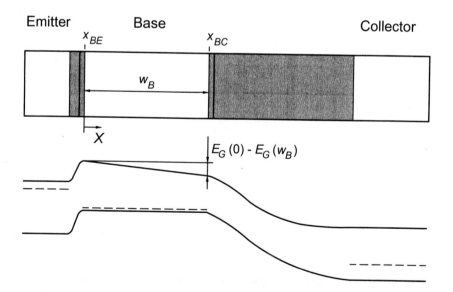

Figure 8.7. Energy band diagram of a graded base HBT.

To calculate the position-dependent intrinsic carrier concentration in the graded base, the expression describing the effect of bandgap narrowing given in Eq. (7-53) is used:

$$n_{iB}^2(X) = n_{iB}^2(0) \exp\left[\frac{E_G(w_B) - E_G(0)}{k_B T} \frac{X}{w_B}\right] \tag{8-19}$$

Putting Eq. (8-19) into the integral in Eq. (8-18) yields

$$\int_0^{w_B} \frac{dX}{n_{iB}^2(X)} = \frac{1}{n_{iB}^2(0)} \frac{k_B T w_B}{E_G(w_B) - E_G(0)} \left[1 - \exp - \frac{E_G(w_B) - E_G(0)}{k_B T}\right] \tag{8-20}$$

Finally, combining Eqs. (8-20) and (8-18), the collector current density of a graded base HBT is given by

$$J_C = -\frac{qD_{nB} n_{iB}^2(0)}{N_{AB} w_B} \frac{E_G(w_B) - E_G(0)}{k_B T} \frac{\exp\dfrac{V_{BE}}{V_T}}{1 - \exp\dfrac{E_G(w_B) - E_G(0)}{k_B T}} \tag{8-21}$$

This approach has been used extensively for the optimization of graded base SiGe HBTs [24, 25]. It can, however, be applied to other graded base HBTs as well.

Under the assumption that the base grading has only a minor influence on the base current, Eq. (8-21) together with the base current model derived earlier, describes the common–emitter current gain of graded base HBTs. The equations presented above are valid for a uniform base grading throughout the base, a case shown in Fig. 8.8(a) for SiGe HBTs. The method has been extended for the more complex piecewise base grading cases shown in Fig. 8.8(b) [26, 27].

8.2.3 HBTs with Spike Emitter–Base Heterojunction

The energy band diagram of a spike heterojunction is plotted in Fig. 8.1(c). The spike at the heterointerface is the result of a noticeable conduction band offset ΔE_C. This is the case for abrupt AlGaAs/GaAs, AlInAs/InGaAs, and InP/InGaAs HBTs, where the material composition changes abruptly from the base to emitter.

In Section 3.3.1 a method to calculate the current across spike heterojunctions was presented. This approach can be adopted to calculate the collector current of spike HBTs. What we have to do is to repeat the derivation from Section 3.3.1 starting with Eq. (3-26). The quantities with the subscript 1 are now related to the emitter, and those with the subscript 2 to the base of the HBT. Furthermore, we have to replace

- $n(x_j^-)$ and $n(x_j^+)$ in Eq. (3-26) by $n(x_{jEB}^-)$ and $n(x_{jEB}^+)$
- $n(x_n)$ and $n(x_p)$ in Eqs. (3-28) and (3-29) by $n(x_{EB})$ and $n(x_{BE})$
- V_{pN} in Eqs. (3-30) and (3-31) by V_{BE}
- x_{Bp} and $(x_{Bp}-x_p)$ in Eq. (3-32) by x_{BC} and w_B

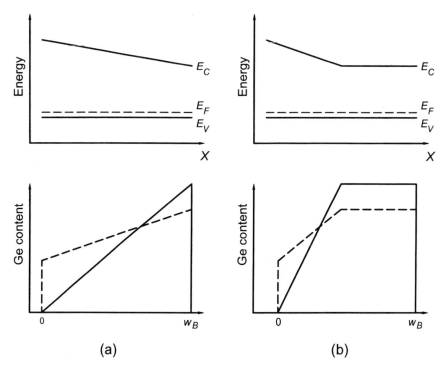

Figure 8.8. Energy band diagrams and Ge profiles in the base of SiGe HBTs for (a) triangular Ge profile and simple trapezoid Ge profile with a constant Ge grading and (b) piecewise linear Ge profile with nonconstant grading.

Then, the electron density at x_{BE}, i.e., at the emitter side of the quasineutral base, becomes

$$n(x_{BE}) = \frac{v_n w_B N_{DE} \exp\left(-\dfrac{V_{B1}}{V_T}\right) + D_{nB} n(x_{BC})}{D_{nB} + v_n w_B \exp\dfrac{V_{B2} - \Delta E_C/q}{V_T}} \quad (8\text{-}22)$$

and the electron current density at x_{BE} is obtained as

$$J_{nB}(x_{BE}) = qD_{nB}\frac{n(x_{BC}) - n(x_{BE})}{w_B} \quad (8\text{-}23)$$

where v_n, V_{B1}, and V_{B2} have the same meanings as described in Section 3.3.1. Neglecting recombination in the quasineutral base, we come to

$$J_C = J_{nB}(x_{BE}) \quad (8\text{-}24)$$

Equations (8-23) and (8-24) give the collector current density of a spike HBT for the case that thermionic emission is assumed to be the only transport mechanism across the emitter–base heterojunction. However, as discussed in Section 3.3.1, the field emission (tunneling) is a second transport mechanism prominent in spike heterojunctions. The thermionic-field-emission current is obtained from Eqs. (3-34)–(3-38).

This collector current model has been applied to investigate spike (abrupt) AlGaAs/GaAs HBTs [28, 29]. Figure 8.9 illustrates the collector currents for an abrupt AlGaAs/GaAs HBT calculated using the conventional drift–diffusion model neglecting the spike, the thermionic-emission model, and the thermionic-field-emission model. Experimental data are also shown. As can be seen, the thermionic-emission model underestimates the collector current. Surprisingly, the drift–diffusion model appears to be a good approximation except for very large voltages. Among the three models considered, the thermionic-field-emission is the most accurate one. Note that the drift–diffusion model is valid for a junction with a smooth conduction band, and the fact that such a model overestimates the collector current in a spike HBT suggests that the spike degrades the emitter injection efficiency.

Apparently, the conduction band spike is not desired, and transistor designs have been developed to reduce or eliminate the effect of spike in GaAs- and InP-based HBTs.

8.2.4 HBT Structures with a Reduced Spike

A conduction band spike at the emitter–base heterointerface necessitates the flow of electrons by means of thermionic emission and tunneling and consequently reduces

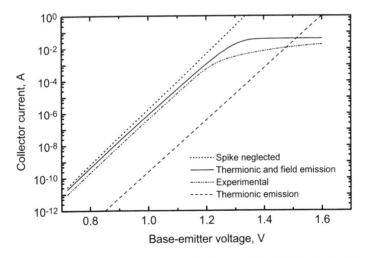

Figure 8.9. Collector current of an abrupt $Al_{0.3}Ga_{0.7}As$/GaAs HBT calculated using the drift–diffusion model, thermionic emission model, and thermionic field emission model. Also shown are experimental results. Data taken from [29].

the collector current. The undesirable effect associated with the spike can be suppressed with additional features added to the emitter–base heterojunction. Two options are frequently used for this purpose for AlGaAs/GaAs and InP HBTs and will be briefly described below.

A. Graded Layer. The first option is to use a graded layer in the emitter. Consider an $Al_{0.3}Ga_{0.7}As$/GaAs HBT. If a thin emitter region (a few hundred Å) adjacent to the heterojunction is graded, i.e., the Al content within this region decreases from $x = 0.3$ in the bulk of emitter to $x = 0$ at the heterointerface, the spike can be effectively removed. A detailed analysis of the dc currents in HBTs having a graded layer can be found in [28, 29].

Figure 8.10 shows the Gummel plots, i.e., the base and collector currents versus the base–emitter voltage, base and collector currents of two AlGaAs/GaAs HBTs with and without a graded layer (the graded layer thicknesses w_G are 150 Å and 0). Due to grading, the collector current increases, which is desirable. The base current, however, increases as well, which is not desirable. A closer look at the currents reveals, however, that the increase of the collector current is larger than that of the base current. Thus, the common–emitter current gain is increased when the graded layer is incorporated into the heterojunction.

B. Setback Layer. Inserting a thin, lightly doped or undoped layer of the same material as the base between the base and heterointerface is the second design option to suppress the effect of the spike. Such a layer is called a setback layer. It cannot fully remove the conduction band spike but makes it electrically less active. This can be seen in Fig. 8.11, which illustrates schematically the band diagram of the AlGaAs/GaAs heterojunction with and without a setback layer. Due to the pres-

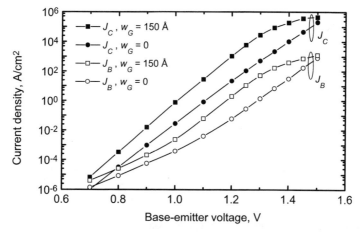

Figure 8.10. Calculated Gummel plots of abrupt and graded AlGaAs/GaAs HBTs. Both transistors are identically designed except for the different grading. Data taken from [29].

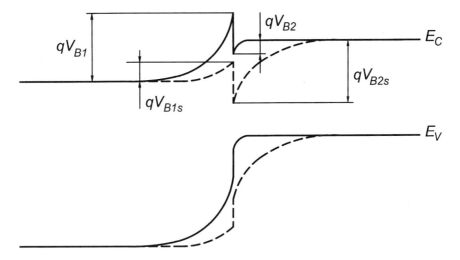

Figure 8.11. Energy band diagrams of AlGaAs/GaAs heterojunctions with (dashed lines, index s) and without a setback layer (solid lines).

ence of the lightly doped setback layer, the space–charge region extends deeper into the base and the potential distribution in the heterojunction changes. This leads to a downward shift of the spike, and the tip of the spike is no longer much higher than the conduction band edge in the emitter. Consequently, the spike presents less of an obstacle for the electron transport across the heterointerface, thereby moving the charge transport mechanism to some extent from the thermionic tunneling to the more efficient drift–diffusion. An analytical model describing the influence of the setback layer on the dc behavior of HBTs, and detailed information are presented in [29, 30]. Figure 8.12 shows the calculated Gummel plots of two AlGaAs/GaAs HBTs with and without a setback layer (the setback layer thickness w_i are 200 Å and 0). Similar to the case of the graded layer, both the collector and base currents increase when a setback layer is used, but the increase in the collector current prevails and the common–emitter current gain is improved.

8.2.5 Other Issues Related to HBT DC Behavior

A. Collector–Emitter Offset Voltage. The common–emitter output characteristics of HBTs often exhibit a nonnegligible offset voltage V_{off} as illustrated in Fig. 8.13. A positive collector current flows only when the applied collector–emitter voltage exceeds V_{off}, whereas for $V_{CE} < V_{off}$ small negative collector currents can be observed. In general, a low offset voltage is desirable for HBTs, as a high offset voltage leads automatically to a high knee voltage V_k. From the discussion of the power analysis of BJTs in Section 7.5 we know that a high knee voltage deteriorates the output power and the power-added efficiency of the transistor.

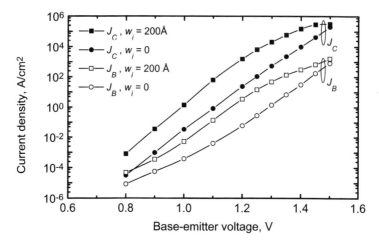

Figure 8.12. Calculated Gummel plots of abrupt AlGaAs/GaAs HBTs with and without a setback layer. Both transistors are identically designed except for the different setback layers. Data taken from [29].

The origin of the offset voltage has been investigated extensively [33–36], and is commonly attributed to the different turn-on voltages of the emitter–base and collector–base junctions. The turn-on voltage of a pn junction can be regarded as the forward voltage required to generate a certain current passing through the junction. This voltage is related to the built-in voltage, and a junction with a larger built-in voltage possesses a larger turn-on voltage as well. In the region of the output characteristics where the collector current rises sharply versus the collector–emitter

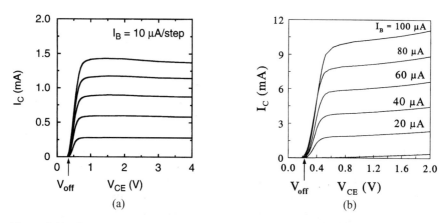

Figure 8.13. Common–emitter output characteristics of experimental (a) InGaP/GaAs HBT [31] and (b) InP/InGaAs HBT [32]. The offset voltages V_{off} in the two devices are illustrated. © 1999, 2001 IEEE.

voltage V_{CE} (i.e., slightly above V_{off} in Fig. 8.13), the transistor operates in the saturation mode. Here, both the emitter–base and the collector–base junctions are forward-biased (see Table 7.1). This means that electrons are injected into the base from both the emitter–base and base–collector junctions. At the same time, both junctions also collect electrons, since the electric field still exists in a forward-biased pn junction and electrons arriving at the junction will be swept across. Thus, if the emitter injects less electrons to the base than the collector does, less electrons will be collected by the collector than those injected from the collector. The result is a negative collector current. This occurs when the turn-on voltage of the emitter–base junction is higher than that of the collector–base junction. The offset voltage is the difference of the two turn-on voltages.

The turn-on voltage of a junction depends on the junction area, built-in voltage, bandgap difference, and, if a spike is present, height of the spike. Even if the emitter–base and the collector–base junctions are made from the same materials and have equal doping densities, a notable offset voltage might still be observed if the emitter area is smaller than the collector area.

B. Effects of the Collector–Base Junction in DHBTs. SiGe HBTs and InP-based HBTs having an InP collector are DHBTs (see Table 8.1). These HBTs have a collector–base heterojunction and a wider bandgap collector. Such a heterojunction affects the HBT behavior in the following two ways.

First, the effect of base pushout (discussed in Section 7.2.2) is altered due to the heterojunction. Let us first consider a transistor with a collector–base homojunction. If this transistor is operated in the forward-active mode at low collector current densities, the electric field in the collector–base space–charge region is negative (see Fig. 7.16) because the collector is more positively biased compared to the base. Thus, the field is directed from the collector to the base in the negative x-direction. If the external collector–base voltage is fixed, the magnitude of the field at the collector–base junction decreases with increasing collector current. The field eventually reduces to zero at a certain collector current density J_K. For a current density higher than J_K, the field becomes positive and holes are injected from the base to collector. This leads to an apparent widening of the quasineutral base or the so-called base pushout (see Section 7.2.2).

The situation is much different in DHBTs. Here, the barrier for hole injection from the base to the collector is enlarged by the valence band discontinuity ΔE_V. Holes can only be injected in the collector if the field in the collector–base heterojunction is positive *and* if the resulting potential overcomes the valence band offset. At the first sight this sounds good, because the undesired base pushout is supressed. However, for collector current densities exceeding J_K, the potential and field distributions in the collector–base heterojunction change in such a manner that the electron transport across the heterojunction is hampered. This effect has been investigated in [19, 37] for SiGe HBTs using numerical device simulations. Taking the results from [37], we show in Figure 8.14 the collector current density J_C of SiGe HBTs having different Ge contents in the base (and thus having different valence band offsets at the collector–base heterointerface) as a function of the base–emitter

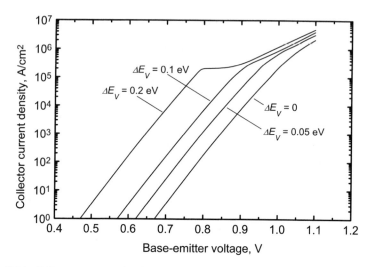

Figure 8.14. Collector current densities of SiGe HBTs calculated for different collector–base valence band offsets. Data taken from [37].

voltage V_{BE}. In the low V_{BE} range, J_C increases linearly in the log-linear plot. Above a certain critical current density J_{Ccrit} (about 2×10^5 A/cm^2), however, J_C increases much slower with increasing V_{BE}. This effect becomes more pronounced for increasing valence band offsets. This phenomenon is frequently called the modified Kirk effect.

The cutoff frequency of DHBTs at current levels above J_{Ccrit} suffers as well, as demonstrated in Fig. 8.15. The falloff of f_T becomes steeper if the valence band offset increases. It should be noted that the transistor structures used in [37] for this study have been deliberately chosen to emphasize the effect of the valence band offset. (Unfortunately, in [37] no detailed information on transistor attributes such as lateral structures and dimensions, doping, and Ge profiles are given.) A further demonstration of the influence of the modified Kirk effect on DHBTs behavior is presented in Fig. 8.16. Here, the cutoff frequencies of two experimental transistors, a Si BJT and a SiGe HBT, are shown. Since both transistors are fabricated in the same lot of wafers and are processed identically [25], a more reasonable and useful comparison can be made. Clearly, the HBT shows a higher peak cutoff frequency, but the degradation of f_T of this device at high collector currents (> 1 mA) is more significant.

The second effect associated with the collector–base heterojunction is related to the conduction band discontinuity ΔE_C. Particularly for the InP DHBT, this effect is substantial due to the large conduction band offset at the collector–base junction. The conduction band edge in the collector–base heterojunction of an InP DHBT is shown in Fig. 8.17. Obviously, the conduction band offset at the collector–base het-

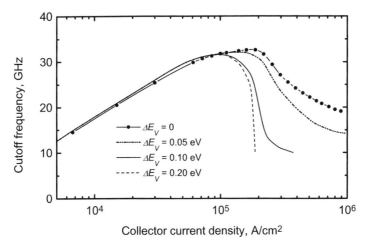

Figure 8.15. Cutoff frequency of SiGe HBTs calculated as a function of the collector current density. Data taken from [37].

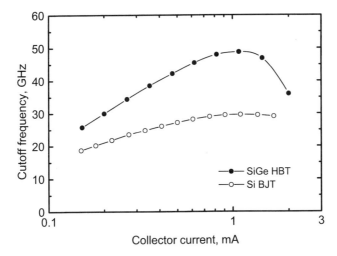

Figure 8.16. Comparison of cutoff frequencies of a SiGe HBT and a Si BJT as a function of the collector current. Both transistors were fabricated in the same wafer lot and processed identically. Data taken from [25].

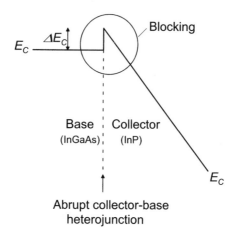

Figure 8.17. Schematic of the conduction band edge in the collector–base heterojunction of an InP DHBT.

erointerface imposes a barrier hindering the electron flow. The undesirable blocking effect of ΔE_C can be reduced using a graded layer in the collector–base heterojunction.

8.3 SMALL-SIGNAL, NOISE, AND POWER ANALYSIS OF HBTs

The basics of the small-signal, noise, and power analysis of HBTs are very similar to those of BJTs. In this section, we will not reiterate the details but briefly discuss the issues pertinent and important to HBTs.

For the small-signal analysis of HBTs, small-signal equivalent circuits based either on the π-model or the T-model can be used. These models have already been discussed in Section 7.3.1 (see Figs. 7.25 and 7.26). The elements of the equivalent circuit for the HBT π-model are calculated as follows. For HBTs with a smooth emitter–base heterojunction, the transconductance g_m defined by Eq. (7-78) for the BJT is still applicable. Carrying out the derivative of Eqs. (8-9) and (8-21) with respect to the base–emitter voltage leads to

$$g_m = \frac{I_C}{V_T} \tag{8-25}$$

which is identical to Eq. (7-79). In the case of HBTs with a spike emitter–base heterojunction, the expressions for the collector current are more complicated (see Section 8.2.3) and an analytical expression for the transconductance is difficult to obtain. Therefore, we recommend that the transconductance of this device be determined numerically. This is done by calculating the collector currents I_{C1} and I_{C2} for two slightly different base–emitter voltages V_{BE1} and V_{BE2} at a constant collector–emitter voltage and then computing the transconductance using

$$g_m = \left.\frac{dI_C}{dV_{BE}}\right|_{V_{CE}=\text{const.}} \approx \frac{I_{C2} - I_{C1}}{V_{BE2} - V_{BE1}} \quad (8\text{-}26)$$

Like the BJT, the collector–emitter resistance r_{ce} of the HBT is obtained from the Early voltage and Eq. (7-81). For the emitter–base resistance r_{eb}, a numerical approach similar to that used to calculate the transconductance, together with the definition of r_{eb} given in Eq. (7-82), is recommended.

Equation (7-84) is applicable for calculating the collector–base capacitance C_{cb}. For SHBTs with a collector–base homojunction, the thickness of the space–charge region is given by Eq. (2-98). For DHBTs with a collector–base heterojunction, the space–charge region thickness is the sum of x_n and x_p expressed in (3-17) and (3-18).

To find the emitter–base capacitance C_{eb}, the space–charge region capacitance C_{dEB} and the diffusion capacitance C_{DEB} have to be determined [see Eqs. (7-85)–(7-87)]. Because of the emitter–base heterojunction, the thickness of the space–charge region needed in Eq. (7-86) is the sum of x_{nEB} and x_{pEB} from Eqs. (8-11) and (8-12). The diffusion capacitance defined in Eq. (7-87), on the other hand, can be more easily obtained numerically. To this end, we first determine the minority carrier charges in the quasineutral emitter and the quasineutral base, Q_{pE} and Q_{nB}, respectively, for two slightly different base–emitter voltages V_{BE1} and V_{BE2}. Then C_{DEB} is given by

$$C_{DEB} = \frac{Q_{pE2} - Q_{pE1}}{V_{BE2} - V_{BE1}} + \frac{|Q_{nB2} - Q_{nB1}|}{V_{BE2} - V_{BE1}} \quad (8\text{-}27)$$

The different delay times and the frequency limits f_T and f_{max} of the HBT can be modeled analogously to those of the BJT described in Sections 7.3.2 and 7.3.3.

The origins of the high-frequency noise in HBTs and BJTs are in general the same. Thus, the noise model by Hawkins described in Section 7.4 can be applied to HBTs as well, and it has been done successfully for GaAs HBTs [38], InP HBTs [39], and SiGe HBTs [40, 41]. The minimum noise figure of the HBT exhibits the same collector current dependence as the Si BJT, i.e., a pronounced U-shape in the NF_{min} versus I_C curve, as shown in Fig. 7.41.

The power characteristics of HBTs and BJTs are also alike, and the treatment given in Section 7.5 is applicable for HBTs. For example, for HBTs used in a Class A power amplifier, Eqs. (7-126)–(7-129) can be utilized to calculate the maximum output power $P_{out,max}$, dc power dissipation P_{dc}, power-added efficiency PAE, and total power (dc power plus RF power) dissipation in the HBT.

As in power BJTs, self-heating is an important factor in designing power HBTs, and this thermal effect is even more significant in III–V HBTs due to the poor thermal conductivity of the III–V compound materials. Issues of self-heating and its effects on the HBT performance will be discussed in detail in the next section.

8.4 SELF-HEATING OF HBTs

It has been shown in Section 2.7 that the thermal conductivity of III–V compound semiconductors is smaller than that of Si. At room temperature, the values are 1.3 W/cm-K, 0.7 W/cm-K, and 0.5 W/cm-K for Si, InP, and GaAs, respectively. Thus, the removal of heat is more difficult and the effect of self-heating is more significant in III–V HBTs. The topic of self-heating will be covered in several sequential steps. First, we will examine self-heating in multifinger HBTs and investigate the temperature and current distributions among the fingers. This is followed by discussions of the temperature dependence of the current gain in HBTs. Finally, the phenomenon of the current gain collapse as a result of self-heating will be addressed.

8.4.1 Temperature-Dependent Collector Current in Multifinger HBTs

Let us consider a multifinger HBT. Due to the power dissipation in the transistor unit cells and the thermal coupling among the fingers, a nonuniform temperature distribution, with the center fingers being hotter than the outer fingers, is commonly seen in this device. Such a temperature distribution can be calculated by solving Eq. (2-129), as has been done in several HBT studies [14, 42–44]. Under dc operation, the power dissipated in the ith finger, $P_{diss,i}$, is equal to $V_{CE} \times I_{Ci}$, where I_{Ci} is the dc collector current of the ith finger. For microwave operation, the dissipated power depends on both the amplifier class and dc bias conditions.

The total collector current I_C of the transistor is the sum of all the collector currents I_{Ci} in the individual fingers. The current of the ith finger strongly depends on the temperature in the emitter–base junction of the particular finger. Typically, because the finger size is small, the temperature within the individual finger can be assumed constant, but Eq. (2-129) permits a nonuniform temperature distribution within the individual finger as well. This can be carried out by dividing the finger into several subregions.

To simplify the analysis, we will consider the case of constant temperature within the individual finger. The collector current of the ith finger for a given finger temperature can be calculated using one of the HBT dc models described in Section 8.2, taking into account the temperature-dependent material properties described in Chapters 2 and 3. Another approach is to write the collector current flowing through the ith finger as [14, 43]

$$I_{Ci} = I_{C0} \exp\left\{\frac{q}{k_B T_A}[V_{BE,ji} - \phi(T_i - T_A)]\right\} \qquad (8\text{-}28)$$

where I_{C0} is the saturation current at ambient temperature, $V_{BE,ji}$ is the emitter–base junction voltage (i.e., the intrinsic emitter–base voltage), ϕ is the thermal-electric feedback coefficient, T_A is the ambient temperature, and T_i is the finger temperature. In Eq. (8-28), all temperature dependences associated with the various parameters have merged into the thermal-electric feedback coefficient. Note that ϕ is negative [see Eq. (8-29)]. The equation describes the important effect that the intrinsic base–emitter voltage required to produce a certain collector current in a finger decreases with increasing temperature. Fig. 8.18 shows the base–emitter voltage, hereafter called the turn-on voltage, necessary to generate a certain collector current in an abrupt AlGaAs/GaAs HBT as a function of the temperature. As expected, the turn-on voltage increases with increasing collector current. An important characteristic is the fact that the temperature dependence of the turn-on voltage is linear. Detailed investigations of different HBTs showed that the base–emitter turn-on voltage decreases by a rate between –3 and –1 mV/°C over a wide range of collector current densities [45, 46]. This rate is the thermal-electric feedback coefficient ϕ. Evaluating the data presented in [45, 46] leads to the following approximate expression for ϕ:

$$\phi(J_C)[V/°C] = -[a \log (J_C) + b] \quad (8\text{-}29)$$

where a and b are fitting parameters and J_C is the collector current density in A/cm^2. For abrupt Al$_{0.35}$Ga$_{0.65}$As/GaAs HBTs, a and b are -1.93×10^{-4} and 1.97×10^{-3}, respectively, and for InP/InGaAs/InP HBTs the corresponding values are -1.87×10^{-4} and 1.89×10^{-4}.

Equations (8-28) and (8-29) suggest that under fixed bias conditions, the current passing through the ith finger increases considerably when the temperature is increased. For discussion, let us consider a 3-finger HBT. At a low total collector current, and thus a low current in the individual finger, self-heating is insignificant and the three fingers carry about the same portion of the total current (one third of the total current). At a high current level, however, self-heating becomes more prominent, the temperatures in all three fingers increase, and the center finger becomes hotter than the two outer fingers. Thus, the current passing through the center finger

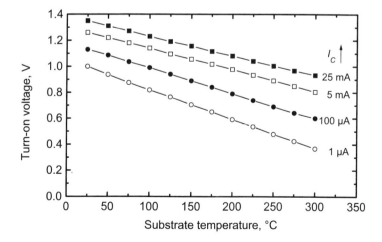

Figure 8.18. Turn-on voltages of the emitter–base heterojunction of an AlGaAs/GaAs HBT as a function of the temperature. Data taken from [45].

increases more than those in the other two fingers, and the two outer fingers contribute less to the total current. Eventually, the center finger can become so hot that this finger carries the entire total collector current and the two outer fingers become inactive.

As mentioned in Section 7.5, incorporating emitter ballast resistors can keep the current distribution in a multifinger transistor more homogeneous. With a ballast resistor, the intrinsic base–emitter voltage in Eq. (8-28) becomes

$$V_{BE,ji} = V_{BE} - R_{E,ball}I_{Ci} \qquad (8\text{-}30)$$

where $R_{E,ball}$ is the resistance of the emitter ballast resistor and V_{BE} in the applied (extrinsic) base–emitter voltage. Strictly speaking, the emitter contact resistance automatically serves as a ballast resistor, but an additional resistance is frequently added to boost the ballasting effect.

Using Eq. (2-129) to model the effect of self-heating and Eq. (8-28) to calculate the temperature-dependent current in the fingers, the current and temperature distributions in multifinger GaAs HBTs have been investigated [43]. The finger currents in a five-finger HBT with and without emitter ballast resistors are shown in Fig. 8.19. For the HBT with ballast resistors, the current distribution among the fingers is quite uniform, even at high current levels. This is not the case for the HBT without ballast resistors, and a high degree of nonunifomity is observed. Figure 8.20 shows the finger temperatures in the HBT with and without emitter ballasting. With ballast resistors, the temperature distribution among the fingers is more homoge-

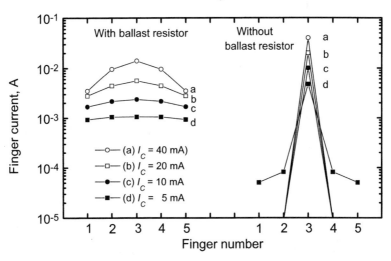

Figure 8.19. Calculated current distribution among the fingers of a five-finger HBT with an emitter ballast resistor of 20 Ω and without an emitter ballast resistor. The curves (a), (b), (c), and (d) correspond to the collector currents of 40, 20, 10, and 5 mA, respectively. Data taken from [43].

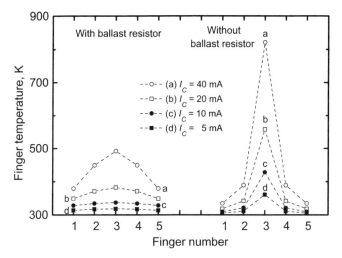

Figure 8.20. Calculated temperature distribution among the fingers of the same HBT in Figure 8.19 with an emitter ballast resistor of 20 Ω and without an emitter ballast resistor. The temperatures are those at the centers of the individual fingers. Data taken from [43].

neous and the peak temperature in the fingers is reduced, particularly in the center finger, due to the much less significant thermal coupling among the fingers.

8.4.2 Temperature Dependence of Current Gain

The current gain of a BJT is an increasing function of temperature, as discussed in Section 7.5. The situation is quite different in HBTs. Let us first concentrate on AlGaAs/GaAs HBTs. In Fig. 8.21, the common–emitter current gains of three different abrupt AlGaAs/GaAs HBTs having a single emitter finger compiled from [14, 47, 48] are shown. Obviously the current gain of these devices decreases with increasing temperature. This behavior is different from Si BJTs and can be explained as follows. In the AlGaAs/GaAs HBT at room temperature, the hole injection from the base into the emitter is only a negligible portion of the total base current due to the presence of the valence band offset. This situation changes when the temperature increases. A higher temperature leads to an increase in the energy of the holes in the base. Figure 8.22 shows qualitatively the hole energies in the valence band of the base at two different temperatures. The low-energy hole corresponds to a representative hole at room temperature, and the high-energy hole corresponds to a hole at an elevated temperature. The high-energy hole is located "deeper" in the valence band and has to surmount a lower energy barrier to move to the emitter. Thus, the blocking of holes resulting from the valence band offset becomes less effective when the temperature is increased, the hole injection current becomes more significant, and the current gain decreases.

Figure 8.23 shows the common–emitter current gains of InGaP/GaAs HBTs,

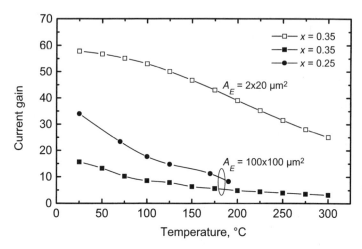

Figure 8.21. Common–emitter current gains of abrupt Al$_x$Ga$_{1-x}$As/GaAs HBTs as a function of the temperature. Data taken from [14, 47, 48].

InAlAs/InGaAs HBTs, and InP/InGaAs HBTs as a function of temperature compiled from [48–50]. Let us first compare the current gains of InGaP/GaAs HBTs in Fig. 8.23 and those of abrupt Al$_{0.35}$Ga$_{0.65}$As/GaAs HBT in Fig. 8.21. The absolute values of the current gains are of no interest here, but only their temperature dependence. The bandgap difference ΔE_G for the AlGaAs/GaAs and InGaP/GaAs pairs are 0.46 eV and 0.43 eV, respectively (i.e., the bandgap differences are almost the same). However, the AlGaAs/GaAs HBT possesses a more pronounced conduction band spike and only about 0.19 eV goes to the valence band offset. On the other hand, the InGaP/GaAs HBT has a ΔE_G of about 0.43 eV and it goes almost entirely to the valence band offset ΔE_V of 0.4 eV. It is shown in Eq. (8-3) that the emitter injection efficiency is proportional to $\exp(\Delta E_V/k_B T)$ for an HBT with a spike emitter–base heterojunction. Thus, it is easy to understand why a much higher tempera-

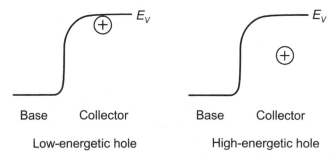

Figure 8.22. Schematic showing the position of low- and high-energy holes in the valence band of the base.

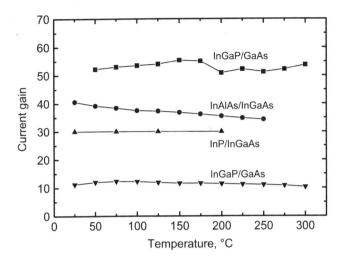

Figure 8.23. Common–emitter current gains of InGaP/GaAs and InP HBTs as a function of the temperature. Data taken from [48–50].

ture is needed in InGaP/GaAs HBT to increase the hole injection from the base to emitter to a degree that it becomes a noticeable share of the total base current. The data given in Fig. 8.23 confirm the above discussions and show that the current gain of InGaP/GaAs HBT is insensitive to the temperature.

The current gain of the InP/InGaAs HBT, which has a large valence band offset of about 0.35 eV, is also not a function of the temperature (see Fig. 8.23) because the hole injection from the base to emitter is again negligibly small even at elevated temperatures. The current gain of the InAlAs/InGaAs HBT, on the other hand, is temperature dependent, but to a lesser extent compared to the $Al_{0.35}Ga_{0.65}As/GaAs$ HBT. This seems to be reasonable because the valence band offset at the InAlAs/InGaAs heterointerface is a bit larger than that at the $Al_{0.35}Ga_{0.65}As/GaAs$ heterointerface (0.195 eV versus 0.19 eV) but smaller than the InP/InGaAs HBT. It should be noted that the temperature dependence of the current gain is not only a function of the band alignment at the emitter–base heterointerface, but depends to some extent on the temperature dependence of other components of the collector and base current as well. As such, the HBT's current gain temperature dependence can sometimes not be explained soley by the valence band offset and has to do with the type and the specific design of the device.

Figure 8.24 shows the variation of the current gains when the temperature changes from 0 to 100 °C for a Si BJT and several SiGe HBTs having three different Ge profiles in the base. Among the three Ge profiles considered, the Ge content at the emitter–base interface is nearly zero for the triangle Ge profile, larger for the trapezoid Ge profile, and largest for the digital (uniform) Ge profile. The results in Fig. 8.24 indicate that when moving from triangle to trapezoid to digital Ge profiles (valence band offset is increasing), the change in temperature-induced variation of

430 HETEROJUNCTION BIPOLAR TRANSISTORS

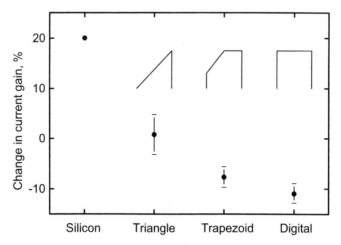

Figure 8.24. Changes in the current gains at temperatures varying from 0 to 100 °C for a Si BJT and SiGe HBTs having three different Ge profiles. Data taken from [25].

the current gain becomes larger. One reason for this occurrence is that relatively low Ge contents are used in these devices, and the three SiGe HBTs are changing more and more from BJT-like to HBT-like. As a result, the current gain of these devices is becoming a decreasing function of the temperature. In other words, the valence band offsets in the trapezoid and digital profiles are not large enough, due to the low Ge contents, to allow for temperature-independent current gains in these devices. Material parameters of SiGe may also play a role in the observed characteristics. Table 8.3 summarizes our discussion on the temperature dependence of the current gain of single-finger BJTs and HBTs. A more in-depth discussion on this topic can be found in [14].

8.4.3 Current Gain Collapse

The collapse of current gain is a phenomenon observed in abrupt multifinger AlGaAs/GaAs HBTs. It is a sudden fall-off of the total collector current when, for a

Table 8.3. Temperature dependence of the current gains of BJT and HBTs[a]

Transistor type	Values of current gain β as $T \uparrow$
Abrupt AlGaAs/GaAs HBT	$\downarrow$
InGaP/GaAs HBT	—
InP/InGaAs HBT	—
InAlAs/InGaAs HBT	weak $\downarrow$
SiGe HBT	— ... $\downarrow$
Si BJT	$\uparrow$

[a]The idea of this table has been taken from [14].

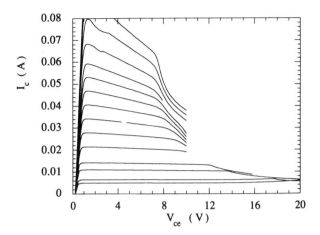

Figure 8.25. Measured I_C versus V_{CE} characteristics of a four-finger AlGaAs/GaAs HBT (parameter is the base current). The current collapse at high V_{CE} is illustrated. Taken from [51] ©1993 IEEE.

given base current, the collector–emitter voltage V_{CE} exceeds a critical value. This is illustrated from the measured I_C–V_{CE} characteristics of a four-finger AlGaAs/GaAs HBT given in Fig. 8.25. At small collector–emitter voltages above 0.5 V, the total collector current shows a gradual decrease with increasing V_{CE}. This operation region is frequently called the negative differential resistance (NDR) region. In the NDR region, all fingers still contribute to the total collector current, but the contribution of the center fingers is larger than that of the outer fingers. The slight decrease of the total collector current is due to the increasing temperature of the transistor when V_{CE} is increased (a larger V_{CE} leads to a larger power dissipation and thus to a temperature rise). This results in a decrease in the current gain for AlGaAs/GaAs HBTs, as discussed in the previous section. As V_{CE} is increased further, the current distribution among the fingers becomes more nonuniform. Eventually, one of the center fingers becomes so hot that it is the only finger conducting current. This gives rise to a large drop in the total collector current and also a sharp fall-off in the current gain because the base current fed to the transistor is fixed. This behavior is totally different from Si BJTs, in which thermal runaway, i.e., a steady increase of the collector current, occurs. Although thermal runaway will destroy the BJT, the HBTs can survive the current gain collapse because it is a nondestructive event.

Single-finger AlGaAs/GaAs HBTs do not show a collapse of current gain. A single-finger transistor continues to operate in the NDR region until the very large applied voltage activates the avalanche breakdown. One would expect a two-finger AlGaAs/GaAs HBT to behave in the same manner as a single-finger HBT, because the symmetry will lead to uniform current and temperature distributions. This, however, is not the case, and current gain collapse can take place in two-finger transistors as well [51, 52]. The reason is that it is not possible to fabricate an HBT with an exactly symmetrical two-finger structure. Thus, even a small temperature difference between the two fingers will eventually lead to current gain collapse.

8.5 ISSUES OF GaAs-BASED HBTs

8.5.1 Transistor Structures

A. AlGaAs/GaAs HBTs. The early experimental and commercial GaAs HBTs employed AlGaAs emitters, and these devices are still being widely used for power amplifiers operating at microwave frequencies. The cross section of a typical AlGaAs/GaAs HBT has been been shown in Figs. 1.22 and 8.6. In principle, it consists of an AlGaAs emitter followed by a GaAs base, a GaAs collector, and a GaAs subcollector.

The exact layer sequence of practical AlGaAs/GaAs HBTs is, however, more complicated. A typical layer sequence, from the top to bottom, for AlGaAs/GaAs HBTs is shown in Table 8.4. The cap layer underneath the emitter contact is n-type (heavily doped) and made of either GaAs or, in most cases, an InGaAs/GaAs sequence. The aim of the cap is to minimize the emitter contact resistance. The n-type AlGaAs layer is followed by the wide-bandgap emitter. Below the emitter is a graded and/or setback layer serving to reduce the conduction band spike at the emitter–base heterointerface. Most HBT designs employ an n-type graded $Al_xGa_{1-x}As$ layer in which the Al content decreases from the value in the emitter down to zero at the heterointerface. A setback layer with an undoped GaAs can also be added after the graded layer. Then the p-type base follows. Most experimental and commercial transistors use GaAs as the base material, but strained InGaAs has been used as well [38, 53]. In any case, the base is doped much higher than the emitter. The base sheet resistance R_{sh} [see Eq. (7-34)] of GaAs HBTs is typically in the range of 100–600 $\Omega/\square$, and thus about one order of magnitude less than in Si BJTs. Below the base, the n-type GaAs collector is located. The collector doping is typically sev-

Table 8.4. Typical layer consequence and make-up of AlGaAs/GaAs HBTs

Layer	Conductivity	Material composition	Thickness, nm	Doping, cm^{-3}
Cap	n$^+$	$In_xGa_{1-x}As$, $x = 0.25-1$	20–100	$1-5 \times 10^{19}$
Cap	n$^+$	$In_xGa_{1-x}As$, x graded	10–100	$1-5 \times 10^{19}$
Cap	n$^+$	GaAs	50–200	$2-5 \times 10^{19}$
Emitter	n	$Al_xGa_{1-x}As$, $x =$ 0.25–0.35	50–200	$1-5 \times 10^{17}$
Graded/setback	n/i	$Al_xGa_{1-x}As$, $x =$ 0.25–0.35/GaAs	20–50/10–20	$1-5 \times 10^{17}/0$
Base	p$^+$	GaAs or InGaAs	30–150	$1-5 \times 10^{19}$
Collector	n	GaAs	500–1000	$5 \times 10^{15} - 2 \times 10^{17}$
Subcollector	n$^+$	GaAs	500–1000	$2-5 \times 10^{18}$

eral orders of magnitude lower than the base doping so that a high breakdown voltage can be achieved. Finally the n-type GaAs subcollector having a high doping density is placed underneath the collector to minimize the parasitic collector resistance. The entire structure is grown epitaxially on a semi-insulating GaAs substrate.

The impurity dopant for the n-type InGaAs, GaAs, and AlGaAs layers is typically Si, and for the p-type base Be or C is used. Several metallization systems are available for the realization of the ohmic emitter, base, and collector contacts of GaAs-based HBTs. The performance of bipolar transistors is sensitive to the quality of the ohmic contacts, and low contact resistances are normally desirable, except for the case of ballasting. For the contacts on n-type GaAs (or InGaAs), i.e., for the emitter and collector contacts, alloyed AuGe/Ni with a contact resistance ρ_{co} of 10^{-6}–10^{-7} Ωcm^2 is frequently used. Nonalloyed WSi/Ti/Pt/Au or Ti/Pt/Au is also a good candidate for the emitter contact. The base metallization consists of Ti/Pt/Au, Au/Zn/Au, or AuBe. The contact resistance depends on the doping density of the underlying semiconductor layer. Because the base is heavily doped, low base contact resistances can easily be obtained. For Ti/Pt and Ti/Pt/Au base contacts, ρ_{co} values are typically between 10^{-6} and 10^{-7} Ωcm^2, and ρ_{co} of 4–7 × 10^{-7} Ωcm^2 for Au/Zn/Au contacts have also been reported [54].

The layer sequence listed in Table 8.4 is first grown on the entire GaAs wafer. To arrive at the HBT structure shown in Fig. 8.6, the cap and the emitter have to be removed outside the emitter region by etching. This etching must be stopped after the AlGaAs layer is removed. Unfortunately there is no etching solution that removes AlGaAs but does not attack GaAs. Thus, selective etching is a technical challenge. Another difficulty of processing the AlGaAs/GaAs HBT is the material quality of the AlGaAs layer. Aluminium is extremely reactive and, during growth, background impurities are likely to be incorporated into the AlGaAs layers, thereby leading to a high density of traps. Moreover, donors used to dope the AlGaAs can create a large number of deep-level traps such as the so-called DX centers. Because of these shortcomings, a different emitter material for GaAs-based HBTs is desirable, and InGaP has been shown to be an excellent alternative.

B. InGaP/GaAs HBTs. The first InGaP/GaAs HBT was reported in 1985 [55], and in the early 1990s the suitability of these transistors for microwave operation has been established. Since then, the InGaP/GaAs HBT has gained increasing popularity. It reached commercial status in the second half of the 1990s and has competed succesfully with the AlGaAs/GaAs HBT ever since. The transistor structures of InGaP/GaAs and AlGaAs/GaAs HBTs are very similar, only the emitter in the InGaP/GaAs HBT is made of $In_{0.49}Ga_{0.51}P$, which is lattice-matched to GaAs.

InGaP/GaAs HBTs have several advantages over AlGaAs/GaAs HBTs. First, the etching solutions for InGaP are selective and do not attack GaAs. Thus, the etching process is automatically stopped once the InGaP emitter is removed. Second, InGaP layers contain far fewer traps and DX centers than AlGaAs layers. The third advantage is that the conduction band offset at the ordered InGaP/GaAs heterointerface is almost zero, and most of the bandgap difference of about 0.43 eV appears as the valence band offset. Thus, the ordered InGaP/GaAs heterojunction is naturally smooth and no graded or setback layers at the emitter–base junction are required.

The designation "ordered" requires a brief explanation. It has been observed that not all InGaP/GaAs heterojunctions are smooth heterojunctions [47]. Instead, conduction band offsets between zero and more than 0.2 eV have been found. This is attributed to the crystalline structure of the InGaP. The structure of ordered InGaP is such that (001) lattice planes of exclusively Ga, P, In, and P atoms are stacked on top of one another, and no intermixing of Ga and In atoms in the plane takes place. Growing an ordered InGaP layer on GaAs yields a smooth heterojunction. InGaP layers in which the Ga, In, and P atoms are randomly distributed on the (001) plane are called the disordered. Growing a disordered InGaP layer on GaAs results in a spike heterojunction. Thus, only the ordered InGaP emitter is used in InGaP/GaAs HBTs.

C. Vertical Design. Power amplification is a main application of GaAs-based HBTs with either an AlGaAs or InGaP emitter. The basic design rules are the same for AlGaAs/GaAs and InGaP/GaAs HBTs, and the design targets are high operating frequency, high power gain, and high output power. As such, transistors with sufficiently high current gain, with both high f_T and f_{max}, high breakdown voltage, and low self-heating are desired. The first step of HBT design is the optimization of the vertical layer structure. A typical guide for the vertical structure has been given in Table 8.4.

In general, a thin and highly doped base is needed. A small base thickness ensures a low base transit time and thus a high f_T, as demonstrated in Fig. 8.26, which shows the reported cutoff frequencies of GaAs-based HBTs as a function of the base thickness. An empirical upper limit fit to the best f_T is also included in the fig-

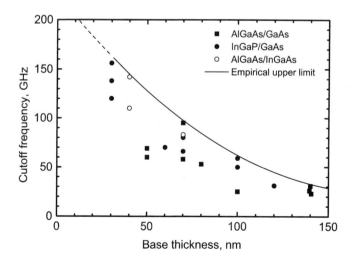

Figure 8.26. Reported cutoff frequencies of GaAs HBTs as a function of the base thickness.

ure. A high base doping density is used so that base punch-through is avoided and a low base resistance is achieved. A low base resistance is imperative for two reasons. First, it leads to a high power gain and a high maximum frequency of oscillation f_{max}. Second, the current crowding is minimized and a high emitter utilization factor is guaranteed (see Section 7.2.2).

There is a trade-off in optimizing the collector thickness and doping density, as discussed in Chapter 7. The requirements of a high cutoff frequency f_T and a simultaneously high collector–emitter breakdown voltage BV_{CE0} are often conflicting. For a high f_T, the collector should be thin and/or highly doped to minimize the transit time related to the collector–base space–charge region, τ_{BC} [see Eq. (7-97)]. A high breakdown voltage, on the other hand, requires a thick and lightly doped collector. It is well known that a GaAs HBT cannot be designed to simultaneously satisfy both high f_T and BV_{CE0}. This is not only true for GaAs HBTs but also for bipolar transistors in general. The reported cutoff frequency of GaAs HBTs and InP HBTs as a function of BV_{CE0} will be shown later in Section 8.6.2. Note that the breakdown voltage of GaAs HBTs is in general considerably higher than that of Si BJTs.

An experimental design study carried out in [56] gives further insight into the optimization of the HBT layer structure. Different abrupt AlGaAs/GaAs HBTs were fabricated and compared in terms of BV_{CE0}, f_T, and f_{max}. These transistors have the same emitter and subcollector design, the same collector doping ($N_{DC} = 7.5 \times 10^{15}$ cm^{-3}), but different collector thicknesses w_C. Furthermore, different base doping N_{AB} and base thicknesses w_B have been considered as well. Table 8.5 shows the outcomes of these HBTs.

The important conclusions one can draw from Table 8.5, which are consistent with the physics discussed above, are:

- An increase in the collector thickness leads to an improvement in the breakdown voltage (compare devices 1, 2, and 3). Inserting an undoped (i) layer between the p$^+$-type base and the n-type collector, as has been done in device

Table 8.5. Measured BV_{CE0}, f_T, and f_{max} for AlGaAs/GaAs HBTs with different base and collector designs

Device	w_C, μm	w_B, μm	N_{AB}, cm^{-3}	BV_{CE0}, V	f_T, GHz	f_{max}, GHz
1	0.7	0.14	1×10^{19}	13.5	21	42
2	1.0	0.14	1×10^{19}	18	19	45
3	1.5	0.14	1×10^{19}	25	—	—
4	0.7 + 0.3i	0.14	1×10^{19}	20.5	20	47
5	1.0[a]	0.14	1×10^{19}	28	—	—
6	1.0	0.14	$5 \times 10^{18} - 5 \times 10^{19}$	17.5	26	66
7	1.0	0.08	2×10^{19}	17	27	52
8	1.0	0.08	4×10^{19}	17.5	28	60

[a]Device 5 has a graded Al$_{0.3}$Ga$_{0.7}$As collector.

4, leads to a reduced field at the collector–base junction and thus an improved breakdown behavior (compare devices 2 and 4).

- A collector with a wider bandgap leads to a higher breakdown voltage. This is the case for device 5, having an AlGaAs collector (compare devices 2 and 5).
- A built-in field in the base results in a drift tendency and thus an additional accelerating force driving the electrons across the base. A built-in field can be obtained by a doping gradient in the base (higher doping at the collector–base junction and lower doping at the emitter–base junction) as in device 6, or by varying the material composition in the base. The result is a smaller base transit time and a higher cutoff frequency (compare devices 2 and 6). It should be noted, however, that most GaAs HBTs use a homogeneous base without any grading of the doping or material composition.
- A thinner base leads to a smaller base transit time and a higher cutoff frequency (compare devices 2, 7, and 8).
- A higher base doping density leads to a lower base resistance and a higher f_{max} (compare devices 7 and 8).

D. Lateral Design. Although the cutoff frequency of an HBT is mainly determined by the vertical layer design, the maximum frequency of oscillation f_{max} is sensitive to both the vertical and lateral structures of the HBT. The lateral HBT design comprises several aspects. First, the design should minimize the extrinsic collector–base capacitance $C_{cb,ext}$ and the extrinsic base resistance $R_{B,ext}$ to increase the power gain and f_{max}. Second, the emitter width W_E (and thus the base width W_B) should be small enough to ensure a low intrinsic base resistance. This is important to obtain a high f_{max} and a high emitter utilization factor EU. Third, the thermal aspects should be considered (this will be discussed in the next section).

A low $C_{cb,ext}$ can be obtained by minimizing the area of the collector–base junction not located underneath the emitter. In addition, a small distance between the metallic base and emitter contacts results in a low $R_{B,ext}$. If the base contacts are

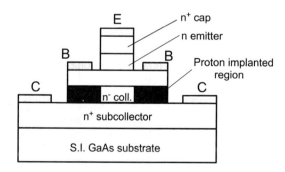

Figure 8.27. GaAs HBT with proton implanted regions to reduce the extrinsic collector–base capacitance. After [58].

self-aligned to the emitter, the distance between the emitter and base contacts can be made small and, at the same time, the size of the extrinsic transistor reduced. Thus, both $C_{cb,ext}$ and $R_{B,ext}$ are minimized in a self-aligned structure, as reported in [57–59]. A further reduction of $C_{cb,ext}$ can be achieved when self alignment is combined with proton implantation (as shown in Fig. 8.27), which changes the extrinsic collector from n-type to insulating.

The emitter width W_E is a relatively uncritical parameter in regard to the high-frequency performance of HBTs. In fact, the requirements on the minimum lateral dimensions of III–V HBTs are much more relaxed than those for microwave FETs and Si BJTs. In the higher GHz applications, the critical lateral dimension of FETs (gate length) and BJTs (emitter width) must be in the deep submicron range, whereas III–V HBTs with an emitter width in the μm range are very common. The emitter width should, however, not be too large. Although the base doping is quite high, a large emitter width will result in an unacceptably large intrinsic base resistance and thus a low f_{max}. A large intrinsic base resistance would also lead to considerable current crowding and low emitter utilization factor (see Section 7.2.2 for discussions of current crowding and the dc emitter utilization factor EU_{dc}). At microwave frequencies, the high-frequency emitter utilization factor EU_{ac} is also important. These two emitter utilization factors have been modeled in [14], and the main results of the study are:

- Both EU_{dc} and EU_{ac} depend on the dc operating conditions, and they decrease with increasing collector current and increasing base current. The reason for this is the increasing lateral voltage drop in the intrinsic base caused by the increasing base current.

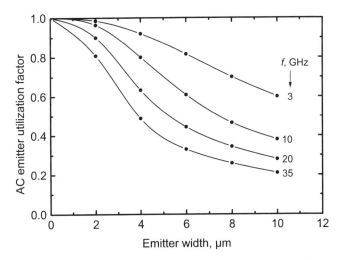

Figure 8.28. High-frequency emitter utilization factor EU_{ac} plotted as a function of emitter width at different frequencies. Data taken from [14].

- Both EU_{dc} and EU_{ac} depend on the base sheet resistance, and they decrease with increasing sheet resistance.
- EU_{ac} becomes smaller when the operating frequency increases.

Figure 8.28 illustrates EU_{ac} calculated as a function of emitter width at different frequencies. As a rule of thumb, the emitter utilization factor should be larger than 0.8 at the operating frequency. Typical emitter widths of GaAs HBTs are in the range of 1–3 μm.

E. Thermal Design. The great majority of HBTs have multifinger, multicell structures and are used in power amplifiers. Thus, the thermal design plays an important role in these devices. The main goal of the design is to minimize self-heating and to strive for uniform temperature and current distributions among the emitter fingers.

The choice of the emitter dimensions, i.e., of the emitter length L_E (the length in the third dimension) and emitter width W_E, depend on the operating frequency. Figure 8.29 shows the emitter dimensions of GaAs power HBTs as a function of the operating frequency. Although the data scatters, the general trend of both L_E and W_E to decrease with increasing operating frequency is clear. A comparison of the emitter length for the HBT from Fig. 8.29 with the gate width W of power GaAs MESFETs (the gate width of a FET corresponds to the emitter length of a bipolar transistor, as both dimensions govern the total output power of the device) reveals that $W \gg L_E$ for a given frequency. For GaAs HBTs with operating frequencies in the range of 1 to 35 GHz, typical values for L_E and W_E are 10–20 μm and 2–4 μm, respectively.

Three different cell designs for power HBTs are shown in Fig. 8.30 [60]. Design (a) consists of one emitter finger and two base contacts located between the two

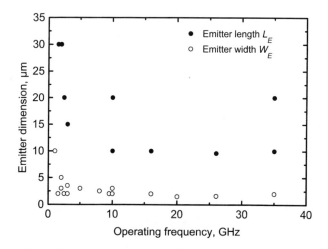

Figure 8.29. Emitter length and emitter width of fabricated GaAs HBTs as a function of the targeted operating frequency.

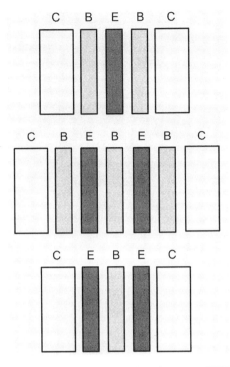

Figure 8.30. Different cell designs for power HBTs [60].

collector contacts. It is a typical cell design and is preferable from the thermal management point of view. This is because, for given W_E and L_E, the power dissipation is low and self-heating is less significant in this structure. Design (b) consists of two emitter fingers and three base contacts. This option is more compact and leads to a higher cell output power than Design (a). The heat removal, however, is more difficult, and more notable self-heating can be expected because of the higher power density in such a cell structure. Design (c) has a structure of two emitter fingers and one base contact and is also more compact than Design (a).

There is another implication associated with the different cell designs. The active transistor area, which influences the intrinsic collector–base capacitance $C_{cb,int}$, is equal to the emitter area $A_E = L_E \times W_E$ for one transistor cell. If we assume the spacings between the emitter, base, and collector contacts are negligibly small compared to the emitter, base, and collector widths (this is a reasonable assumption for a self-aligned technology), the total (intrinsic and extrinsic) collector–base capacitance is proportional to the sum of the areas of the emitter, base, and collector of the cell. The ratios of the intrinsic collector–base capacitance to the total collector–base capacitance are 1/3, 2/5, and 2/3 for Designs (a), (b), and (c), respectively. Thus, Design (c) is the most preferable in terms of f_{max}. To achieve a certain level of output power, a large number of parallel-connected cells is commonly necessary.

A possible way to reduce the current and temperature nonuniformity in the HBT

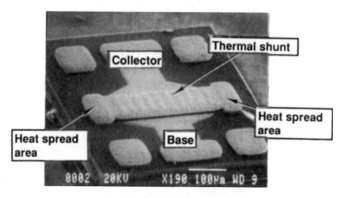

Figure 8.31. Topology of a GaAs power HBT with a thermal shunt. The total emitter area of the transistor is 600 μm². Taken from [62] © 1993 IEEE.

is to use ballast resistors. The effect of emitter ballast resistors on the BJT and GaAs HBT dc behavior has been discussed in Sections 7.5 and 8.4.1, respectively. Another option is to use base ballast resistors, which is particularly effective for GaAs HBTs [14, 61]. It should be noted that base ballasting is not suited for preventing thermal runaway in Si BJTs. A further design option to minimize the temperature inhomogeneity in bipolar transistors is to use thermal shunts. A thermal shunt is a metal bridge that connects all the emitter fingers of a multifinger transistor, thereby creating an equalized temperature on the emitter fingers. Figure 8.31 shows the topology of an AlGaAs/GaAs power HBT having the thermal shunt, and Figs. 8.32(a) and (b) show the cross section and thermal equivalent circuit, respectively, for such a device.

Finally, the design of the substrate offers another important opportunity to reduce the effect of self-heating. The basic rules for the substrate design of power FETs discussed in Section 4.6.3 are applicable for power HBTs. The design options include substrate thinning (a typical substrate thickness is about 30 μm), covering the back side of substrate with a metallic plated heat sink, and using via holes filled with metal.

8.5.2 GaAs-Based HBT Performance

GaAs HBTs with cutoff frequencies f_T in the range of 100–150 GHz (see Fig. 8.26) and maximum frequencies of oscillation f_{max} in the range of 200–350 GHz have been realized. Figure 8.33 shows the reported f_{max} versus f_T data for AlGaAs/GaAs, InGaP/GaAs, and InGaAs-based HBTs. Although the figure does not contain any information on transistor design, it reveals that f_{max} exceeds f_T in most GaAs HBTs. Table 8.6 summarizes the state of the art of GaAs HBTs in terms of f_T and f_{max}. Also included are two transistors with high collector doping densities N_{DC} to demonstrate the influence of N_{DC} on f_{max}. These two transistors do not possess record f_T or f_{max}, but rather the highest reported f_{max} at these high collector doping densities. As ex-

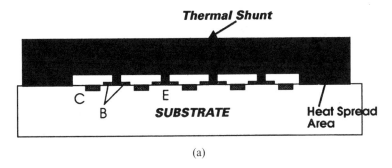

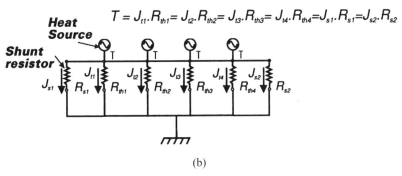

Figure 8.32. (a) Schematic cross section and (b) thermal equivalent circuit of a power HBT with a thermal shunt. Taken from [62] © 1993 IEEE.

pected, an HBT with a lower collector doping density possesses a higher f_{max}. No clear trend of f_T for different N_{DC} can be observed, however.

The reported total output power data of GaAs HBTs as a function of the operating frequency are given in Fig. 8.34, indicating that, as expected, the output power decreases with increasing frequency. The same trend can be found in microwave FETs. The state of the art is demonstrated by GaAs HBTs delivering output powers of 10 W at 1 GHz and 1 W at 35 GHz. The output power density of bipolar transistors is frequently given in µW/mm² (mm² is the emitter area), which is also a decreasing function of the frequency, as shown in Fig. 8.35. For microwave FETs, the output power density is commonly given in W/mm (mm is the gate width). Note that the lateral HBT dimension corresponding to the gate width in FETs is the emitter length L_E. As can be seen in Fig. 1.27, wide-bandgap FETs (AlGaN/GaN HEMTs) possess a maximum output power density of about 10 W/mm, and the output power density of III–V FETs (MESFETs and HEMTs) does not exceed 2 W/mm even at low frequencies. GaAs HBTs, on the other hand, with a record power density of 30 W/mm (mm is the emitter length) at 10 GHz [62] have been reported. These high power densities per mm emitter length, along with the results in Fig. 8.35, demonstrate that HBTs are much more compact devices compared to FETs for power applications. Nevertheless, the total output power of HBTs in commonly smaller than that of high-power FETs. The high

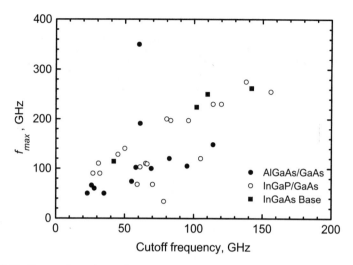

Figure 8.33. Reported maximum frequency of oscillation versus cutoff frequency for GaAs HBTs.

power densities in HBTs give rise to considerable self-heating, and a proper thermal design is critical for such devices.

GaAs HBTs are much less popular for low-noise applications. Nonetheless, there are several studies on the high-frequency noise performance of these devices and on their usage as low-noise transistors. Fig. 8.36 shows the reported minimum noise figures of experimental GaAs HBTs as a function of the frequency. Noise figures below 1 dB at 2 GHz and below 2 dB up to 18 GHz have been demonstrated [38, 68]. Also included are the reported noise data for InP-based HBTs. The noise performance of InP HBTs is superior to that of GaAs HBTs for frequencies up to 5 GHz. Above this frequency, the noise figures InP HBTs are considerably higher than those of GaAs HBTs.

Table 8.6. Record performance of GaAs HBTs in terms of f_T and f_{max}

HBT type (emitter–base junction)	f_T, GHz	f_{max}, GHz	N_{DC}, cm^{-3}	Ref.
AlGaAs/GaAs	60	350	3×10^{16}	[6]
AlGaAs/GaAs	114	148	—	[63]
InGaP/GaAs	156	255	3×10^{16}	[5]
InGaP/GaAs	138	275	2×10^{16}	[64]
InGaP/GaAs	120	230	undoped	[65]
AlGaAs/InGaAs	142	262	5×10^{16}	[53]
AlGaAs/GaAs	95	105	1×10^{17}	[66]
InGaP/GaAs	78	34	2×10^{17}	[67]

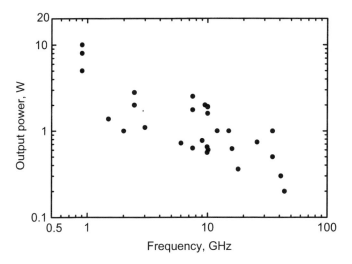

Figure 8.34. Reported output powers of GaAs HBTs as a function of the frequency.

8.6 ISSUES OF InP-BASED HBTs

8.6.1 Transistor Structures

A. Conventional InP HBTs. InP HBTs are typically formed of lattice-matched layers epitaxially grown on InP substrates. Materials lattice-matched to InP are

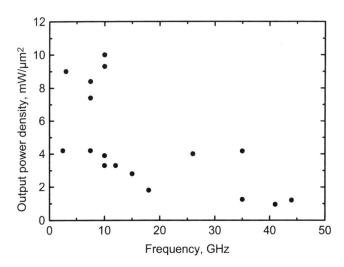

Figure 8.35. Reported output power densities (in mW/μm^2) of GaAs HBTs as a function of the frequency.

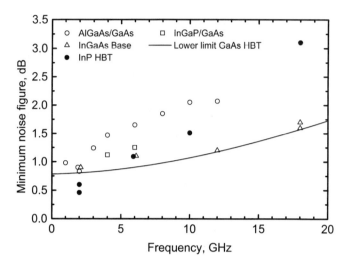

Figure 8.36. High-frequency noise performance of GaAs and InP HBTs. Also included is an empirical lower limit fit.

$In_{0.53}Ga_{0.47}As$ and $In_{0.52}Al_{0.48}As$. Thus, one can combine InP (E_G = 1.35 eV) or $In_{0.52}Al_{0.48}As$ (E_G = 1.45 eV) and the narrower-bandgap material $In_{0.53}Ga_{0.47}As$ (E_G = 0.74 eV) to obtain different versions of SHBTs and DHBTs. Possible options for the layer sequence of InP HBTs have been given in Table 8.1. Although InAlAs/InGaAs SHBTs, InP/InGaAs SHBTs, and InP/InGaAs DHBTs are quite common, only few reports on the InAlAs/InGaAs DHBTs are available. Note that InP HBT is the general designation for HBTs on InP substrates, and the terms InP/InGaAs and InAlAs/InGaAs describe the emitter–base heterostructure pairs of InP HBTs. Table

Table 8.7. Typical layer sequences and make-ups of InP HBTs

Layer	InP/InGaAs DHBT	InP/InGaAs SHBT	InAlAs/InGaAs SHBT
Cap	50–200 nm n^+ InGaAs $1–3 \times 10^{19}$ cm^{-3}	50–200 nm n^+ InGaAs $1–3 \times 10^{19}$ cm^{-3}	50–100 nm n^+ InGaAs $1–3 \times 10^{19}$ cm^{-3}
n^+ Emitter	50–200 nm n^+ InP $1–3 \times 10^{19}$ cm^{-3}	50–200 nm n^+ InP $1–3 \times 10^{19}$ cm^{-3}	$\approx$ 50 nm n^+ InAlAs 10^{19} cm^{-3}
Emitter	50–200 nm n InP $2–8 \times 10^{17}$ cm^{-3}	50–200 nm n InP $2–8 \times 10^{17}$ cm^{-3}	50–200 nm n InAlAs $2–8 \times 10^{17}$ cm^{-3}
Base	30–100 nm p^+ InGaAs $2–8 \times 10^{19}$ cm^{-3}	30–100 nm p^+ InGaAs $2–8 \times 10^{19}$ cm^{-3}	50–100 nm p^+ InGaAs $\approx 3 \times 10^{19}$ cm^{-3}
Collector	200–1000 nm composite n^--InGaAs/n-InP/n^--InP	200–1000 nm InGaAs $1–3 \times 10^{16}$ cm^{-3}	200–1000 nm InGaAs $1–3 \times 10^{16}$ cm^{-3}
Subcollector	100–400 nm n^+ InGaAs $\approx 10^{19}$ cm^{-3}	100–400 nm n^+ InGaAs $\approx 10^{19}$ cm^{-3}	100–700nm n^+ InGaAs $\approx 10^{19}$ cm^{-3}

8.7 lists the typical layer sequence and make-ups (material composition, thickness, doping concentration) of InP HBTs.

The n^+ InGaAs cap layer on the top of the structure ensures a low emitter contact resistance. The emitter layer can be InP or InAlAs, but InP is normally preferred for the following electrical and technological reasons. The bandgap difference ΔE_G of the $In_{0.52}Al_{0.48}As/In_{0.53}Ga_{0.47}As$ pair (i.e., $In_{0.53}Ga_{0.47}As$ is the base layer) is larger than that of $InP/In_{0.53}Ga_{0.47}As$ pair by about 100 meV (see Table 3.4). However, in an abrupt $In_{0.52}Al_{0.48}As/In_{0.53}Ga_{0.47}As$ heterojunction, a large portion of ΔE_G goes to the conduction band offset ΔE_C and only a small portion goes to the valence band offset ΔE_V ($\Delta E_C/\Delta E_V \approx 0.52$ eV/0.19 eV). As such, the InAlAs/InGaAs heterojunction is frequently graded to reduce the spike. The values of ΔE_G, ΔE_C, and ΔE_V for the $InP/In_{0.53}Ga_{0.47}As$ heterojunction, on the other hand, are about 0.62, 0.27, and 0.34 eV, respectively. From this perspective, InP is a better material for the emitter. The second reason has to do with the etching process. We have mentioned earlier that after epitaxial growth of the entire structure, the cap and emitter layers outside the emitter contact have to be etched away. For the InAlAs/InGaAs system, there is no etching solution that can remove the InAlAs emitter but not attack the InGaAs base, a problem similar to the etching of the AlGaAs/GaAs system. For the InP/InGaAs structure, highly selective etching solutions do exist. As a result, from the processing point view, InP/InGaAs is again a better emitter–base heterojunction than InAlAs/InGaAs.

The InGaAs base is thin and heavily doped to simultaneously minimize the base transit time and the intrinsic base resistance. The material composition may be graded to create a built-in field. Note that a graded base is strained, and a homogeneous $In_{0.53}Ga_{0.47}As$ layer is lattice-matched to the underlying collector.

The design of the collector–base junction and the collector is a complex task. The first issue is to decide whether a homojunction or a heterojunction should be used, i.e., whether SHBTs or DHBTs are preferable. The advantage of DHBTs is the wider bandgap of the collector resulting in a larger Early voltage, which is equivalent to a smaller output conductance, and a higher breakdown voltage. The drawback of a collector–base heterojunction is the presence of the conduction band spike hindering the electron transport (see Fig. 8.17). The conduction band spike can be minimized using the so-called composite collector consisting of one of the following layer sequences:

- p^+-InGaAs (base)/n-InGaAs (low doped)/n-InP [69, 70]
- p^+-InGaAs (base)/n-InGaAs (low doped)/n^+-InP (thin)/n-InP [71, 72]
- p^+-InGaAs (base)/n-InGaAsP (composite)/n-InP [73, 74]
- p^+-GaAsSb (base)/n-InP [8]

In principle, either InP or InAlAs is suited for the collector of InP DHBTs, but only the InP collector is commonly used for the following two reasons. First, the undesired conduction band offset at the InGaAs/InAlAs heterointerface is almost twice as much as that at the InGaAs/InP heterointerface. Thus, the suppression of current blocking is more difficult in InGaAs/InAlAs junctions. The second reason is also related to the carrier transport. The collector–base junction is a high-field region, and a high electron saturation velocity in the collector is desirable. The electron saturation velocity in InAlAs is 50% smaller than that in InP.

Finally, the subcollector of InP HBTs consists of heavily doped InGaAs, to reduce the parasitic collector resistance. Since the subcollector is a low-field region, a high low-field electron mobility μ_0 is beneficial, whereas the bandgap of the material is of no concern. InGaAs has a much higher μ_0 than InP and thus is the material of choice.

Typical dopants are Si for the n-type layers and either C or Be for the p-type base. The ohmic emitter and collector contacts can be either alloyed Ni/Ge/Ni/Au or nonalloyed Ti/Pt/Au, whereas for the ohmic base contact Pt/Ti/Pt/Au is used. The specific contact resistances are of the same order as those for GaAs HBTs, i.e., 10^{-6}–10^{-7} Ωcm^2.

The lateral design guidelines for the GaAs HBTs are applicable for the InP HBTs as well.

B. Transferred Substrate InP HBTs. The power gain and maximum frequency of oscillation of bipolar transistors are closely related to the collector–base capacitance C_{cb} [see Eqs. (7-103) and (7-104)]. The intrinsic portion of C_{cb} can be influenced to a certain degree by the doping of the collector. A lower collector doping leads to a widening of the collector–base space–charge region and thus a reduction of C_{cb}. Nonetheless, a wider collector–base space–charge region increases the delay time τ_{BC} [see Eq. (7-97)], and consequently deteriorates the cutoff frequency. Thus, the extent to which C_{cb} could be optimized by the collector doping is quite limited. A more effective way is the reduction of the extrinsic portion of C_{cb}, $C_{cb,ext}$. Conventional design options to reduce $C_{cb,ext}$ have been presented in Section 8.5.1. The ultimate reduction of $C_{cb,ext}$ could be achived by having the same collector–base and emitter–base junction areas. This idea has been adopted in an advanced HBT design called the transferred substrate (TS) HBT [9, 75, 76].

The main processing steps for the InP TS HBT are as follows. First, the layer sequence of the transistor (similar to one of those shown in Table 8.7) is grown epitaxially on the InP substrate. Then the substrate transfer process shown in Fig. 8.37 is carried out. It starts with the deposition of a Si$_3$N$_4$ insulator layer and a thick dielectric layer (benzocyclobutene, BCB). Vias are etched in the BCB and the wafer is electroplated with gold, which also fills the vias. Next the wafer is solder-bonded to a GaAs substrate, and the InP substrate is removed by etching. Finally, the collector contact is formed. The cross section of the TS HBT is shown in Fig. 8.38.

The reduction of $C_{cb,ext}$ in the TS HBT can be illustrated using the schematic given in Fig. 8.39. The collector–base junction area A_{CBj} of this device is equal to $L_E \times (W_E + 2W_{EB} + 2 \times W_B)$. Although it is larger than the collector contact area A_{Cco} given by $A_{Cco} = L_E \times (W_E + 2W_{EB} + 2 \times W_{BC})$, only A_{Cco} contributes to C_{cb}. The reason is that the collector region outside the collector contact is thinned and fully depleted, and a variation of the collector–base voltage does not change the charge storage

8.6 ISSUES OF InP-BASED HBTs 447

1) Normal emitter, base processes. Deposit silicon nitride insulator.

2) Coat with BCB polymer. Etch vias.

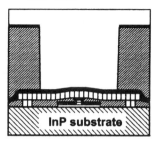

3) Electroplate with gold. Die attach to carrier substrate.

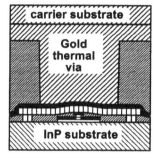

4) Invert wafer. Remove InP substrate. Deposit collector.

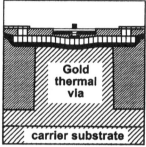

Figure 8.37. Main processing steps for the TS InP HBT. Taken from [9] © 2001 IEEE.

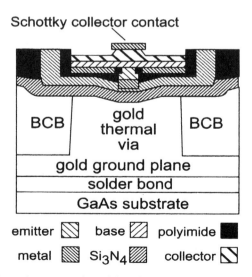

Figure 8.38. Schematic cross section of the TS InP HBT. Taken from [9] © 2001 IEEE.

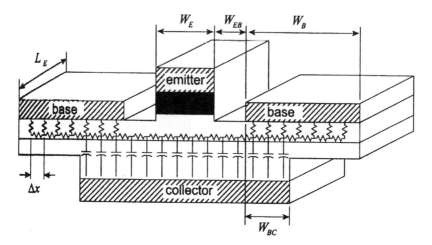

Figure 8.39. Equivalent circuit of the TS HBT showing the resistive and capacitive components in the collector–base junction. Taken from [9] © 2001 IEEE.

in this region. Thus, the collector–base capacitance in the TS HBT does not depend on the collector–base junction area but on the collector contact area only. Although the current TS technology is certainly not suited for the cost-effective mass production, it nonetheless clearly demonstrates the ultimate potential of InP HBTs and has resulted in HBTs with record f_{max} considerably exceeding the f_{max} obtained with any other type of microwave transistors.

8.6.2 InP HBT Performance

We first discuss the performance of conventional InP HBTs and then present data on InP TS HBT performance. Conventional InP HBTs show extremely high frequency limits. Transistors with f_T exceeding 300 GHz and a record f_{max} of 300 GHz have been reported. In general, InP HBTs show higher f_T and f_{max} (if the 350 GHz f_T of the GaAs HBT from [6] is excluded) than all other classes of bipolar transistors. Figure 8.40 shows the reported maximum frequency of oscillation versus cutoff frequencies for various InP HBTs. Impressive frequency performances are found. InP/InGaAs HBTs show higher f_T and f_{max} compared to InAlAs/InGaAs transistors. Table 8.8 summarizes the state of the art of InP HBTs in terms of f_T and f_{max}.

The reported cutoff frequency of InP HBTs as a function of the emitter–collector breakdown voltage is shown in Fig. 8.41. Included is also data for GaAs HBTs. Again, the tradeoff between the speed and breakdown voltage is clearly illustrated. As expected, DHBTs exhibit a larger breakdown voltage for a given cutoff frequency compared to SHBTs, as the breakdown voltage is related to the bandgap of the collector material. In general, the InP SHBT breaks down at a smaller voltage than InP DHBTs and GaAs HBTs.

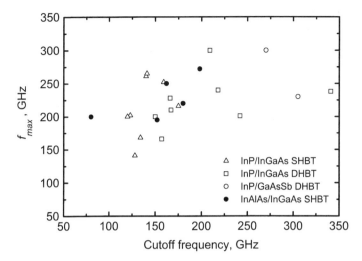

Figure 8.40. Reported maximum frequency of oscillation versus cutoff frequency for InP HBTs.

The minimum noise figures of InP HBTs are shown in Fig. 8.36. A minimum noise figure of about 0.5 dB at 2 GHz has been reported [39, 74], which is superior to the noise figures of GaAs HBTs. Very little data has been reported on the power performance of InP HBTs so far [78, 79] but, generally, the maximum output power obtainable from InP HBTs is lower than that from GaAs HBTs.

Finally, we take a look at the frequency limits of experimental transferred substrate InP HBTs. Table 8.9 shows the reported f_T and f_{max}, together with the design details, of TS SHBTs and DHBTs. An astonishing f_{max} of larger than 1000 GHz is demonstrated.

Table 8.8. Cutoff frequency, maximum frequency of oscillation, and make-ups of high-performance InP HBTs

Type	f_T, GHz	f_{max}, GHz	Emitter area, ($W_E \times L_E$), μm²	Base, nm	Collector, nm	Ref.
InP/InGaAs SHBT	175	216	0.6 × 9.1	50	300	[72]
InP/InGaAs SHBT	140	261	0.6 × 9.1	50	300	[72]
InP/InGaAs DHBT	341	238	0.8 × 3.0	30	150	[7]
InP/InGaAs DHBT	209	300	0.6 × 9.1	50	300	[72]
InP/InGaSb DHBT	270	300	0.4 × 11	25	200	[8]
InP/InGaSb DHBT	305	230	0.4 × 11	20	200	[8]
InAlAs/InGaAs SHBT	198	272	1 × 3	?	200	[77]

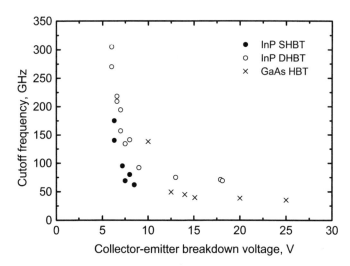

Figure 8.41. Reported cutoff frequency versus the collector–emitter breakdown voltage for InP SHBTs, InP DHBTs, and GaAs HBTs.

8.7 ISSUES OF SiGe HBTs

8.7.1 Transistor Structures

A. SiGe Base HBT and True SiGe HBT. In general, the vertical structure of a SiGe HBT consists of, from bottom up, a Si substrate, an n^+-type Si subcollector, an n-type Si collector, a p-type strained SiGe base, and an n-type Si emitter. From the very beginning of SiGe HBT research, two different design philosophies had been developed and investigated, and SiGe HBTs based on these two concepts are currently commercially available.

One of the concepts was introduced by IBM researchers and is frequently called SiGe base HBT [25, 81]. It employs the same basic design and doping profile as Si BJTs, i.e., an n^+-type polysilicon emitter, followed by a n^+-type emitter formed by outdiffusion of As from the polysilicon emitter to the underlying Si, and a p-type base. The main difference of the SiGe-based HBT compared to a Si BJT is the base.

Table 8.9. Frequency limits and design details of InP transferred substrate HBTs

Type	f_T, GHz	f_{max}, GHz	Emitter/collector area, $W_E \times L_E, W_C \times L_C$, μm²	Base, nm	Collector, nm	Ref.
InAl/InGaAs SHBT	204	1080	0.4 × 6/0.7 × 10.0	40	300	[9]
InAlAs/InGaAs SHBT	295	295	1 × 8/2 × 8.5	30	200	[76]
InAlAs/InGaAs SHBT	162	820	0.4 × 6/0.4 × 6	40	270	[75]
InP/InGaAs DHBT	141	425	0.5 × 8/1.2 × 8.75	40	300	[80]

It consists of strained narrow bandgap p-type SiGe with a doping level comparable to Si BJTs, i.e., less than the emitter doping. The Ge content x in the base in nonuniform. At the emitter–base junction it is low (commonly $x = 0$ to 0.1) and increases toward the collector–base junction. The main effect of the Ge is the decreasing bandgap toward the collector and the resulting built-in field, whereas the effect of the heterojunction between emitter and base is only of secondary importance.

The second concept, pioneered by a Daimler Benz group, fully exploits the properties of the emitter–base heterojunction [82, 83]. To distinguish this concept from the SiGe base HBT we will call it True SiGe HBT. The main feature of the true SiGe HBT is a base doping considerably exceeding the emitter doping, similar to III–V HBTs. The Ge content in the base of these transistors is much higher than in SiGe base HBTs and is in the range of $x = 0.2$ to 0.4. Although the Ge profile may be graded, the grading is of secondary importance. Because the base doping is very high, the base can be made extremely thin and the base sheet resistance is lower compared to SiGe-based HBTs. The sheet resistance of the intrinsic base of true SiGe HBTs is around 1 kΩ/□ compared to 4–10 kΩ/□ for SiGe base HBTs.

In Figs. 8.42 and 8.43, the doping and Ge profiles characteristic for SiGe base HBTs and true SiGe HBTs are shown. Table 8.10 lists the typical layer sequence and design of the two SiGe HBT concepts. It goes without saying that features of both basic concepts can be combined in the design of a HBT.

As in the case of Si BJTs and III–V HBTs, optimizing the collector doping concentration N_{DC} is a challenging task. First, one has to find a compromise between the cutoff frequency (requiring high N_{DC}) and breakdown voltage (requiring low N_{DC}). Furthermore, N_{DC} affects both the intrinsic and the extrinsic collector–base capacitances $C_{cb,int}$ and $C_{cb,ext}$. If a high f_T and thus a high N_{DC} in the intrinsic collector is required, then an option to minimize $C_{cb,ext}$ is to use the so-called selectively implanted collector (SIC). It is realized before the emitter metallization by ion

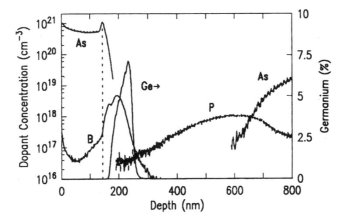

Figure 8.42. Typical doping and Ge profiles of the SiGe base HBT. The polysilicon emitter extends from depth = 0 to the dashed vertical line. Taken from [25] © 1995 IEEE.

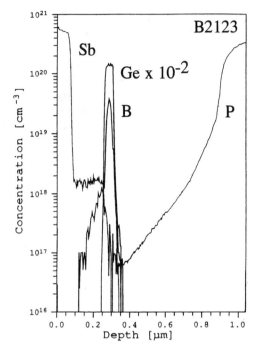

Figure 8.43. Typical doping and Ge profiles of the true SiGe HBT. Taken from [82] © 1992 IEEE.

Table 8.10. Typical layer sequence and make-ups of SiGe base and true SiGe HBTs

Layer	SiGe base HBT	True SiGe HBT
Cap	150–250 nm n^+-polysilicon $\approx 10^{21}$ cm^{-3} (dopant As or P)	50–300 nm n^+-Si 2–3×10^{20} cm^{-3} (dopant Sb)
Emitter	20–40 nm n^+-Si $\approx 10^{21}$ cm$^{-3} \to N_{AB}$ (dopant As or P)	50–250 nm n-Si 3×10^{17}-2×10^{18} cm^{-3} (dopant Sb)
Spacer (optional)	2–5 nm i-SiGe	2–5 nm i-SiGe
Base	30–50 nm strained p-SiGe 2×10^{18}-3×10^{19} cm^{-3} (dopant B)	8–40 nm strained p^+-SiGe 2–8×10^{19} cm^{-3} (dopant B)
Spacer (optional)	5–15 nm i-SiGe	5–15 nm i-SiGe
Collector	100–400 nm n-Si 10^{16}–7×10^{17} cm^{-3} (dopand P)	100–400 nm n-Si 10^{16}–7×10^{17} cm^{-3} (dopand Sb)
Subcollector	n^+-Si	n^+-Si

implantation into the intrinsic collector through the emitter window and the SiGe base. Thus, the active collector region, i.e., the region underneath the emitter, becomes highly doped, whereas the extrinsic collector remains lightly doped.

The subcollector of SiGe HBTs can either be ion-implanted into the Si substrate and subsequently diffused during annealing, or be grown epitaxially. After the formation of the subcollector, all other layers (except the polysilicon emitter) are grown epitaxially either by chemical vapor deposition (CVD) or by MBE. While the CVD is common for SiGe base HBTs, the MBE is more suitable for true SiGe HBTs. For the ohmic collector contact Ti/Au or Ti/Pt/Au metallizations are common and for the base contact typically Pt/Au is used. In true SiGe HBTs the emitter contact materials are Pt/Au or Cr/Au, whereas for the SiGe-based HBT the same emitter contacts as in conventional Si BJTs are used.

During the post-epitaxial processing, boron diffuses out of the thin and heavily doped SiGe base into the collector and emitter. This can lead to a shift of the Si/SiGe interfaces into the Si emitter and Si collector. Then the pn junctions do no longer coincide with the Si/SiGe heterojunctions, but n-Si/p-Si homojunctions will occur [84]. This will lead to conduction band barriers at the emitter–base and base collector junctions. At the emitter side, the energy barrier for electron injection will increase and part the advantage of the heterojunction will be lost. The barrier at the collector–base junction, on the other hand, has an effect like the barrier shown in Fig. 8.17 and leads to current blocking, increased electron storage in the base, and a decrease of the cutoff frequency. Inserting undoped spacer layers can be used to guarantee the location of the pn junctions inside the SiGe layer even if boron outdiffusion occurs. Another effective means to suppress base dopant out-diffusion in SiGe HBTs is to add carbon to the SiGe base during growth [85]. Within only a few years, such SiGe:C HBTs have become very popular and are fabricated by several companies [86–88]. Typical carbon concentrations are in the range from 5×10^{19} to 5×10^{20} cm^{-3}.

B. Structure Optimization Since the early 1990s, a lot of work has been done to develop reliable device models and simulation tools for SiGe HBTs (see, e.g., [89–95]). These models and tools have been extensively used to investigate the physics of SiGe HBTs and to optimize their design. In the following, we will discuss several issues related to the design of SiGe HBTs with high values of the characteristic frequencies f_T and f_{max}.

In Fig. 8.44 the influence of the Ge profile on the cutoff frequency of two basic SiGe HBT structures is illustrated. The cutoff frequencies have been obtained from numerical device simulations. Details on the device simulator used can be found in [96]. The Ge profiles investigated have a trapezoid shape as shown in Fig. 8.8(b). The Ge content at the emitter–base junction (in %) is designated as γ and the length of the Ge grading is expressed by θ, where θ is the length of the graded region divided by the base thickness in %. Details on the structures are given in Table 8.11.

Clearly, the transistor structure with the narrow base and collector layers shows a higher maximum cutoff frequency than the transistor with the more relaxed design. A Ge grading is beneficial for both structures. It can be seen from Fig. 8.44,

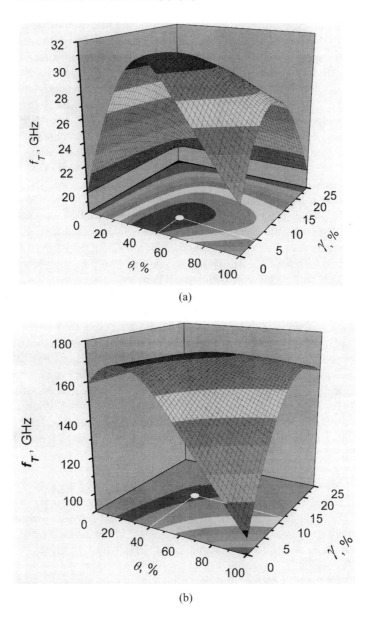

Figure 8.44. Cutoff frequency of two different basic SiGe HBT structures as a function of the Ge profile in the base. The Ge profile is expressed by the parameters Θ and γ; see text. The maximum Ge content in both structures is 25%. (a) Structure 1 from Table 8.11. (b) Structure 2 from Table 8.11.

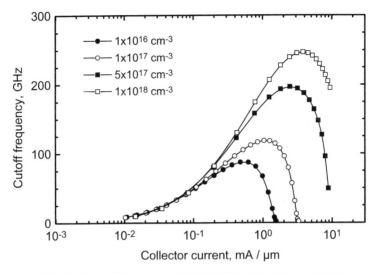

Figure 8.45. Calculated cutoff frequency of SiGe HBTs with different SIC dopings versus collector curent.

however, that the dependence of the cutoff frequency on variations of the Ge profile is different for the two structures.

The frequency limits of SiGe HBTs can be estimated from the simulation results in Figs. 8.45 and 8.46 [97]. Here, extremely scaled SiGe HBT structures with an emitter width W_E of 0.2 μm, a base thickness w_B of 10 nm, a collector thickness w_C of 150 nm, and with a selectively implanted collector (SIC) have been investigated. The doping concentration of the extrinsic collector is 10^{16} cm^{-3} in all cases, and the SIC doping has been varied. Figure 8.45 shows the cutoff frequency of the scaled HBT as a function of the collector current. An increasing doping of the the SIC results in a continous increase of f_T. Because the base thickness of the HBT is very small, the total emitter–collector transit time τ_{EC} is dominated by the collector–base space–charge region transit time τ_{BC} [see Eq. (7-97)]. A high doping of the SIC results in a narrow collector–base space–charge region, and thus in a small τ_{BC} and in a high f_T. The maximum frequency of oscillation shown in Fig. 8.46 reveals a different behavior. If the SIC doping is increased from 10^{16} cm^{-3} up to about 5×10^{17} cm^{-3}, f_{max} increases as well. For SIC dopings above 5×10^{17} cm^{-3}, however, f_{max}

Table 8.11. Design information on the SiGe HBTs from Fig. 8.44

Structure	Base, w_B, nm	Collector, w_C, nm	N_{DC}, cm^{-3}	Maximum f_T, GHz	Optimum Θ, %	Optimum γ, %
1	60	400	1.5×10^{16}	30	50	4.8
2	10	100	2.5×10^{17}	169	40	12

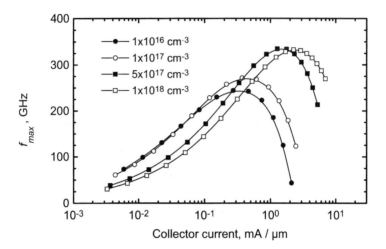

Figure 8.46. Calculated maximum frequency of oscillation of SiGe HBTs with different SIC dopings versus collector curent.

does not increase further but rather slightly drops. Such a behavior can is attributed to the competition of two effects [see Eq. (7-104)]. First, a higher SIC doping will lead to an enhancement of f_T as discussed above. At the same time, however, the space–charge region of the collector–base junction becomes narrower, leading to a rise of the collector–base capacitance C_{cb}. The first trend tends to increase f_{max}, but

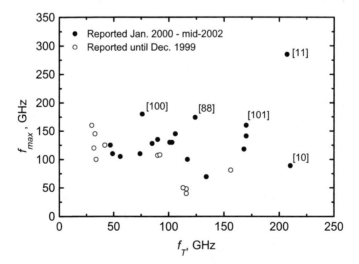

Figure 8.47. Reported maximum frequency of oscillation versus cutoff frequency for SiGe HBTs. Note the considerable progress made in the years 2000 and 2001.

the second one has the opposite effect. It should be noted that SiC doping concentrations of 10^{18} cm^{-3} will lead to very low breakdown voltages.

The data shown in Figs. 8.45 and 8.46 suggest that a properly designed SiGe could achieve both f_T and f_{max} values around 300 GHz. The progress of SiGe HBT performance obtained during the late 1990s and the state of the art transistors reported until early 2002 confirm this statement. In the following section we will discuss the state of the art of SiGe HBTs in more detail.

8.7.2 SiGe HBT Performance

The first SiGe HBT was reported in 1987 [98]. Within the next three years, the cutoff frequency of SiGe HBTs had reached 75 GHz, which clearly exceeded the speed obtainable from Si BJTs at that time [99]. Since then, a steady improvement of SiGe HBT high-frequency performance has been made. Figure 8.47 shows the reported maximum frequencies of oscillation f_{max} versus cutoff frequencies f_T of SiGe HBTs. Note the remarkable progress made in the period from 2000 to 2002. The state of the art of SiGe HBTs in terms of f_T and f_{max} is summarized in Table 8.12.

In Fig. 8.48, the reported cutoff frequencies of SiGe HBTs are shown as a function of the base thickness. Clearly, a thin base is critical for achieving a high f_T. Figure 8.49 compiles the reported f_T and f_{max} of SiGe HBTs versus the collector doping density. It confirms a design issue we have already known from Si BJTs and III–V HBTs, namely the tradeoff between f_T and f_{max} in regard to the collector doping. A high collector doping density leads to a high cutoff frequency, but at the expense of a reduced maximum frequency of oscillation.

The minimum noise figures verus the frequency of SiGe HBTs are shown in Fig. 8.50. The high-frequency noise behavior of SiGe HBTs is the best among all bipolar transistors. This is the reason why SiGe HBTs compete very well with GaAs MESFETs and GaAs HEMTs for low-noise applications up to 10 GHz.

Another important issue is that the potential of SiGe HBTs can be more fully exploited if they are integrated into an existing CMOS process. This integration will allow for the manufacturing of cost-effective, high-performance BiCMOS ICs for mass consumer markets such as mobile communications. Much work has been done since the late 1990s to develop a CMOS-compatible SiGe HBT technology. Figure

Table 8.12. Cutoff frequency, maximum frequency of oscillation, emitter–collector breakdown voltage, and emitter width of high-performance SiGe HBTs

f_T, GHz	f_{max}, GHz	BV_{CE0}, V	W_E, μm	Ref.
210	89	1.8	0.2	[10]
76	180	2.5	0.2	[100]
124	174	2.3	0.2	[88]
170	160	2	0.15	[101]
207	285	1.7	0.12	[11]

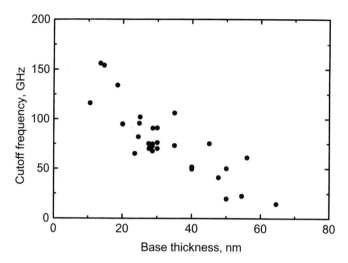

Figure 8.48. Reported cutoff frequency of SiGe HBTs as a function of the base thickness.

8.51 shows f_T and f_{max} obtained from SiGe HBTs fabricated using a BiCMOS process as a function of emitter width, which is the smallest lateral dimension of an HBT. If the rule of thumb that $f_T \approx f_{max}$ should be 10 times the operating frequency of the system is adopted, then it becomes clear from Fig. 8.51 that CMOS-compatible SiGe HBTs with emitter widths of 0.15, 0.4, and 0.8 μm are well suited for 15, 10, and 5 GHz operations.

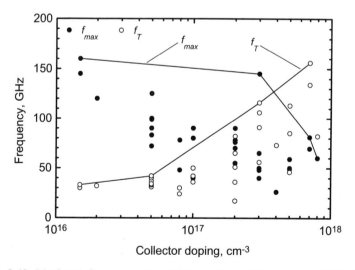

Figure 8.49. Maximum frequency of oscillation and cutoff frequency of SiGe HBTs as a function of the collector doping.

8.7 ISSUES OF SiGe HBTs   459

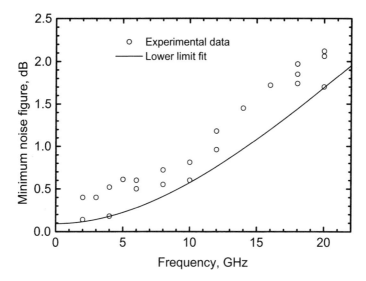

Figure 8.50. Reported minimum noise figures of SiGe HBTs as a function of the frequency. Also shown is an empirical lower limit fit.

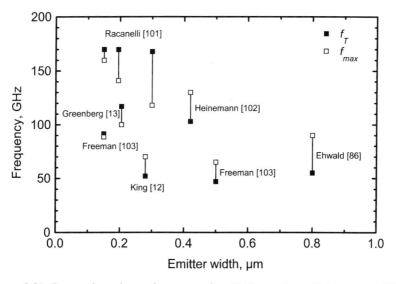

Figure 8.51. Reported maximum frequency of oscillation and cutoff frequency of HBTs based on the Si BiCMOS technology.

Prior to the mid 1990s, SiGe HBTs were not considered suitable for power amplifiers due to their low breakdown voltages. Figure 8.52 shows the reported cutoff frequency of SiGe HBTs versus the collector–emitter breakdown voltage. Also included are data for Si BJTs. The breakdown voltage of SiGe HBTs is normally not higher than that of Si BJTs having a comparable collector design. However, for a given BV_{CE0}, SiGe HBTs exhibit a considerably higher cutoff frequency. Also note that the breakdown voltages in Fig. 8.52 are considerably lower than those of GaAs HBTs. Regardless of this drawback, the advantage of being able to integrate high-speed Si and SiGe components onto a single chip has led to increased research activities on SiGe power HBTs in the late 1990s and early 2000s. A highlight is a SiGe power HBT delivering an enormous 230 W output power at 2.8 GHz [104]. This transistor has a very low collector doping concentration of only 3×10^{15} cm^{-3}, thereby resulting in high BV_{CE0} of 39 V and BV_{CB0} of 62 V (unfortunately, no information on f_T is given). Table 8.13 summarizes the state of the art of SiGe power HBTs.

The development of SiGe HBTs was no doubt one of the hottest and most active topics in the microwave semiconductor electronics industry during the late 1990s and early 2000s. SiGe HBTs clearly outperformed Si BJTs in terms of frequency performance, and they are serious competitors to III–V transistors in low-noise applications as well. Further advances in the field of SiGe power HBTs can be expected.

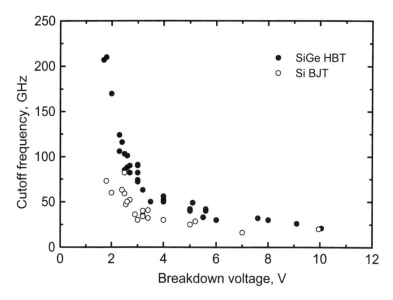

Figure 8.52. Reported cutoff frequency versus collector–emitter breakdown voltage for Si BJTs and SiGe HBTs.

Table 8.13. State of the art of SiGe power HBTs

Frequency, GHz	Output power, W	Output power, mW/μm^2	Density, W/mm	Ref.
2.4	1	2.4	6.1	[105]
2.8	230	—	—	[104]
5.7	0.1	0.79	1	[106]
8.4	0.32	—	—	[107]
12.6	0.28	0.4	—	[108]

REFERENCES

1. W. Shockley, Claim 2 of U.S. Patent 2.569.347, Filed 26 June 1948, Issued 25 September 1951, Expired 24 September 1968.
2. H. Kroemer, Theory of a Wide-Gap Emitter for Transistors, *Proc. IRE, 45,* pp. 1535–1537, 1957.
3. W. P. Dumke, J. M. Woodall, and V. L. Rideout, GaAs-GaAlAs Heterojunction Transistor for High Frequency Operation, *Solid-State Electron., 15,* pp. 1339–1343, 1972.
4. J.-P. Bailbe, A. Marty, P. H. Hiep, and G. E. Rey, Design and Fabrication of High-Speed GaAlAs/GaAs Heterojunction Transistors, *IEEE Trans. Electron Devices, 27,* pp. 1160–1164, 1980.
5. T. Oka, K. Hirata, K. Ouchi, H. Uchiyama, T. Taniguchi, K. Mochizuki, and T. Nakamura, Advanced Performance of Small-Scaled InGaP/GaAs HBT's with f_T over 150 GHz and f_{max} over 250 GHz, *IEDM Tech. Dig.,* pp. 653–656, 1998.
6. W. J. Ho, N. L. Wang, M. F. Chang, A. Sailer, and J. A. Higgins, Self-Aligned, Emitter-Edge-Passivated AlGaAs/GaAs Heterojunction Bipolar Transistors with Extrapolated Maximum Oscillation Frequency of 350 GHz, *Dev. Res. Conf.,* paper IVA-1, 1992. See also *IEEE Trans. Electron Devices, 39,* p. 2655, 1992.
7. M. Ida, K. Kurishima, N. Watanabe, and T. Enoki, InP/InGaAs DHBTs with 341-GHz f_T at High Current Density of Over 800 kA/cm^2, *IEDM Tech. Dig.,* pp. 776–779, 2001.
8. M. W. Dvorak, O. J. Pitts, S. P. Watkins, and C. R. Bolognesi, Abrupt Junction InP/GaAsSb/InP Double Heterojunction Bipolar Transistors with F_T as High as 250 GHz and BV_{CE0} > 6 V, *IEDM Tech. Dig.,* pp. 178–181, 2000.
9. M. Rodwell, M. Urteaga, T. Mathew, D. Scott, D. Mensa, Q. Lee, J. Guthrie, Y. Betser, S. C. Martin, R. P. Smith, S. Janagathan, S. Krishnan, S. I. Long, R. Pullela, B. Agarwal, U. Bhattacharya, L. Samoska, and M. Dahlstrom, Submicron Scaling of HBTs, *IEEE Trans. Electron Devices, 48,* pp. 2606–2624, 2001.
10. S. J. Jeng, B. Jagannathan, J.-S. Rieh, J. Johnson, K. T. Schonenberg, D. Greenberg, A. Stricker, H. Chen, M. Khater, D. Ahlgreen, G. Freeman, K. Stein, and S. Subbanna, A 210-GHz f_T SiGe HBT with a Non-Self-Aligned Structure, *IEEE Electron Device Lett., 22,* pp. 542–544, 2001.
11. B. Jagannathan, M. Khater, F. Pagette, J.-S. Rieh, D. Angell, H. Chen, J. Florkey, F. Golan, D. R. Greenberg, R. Groves, S. J. Jeng, J. Johnson, E. Mengistu, K. T. Schonen-

berg, C. M. Schnabel, P. Smith, A. Stricker, D. Ahlgreen, G. Freeman, K. Stein, and S. Subbanna, Self-Aligned SiGe NPN Transistors with 285 GHz f_{MAX} and 207 GHz f_T in a Manufacturable Technology, *IEEE Electron Device Lett. 23,* pp. 258–260, 2002.

12. C. A. King, M. R. Frei, M. Mastrapasqua, K. K. Ng, Y. O. Kim, R. W. Johnson, S. Moinian, S. Martin, H.-I. Cong, F. P. Klemens, R. Tang, D. Nguyen, T.-I. Hsu, T. Campbell, S. J. Molloy, L. B. Fritzinger, T. G. Ivanov, K. K. Bourdelle, C. Lee, Y.-F. Chyan, M. S. Carroll, and C. W. Leung, Very Low Cost Graded SiGe Base Bipolar Transistors for a High Performance Modular BiCMOS Process, *IEDM Tech. Dig.,* pp. 565–568, 1999.

13. D. Greenberg, S. Sweeney, C. LaMothe, K. Jenkins, D. Friedman, B. Martin Jr., G. Freeman, D. Ahlgren, S. Subbanna, and A. Joseph, Noise Performance and Considerations for Integrated RF/Analog/Mixed-Signal Design in a High-Performance SiGe BiCMOS Technology, *IEDM Tech. Dig.,* pp. 495–498, 2001.

14. W. Liu, *Handbook of III–V Heterojunction Bipolar Transistors,* Wiley, New York, 1998.

15. H. Kroemer, Heterostructure Bipolar Transistors and Integrated Circuits, *Proc. IEEE, 70,* pp. 13–25, 1982.

16. S. Tiwari and S. L. Wright, Material Properties of p-type GaAs at Large Dopings, *Appl. Phys. Lett., 56,* pp. 563–565, 1990.

17. F. M. Bufler, P. Graf, B. Meinerzhagen, H. Kibbel, and G. Fischer, A New Comprehensive and Experimentally Verified Electron Transport Model for Strained SiGe, *Proc. SISPAD,* pp. 57–58, 1996.

18. B. Meinerzhagen and F. M. Bufler, Private Communication.

19. D. M. Nuernbergk, *Simulation des Transportverhaltens in Si/Si$_{1-x}$Ge$_x$/Si-Heterobipolartransistoren,* H. Utz Verlag, München 1999.

20. C. H. Henry, R. A. Logan, F. R. Merrit, and C. G. Betha, Radiative and Nonradiative Lifetimes in n-Type and p-Type 1.6 μm InGaAs, *Electron. Lett., 20,* pp. 358–359, 1984.

21. S. C. Jain and D. J. Roulston, A Simple Expression for Bandgap Narrowing (BGN) in Heavily-Doped Si, Ge, GaAs, and Ge$_x$Si$_{1-x}$ Strained Layers, *Solid-State Electron., 34,* pp. 453–465, 1991.

22. S. C. Jain, J. M. McGregor, and D. J. Roulston, Band-Gap Narrowing in Novel III–V Semiconductors, *J. Appl. Phys. 68,* pp. 3747–3749, 1990.

23. C. D. Parikh and F. A. Lindholm, Space-Charge Region Recombination in Heterojunction Bipolar Transistors, *IEEE Trans. Electron Devices, 39,* pp. 2197–2205, 1992.

24. D. L. Harame, J. M. C. Stork, B. S. Meyerson, K. Y.-J. Hsu, J. Cotte, K. A. Jenkins, J. D. Cressler, P. Restle, E. F. Crabbe, S. Subbanna, T. E. Tice, B. W. Scharf, and J. A. Yasaitis, Optimization of SiGe HBT Technology for High Speed Analog and Mixed-Signal Applications, *IEDM Tech. Dig.,* pp. 71–74, 1993.

25. D. L. Harame, J. H. Comfort, J. D. Cressler, E. F. Crabbe, J. Y.-C. Sun, B. S. Meyerson, and T. Tice, Si/SiGe Epitaxial-Base Transistors–Part I: Materials, Physics, and Circuits, *IEEE Trans. Electron Devices, 42,* pp. 455–468, 1995.

26. J. D. Cressler, J. H. Comfort, E. F. Crabbe, G. L. Patton, J. M. C. Stork, J. Y.-C. Sun, and B. S. Meyerson, On the Profile Design and Optimization of Epitaxial Si- and SiGe-Base Bipolar Technology for 77 K Applications–Part I: Transistor DC Design Considerations, *IEEE Trans. Electron Devices, 40,* pp. 525–541, 1993.

27. J. D.Cressler, E. F. Crabbe, J. H. Comfort, J. M. C. Stork, and J. Y.-C. Sun, On the Profile Design and Optimization of Epitaxial Si- and SiGe-Base Bipolar Technology for 77 K Applications–Part II: Circuit Performance Issues, *IEEE Trans. Electron Devices, 40,* pp. 542–556, 1993.
28. A. A. Grinberg, M. S. Shur, R. J. Fischer, and H. Morkoc, An Investigation of the Effect of Graded Layers and Tunneling on the Performance of AlGaAs/GaAs Heterojunction Bipolar Transistors, *IEEE Trans. Electron Devices, 31,* pp. 1758–1765, 1984.
29. J. J. Liou, *Principles and Analysis of AlGaAs/GaAs Heterojunction Bipolar Transistors,* Artech House, Norwood, MA 1996.
30. J. J. Liou, C. S. Ho, L. L. Liou, and C. I. Huang, An Analytical Model for Current Transport in AlGaAs/GaAs Abrupt HBTs with a Setback Layer, *Solid-State Electron., 36,* pp. 819–825, 1993.
31. T. Oka, K. Hirata, H. Suzuki, K. Ouchi, H. Uchiyama, T. Taniguchi, K. Mochizuki, and T. Nakamura, High-Speed Small-Scale InGaP/GaAs HBT Technology and its Application to Integrated Circuits, *IEEE Trans. Electron Devices, 48,* pp. 2625–2630, 2001.
32. A. Bandyopadhyay, S. Subramanian, S. Chandrasekhar, A. G. Dentai, and S. M. Goodnick, Degradation of DC Characteristics of InGaAs/InP Single Heterojunction Bipolar Transistors Under Electron Irradiation, *IEEE Trans. Electron Devices, 46,* pp. 840–849, 1999.
33. S. C. Lee, J. N. Kau, and H.-H. Lin, Origin of High Offset Voltage in an AlGaAs/GaAs Heterojunction Bipolar Transistor, *Appl. Phys. Lett., 45,* pp. 1114–1116, 1984.
34. N. Chand, R. Fischer, and H. Morkoc, Collector-Emitter Offset Voltage in AlGaAs/GaAs Heterojunction Bipolar Transistor, *Appl. Phys. Lett., 47,* pp. 313–315, 1985.
35. T. Won, S. Iyer, S. Agarwal, and H. Morkoc, Collector Offset Voltage of Heterojunction Bipolar Transistor Grown by Molecular Beam Epitaxy, *IEEE Electron Device Lett., 10,* pp. 274–276, 1989.
36. L. L. Liou, J. Ebel, and C. I. Huang, The Offset Voltage of Heterojunction Bipolar Transistors Using Two-Dimensional Numerical Simulation with Current Boundary Conditions, *IEEE Trans. Electron Devices, 39,* pp. 742–745, 1992.
37. P. E. Cottrell and Z. Yu, Velocity Saturation in the Collector of $Si/Ge_xSi_{1-x}/Si$ HBT's, *IEEE Electron Device Lett., 11,* pp. 431–433, 1990.
38. H. Dodo, Y. Amamiya, T. Niwa, M. Mamada, N. Goto, and H. Shimawaki, Microwave Low-Noise AlGaAs/InGaAs HBT's with p^+-Regrown Base Contacts, *IEEE Electron Device Lett., 19,* pp. 121–123, 1998.
39. Y. K. Chen, R. Nottenburg, M. B. Panish, R. A. Hamm, and D. A. Humphrey, Microwave Noise Performance of InP/InGaAs Heterostructure Bipolar Transistors, *IEEE Electron Device Lett., 10,* pp. 470–472, 1989.
40. H. Schumacher, U. Erben, and A. Gruhle, Noise Characterization of Si/SiGe Heterojunction Bipolar Transistors at Microwave Frequencies, *Electron. Lett., 28,* pp. 1167–1168, 1992.
41. W. E. Ansley, J. D. Cressler, and D. M. Richey, Base-Profile Optimization for Minimum Noise Figure in Advanced UHV/CVD SiGe HBT's, *IEEE Trans. Microwave Theory Tech., 46,* pp. 653–660, 1998.
42. G.-B. Gao, M.-Z. Wang, X. Gui, and H. Morkoc, Thermal Design Studies of High-

Power Heterojunction Bipolar Transistors, *IEEE Trans. Electron Devices, 36,* pp. 854–863, 1989.

43. W. Liu and B. Bayraktaroglu, Theoretical Calculations of Temperature and Current Profiles in Multi-Finger Heterojunction Bipolar Transistors, *Solid-State Electron., 36,* pp. 125–132, 1993.
44. C-W. Kim, N. Goto, and K. Honjo, Thermal Behavior Dependending on Emitter Finger and Substrate Configurations in Power Heterojunction Bipolar Transistors, *IEEE Trans. Electron Devices, 45,* pp. 1190–1195, 1998.
45. W. Liu and A. Yuksel, Thermal-Electric Feedback Coefficients in GaAs-Based Heterojunction Bipolar Transistors, *Solid-State Electron., 38,* pp. 407–411, 1995.
46. W. Liu, Thermal Management to Avoid the Collapse of Current Gain in Power Heterojunction Bipolar Transistors, *Dig. GaAs IC Symp.,* pp. 147–150, 1995.
47. W. Liu, E. Beam, T. Kim, and A. Khatibzadeh, Recent Developments in GaInP/GaAs Heterojunction Bipolar Transistors, in M. F. Chang (ed.), *Current Trends in Heterojunction Bipolar Transistors,* World Scientific, Singapore, 1996.
48. W. Liu, S. K. Fan, T. Henderson, and D. Davito, Temperature Dependence of Current Gains in GaInP/GaAs and AlGaAs/GaAs Heterojunction Bipolar Transistors, *IEEE Trans. Electron Devices, 40,* pp. 1351–1353, 1993.
49. M. Hafizi, W. E. Stanchina, R. A. Metzger, P. A. Mcdonald, and F. Williams, Jr., Temperature Dependence of DC and RF Characteristics of AlInAs/GaInAs HBTs, *IEEE Trans. Electron Devices, 40,* pp. 1583–1588, 1993.
50. W. Liu, H. Chan, and Y. Kao, Thermal Properties and Thermal Instabilities in InP-Based Heterojunction Bipolar Transistors, *IEEE Trans. Electron Devices, 43,* pp. 388–395, 1996.
51. W. Liu, S. Nelson, D. Hill, and A. Khatibzadeh, Current Gain Collapse in Microwave Multi-Finger Heterojunction Bipolar Transistors Operated at Very High Power Density, *IEEE Trans. Electron Devices, 40,* pp. 1917–1927, 1993.
52. W. Liu and A. Khatibzadeh, The Collapse of Current gain in Multi-Finger Heterojunction Transistors: Its Substrate Temperature Dependence, Instability Criteria, and Modeling, *IEEE Trans. Electron Devices, 41,* pp. 1698–1707, 1994.
53. Y. Amamiya, T. Niwa, N. Nagano, M. Mamada, Y. Suzuki, and H. Shimawaki, 40-GHz Frequency Dividers with Reduced Power Dissipation Fabricated Using High-Speed Small-Emitter-Area AlGaAs/InGaAs HBTs, *Dig. GaAs IC Symp.,* pp. 121–124, 1998.
54. M. F. Chang and P. M. Asbeck, III–V Heterojunction Bipolar Transistors for High-Speed Applications, in M. F. Chang (ed.), *Current Trends in Heterojunction Bipolar Transistors,* World Scientific, Singapore, 1996.
55. M. J. Mondry and H. Kroemer, Heterojunction Bipolar Transistor Using a (Ga,In)P Emitter on a GaAs Base, Grown by Molecular Beam Epitaxy, *IEEE Electron Device Lett., 6,* pp. 175–177, 1985.
56. M. Hafizi, D. C. Streit, L. T. Tran, K. W. Kobayashi, D. K. Umemoto, A. K. Oki, and S. K. Wang, Experimental Study of AlGaAs/GaAs HBT Device Design for Power Applications, *IEEE Electron Device Lett., 12,* pp. 581–583, 1991.
57. B. Bayraktaroglu, N. Camilleri, H. D. Shih, and H. Q. Tserng, AlGaAs/GaAs Heterojunction Bipolar Transistors with 4W/mm Power Density at X-Band, *Dig. IEEE MTT-S.,* pp. 387–390, 1987.

58. K. Nagata, O. Nakajima, Y. Yamauchi, T. Nittono, H. Ito, and T. Ishibashi, Self-Aligned AlGaAs/GaAs HBT with Low Emitter Resistance Utilizing InGaAs Cap Layer, *IEEE Trans. Electron Devices, 35,* pp. 2–7, 1988.

59. N. Hayama, C-W. Kim, H. Takahashi, N. Goto, and K. Honjo, High-Efficiency, Small-Chip AlGaAs/GaAs Power HBTs for Low-Voltage Digital Cellular Phones, *IEEE MTT-S Dig.,* pp. 1307–1310, 1997.

60. B. Bayraktaroglu and J. A. Higgins, HBTs for Microwave Power Applications, in M. F. Chang (ed.), *Current Trends in Heterojunction Bipolar Transistors,* World Scientific, Singapore, 1996.

61. W. Liu, A. Khatibzadeh, J. Sweder, and H. Chau, The Use of Base Ballasting to Prevent the Collapse of Current Gain in AlGaAs/GaAs Heterojunction Bipolar Transistors, *IEEE Trans. Electron Devices, 43,* pp. 245–250, 1996.

62. B. Bayraktaroglu, J. Barette, L. Kehias, C. I. Huang, R. Fitch, R. Neidhard, and R. Scherer, Very High-Power-Density CW Operation of GaAs/AlGaAs Microwave Heterojunction Bipolar Transistors, *IEEE Electron Device Lett., 14,* pp. 493–495, 1993.

63. Y. Matsuoka, S. Yamahata, S. Yamaguchi, K. Murata, E. Sano, and T. Ishibashi, IC-Oriented Self-Aligned High-Performance AlGaAs/GaAs Ballistic Collection Transistors and Their Application to High Speed ICs, *IEICE Trans. Electronics E-76-C,* pp. 1392–1399, 1993.

64. T. Oka, K. Hirata, K. Ouchi, H. Uchiyama, K. Mochizuki, and T. Nakamura, InGaP/GaAs HBT's with High-Speed and Low-Current Operation Fabricated Using WSi/Ti as the Base Electrode and Burying SiO_2 in the Extrinsic Collector, *IEDM Tech. Dig.,* pp. 739–742, 1997.

65. K. Mochizuki, K. Ouchi, K. Hirata, T. Tanoue, T. Oka, and H. Masuda, Polycrystal Isolation of InGaP/GaAs HBT's to Reduce Collector Capacitance, *IEEE Electron Device Lett., 19,* pp. 47–49, 1998.

66. Y. Ota, T. Hirose, A. Ryoji, and M. Inada, AlGaAs/GaAs Heterojunction Bipolar Transistors Fabricated with Various Collector-Carrier-Concentrations, *Electron. Lett., 26,* pp. 203–205, 1990.

67. K. Mochizuki, T. Tanoue, T. Oka, K. Ouchi, K. Hirata, and T. Nakamura, High-Speed InGaP/GaAs Transistors with a Sidewall Base Contact Structure, *IEEE Electron Device Lett., 18,* pp. 562–564, 1997.

68. K. W. Kobayashi, L. T. Tran, A. K. Oki, and D. C. Streit, Noise Optimization of a GaAs HBT Direct-Coupled Low Noise Amplifier, *IEEE MTT-S Dig.,* pp. 815–818, 1996.

69. H.-F. Chau and E. A. Beam III, High-Speed, High-Breakdown Voltage InP/InGaAs Double-Heterojunction Bipolar Transistors Grown by MOMBE, *IEEE Device Res. Conf.,* paper IVA-1, see also *IEEE Trans. Electron Devices, 40,* p. 2121, 1993.

70. H.-F. Chau, E. A. Beam III, Y.-C. Kao, and W. Liu, InP-Based Heterojunction Bipolar Transistors, in M. F. Chang (ed.), *Current Trends in Heterojunction Bipolar Transistors,* World Scientific, Singapore, 1996.

71. O. Sugiura, A. G. Dentai, C. H. Joyner, S. Chandrasekhar, and J. C. Campbell, High-Current Gain InGaAs/InP Double-Heterojunction Bipolar Transistors Grown by Metal Organic Vapor Phase Epitaxy, *IEEE Electron Device Lett., 9,* pp. 253–255, 1988.

72. A. Fujihara, Y. Ikenaga, H. Takahashi, M. Kawanaka, and S. Tanaka, High-Speed

InP/InGaAs DHBTs with Ballistic Collector Launcher Structure, *IEDM Tech. Dig.*, pp. 772–775, 2001.

73. K. Kurishima, H. Nakayima, T. Kobayashi, Y. Matsuoka, and T. Ishibashi, Fabrication and Characterization of High-Performance InP/InGaAs Double-Heterojunction Bipolar Transistors, *IEEE Trans. Electron Devices, 41,* pp. 1319–1326, 1994.

74. V. Danelon, F. Aniel, J. L. Benchimol, J. Mba, M. Riet, P. Crozat, G. Vernet, and R. Adde, Noise Parameters of InP-Based Double Heterojunction Base-Collector Self-Aligned Bipolar Transistors, *IEEE Microwave and Guided Wave Lett., 9,* pp. 195–197, 1999.

75. Q. S. Lee, S. C. Martin, D. Mensa, R. P. Smith, J. Guthrie, and M. J. W. Rodwell, Submicron Transferred-Substrate Bipolar Transistors, *IEEE Electron Device Lett., 20,* pp. 396–398, 1999.

76. Y. Betser, D. Scott, D. Mensa, S. Jaganathan, T. Mathews, and M. J. Rodwell, InAlAs/InGaAs HBTs with Simultaneously High Values of F_τ and F_{max} for Mixed Signal Analog/Digital Applications, *IEEE Electron Device Lett., 22,* pp. 56–58, 2001.

77. M. Sokolich, C. H. Fields, S. Thomas III, B. Shi, Y. K. Boegeman, M. Montes, R. Martinez, A. R. Kramer, and M. Madhav, A Low-Power 72.8-GHz Static Frequency Divider in AlInAs/InGaAs HBT Technology, *IEEE J. Solid-State Circuits, 36,* pp. 1328–1334, 2001.

78. R. S. Virk, M. Y. Chen, C. Nguyen, T. Liu, M. Matloubian, and D. B. Rensch, A High-Performance AlInAs/InGaAs/InP DHBT K-Band Power Cell, *IEEE Microwave and Guided Wave Lett., 7,* pp. 323–325, 1997.

79. D. Sawdai, K. Yang, S. S.-H. Hsu, D. Pavlidis, and G. I. Haddad, Power Performance of InP-Based Single and Double Heterojunction Bipolar Transistors, *IEEE Trans. Microwave Theory Tech., 47,* pp. 1449–1456, 1999.

80. S. Lee, H. J. Kim, M. Urteaga, S. Krishnan, Y. Wie, M. Dahlstrom, and M. Rodwell, Transferred Substrate InP/InGaAs/InP Double Heterojunction Bipolar Transistors with f_{max} = 425 GHz, *Electron. Lett., 37,* pp. 1096–1098, 2001.

81. D. L. Harame, J. H. Comfort, J. D. Cressler, E. F. Crabbe, J. Y.-C. Sun, B. S. Meyerson, and T. Tice, Si/SiGe Epitaxial-Base Transistors–Part II: Process Integration and Analog Applications, *IEEE Trans. Electron Devices, 42,* pp. 469–482, 1995.

82. A. Gruhle, H. Kibbel, U. König, U. Erben, and E. Kasper, MBE-Grown Si/SiGe HBT's with High β, f_T, and f_{max}, *IEEE Electron Device Lett., 13,* pp. 206–208, 1992.

83. A. Schüppen, A. Gruhle, H. Kibbel, and U. König, Mesa and Planar SiGe-HBTs on MBE-Wafers, *J. Mat. Sci. Electron. 6,* pp. 298–305, 1995.

84. E. J. Prinz, P. M. Garone, P. V. Schwartz, X. Xiao, and J. C. Sturm, The Effects of Base Dopant Outdiffusion and Undoped $Si_{1-x}Ge_x$ Junction Spacer Layers in $Si/Si_{1-x}Ge_x/Si$ Heterojunction Bipolar Transistors, *IEEE Electron Device Lett., 12,* pp. 42–44, 1991.

85. H. J. Osten, G. Lippert, D. Knoll, R. Barth, B. Heinemann, H. Rücker, and P. Schley, The Effect of Carbon Incorporation on SiGe Heterobipolar Transistor Performance and Process Margin, *IEDM Tech. Dig.,* pp. 803–806, 1997.

86. K. E. Ewald, D. Knoll, B. Heinemann, K. Chang, J. Kirchgessner, R. Mauntel, I. S. Lim, J. Steele, P. Schley, B. Tillack, A. Wolff, K. Blum, W. Winkler, M. Pierschel, U. Jagdhold, R. Barth, T. Grabolla, H. J. Erzgräber, B. Hunger, and H. J. Osten, Modular Integration of High-Performance SiGe:C HBTs in a Deep Submicron, Epi-Free CMOS Process, *IEDM Tech. Dig.,* pp. 561–564, 1999.

87. J. Böck, H. Schäfer, H. Knapp, D. Zschög, K. Aufinger, M. Wurzer, S. Boguth, R. Stengl, R. Schreiter, and T. F. Meister, High-Speed SiGe:C Bipolar Technology, *IEDM Tech. Dig.*, pp. 344–347, 2001.
88. K. Oda, E. Ohue, I. Suzumura, R. Hayami, A. Kodoma, H. Shimamoto, and K. Washio, Self-Aligned Selective-Epitaxial-Growth $Si_{1-x-y}Ge_xC_y$ HBT Technology Featuring 170-GHz f_{max}, *IEDM Tech. Dig.*, pp. 332–335, 2001.
89. M. Anderson, Z. Xia, P. Kuivalaien, and H. Pohjonen, Compact $Si_{1-x}Ge_x/Si$ Heterojunction Bipolar Transistor Model for Device and Circuit Simulation, *IEE Proc. Circuits Devices Syst. 142*, pp. 1–7, 1995.
90. D. Rosenfeld and S. A. Alterowitz, The Composition Dependence of the Cut-Off Frequency of Ungraded $Si_{1-x}Ge_x/Si_{1-y}Ge_y/Si_{1-x}Ge_x$ HBTs, *Solid-State Electron., 38*, pp. 641–651, 1995.
91. C. Jungemann, S. Keith, and B. Meinerzhagen, Full-Band Monte Carlo Device Simulation of a Si/SiGe HBT with a Realistic Doping Profile, *IEICE Trans. Electron. E83-C*, pp. 1228–1234, 1999.
92. G. Niu, W. E. Ansley, S. Zhang, J. D. Cressler, C. S. Webster, and R. A. Groves, Noise Parameter Optimization of UHV/CVD SiGe HBT's for RF and Microwave Applications, *IEEE Trans. Electron Devices, 46*, pp. 1589–1598, 1999.
93. J. Geßner, F. Schwierz, H. Mau, D. Nuernbergk, M. Roßberg, and D. Schipanski, Simulation of the Frequency Limits of SiGe HBTs, *Proc. MSM'99*, pp. 407–410, 1999.
94. F. Schwierz, J. Geßner, and D. Schipanski, Design of SiGe HBTs for High Frequency Operation, *Ext. Abstr. SSDM '99*, pp. 130–131, 1999.
95. J. B. Johnson, A. Stricker, A. J. Joseph, and J. A. Slinkman, A Technology Simulation Methodology for AC Performance Optimization of SiGe HBTs, *IEDM Tech. Dig.*, pp. 489–492, 2001.
96. H. Mau, *Anpassung und Implementation des Energietransportmodells zur vergleichenden Simulation mit dem Drift-Diffusions-Modell an SiGe-Heterobipolartransistoren*, Ph.D. Thesis, TU Ilmenau, 1997.
97. J. Geßner and F. Schwierz, Vertical and Lateral Design of SiGe HBTs for Very-High Frequency Operation, Unpublished, TU Ilmenau 2002.
98. S. S. Iyer, G. L. Patton, S. L. Delage, S. Tiwari and J. M. C. Storck, Silicon-Germanium Base Heterojunction Bipolar Transistors by Molecular Beam Epitaxy, *IEDM Tech. Dig.*, pp. 874–877, 1987.
99. G. L. Patton, J. H. Comfort, B. S. Meyerson, E. F. Crabbe, G. J. Scilla, E. De Fresart, J. M. C. Storck, J. Y.-C. Sun, D. L. Harame, and J. N. Burghartz, 75-GHz f_T SiGe-Base Heterojunction Bipolar Transistors, *IEEE Electron Device Lett., 11*, pp. 171–173, 1990.
100. K. Washio, E. Ohue, H. Shimamoto, K. Oda, R. Hayami, Y. Kiyota, M. Tanabe, M. Kondo, T. Hashimoto, and T. Harada, A 0.2-μm 180-GHz-f_{max} 6.7-ps-ECL SOI/HRS Self-Aligned SEG SiGe HBT/CMOS Technology for Microwave and High-Speed Digital Applications, *IEDM Tech. Dig.*, pp. 741–744, 2000.
101. M. Racanelli, K. Schuegraf, A. Kalburge, A. Kar-Roy, B. Shen, C. Hu, D. Chapek, D. Howard, D. Quon, F. Wang, G. U'ren, L. Lao, H. Tu, J. Zheng, J. Zhang, K. Bell, K. Yin, P. Joshi, S. Akhtar, S. Vo, T. Lee, W. Shi, and P. Kempf, Ultra High Speed SiGe NPN for Advanced BiCMOS Technology, *IEDM Tech. Dig.*, pp. 336–339, 2001.
102. B. Heinemann, D. Knoll, R. Barth, D. Bolze, K. Blum, J. Drews, K.-E. Ehwald, G. G. Fischer, K. Köpke, D. Krüger, R. Kurps, H. Rücker, P. Schley, W. Winkler, and H.-E.

Wulf, Cost-Effective High-Performance High-Voltage SiGe:C HBTs with 100 GHz f_T and $BV_{CE0} \times f_T$ Products Exceeding 220 VGHz, *IEDM Tech. Dig.*, pp. 348–351, 2001.

103. G. Freeman, D. Ahlgren, D. R. Greenberg, R. Groves, F. Huang, G. Hugo, B. Jagannathan, S. J. Jeng, J. Johnson, K. Schonenberg, K. Stein, R. Volant, and S. Subbanna, A 0.18μm 90 GHz f_T SiGe HBT BiCMOS, ASIC-Compatible, Copper Interconnect Technology for RF and Microwave Applications, *IEDM Tech. Dig.*, pp. 569–572, 1999.

104. P. A. Potyrai, K. J. Petrosky, K. D. Hobart, F. J. Kub, and P. E. Thompson, A 230 Watt S-Band SiGe HBT, *Dig. IEEE MTT-S*, pp. 673–676, 1996.

105. J. N. Burghartz, J.-O. Plouchart, K. A. Jenkins, C. S. Webster, and M. Soyuer, SiGe Power HBT's for Low-Voltage, High-Performance RF Applications, *IEEE Electron Device Lett., 19,* pp. 103–105, 1998.

106. U. Erben, M. Wahl, A. Schüppen, and H. Schumacher, Class-A SiGe HBT Power Amplifiers at C-Band Frequencies, *IEEE Microwave and Guided Wave Letters, 5,* pp. 435–436, 1995.

107. Z. Ma, S. Mohammadi, L.-H. Lu, P. Bhattacharya, L. P. B. Katehi, S. A. Alterowitz, and G. E. Ponchak, An X-Band High-Power Amplifier Using SiGe/Si HBT and Lumped Passive Components, *IEEE Microwave and Wireless Components Lett. 11,* pp. 287–289, 2001.

108. Z. Ma, S. Mohammadi, P. Bhattacharya, L. P. B. Katehi, S. A. Alterowitz, G. E. Ponchak, K. M. Strohm, and J.-F. Luy, Ku-Band (12.6 GHz) SiGe/Si High-Power Heterojunction Bipolar Transistors, *Electron. Lett., 37,* pp. 1140–1142, 2001.

APPENDIX 1

FREQUENTLY USED SYMBOLS

Symbol	Description	Unit
a	Lattice constant	Å
a	Layer thickness	cm or μm
A	Area	cm^2 or μm^2
BV	Breakdown voltage	V
C	Capacitance	F
C_d	Space-charge capacitance	F
C_D	Diffusion capacitance	F
d_{sc}	Thickness of space-charge region	cm or μm
$\mathcal{D}$	Dielectric displacement	As/cm^2
D_n, D_p	Diffusion coefficient for electron, holes	cm^2/s
$\mathcal{E}$	Electric field	V/cm
$\mathcal{E}_S$	Saturation field	V/cm
E	Energy	eV
E_A, E_D	Acceptor, donor level	eV
E_C	Conduction band edge	eV
E_F	Fermi level	eV
E_G	Bandgap	eV
E_i	Intrinsic Fermi level	eV
EU	Emitter utilization factor	—
E_V	Valence band edge	eV
f	Frequency	Hz
$f(E)$	Fermi-Dirac distribution function	—
f_{max}	Maximum frequency of oscillation	Hz or GHz
f_T	Cutoff frequency	Hz or GHz
g_m	Transconductance	A/V
G_e	Emitter injection efficiency	—
h	Planck constant	eV-s or J-s
h_{21}	Short-circuit current gain	dB
i	ac current	A
I	dc Current	A
I_C, I_B, I_E	Collector, base, emitter current	A
I_D	Drain current	A

Symbol	Description	Unit
j	Imaginary unit	—
J	Current density	A/cm^2
J_{Di}	Diffusion current density	A/cm^2
J_{Dr}	Drift current density	A/cm^2
k	Stability factor	—
k_B	Boltzmann constant	eV/K or J/K
L, L_{ch}	Gate, channel length	μm
L_n, L_p	Minority electron, hole diffusion length	μm
L_1, L_2	Length of region 1, 2 in FETs	μm
$m_{n,ds}, m_{h,ds}$	Electron, hole density of states efective mass	kg
m^*	Effective electron mass (transport, conductivity)	kg
MAG	Maximum available gain	dB
MSG	Maximum stable gain	dB
n	Electron density	cm^{-3}
n_i	Intrinsic carrier concentration	cm^{-3}
n_s	Electron sheet density	cm^{-2}
n_0	Electron density (thermal equilibruim)	cm^{-3}
N_A, N_D	Acceptor, donor concentration	cm^{-3}
N_C, N_V	Effective density of states in the conduction, valence band	cm^{-3}
NF, NF_{min}	Noise figure, minimum noise figure	dB
p	Hole density	cm^{-3}
p_0	Hole density (thermal equilibrium)	cm^{-3}
p	Normalized potential (at the boundary between regions 1 and 2 in FETs)	—
P	Momentum	kg-m/s
P	Electric power	W
PAE	Power added efficiency	%
P_{out}	Output power	W
q	Elementary charge	As
Q	Charge	As
R	Resistance	Ω
R_{co}	Contact resistance	Ω
s	Normalized potential (at source)	—
t	Time	s
t_c	Critical thickness	μm
t_{ox}	Oxide thickness	nm or μm
T	Temperature	K
U	Unilateral power gain	dB
v	ac voltage	V
v_n, v_p	Electron, hole velocity	cm/s
v_S	Saturation velocity	cm/s
v_{th}	Thermal velocity	cm/s

FREQUENTLY USED SYMBOLS

Symbol	Description	Unit
V	dc voltage	V
V_A	Early voltage	V
V_{bi}	Built-in voltage	V
V_{GS}, V_{DS}	Gate-source, drain-source voltage	V
V_{BE}, V_{CE}	Base-emitter, collector-emitter voltage	V
V_k	Knee voltage	V
V_{th}	Threshold voltage	V
V_T	Thermal voltage	V
w	Normalized potential	—
w	Thickness	cm or μm
w_B, w_C, w_E	Base, collector, emitter thickness	cm or μm
W	Gate width	cm or μm
W_B, W_E	Base, emitter width	cm or μm
α	Common base current gain	—
α_B	Base transport factor	—
β	Common emitter current gain	—
χ	Electron affinity	eV
ε	Dielectric constant	As/Vcm
κ	Thermal conductivity	W/cm-°C
φ	Potential	V
μ_n, μ_p	Electron, hole mobility	cm^2/Vs
μ_0	Low-field mobility	cm^2/Vs
ρ	Specific resistance, resistivity	Ω cm
τ_E, τ_P	Energy, momentum relaxation time	s
τ_T	Transit time	s
ΔE_{BGN}	Bandgap narrowing	eV
$\Delta E_C, \Delta E_V$	Conduction band, valence band offset	eV
ΔE_G	Bandgap difference	eV
ϕ	Work function	eV
ϕ_b	Schottky barrier height	eV
ϖ	Angular frequency	Hz

APPENDIX 2

PHYSICAL CONSTANTS AND UNIT CONVERSIONS

Quantity	Symbol	Value in SI Units	Converted Value
Boltzmann constant	k_B	1.381×10^{-23} J/K	8.620×10^{-5} eV/K
Electron rest mass	m_0	9.110×10^{-28} g	5.686×10^{-16} eVs^{2}cm^{-2}
Elementary charge	q	1.602×10^{-19} C	1.602×10^{-19} As
Dielectric constant in vacuum	ε_0	8.854×10^{-12} F/m	8.854×10^{-14} As/Vcm
Planck constant	h	6.626×10^{-34} Js	4.136×10^{-15} eVs
Speed of light in vacuum	c	2.998×10^8 m/s	2.998×10^{10} cm/s
Thermal voltage ($T = 300$ K)	$k_B T/q$	0.02586 V	25.86 mV

In semiconductor electronics it is common to work with units different from the SI units. The conversion relations between SI units and the units used in semiconductor electronics are given below:

$$1 \text{ J} = 1 \text{ VAs} = 1 \text{ Ws}$$

$$1 \text{ eV} = 1.602 \times 10^{-19} \text{ Ws}$$

$$1 \text{ Ws} = \frac{1}{1.602 \times 10^{-19}} \text{ eV} = 6.242 \times 10^{18} \text{ eV}$$

$$1 \text{ g} = 10^{-7} \text{ Ws}^3 \text{ cm}^{-2} = 6.242 \times 10^{11} \text{ eV s}^2 \text{ cm}^{-2}$$

$$0 \text{ K} = -273.2 \text{ °C}$$

$$0 \text{ °C} = 273.2 \text{ K}$$

$$1 \text{ Å} = 10^{-8} \text{ cm}$$

$$1 \text{ nm} = 10 \text{ Å} = 10^{-7} \text{ cm}$$

$$1 \text{ μm} = 10^{-4} \text{ cm}$$

$$1 \text{ mil} = 25 \text{ μm}$$

$$1 \text{ in.} = 2.54 \text{ cm}$$

Prefixes

Multiple	Prefix	Symbol	Multiple	Prefix	Symbol
10^{12}	Tera	T	10^{-15}	Femto	f
10^{9}	Giga	G	10^{-12}	Pico	p
10^{6}	Mega	M	10^{-9}	Nano	n
10^{3}	Kilo	k	10^{-6}	Micro	μ
			10^{-3}	Milli	m

APPENDIX 3

MICROWAVE FREQUENCY BANDS

IEEE Radar Band Designation [1]

Designation	Frequency Range, GHz	Wave Length, cm
L band	1–2	30–15
S band	2–4	15–7.5
C band	4–8	7.50–3.75
X band	8–12	3.75–2.50
Ku band	12–18	2.50–1.67
K band	18–27	1.67–1.11
Ka band	27–40	1.11–0.75
V band	40–75	0.75–0.40
W band	75–110	0.40–0.27
mm band	110–300	0.27–0.10
sub-mm band	300–3000	0.10–0.01

Old U.S. Military Microwave Bands [2]

Designation	Frequency Range, GHz
L band	0.39–1.55
S band	1.55–3.90
C band	3.90–6.20
X band	6.20–10.90
K band	10.90–36.00
Q band	36.00–46.00
V band	46.00–56.00
W band	56.00–100.00

REFERENCES

1. The IEEE Radar Band Designation, http://www.naval.com/radio-bands.htm.
2. S. Y. Liao, *Microwave Devices and Circuits*, Prentice-Hall, Englewood Cliffs, 1990.

APPENDIX 4

TWO-PORT CALCUATIONS

The small-signal behavior of a two-port can be described by different sets of so-called small-signal parameters. Commonly the scattering parameters (S parameters), the admittance parameters (Y parameters) and the hybrid parameters (H parameters) are used to represent microwave transistors. For the experimental characterization of microwave transistors, the S parameters are measured, while for physically based small-signal eqivalent circuits, Y or H parameters are typically used. Each parameter set consists of four parameters and can be written in matrix form as:

$$\|X\| = \begin{Vmatrix} x_{11} & x_{12} \\ x_{12} & x_{22} \end{Vmatrix} \tag{A4-1}$$

where X is S, Y, or H.

When two two-ports are connected to each other and the small-signal parameters of the entire circuit are to be determined, two further sets of small-signal parameters are helpful, namely the impedance parameters (Z parameters) and the chain parameters (A parameters). The different types of small-signal parameters can be converted into each other by applying certain conversion rules. In the following, the definitions of the Y, H, Z, and A parameters, the conversion rules, and the rules for calculating the small-signal parameters of connected two-ports are given. S parameters have already been defined in Section 1.2. Important for the definition of the parameters are the directions in which currents and voltages are assumed to be positive. The equations and conversion rules in this appendix are based on the notation given in Section 1.2, Fig. 1.2.

Definition of the Y, H, Z, and A Parameters

Y parameters $\qquad$ H parameters $\qquad$ Z parameters $\qquad$ A parameters

$$y_{11} = \frac{i_1}{u_1}\bigg|_{u_2=0} \quad h_{11} = \frac{u_1}{i_1}\bigg|_{u_2=0} \quad z_{11} = \frac{u_1}{i_1}\bigg|_{i_2=0} \quad a_{11} = \frac{u_1}{u_2}\bigg|_{i_2=0}$$

$$y_{12} = \frac{i_1}{u_2}\bigg|_{u_1=0} \quad h_{12} = \frac{u_1}{u_2}\bigg|_{i_2=0} \quad z_{12} = \frac{u_1}{i_2}\bigg|_{i_1=0} \quad a_{12} = \frac{u_1}{-i_2}\bigg|_{u_2=0} \tag{A4-2}$$

$$y_{21} = \left.\frac{i_2}{u_1}\right|_{u_2=0} \quad h_{21} = \left.\frac{i_2}{i_1}\right|_{u_2=0} \quad z_{21} = \left.\frac{u_2}{i_1}\right|_{i_2=0} \quad a_{21} = \left.\frac{i_1}{u_2}\right|_{i_2=0}$$

$$y_{22} = \left.\frac{i_2}{u_2}\right|_{u_1=0} \quad h_{22} = \left.\frac{i_1}{u_2}\right|_{i_2=0} \quad z_{22} = \left.\frac{u_2}{i_2}\right|_{i_1=0} \quad a_{22} = \left.\frac{i_1}{-i_2}\right|_{u_2=0}$$

Rules for the Conversion of S Parameters to Y and H Parameters

$$\|S\| = \left\| \begin{array}{cc} \dfrac{1-y'_{11}+y'_{22}-\Delta y'}{1+y'_{11}+y'_{22}+\Delta y'} & \dfrac{-2y'_{12}}{1+y'_{11}+y'_{22}+\Delta y'} \\ \dfrac{-2y'_{12}}{1+y'_{11}+y'_{22}+\Delta y'} & \dfrac{1+y'_{11}-y'_{22}+\Delta y'}{1+y'_{11}+y'_{22}+\Delta y'} \end{array} \right\| = \left\| \begin{array}{cc} \dfrac{-1+h'_{11}-h'_{22}+\Delta h}{1+h'_{11}+h'_{22}+\Delta h'} & \dfrac{2h'_{12}}{1+h'_{11}+h'_{22}+\Delta h'} \\ \dfrac{-2h'_{21}}{1+h'_{11}+h'_{22}+\Delta h'} & \dfrac{1+h'_{11}-h'_{22}-\Delta h'}{1+h'_{11}+h'_{22}+\Delta h'} \end{array} \right\|$$

(A4-3)

where y'_{ij} and and h'_{ij} are the normalized y and h parameters, and $\Delta y'$ and $\Delta h'$ are the determinants given by $\Delta y'_{ij} = y'_{11}y'_{22} - y'_{12}y'_{21}$ and $\Delta h' = h'_{11}h'_{22} - h'_{12}h'_{21}$. Measured S parameters are always related to a microwave system (a transistor, an amplifier, etc.) with a certain characteristic impedance Z_0. Most systems have a characteristic impedance of 50Ω. Therefore the conversion of Y and H parameters to S parameters, and vice versa, must take Z_0 into account. This is done using the normalized Y and H parameters, Y' and H', respectively, in (A1-3) given by

$$\|Y'\| = Z_0 \left\| \begin{array}{cc} y_{11} & y_{12} \\ y_{21} & y_{22} \end{array} \right\| \text{ and } \|H'\| = \left\| \begin{array}{cc} h_{11}/Z_0 & h_{12} \\ h_{21} & y_{22}Z_0 \end{array} \right\| \quad (A4\text{-}4)$$

Rules for the Conversion of Y and H Parameters to S Parameters

$$\|Y'\| = \left\| \begin{array}{cc} \dfrac{1-s_{11}+s_{22}-\Delta s}{1+s_{11}+s_{22}+\Delta s} & \dfrac{-2s_{12}}{1+s_{11}+s_{22}+\Delta s} \\ \dfrac{-2s_{21}}{1+s_{11}+s_{22}+\Delta s} & \dfrac{1+s_{11}-s_{22}-\Delta s}{1+s_{11}+s_{22}+\Delta s} \end{array} \right\|$$

(A4-5)

$$\|H'\| = \left\| \begin{array}{cc} \dfrac{1+s_{11}+s_{22}+\Delta s}{1-s_{11}+s_{22}-\Delta s} & \dfrac{2s_{12}}{1-s_{11}+s_{22}-\Delta s} \\ \dfrac{2s_{21}}{1-s_{11}+s_{22}-\Delta s} & \dfrac{1-s_{11}-s_{22}+\Delta s}{1-s_{11}+s_{22}-\Delta s} \end{array} \right\|$$

Rules for the Conversion Between Y, H, Z, and A Parameters

	$\|y\|$	$\|h\|$	$\|z\|$	$\|a\|$
$\|y\|$	$\begin{Vmatrix} y_{11} & y_{12} \\ y_{21} & y_{22} \end{Vmatrix}$	$\dfrac{1}{h_{11}}\begin{Vmatrix} 1 & -h_{12} \\ h_{21} & \Delta h \end{Vmatrix}$	$\dfrac{1}{\Delta z}\begin{Vmatrix} z_{22} & -z_{12} \\ -z_{21} & z_{11} \end{Vmatrix}$	$\dfrac{1}{a_{12}}\begin{Vmatrix} a_{22} & -\Delta a \\ -1 & a_{11} \end{Vmatrix}$
$\|h\|$	$\dfrac{1}{y_{11}}\begin{Vmatrix} 1 & -y_{12} \\ y_{21} & \Delta y \end{Vmatrix}$	$\begin{Vmatrix} h_{11} & h_{12} \\ h_{21} & h_{22} \end{Vmatrix}$	$\dfrac{1}{z_{22}}\begin{Vmatrix} \Delta z & z_{12} \\ -z_{21} & 1 \end{Vmatrix}$	$\dfrac{1}{a_{22}}\begin{Vmatrix} a_{12} & \Delta a \\ -1 & a_{21} \end{Vmatrix}$
$\|z\|$	$\dfrac{1}{\Delta y}\begin{Vmatrix} y_{22} & -y_{12} \\ -y_{21} & -y_{11} \end{Vmatrix}$	$\dfrac{1}{h_{22}}\begin{Vmatrix} \Delta h & h_{12} \\ -h_{21} & 1 \end{Vmatrix}$	$\begin{Vmatrix} z_{11} & z_{12} \\ z_{21} & z_{22} \end{Vmatrix}$	$\dfrac{1}{a_{21}}\begin{Vmatrix} a_{12} & \Delta a \\ 1 & a_{22} \end{Vmatrix}$
$\|a\|$	$\dfrac{1}{y_{21}}\begin{Vmatrix} -y_{22} & -1 \\ -\Delta y & -y_{11} \end{Vmatrix}$	$-\dfrac{1}{h_{21}}\begin{Vmatrix} \Delta h & h_{11} \\ h_{22} & 1 \end{Vmatrix}$	$\dfrac{1}{z_{21}}\begin{Vmatrix} z_{11} & \Delta z \\ 1 & z_{22} \end{Vmatrix}$	$\begin{Vmatrix} a_{11} & a_{12} \\ a_{21} & a_{22} \end{Vmatrix}$

$$(\text{A4-6})$$

Rules for the Calculation of the Small-Signal Parameters of Connected Two-Ports

For the analysis of small-signal equivalent circuits of microwave transistors, the following three types of connected two-ports are of interest: series, parallel, and chain connection. Figure A4.1 shows the circuit diagrams for these three cases. The total small-signal parameters $\|X_{\text{tot}}\|$ of the circuit consisting of the two-ports $TP1$ and $TP2$ can be calculated from the parameters of the individual two-ports, $\|X_1\|$ and $\|X_2\|$, as follows:

Series circuit	$\|Z_{\text{tot}}\|$	$=$	$\|Z_{TP1}\| \quad +$	$\|Z_{TP2}\|$
Parallel circuit	$\|Y_{\text{tot}}\|$	$=$	$\|Y_{TP1}\| \quad +$	$\|Y_{TP2}\|$
Chain circuit	$\|A_{\text{tot}}\|$	$=$	$\|A_{TP1}\| \quad \times$	$\|A_{TP2}\|$

$$(\text{A4-7})$$

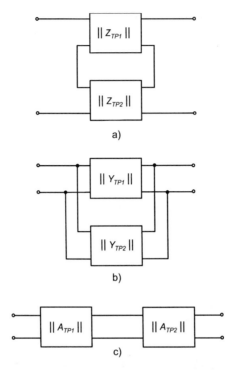

Figure A4.1. Circuit diagram of connected two-ports. (a) Series connection. (b) Parallel connection. (c) Chain connection.

APPENDIX 5

IMPORTANT MATERIAL PROPERTIES OF SELECTED MATERIALS (AT ROOM TEMPERATURE)

	Si	GaAs	$Al_{0.3}Ga_{0.7}As$	$In_{0.53}Ga_{0.47}As$	$In_{0.52}Al_{0.48}As$
E_G (eV)	1.124	1.422	1.80	0.737	1.451
N_C (cm^{-3})	3.22×10^{19}	3.97×10^{17}	7.0×10^{17}	2.08×10^{17}	5.16×10^{17}
N_V (cm^{-3})	1.82×10^{19}	9.4×10^{18}	1.44×10^{18}	9.16×10^{18}	7.34×10^{18}
n_i (cm^{-3})	8.82×10^{10}	2.21×10^{6}	7.71×10^{2}	8.94×10^{11}	1.27×10^{6}
ε_r	11.9	12.85	12.03	14.07	12.73
κ (W/cm-°C)	1.3	0.5	0.137	0.05	—

	$Al_{0.3}Ga_{0.7}As$/ GaAs	InP/ $In_{0.53}Ga_{0.47}As$	$In_{0.52}Al_{0.48}As$/ $In_{0.53}Ga_{0.47}As$	$In_{0.59}Ga_{0.51}P$/ GaAs (ordered)
ΔE_G (eV)	0.378	0.616	0.714	0.438
ΔE_C (eV)	0.219	0.271	0.52	0.034
ΔE_V (eV)	0.159	0.345	0.194	0.404

INDEX

1/f noise, 194
2DEG, *see* Two-dimensional electron gas

A parameters, 477
Acceptor, 65, 97
Accumulation, 296–298, 306
Accumulation layer, 306
Activation energy, 65–66, 73, 224, 233
Amplifier class, 204
Associated gain, 11, 214–215, 268, 278, 336
Atmospheric window, 39
Avalanche breakdown, *see* Impact ionization

Back injection, 342
Ballast resistor, 386, 426, 440
Band diagram, 63–66, 70, 79
    of BJTs, 343
    of HEMTs, 16, 232
    of HBTs, 17, 402–403, 412, 414, 416–417
    of pn junctions, 96, 101
    of Schottky junctions, 106
    of the two-terminal MOS structure, 295–298
Band offset, 16–20, 23–24, 121–123, 125, 131–134, 150, 155, 233, 262, 271, 402–404, 420–422, 428–429, 434, 445, 481
Band structure, 66, 85–86
Bandgap, 19, 34, 63, 76–77, 122–123, 130–131, 271, 296, 481
    narrowing, 356–358, 365, 407
    temperature dependence, 76–77
Barrier transparency, 138
Base current, 342–343, 346, 356–357, 390–391, 409–411
Base pushout, 359–363, 419–420
Base resistance, 353–355
Base station, 41–42
Base transport factor, 345
BJT:
    GaAs BJT, 334
    Ge BJT, 334
    Si BJT, 334
Boltzmann distribution, 69, 75
Breakdown field, 111
Breakdown voltage, 108, 111–112
    of bipolar transistors, 29, 32–33, 366–368, 393–395, 434–435, 448–450, 457, 460
    of field-effect transistors, 38, 42, 206, 217–218, 223, 269, 271, 276–278, 282, 324–327
Built-in field, 405, 412, 436, 445, 451
Built-in voltage:
    of heterojunctions, 128
    of pn junctions, 98
    of Schottly junctions, 106

Cappy model, 176–180, 187–190, 199, 203
Cappy-like model, 246–253, 259
Cellular phone, 2, 39, 41
Channel length, 294
Charging time, *see* Delay time
Collector-base capacitance, 28, 374, 423, 436–437, 446, 451
Collector resistance, 352–356
Common-base, 347
Common-emitter, 347
Comparison:
    BJT vs. HBT, 29–34
    GaAs MESFET vs. Si BJT, 15
    of different microwave bipolar transistors, 29–32
    of different microwave FETs, 25–28, 36–37
    of different microwave transistor types, 22, 33–35, 45–52
    of wide bandgap FETs, 35
Complete ionization, 72
Conduction band, 63–64
Contact resistance, 173–174, 223, 242–244, 306–307, 352–353, 433, 446
Critical thickness, 18, 123–124

483

Current crowding, 368
Current gain, 190, 194
  collapse, 430 432
  common-base, 347
  common-emitter, 347, 391
  temperature dependence, 385, 427–430
Cutoff frequency, 9–11, 190–192, 259–265, 268, 378–379
  data for bipolar transistors, 22, 29–30, 32, 49,392–394, 442, 449–450, 454–460
  data for FETs, 22, 24–25, 35, 37, 212–213, 224–225, 267–268,   271–272, 276, 278–279, 282, 321–323, 326–327
CVD, 453

Decibel, 8, 11, 190
Degenerate semiconductor, 70
Delay time:
  ECL gate delay, 335–336
  in bipolar transistors, 376–379, 423
  in FETs, 182–183, 262–265
Density of states, 67–68, 142
Depletion, 296–298
Depletion approximation, 99, 107, 167
Depletion capacitance, 374
Depletion-mode FET, 164–165, 210, 236
Dielectric constant, 98–99
Diffusion capacitance, 374
Diffusion coefficient, 89–90
Diffusion current, 89–90
Diffusion length, 104
Diffusion noise, 198
Donor, 65
Double heterojunction HBT, 401
Drain conductance, 183
Drain resistance (small-signal), 183, 185, 188, 255, 314
Drift current, 86–87
Drift-diffusion model, 106
DX center, 18, 233, 271

Early effect, 349–350
Early voltage, 350–351
Effective density of states, 67–68, 76
Einstein relation, 90
Electron affinity, 97, 125, 295–297
Emitter–base resistance, 374
Emitter finger (see also multifinger), 387, 427, 438–440
Emitter injection efficiency, 17, 404, 410
Emitter resistance, 353
Emitter utilization, 396, 437–438
Energy balance equation, 91–92, 178–179
Enhancement-mode FET, 164–165, 210, 236

f_{max}, see Maximum frequency of oscillation
f_T, see, cutoff frequency
Fermi-Dirac distribution, 67
Fermi integral, 68
Fermi level, 65
Field emission, 135–136, 138
Figure of merit, 4
Flat band condition, 297
Forward active mode, 338–339
Fringing capacitance, 186, 212

Gate-drain capacitance, 183, 186, 188, 256–257, 314
Gate finger width, 211, 216–221, 259, 315
Gate oxide, 293, 321, 326
Gate resistance, 184, 187, 211, 257–259, 315–316, 322
Gate-source capacitance, 183, 186, 188, 192, 256, 314, 316
Generation, 65
Graded base, 411–413
Graded layer, 416
Gradual channel approximation, 167, 238
Gummel number, 65
Gummel plot, 56–357, 390–391, 416–418

H parameters, 477–479
Handset, 41–42
HBT, 400
  GaAs HBT, 432
  InP HBT, 443
  SiGe HBT, 450
  transferred substrate HBT, 446
Heat flow equation, 113–114
HEMT, 231
  AlSb/InAs HEMT, 234
  AlGaN/GaN HEMT, 280
  conventional AlGaAs/GaAs, 266
  GaAs mHEMT, 273
  GaAs pHEMT, 270
  InP HEMT, 276
  Si-based, 233
Heterojunction, 16, 23
  band diagram, 16, 17, 126, 135, 136, 141, 147, 403, 417
  smooth, 135, 139, 401–404
  spike, 134–139, 401–404, 413–417
  type I, II, 121–123
Hexagonal lattice, 34, 61, 121–122
High-level injection, 103, 348
Horizontal-flow contact, 353

Impact ionization, 108–112
Incomplete ionization, 73, 224

Intrinsic carrier concentration, 71
Intrinsic Fermi level, 65, 80
Inversion, 298
Inversion channel, 291
Ionization coefficient, 100–110

Joyce–Dixon approximation, 75

Kirk effect, *see* Base pushout
Knee voltage, 205–207, 217, 223, 384, 417

Lattice constant, 18–19, 121–122
Lattice mismatch, 18
LDMOSFET, 37–38, 293, 326–327
Load line, 205–206
Long-base junction, 105
Low-field mobility, *see* Mobility
Low-level injection, 102–103, 344

MAG, see, Maximum available gain
Majority carrier, 71
Mass action law, 71
Maximum available gain, 7, 192–194, 221
Maximum frequency of oscillation, 9, 28, 41, 192–194, 378–379
 data for bipolar transistors, 22, 29, 31, 49, 380, 392–394, 435, 440–442, 448–450, 456–459
 data for FETs, 22, 25–26, 35, 38, 212–214, 224–225, 268, 271–272, 276, 278–279, 282, 323–324
Maximum gate finger width, 218
Maximum stable gain, 8, 192–194, 221
MBE, 16, 400, 453
Mean free path, 78
Mesa, 211, 212
MESFET, 163
 GaAs MESFET, 208
 SiC MESFET, 223
 GaN MESFET, 225
Metamorphic, 20, 273–276
Microwaves, 1
Minimum noise figure, 11, 199–204, 265, 383
 data for bipolar transistors, 15, 31–33, 394–396, 442, 444, 449, 457, 459
 data for FETs, 15, 26–27, 37–38, 214–215, 268–269, 272–273, 280, 283, 325
Minority carriers, 65–66
 Density, 100–103
 diffusion length, 104, 348
 life time, 104, 348, 406–407
 mobility, 81, 348, 406–407
Mobility, 24, 79–84, 154–157
MOCVD, 16, 400

Modulation efficiency, 259–264
Momentum, 66, 91
Momentum balance equation, 91, 178, 251
MOSFET, 292
MSG, *see* Maximum stable gain
Multifinger design, 114, 211, 216, 315, 386, 389, 396, 424–427, 430, 438–439
Mushroom gate, 22–23, 211, 257–259

NF_{min}, *see* Minimum noise figure
Neutrality condition, 71
Noise, 11, 194–197, 379–381
Noise figure, 11
Noise temperature, 11
Nondegenerate semiconductor, 70
Normally-on, normally-off, 42, 165–166
Nyquist formula, 197

Octave, 9
Offset voltage, 417
Ordered InGaP, 433–434
Output power, 12, 205–206, 384–385
 data for bipolar transistors, 13, 31–34, 396, 441–443, 461
 data for FETs, 27–28, 36–38, 217, 222, 224–226, 269, 273–276, 283, 326–327
Output power density, 12
Overlap capacitance, 314
Oxide capacitance, 301

Pad capacitance, 244–245
PHS-like model, 237–241, 254–257, 303–306, 313–314
PHS model, 167–172, 184–187
pinchoff voltage, 148, 169, 239
pn junction, 96
Poisson equation, 98, 107, 127
Polarization, 157–160
Polysilicon, 293
 base, 335, 387–388
 emitter, 335, 353, 364, 369–372, 387, 450
 gate, 292
Potential, relation between potential and energy, 79–80
Power-added efficiency, 12, 207, 385
 data, 35
Power gain, 7
Pseudomorphic, *see also* Strained layer, 18–19, 23, 121
π-model, 371–373

Quasi-Fermi level, 101
Quasi-saturation, 359

Recessed gate, 173, 209
Recombination, 65
Relaxation time, 91–92
Relaxation time approximation, 91–96, 178, 246
Resistivity, *see* specific resistance

S parameters, 5, 478
Satellite communication, 39, 41
Saturation drain-source voltage, 171–172, 241–242
Saturation velocity, 84, 168, 175, 303, 351
Schottky barrier height, 106–107, 234–235
Schottky junction, 106
Schrödinger-Poisson, 142, 143, 146, 151–153, 159, 235, 304
Selectively implanted collector, 335, 387–389, 451, 455
Self-alignment, 335–336, 388, 437
Self-heating, 112–117, 385–387, 424–431, 438–440
Setback layer, 416–417
Sheet resistance, 244, 306–307, 353
  base, 432, 451
Short-base junction, 105
Shot noise, 380
Silicide, 321–322
Single heterojunction HBT, 401
Small-signal condition, 180–182
Small-signal equivalent circuit, 182, 191, 317–320, 373, 376
  for noise considerations, 200, 381
Small-signal parameters, 4–5, 477–480
SOI MOSFET, 294
Source resistance, 172–175, 242–246, 306–307
Space-charge region width
  heterojunction, 129
  pn junction, 100
  Schottky junction, 108
Spacer, 141, 146, 155, 234, 266–267
Stability, 6
Stability factor, 6–7
Strained layer, see also Pseudomorphic, 18, 123–131, 134, 154
Stray capacitance, 314

Subband, 142–144
Surface recombination, 369
System on chip, 40

Thermal conductivity, 115–116
Thermal-electric feedback coefficient, 424
Thermal equilibrium, 63
Thermal noise, 379
Thermal resistance, 117
Thermal runaway, 386, 431
Thermal shunt, 440
Thermal velocity, 78
Thermal voltage, 102
Thermionic emission, 135–138
Thermionic field emission, 136, 138
Threshold voltage, 210, 239, 298–301
Threshold voltage shift (modification), 310–313
T-model, 376, 381
Transconductance, 182
  FET, 184, 188, 254
  bipolar transistor, 373, 423
Transferred substrate HBT, 21, 30, 401, 446–450
Transit time, 187–189, 257
Tunneling, 136, 321
Tunneling resistance, 42
Turn-on voltage, 418–419, 425
Two-dimensional electron gas, 16, 140, 304
  sheet density, 24, 142, 156, 235, 239
  mobility, 24, 154, 156
Two-dimensional hole gas, 141
Two-port, 4–7, 477–480

Unilateral power gain, 8, 192–194, 221, 379
Unit cell, 216

Vacuum level, 97
Valence band, 63–64
Velocity overshoot, 94
Vertical-flow contact, 353
VLSI, 1

Work function, 295–297

Y parameters, 4, 189, 318, 477–480